BEHAVIOUR OF ELECTROMAGNETIC WAVES IN DIFFERENT MEDIA AND STRUCTURES

Edited by **Ali Akdagli**

Behaviour of Electromagnetic Waves in Different Media and Structures
Edited by Ali Akdagli

Published by InTech
Janeza Trdine 9, 51000 Rijeka, Croatia

Publishing Process Manager Iva Lipović
Technical Editor Teodora Smiljanic
Cover Designer Jan Hyrat

Printed in Croatia

Behaviour of Electromagnetic Waves in Different Media and Structures,
Edited by Ali Akdagli
p. cm.
ISBN 978-953-307-302-6

For EU product safety concerns:
IN TECH d.o.o., Prolaz Marije Krucifikse Kozulić 3, 51000 Rijeka, Croatia,
info@intechopen.com or visit our website at intechopen.com.

Meet the editor

Ali Akdagli received the B.S., M.S., and Ph.D. degree from Erciyes University, Kayseri, in 1995, 1997, and 2002, respectively, all in electronic engineering. From 2003 to 2006, he was an assistant professor in the electronic engineering department at Erciyes University. He joined the same department at Mersin University where currently he is an associate professor. He has published more than 70 papers in journals and conference proceedings. His current research interests include, evolutionary optimization techniques (genetic algorithm, ant colony optimization, differential evolution, particle swarm optimization, artificial bee colony algorithms), artificial neural networks and their applications to electromagnetics, microwave circuits, microstrip antennas and antenna pattern synthesis problems. Dr. Akdagli is a editorial advisory board member of Recent Patents on Electrical Engineering which is a journal published by Bentham Science.

Contents

Preface

Lately, there has been a growing interest in electromagnetic wave propagation in complex systems such as modern materials and structures. This book is intended to give the explanatory 20 chapters which consist of original works of the leading scientists in the field of wave propagation which produced theoretical and experimental methods in this field of research and obtained important results.

In chapter 1, the possible procedures for the extraction of electric and magnetic parameters of dispersive materials were outlined, and it was shown that the complex relative permittivity and magnetic permeability can be modeled with either dispersive laws or on a point-by-point basis (at individual frequencies).

In the frame of the Kirchhoff method, the average coefficient of light scattering by surface fractal structures was calculated in chapter 2. A normalized band-limited Weierstrass function was presented for modeling 2D fractal rough surfaces. On the basis of numerical calculation of average scattering coefficient the scattering indicatrises diagrams for various surfaces and falling angles were calculated.

As it is well known the method of moments and finite difference method are widely used as electromagnetic numerical techniques. In the chapter 3, depending on the investigated post geometry, a hybrid method which utilities mode-matching technique, method of moments and finite difference method defined in the frequency domain was proposed for electromagnetic wave scattering from structures containing complex cylindrical or spherical objects.

The standard spaces were constructed under the physical presentation by solving the eigen-value problem of the matrixes of dielectric permittivity and magnetic permeability in chapter 4. Based on the spaces, it was discussed the modal equations of electromagnetic waves for anisotropic media, bi-anisotropic media, dispersive medium and chiral medium, respectively, by converting the classical Maxwell's vector equation to the eigen Maxwell's scalar equation, each of which shows the existence of an electromagnetic sub-wave, and its propagation velocity, propagation direction, polarization direction and space pattern were completely determined in the equations.

In chapter 5, some electromagnetic wave equations to show applications to develop the analytic design formula for the cavity design were presented. Several examples for the rectangular cavity was introduced for atmospheric microwave plasma torch as a rectangular, which has uniform electromagnetic wave distribution to produce wide area plasma in atmospheric pressure environment. The annular cavity for klystrode was introduced for a microwave vacuum oscillator as a circular example, which adapted the grid structure and the electron beam as an annular shape, gives high efficiency compared with conventional klystrode.

Ground penetrating radar (GPR) is based on high-frequency electromagnetic wave propagation and its detecting targets are below the ground surface. In order to understand the performance of GPR, permittivity testing and analysis are critical. In chapter 6, permittivity analysis and measurement methods for rock sample were given for GPR applications.

Focusing on the propagation of high frequency pulse electromagnetic waves in layered lossy and dispersive medium and after the physical forward modeling experiment, in chapter 7, the delamination in polished wall paintings by wall coupling antennas using GPR was explained. It was shown that the ultra-wide band GPR is capable of detecting delamination in vertical resolution of about 5 mm when it is equipped with a transmitting antenna of 1.6 GHz central frequency.

With the help of Maxwell's equations, the laws of formation of the impulse of electromagnetic radiation in dielectric environment for conditions self-induced transparency were considered in chapter 8.

In chapter 9, ultrafast electromagnetic waves emitted from semiconductors under high electric fields, which are closely related with ultrafast nonequilibrium transport of carriers in semiconductor, were investigated.

The behavior of electromagnetic waves emitted from within the ionospheric plasma and the analytical solutions which are necessary to understand the characteristics of the environment were explained in chapter 10. Problems in plasma physics at the conductivity, dielectric constants and refractive index were defined according to the media parameters.

In the recent years, by developing the usage of new popular electronic-communication gadgets and home appliances like mobile phones and microwave ovens which are mostly sources of electromagnetic wave radiation, a severe public concern regarding the side-effects whether positive or negative on human health and environment has arisen. In chapter 11, the different types of electromagnetic field and their characteristics besides of definitions were introduced and investigated separately. The biologic effects on live tissues and human body were also investigated for concerned fields.

Magnetic reconnection is a very important physical process in astrophysical and laboratory plasmas, which enables reconfiguration of the magnetic field topology and converts the magnetic field energy to plasma kinetic and thermal energy. The diffusion region is a crucial region of reconnection where magnetic field and plasma decouple from each other and strong wave activity and complex wave particle interactions occur. In chapter 12, observations of low frequency electromagnetic waves where reconnection signatures are well defined as well as those observations where one only speculates about such relationship were summarized.

The switching mode circuit (SMC) is being applied widely to the power circuit and the signal circuit including the digital circuit now. The SMC works by the transistor or the other switching devices. The SEMW theory was developed for the design and analysis of the traces and interconnects on the SMC. The SEMW theory and its application examples of the analysis of the signal line and the power line on the system on a chip were presented in chapter 13.

In chapter 14, the nonlinear absorption of a strong electromagnetic wave by confined electrons in low-dimensional systems in the presence of an external magnetic field was investigated. By using the method of the quantum kinetic equation for electrons, the expressions for the electron distribution function and the nonlinear absorption coefficient in quantum wells, doped superlattics, cylindrical quantum under the influence of an external magnetic field were obtained.

The conditions under which transverse electromagnetic (TEM) waves exist in a sourceless medium was examined in chapter 15. It was shown that TEM waves could be classified according to whether their Poynting vector is identically zero or nonzero.

Lorentz-Dirac equation (LDE) is the widely accepted classical equation to describe the motion of a scalar point charge acted by external electromagnetic fields and its own radiating fields. A new reduction of order form of LDE, which coincides with that obtained by the method of Landau and Lifshitz in its Taylor series form was presented in chapter 16.

In chapter 17, the polarization effects arisen from perforated metallic plates exhibiting extraordinary transmission were given. Setting aside the state-of-the-art of perforated metallic plates, it was shown that by applying Babinet's principle, subwavelength hole arrays arranged in rectangular lattice can further enhance its potential polarization response.

THz wave technology is remarkable in recent optical engineering area. Many development and application researches are being carried out. In chapter 18, one of such researches, study of internal measurement application for insulating materials, was reviewed. Some measurements were carried out with polyethylene, in which THz wave shows high transmission probability, and internal cavity interface and mechanical stress was detected. The applicability of such internal measurement in polyethylene was presented.

In chapter 19, the aim was to construct the conditions for optical reciprocity in the case with a nonlocal anisotropic magnetic permeability and electric permittivity, motivated by the recent explosion in the research with metamaterials according to two different mathematical viewpoints.

Finally, in the last chapter, beamforming of antenna arrays focused in near field zone or intermediate-field zone was studied and new principles of this process were revealed with a purpose of increasing the 3-dimensional gain performance of antenna arrays at a wide range of angles and improving the quality level of this technique for expanding applicability of the focused antenna.

I hope interested readers have the possibility for finding the important and fundamental results on electromagnetic wave propagation in media that exhibits different electrical and magnetic properties.

I thank to Mr. Željko Špalj and Msc. Iva Lipovic who helped me. I would also like to thank to Prof. Dr. Kerim GUNEY who had a lot of support for my scientific development.

Ali Akdagli
Electrical and Electronics Dept., Mersin University
Turkey

1

Electric and Magnetic Characterization of Materials

Leonardo Sandrolini, Ugo Reggiani and Marcello Artioli
Department of Electrical Engineering, University of Bologna
Italy

1. Introduction

The knowledge of the electric and magnetic properties of materials over a broadband frequency range is an essential requirement for accurate modelling and design in several engineering applications. Such applications span printed circuit board design, electromagnetic shielding, biomedical research and determination of EM radiation hazards (Deshpande et al. (1997); Li et al. (2011); Murata et al. (2005)). The electric and magnetic properties of materials usually depend on several factors: frequency, temperature, linearity, isotropy, homogeneity, and so on. The dispersive behaviour exhibited by these materials can be represented by a complex relative permittivity and magnetic permeability which depend on frequency as

$$\hat{\varepsilon}(\omega) = \varepsilon'(\omega) - j\varepsilon''(\omega) \tag{1}$$

$$\hat{\mu}(\omega) = \mu'(\omega) - j\mu''(\omega) \tag{2}$$

being ω the angular frequency, ε', μ' the real parts and ε'', μ'' the imaginary parts of the complex relative permittivity and magnetic permeability, respectively. The real part takes the ability of the medium to store electrical (or magnetic) energy into account, the imaginary part the dielectric (or magnetic) energy losses. The interaction of incident electromagnetic fields with a material can be successfully investigated only when accurate information on the complex permittivity and magnetic permeability is attained. For example, from the knowledge of the frequency dependence of the complex relative permittivity and magnetic permeability of a material, the shielding effectiveness of a structure made of that material can be predicted; similarly, signal interconnects can be accurately designed when the frequency dependence of the dielectric substrate is known; from dielectric property information of tissues the spatial distribution of an incident electromagnetic field and the absorbed power can be accurately determined. Although the complex relative permittivity and magnetic permeability are quantities not directly measurable, they are reconstructed from the measurement of a sensor reflection coefficient or scattering parameters, which can be obtained with a number of different techniques proposed and developed over the last decades (Afsar et al. (1986); Baker-Jarvis et al. (1995); Faircloth et al. (2006); Ghodgaonkar et al. (1990); Queffelec et al. (1994)). Some of these techniques are: open-ended coaxial probe, free-space measurement, cavity resonator, parallel plate capacitor, transmission-line techniques (microstrip, waveguide, etc.); they may be in time domain or frequency domain and make use of probes with one or two ports. No technique is all-embracing as each is limited by its own constraint to specific frequencies, materials (e.g., liquid, malleable or solid material; isotropic or anisotropic) and applications. Regardless of the technique used for the measurement, the common challenge is the extraction of the complex relative permittivity and/or magnetic permeability from measured data by expressing the measured quantities as a function of these parameters. The inversion problem to be solved is thus affected by the mathematical model, i.e., the theoretical expressions that relate the electrical and/or magnetic parameters to the measured quantities. The inversion problem can be solved with deterministic or stochastic methods. The complex relative permittivity and magnetic permeability can be determined over the whole frequency range of interest or on a point-by-point basis (at individual frequency points). It is assumed that the materials considered in the analysis present a negligible ohmic conductivity ($\sigma = 0$). The chapter is organized as follows. Section 2 of the chapter will briefly cover the most common experimental methods used in the electric and magnetic characterization of materials. The problem formulation is presented in Section 3 and the outline of a proposed procedure for parameter extraction is given in Section 4. Finally, results obtained with the proposed approach are shown and commented in Section 5.

2. Techniques of measurement

There exist a large number of techniques developed for the measurement of the electric and magnetic properties of materials. The most common and widespread one- and two-port techniques are transmission-line techniques (open-ended coaxial probe (Misra et al. (1990); Xu et al. (1991)), rectangular (Deshpande et al. (1997); Faircloth et al. (2006); Jarem et al. (1995)) or cylindrical (Ligthart (1983)) waveguide, microstrip or stripline (Barry (1986); Queffelec et al. (1994))), free-space measurement (Galek et al. (2010); Ghodgaonkar et al. (1990)) and cavity resonator (Yoshikawa and Nakayama (2008)). These techniques are different for accuracy and frequency bandwidth of measurement; some are nondestructive and noncontacting and may require sample preparation. Measurement results can also be different because of the field orientation with respect to the material interface, being the measurement more accurate when the fields are tangential to the interface. Although resonator techniques are recognized as more accurate than transmission-line techniques, they can be applied in a narrow frequency band only. In the next sections an overview of the most common broadband techniques of measurement is given.

2.1 One-port techniques

2.1.1 Open-ended coaxial probe

Open-ended coaxial probes have been used extensively by a number of authors mainly for measuring the complex permittivity of dispersive materials (Misra et al. (1990); Stuchly et al. (Febr. 1994); Xu et al. (1991)). For instance, they have been used to measure the electric properties of biological tissues, soils, food, chemicals. The open end of the probe is put in contact with a specimen of the material and the complex reflection coefficient at the aperture is measured with a vector network analyzer (VNA). The technique is particularly suitable for liquids or malleable solids that make a good contact with the probe face. The measurement of solid materials may be affected by a significant error if there are air gaps between the face of the probe and the sample due to surface roughness of the sample. In fact, the electric field at the probe aperture has both the radial and axial components. Models which keep the lift-off of the probe into account have also been proposed (Baker-Jarvis et al. (1994)). Basically, the technique consists in retrieving the complex permittivity from the measurement by relating it to the coaxial probe aperture admittance which, in turn, is obtained from the measured reflection coefficient. The probe is designed in order to have only the TEM mode propagating along the coaxial line, therefore the upper limit of the frequency range in which this fixture can be employed is determined by the frequency cutoff of the higher order modes created at the discontinuity introduced by the material under test. This cutoff frequency depends on the inner and outer diameters of the probe. Moreover, the cell must be long enough to make the evanescent modes decay appreciably far from the open end of the probe. Another issue of this technique concerns the probe calibration, which is usually carried out in three steps with factory-standard calibration loads (short, open and load terminations) (Blackham and Pollard (1997)). In order to improve the accuracy of the measurement, reference loads with liquids of known permittivity (Marsland and Evans (1987)) or short-cavitiy terminations (Otto and Chew (1991)) have been proposed.

2.2 Two-port techniques

2.2.1 Waveguide

The technique consists in filling completely or in part the cross-section of a waveguide (or TEM transmission line) with a material sample and in measuring the scattering parameters by means of a VNA in a broadband frequency range. The electric and magnetic parameters of the material are found through the discontinuity introduced by the sample inside the waveguide as the scattering parameters are related to the permittivity and magnetic permeability of the material with the scattering equations (Nicolson and Ross (1970)). Although preparation is simple, the sample needs to be machined to be fit into the fixture. The most common geometry for the waveguides is the rectangular one; in particular, the waveguide is designed to have only the dominant mode TE_{10} in order to avoid exciting higher order modes.

2.2.2 Free-space measurement

The technique consists in measuring the insertion loss and phase change of a material sample by means of a couple of antennas. The measurement is carried out in free space in a wide broadband frequency range, which extends from a few tens of MHz to tens of GHz according to the available instrumentation. As the technique is contactless and nondestructing, it can be suitable to high-temperature measurements (Ghodgaonkar et al. (1990)). The two antennas are connected to the two ports of a VNA and the scattering parameters related to transmitted and reflected fields are measured. The permittivity and magnetic permeability of the material are then calculated through the scattering equations.

3. Problem formulation

The extraction of the electric and/or magnetic parameters of materials from measurement is a two-step process. The first step of this inversion problem is to find a mathematical model that relates the electrical and/or magnetic parameters of the material under test to the measured quantities.

3.1 Open-ended coaxial probe formulation

For the open-ended coaxial probe measurement technique the complex relative permittivity is determined by inverting the expression of $\hat{Y}(\hat{\varepsilon})$, where $\hat{Y}$ is the aperture admittance of the probe (Stuchly et al. (Febr. 1994))

$$\hat{Y} = Y_0 \frac{1 - \hat{\Gamma}}{1 + \hat{\Gamma}} \tag{3}$$

where Y_0 is the characteristic admittance of the coaxial line and $\hat{\Gamma}$ is the reflection coefficient at the aperture. There are several analytical expressions for the aperture admittance of open-ended coaxial probes (De Langhe et al. (1993); Misra et al. (1990); Xu et al. (1987; 1991)) which contain the complex permittivity explicitly and that can be compared to the measured admittance. Some are very heavy from the computational point of view and may result in convergence problems when numerically solved, because of the presence of multiple integrals, Bessel functions and sine integrals. The expression for the aperture admittance given by (Marcuvitz (1951)), found by matching the electromagnetic field around the probe aperture, can be adopted

$$\hat{Y} = \frac{\sqrt{\hat{\varepsilon}}Y_0}{\sqrt{\varepsilon_{cl}}\ln(b/a)}\left\{\int_0^{\pi/2}\left[J_0\left(\gamma_0 a\sqrt{\hat{\varepsilon}}\sin\theta\right) - J_0\left(\gamma_0 b\sqrt{\hat{\varepsilon}}\sin\theta\right)\right]^2 \frac{d\theta}{\sin\theta}\right.$$
$$+\frac{j}{\pi}\int_0^{\pi}\left[2Si\left(\gamma_0 a\sqrt{\hat{\varepsilon}}\sqrt{1+\frac{b^2}{a^2}-2\frac{b}{a}\cos\theta}\right)\right.$$
$$\left.\left.-Si\left(2\gamma_0 a\sqrt{\hat{\varepsilon}}\sin\frac{\theta}{2}\right) - Si\left(2\gamma_0 b\sqrt{\hat{\varepsilon}}\sin\frac{\theta}{2}\right)\right]d\theta\right\} \quad (4)$$

where: $\hat{\varepsilon}$ is the complex relative permittivity of the material under test, ε_{cl} is the relative permittivity of the coaxial line, a and b are the inner and outer radii of the coaxial line, respectively, γ_0 is the absolute value of the propagation constant in free space (see (16)), Si and J_0 are the sine integral and the Bessel function of zero order, respectively. This integral expression can be evaluated numerically by means either of series expansion as in (Misra et al. (1990); Xu et al. (1987)) or numerical integration.

3.2 Two-port formulation

In a similar manner, for two-port techniques the theoretical expressions of the scattering parameters as functions of the complex relative permittivity (1) and magnetic permeability (2) have to be found. This can be easily achieved expressing the scattering parameters $\hat{S}_{11}(\omega)$ and $\hat{S}_{21}(\omega)$, which can be measured with a VNA and with a two-port fixture, as a function of the reflection coefficient of the air-sample interface and transmission coefficient, $\hat{\Gamma}$ and $\hat{T}$, respectively. For the TEM propagation mode (free-space measurement system and TEM transmission line) and for waveguides with only the TE_{10} propagation mode these expressions for S_{11} and S_{21} are (Barroso and De Paula (2010); Boughriet et al. (1997); Ghodgaonkar et al. (1990); Ligthart (1983))

$$\hat{S}_{11} = \frac{\hat{\Gamma}\left(1-\hat{T}^2\right)}{1-\hat{\Gamma}^2\hat{T}^2}, \quad (5)$$

$$\hat{S}_{21} = \frac{\hat{T}\left(1-\hat{\Gamma}^2\right)}{1-\hat{\Gamma}^2\hat{T}^2} \quad (6)$$

being

$$\hat{\Gamma} = \frac{\hat{Z}-Z_0}{\hat{Z}+Z_0}, \quad (7)$$

$$\hat{T} = e^{-\hat{\gamma}d} \quad (8)$$

and for a rectangular waveguide (Inan and Inan (2000); Kraus and Fleisch (1999); Sadiku (2007))

$$\hat{Z} = j\frac{\omega\mu_0\hat{\mu}}{\hat{\gamma}}, \tag{9}$$

$$Z_0 = j\frac{\omega\mu_0}{\hat{\gamma}_0}, \tag{10}$$

$$\hat{\gamma} = j\frac{2\pi}{\lambda_0}\sqrt{\hat{\varepsilon}\hat{\mu} - \left(\frac{\lambda_0}{\lambda_{0c}}\right)^2}, \tag{11}$$

$$\hat{\gamma}_0 = j\frac{2\pi}{\lambda_0}\sqrt{1 - \left(\frac{\lambda_0}{\lambda_{0c}}\right)^2} \tag{12}$$

where $\hat{Z}$ and $\hat{\gamma}$ are the intrinsic impedance and propagation constant of the filled waveguide, respectively, Z_0 and $\hat{\gamma}_0$ are the intrinsic impedance and propagation constant of the empty waveguide, respectively, and d is the thickness of the sample. The propagation constants depend on the wavelength in free space λ_0 and the cutoff wavelength in the waveguide λ_{0c} (i.e., the wavelength in free space at the cutoff frequency in the empty waveguide (Kraus and Fleisch (1999))). For a rectangular waveguide $\lambda_{0c} = 2a$, where the width of the waveguide a is chosen to be twice the height of the waveguide in order to have only the TE_{10} propagation mode impinging on the material sample in the frequency range of interest.
For free space-measurement, (9)-(12) become (Galek et al. (2010); Ghodgaonkar et al. (1990))

$$\hat{Z} = Z_0\sqrt{\frac{\hat{\mu}}{\hat{\varepsilon}}}, \tag{13}$$

$$Z_0 = \sqrt{\frac{\mu_0}{\varepsilon_0}}, \tag{14}$$

$$\hat{\gamma} = \hat{\gamma}_0\sqrt{\hat{\varepsilon}\hat{\mu}}, \tag{15}$$

$$\hat{\gamma}_0 = j\frac{2\pi}{\lambda_0} \tag{16}$$

where $\hat{Z}$ and Z_0 are the intrinsic impedances of the material under test and free space, respectively.
Once the scattering parameters (5) and (6) are expressed as functions of $\hat{\varepsilon}$ and $\hat{\mu}$, they must be inverted to yield the complex relative permittivity and magnetic permeability.

4. Procedure for the extraction

The second step of the inversion problem is the extraction of the material parameters from the measured quantities.

4.1 Nicolson-Ross-Weir procedure

In the standard Nicolson-Ross-Weir procedure (Nicolson and Ross (1970); Weir (1974)), the relative complex permittivity and magnetic permeability are obtained explicitly for a rectangular waveguide from (7) to (12) on a point-by-point basis as

$$\hat{\mu} = \frac{\lambda_{0g}}{\hat{\Lambda}}\left(\frac{1+\hat{\Gamma}}{1-\hat{\Gamma}}\right), \tag{17}$$

$$\hat{\varepsilon} = \frac{\lambda_0^2}{\hat{\mu}} \left(\frac{1}{\hat{\Lambda}^2} + \frac{1}{\lambda_{0c}^2} \right) \tag{18}$$

where λ_{0g} is the wavelength in the empty waveguide

$$\lambda_{0g} = \frac{\lambda_0}{\sqrt{1 - (\lambda_0/\lambda_{0c})^2}} \tag{19}$$

and

$$\frac{1}{\hat{\Lambda}^2} = -\left(\frac{1}{2\pi d} \ln \frac{1}{\hat{T}} \right)^2. \tag{20}$$

This requires to express $\hat{\Gamma}$ and $\hat{T}$ from the measured scattering parameters $\hat{S}_{11}$ and $\hat{S}_{21}$: from (5) and (6) one can write

$$\hat{K} = \frac{\hat{S}_{11}^2 - \hat{S}_{21}^2 + 1}{2\hat{S}_{11}} \tag{21}$$

$$\hat{\Gamma} = \hat{K} \pm \sqrt{\hat{K}^2 - 1} \tag{22}$$

$$\hat{T} = \frac{\hat{S}_{11} + \hat{S}_{21} - \hat{\Gamma}}{1 - (\hat{S}_{11} + \hat{S}_{21})\,\hat{\Gamma}}. \tag{23}$$

However, this procedure presents phase ambiguity and suffers instability at frequencies where the sample length is a multiple of one-half wavelength in low-loss materials. To overcome this problem, in (Baker-Jarvis et al. (1990)) an iterative procedure has been proposed, which gives stable solutions over the frequency range. This technique requires setting the relative magnetic permeability to 1 and a good initial guess for the permittivity (usually a solution of the Nicolson-Ross equations).

4.2 Fitting procedure

A different procedure for the extraction of material parameters involves minimizing the distance between the calculated aperture admittance (4) or scattering parameters (5) and (6) and the corresponding measured quantities through fitting algorithms, which may be based either on deterministic or stochastic optimization procedures. The minimization can be carried out over the whole frequency range or on a point-by-point basis (i.e., at individual frequency points).

The former approach, followed by a number of authors, consists in modelling the complex relative permittivity and magnetic permeability with a prespecified functional form whose parameters needs to be determined with an optimization procedure. Laurent series can be used for complex relative permittivity and magnetic permeability models (Domich et al. (1991)), as well as dispersive laws, such as Havriliak-Negami and its special cases Cole-Cole and Debye to model dielectric relaxation (Kelley et al. (2007)), or the Lorentz model for both dielectric and magnetic dispersion (Koledintseva et al. (2002)). The Havriliak-Negami model is an empirical modification of the single-pole Debye relaxation model

$$\hat{\varepsilon}(\omega) = \varepsilon_\infty + \frac{\varepsilon_s - \varepsilon_\infty}{\left[1 + (j\omega\tau)^{1-\alpha}\right]^\beta} \tag{24}$$

where ε_s and ε_∞ are the values of the real part of the complex relative permittivity at low and high frequency, respectively, τ is the relaxation time, and α and β are positive real constants

$(0 \leq \alpha, \beta \leq 1)$. From this model, the Cole-Cole equation can be derived setting $\beta = 1$; the Debye equation is obtained with $\alpha = 0$ and $\beta = 1$. This empirical model has the ability to give a better fit to the behaviour of dispersive materials over a wide frequency range. When multiple relaxation times are needed, the complex relative permittivity can be modelled with a Debye function expansion

$$\hat{\varepsilon}(\omega) = \varepsilon_\infty + \sum_{n=1}^{N} \frac{\Delta\varepsilon_n}{1 + j\omega\tau_n} \tag{25}$$

being $\Delta\varepsilon_n$ and τ_n for $n = 1, 2, \ldots, N$ the strengths and relaxation times of the Debye dispersion, respectively. Modelling the complex relative permittivity with the Lorentz model yields

$$\hat{\varepsilon}(\omega) = \varepsilon_\infty + \frac{(\varepsilon_s - \varepsilon_\infty)\,\omega_0^2}{\omega_0^2 + j\omega\delta - \omega^2} \tag{26}$$

where ω_0 is the resonance frequency and δ is the damping factor. For instance, for open-ended coaxial probe measurements of complex relative permittivity, introducing (24) into (4) we obtain an aperture admittance which depends then on five parameters

$$\hat{Y}\left(\varepsilon_s, \varepsilon_\infty, \tau, \alpha, \beta\right) \tag{27}$$

which reduce to four in the case of the Cole-Cole model

$$\hat{Y}\left(\varepsilon_s, \varepsilon_\infty, \tau, \alpha\right) \tag{28}$$

and to three for the Debye model

$$\hat{Y}\left(\varepsilon_s, \varepsilon_\infty, \tau\right). \tag{29}$$

For the Debye function expansion, the aperture admittance depends on $2N + 1$ parameters

$$\hat{Y}\left(\Delta\varepsilon_1, \ldots, \Delta\varepsilon_N, \varepsilon_\infty, \tau_1, \ldots, \tau_N\right). \tag{30}$$

Eventually, for the Lorentz model, the aperture admittance is

$$\hat{Y}\left(\varepsilon_s, \varepsilon_\infty, \omega_0, \delta\right). \tag{31}$$

With reference to the above listed models (24)-(26) for the dispersion law, this approach is summarized in Fig. 1. The unknown parameters are then extracted by fitting the expressions from (27) to (31) to the measured characteristic data.

It can be observed that this approach can also be applied to measurement techniques of the complex permittivity other than open-ended coaxial probes, provided that the analytical relation between a measurable quantity and the complex permittivity is known. In a similar manner, it can be applied to the simultaneous extraction of the complex permittivity and magnetic permeability of a material by comparing the analytical reflection ($\hat{S}_{11}$) and transmission ($\hat{S}_{21}$) coefficients to the measurements carried out with a VNA in the standard transmission-line or free-space techniques.

The success of the extraction of the model parameters relies on a proper choice of the dispersive laws for the material under test; conversely, the fitting algorithms may experience nonconvergence issues or the parameters of the models may be determined with excessive errors. Especially for newly developed materials, individuating the proper dispersion laws may result in a difficult task. Moreover, the complexity of the models affects also the choice of the fitting algorithm for the parameter extraction.

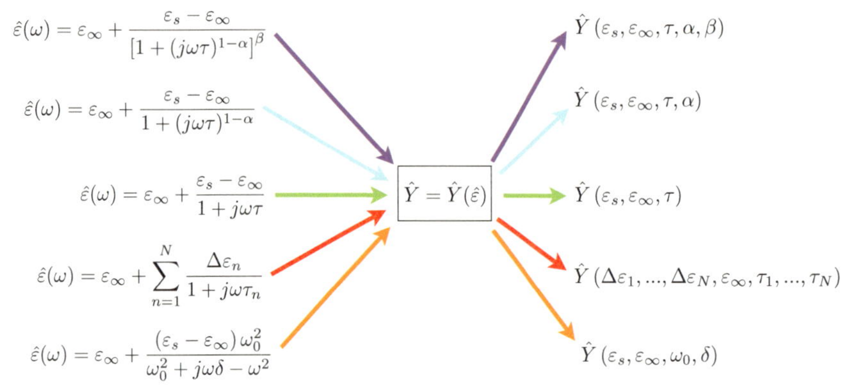

Fig. 1. Aperture admittance of an open-ended coaxial probe as a function of dispersion law parameters.

For these reasons, the latter approach can be followed, which consists in not making any assumption on the dispersive laws and in determining the real and imaginary parts of the material complex parameters at each measurement frequency (on a point-by-point basis). Once the complex relative permittivity and magnetic permeability as a function of frequency are known, proper models can be chosen to represent the material properties over the entire frequency range of measurement. For example, for isotropic materials which exhibit electric and magnetic properties, the inversion process involves the minimization of an objective function φ of the kind

$$\varphi\left(f_i, \varepsilon', \varepsilon'', \mu', \mu''\right) = \sum_{i=1}^{N} \Big[\left(\hat{S}_{11\text{measured}}\left(f_i\right) - \hat{S}_{11\text{calculated}}\left(f_i, \varepsilon', \varepsilon'', \mu', \mu''\right)\right)^2 + \left(\hat{S}_{21\text{measured}}\left(f_i\right) - \hat{S}_{21\text{calculated}}\left(f_i, \varepsilon', \varepsilon'', \mu', \mu''\right)\right)^2 \Big] \tag{32}$$

where f_i is the generic frequency of the N frequencies of the measurement data set. The reflection and transmission coefficients $\hat{S}_{11}$ and $\hat{S}_{21}$ are those of a two-port technique of Sec. 2. Similarly, when the open-ended coaxial probe is used, the determination of the complex permittivity requires to minimize an objective function ψ at each frequency of the measurement data set

$$\psi\left(f_i, \varepsilon', \varepsilon''\right) = \sum_{i=1}^{N} \left(\hat{Y}_{\text{measured}}\left(f_i\right) - \hat{Y}_{\text{calculated}}\left(f_i, \varepsilon', \varepsilon''\right)\right)^2. \tag{33}$$

This approach is handled with the proposed procedure, where a number of optimization algorithms are used to search the minimum values and to overcome possible convergence issues or local minima stalemates. They can be deterministic (e.g., Newton, Interior Point, Quasi-Newton, Levenberg-Marquardt, Gradient, Nonlinear Conjugate Gradient, Principal Axis, Nelder-Mead) or stochastic (e.g., Differential Evolution, Simulated Annealing, Random Search, Particle Swarm). This choice is also motivated as different optimization methods may result in being more appropriate to extract the complex relative permittivity and/or magnetic permeability of the same material at different measurement frequencies. For each measurement frequency, the optimization methods are applied according to a user defined sequence. In case an optimization method does not reach convergence or the desired accuracy within the maximum number of iterations, the minimum search is repeated with the next optimization method in the sequence. The proposed procedure, summarized in the flow-chart of Fig. 2, was implemented in Mathematica (Wolfram Research, Inc. (2008)), a powerful and complex programming environment with the capability of performing both numerical and symbolic calculations. This computational language can be easily extended developing custom algorithms. The optimization methods employed in the parameter extraction procedure are listed in Table 1; of course, other optimization algorithms may be added to this sequence, thus increasing the chances of determining the complex relative permittivity with the requested accuracy. The worst result is given by the failure of all optimization methods.

The interesting aspect of this approach is that the determination of the complex relative permittivity and magnetic permeability may be enhanced with subsequent refinements; more experimental points measured at different frequencies can be added to the first set of measurement data. The procedure can then be run on this additional data set only, extracting the complex relative permittivity at these additional frequencies. Parallelization is another interesting feature of the proposed procedure: in fact, the minimization of the objective function is carried out at each single frequency, and each minimization process is independent from the others. The procedure can then be quite easily implemented on a grid computing system to speed up the extraction process.

5. Results

In order to show the application of the procedure, the complex relative permittivity of methanol was extracted over the frequency range 1–15 GHz. For this nonmagnetic material, the reflection coefficient is sufficient for the complex permittivity determination and thus a single objective function in terms of $\hat{\varepsilon}$ was used. In particular, the complex relative permittivity was obtained from the aperture admittance of an open-ended coaxial probe, that can be related to the reflection coefficient. With reference to (4), the inner and outer radii of the open-ended coaxial probe are $a = 0.04\,cm$ and $b = 0.114\,cm$, respectively, and the dielectric between the inner and outer conductors has a relative permittivity $\varepsilon_{cl} = 1.58$ (Misra et al. (1990)). With the former approach outlined in Sec. 4.2, solving the inverse problem yields a vector of the unknown parameters. The dispersion models considered in the extraction are the Havriliak-Negami, the Cole-Cole, the Debye and Debye function expansion models. As some of the algorithms are stochastic with initial values chosen randomly within a search range, different runs of the procedure may give spread solutions which are averaged out. The dispersion of the parameter values of thee models (Havriliak-Negami, Cole-Cole, and Debye) can be compared graphically, as shown in Fig. 3, by normalizing each parameter value over its own search range. It can be noticed that the fewer the parameters, the less they are spread

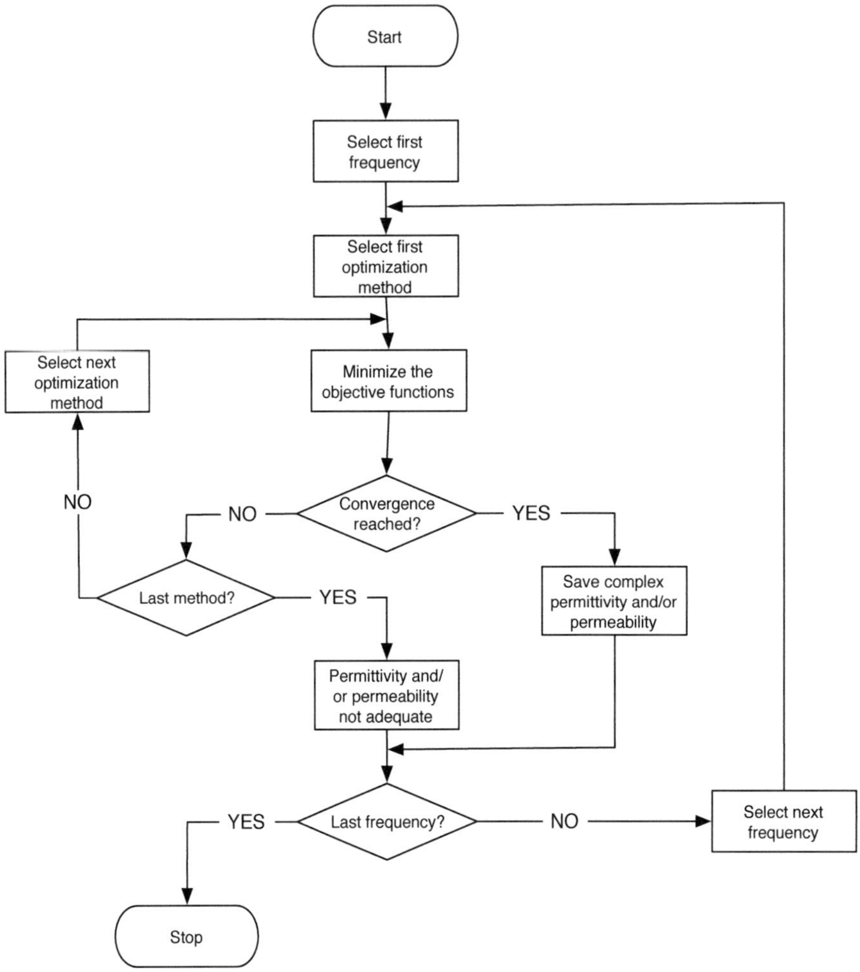

Fig. 2. Flow-chart of the extraction procedure of the complex relative permittivity and/or magnetic permeability.

over the search range: the parameters of the Debye model are less dispersed than those of Havriliak-Negami or Cole-Cole models. Conversely, when the latter approach in Sec. 4.2 is adopted, the complex permittivity is extracted on a point-by-point basis at each frequency of the measurement range. The real and imaginary parts of the complex relative permittivity extracted with both approaches are shown in Figs. 4 and 5, respectively, together with the permittivity values as calculated in (Misra et al. (1990)). The curves in the plots are labelled accordingly. From the two plots it can be noticed that the behaviour versus frequency of the complex relative permittivity extracted at individual frequencies (labelled "No model" in the graphics) is in good agreement with that related to the Havriliak-Negami and Debye function expansion dielectric relaxation models. The agreement is less good with both the Debye and Cole-Cole dispersive models, which give similar values for the permittivity over the frequency range.

The validation of the extraction procedure consists in calculating the aperture admittance of the coaxial probe using (4) and the complex relative permittivity extracted and comparing

	Method
Deterministic	Newton
	Interior Point
	Quasi-Newton
	Levenberg–Marquardt
	Gradient
	Nonlinear Conjugate Gradient
	Principal Axis
	Nelder–Mead
Stochastic	Differential Evolution
	Simulated Annealing
	Random Search
	Particle Swarm

Table 1. Optimization methods employed in the parameter extraction procedure.

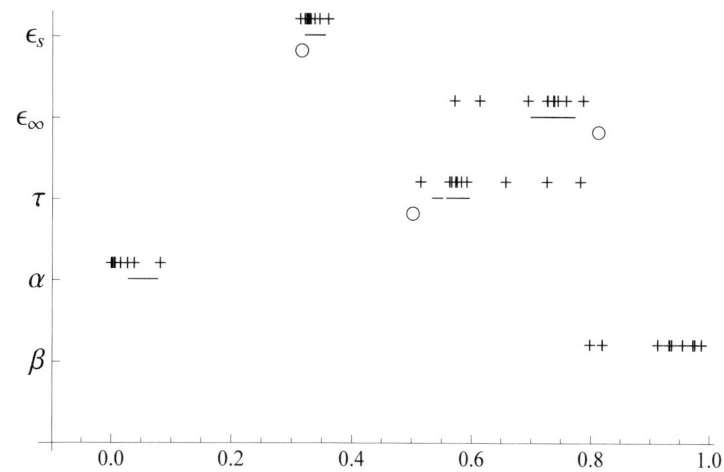

Fig. 3. Normalized parameter values obtained with the Havriliak-Negami (+), Cole-Cole (-) and Debye (○) models.

it to the measured one. The comparison is shown in Figs. 6 and 7 for the real and imaginary parts of the aperture admittance, respectively. The plots show that the aperture admittance calculated with the complex permittivity extracted on a point-by-point basis and that obtained with the Havriliak-Negami and Debye function expansion models are in a very good agreement with the measurement. Differently, the aperture admittance calculated with the permittivity modelled with the Debye and Cole-Cole models differs from the measured one especially at higher frequencies. The complex permittivity extracted on a point-by-point basis shows the best overall result.

6. Conclusion

The chapter outlines possible procedures for the extraction of electric and magnetic parameters of dispersive materials. The complex relative permittivity and magnetic permeability can be modelled with either dispersive laws or on a point-by-point basis (at individual frequencies). The latter approach was implemented in a procedure proposed in

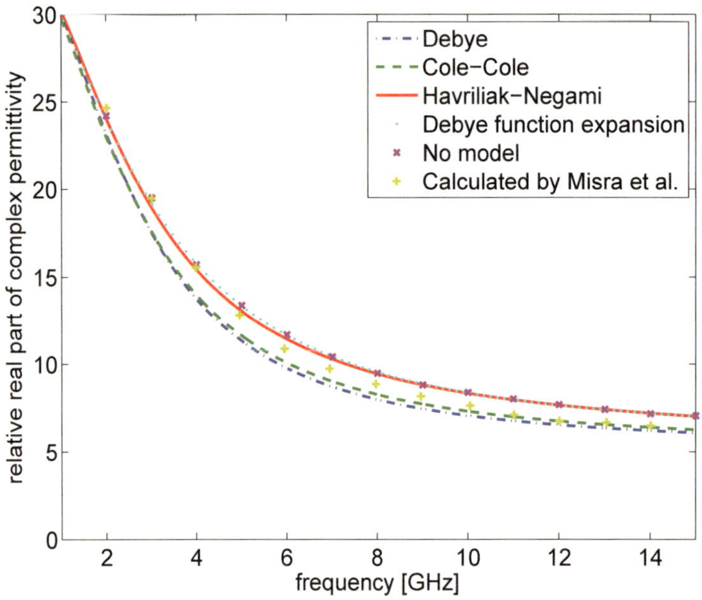

Fig. 4. Comparison of the real part of the complex relative permittivity versus frequency extracted with the Havriliak-Negami, Cole-Cole, Debye and Debye function expansion models and without assuming a dispersive model with that calculated according to the procedure in (Misra et al. (1990)).

this chapter, where a number of optimization algorithms are cycled to extract the complex relative permittivity and to overcome possible convergence issues or local minima stalemates. The assessment and validation of the procedure were carried out against the experimental data for the aperture admittance of an open-ended coaxial probe immersed in methanol. It is found that the frequency behaviour of the complex relative permittivity extracted with the procedure is in a very good agreement with that modelled according to two classic dielectric relaxation models (Havriliak-Negami, and Debye function expansion models), commonly adopted in literature to represent dispersive materials. Furthermore, the calculated aperture admittance was compared with the measured one. The comparison shows that the best level of agreement between the calculated and measured aperture admittance is obtained with the complex relative permittivity extracted with the proposed procedure. This approach is particularly advantageous when applied to new developed materials or materials whose frequency behaviour is not known (materials which cannot be modelled *a priori* with a dielectric relaxation model). In any case, once the complex relative permittivity is known, a proper dielectric relaxation model can be adopted to represent the permittivity over the whole frequency range of measurement. Its parameters can be quickly determined with standard interpolation routines of the complex relative permittivity values previously extracted. Interesting features of the proposed procedure are the possibility of further refining the complex relative permittivity extraction by adding more experimental data points at later times, and its intrinsic parallel nature, being each minimization process carried out at every single frequency independent from the others.

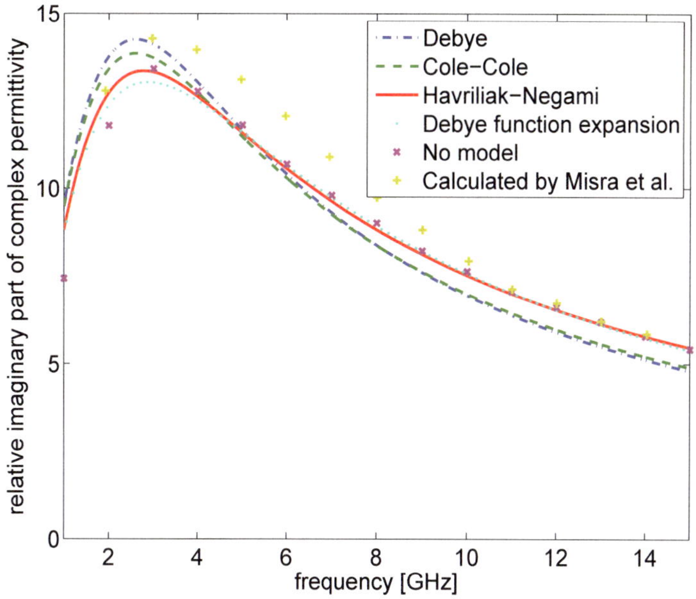

Fig. 5. Comparison of the imaginary part of the complex relative permittivity versus frequency extracted with the Havriliak-Negami, Cole-Cole, Debye and Debye function expansion models and without assuming a dispersive model with that calculated according to the procedure in (Misra et al. (1990)).

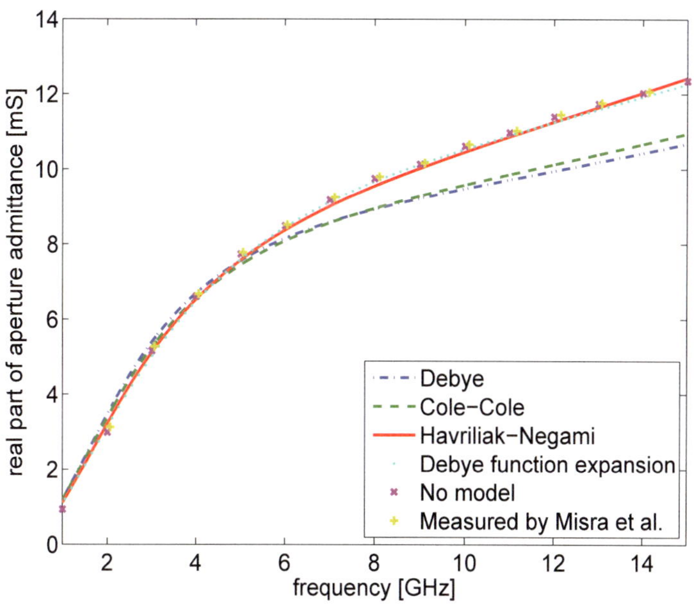

Fig. 6. Comparison of the real part of the calculated and measured (Misra et al. (1990)) aperture admittance versus frequency.

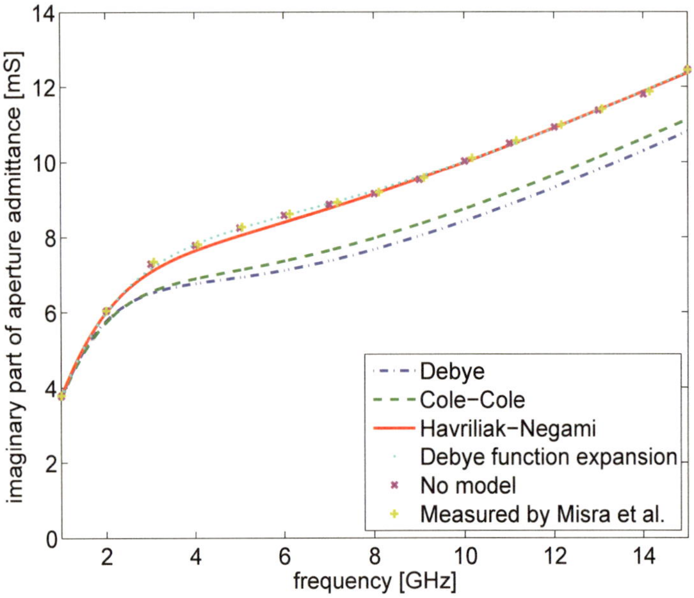

Fig. 7. Comparison of the imaginary part of the calculated and measured (Misra et al. (1990)) aperture admittance versus frequency.

7. References

Afsar, M., Birch, J., Clarke, R. and Chantry, G. (1986). The measurement of the properties of materials, *Proceedings of the IEEE* 74(1): 183–99.

Baker-Jarvis, J., Janezic, M. D., Domich, P. D. and Geyer, R. G. (1994). Analysis of an open-ended coaxial probe with lift-off for nondestructive testing, *IEEE Transactions on Instrumentation and Measurement* 43(5): 711–718.

Baker-Jarvis, J., Jones, C., Riddle, B., Janezic, M., Geyer, R., Grosvenor Jr., J. H. and Weil, C. (1995). Dielectric and magnetic measurements: a survey of nondestructive, quasi-nondestructive, and process-control techniques, *Research in Nondestructive Evaluation* 7(2-3): 117–36.

Baker-Jarvis, J., Vanzura, E. J. and Kissick, W. A. (1990). Improved technique for determining complex permittivity with the transmission/reflection method, *IEEE Transactions on Microwave Theory and Techniques* 38(8): 1096–1103.

Barroso, J. and De Paula, A. (2010). Retrieval of permittivity and permeability of homogeneous materials from scattering parameters, *Journal of Electromagnetic Waves and Applications* 24(11-12): 1563–1574.

Barry, W. (1986). Broad-band, automated, stripline technique for the simultaneous measurement of complex permittivity and permeability., *IEEE Transactions on Microwave Theory and Techniques* MTT-34(1): 80–84.

Blackham, D. V. and Pollard, R. D. (1997). Improved technique for permittivity measurements using a coaxial probe, *IEEE Transactions on Instrumentation and Measurement* 46(5): 1093–1099.

Boughriet, A. H., Legrand, C. and Chapoton, A. (1997). Noniterative stable transmission/reflection method for low-loss material complex permittivity determination, *IEEE Transactions on Microwave Theory and Techniques* 45(1): 52–57.

De Langhe, P., Blomme, K., Martens, L. and De Zutter, D. (1993). Measurement of low-permittivity materials based on a spectral-domain analysis for the open-ended coaxial probe, *IEEE Transactions on Instrumentation and Measurement* 42(5): 879–886.

Deshpande, M., Reddy, C., Tiemsin, P. and Cravey, R. (1997). A new approach to estimate complex permittivity of dielectric materials at microwave frequencies using waveguide measurements, *IEEE Transactions on Microwave Theory and Techniques* 45(3): 359–366.

Domich, P. D., Baker-Jarvis, J. and Geyer, R. G. (1991). Optimization techniques for permittivity and permeability determination, *Journal of Research of the National Institute of Standards and Technology* 96(5): 565–575.

Faircloth, D. L., Baginski, M. E. and Wentworth, S. M. (2006). Complex permittivity and permeability extraction for multilayered samples using s-parameter waveguide measurements, *IEEE Transactions on Microwave Theory and Techniques* 54(3): 1201–1208.

Galek, T., Porath, K., Burkel, E. and Van Rienen, U. (2010). Extraction of effective permittivity and permeability of metallic powders in the microwave range, *Modelling and Simulation in Materials Science and Engineering* 18(2): 1–13.

Ghodgaonkar, D. K., Varadan, V. V. and Varadan, V. K. (1990). Free-space measurement of complex permittivity and complex permeability of magnetic materials at microwave frequencies, *IEEE Transactions on Instrumentation and Measurement* 39(2): 387–394.

Inan, U. S. and Inan, A. S. (2000). *Electromagnetic Waves*, Prentice Hall, Upper Saddle River, NJ, USA.

Jarem, J. M., Johnson Jr., J. B. and Albritton, W. (1995). Measuring the permittivity and permeability of a sample at Ka band using a partially filled waveguide, *IEEE Transactions on Microwave Theory and Techniques* 43(12 pt 1): 2654–2667.

Kelley, D. F, Destan, T. J. and Luebbers, R. J. (2007). Debye function expansions of complex permittivity using a hybrid particle swarm-least squares optimization approach, *IEEE Transactions on Antennas and Propagation* 55(7): 1999–2005.

Koledintseva, M. Y., Rozanov, K. N., Orlandi, A. and Drewniak, J. L. (2002). Extraction of the lorentzian and debye parameters of dielectric and magnetic dispersive materials for FDTD modeling, *Journal Electrical Engineering* 53(9): 97–100.

Kraus, J. D and Fleisch, D. A. (1999). *Electromagnetics with Applications*, 5th edn, Mc Graw-Hill, New York, USA.

Li, X., Han, X., Du, Y. and Xu, P. (2011). Magnetic and electromagnetic properties of composites of iron oxide and cob alloy prepared by chemical reduction, *Journal of Magnetism and Magnetic Materials* 323(1): 14–21.

Ligthart, L. P. (1983). A fast computational technique for accurate permittivity determination using transmission line methods, *IEEE Transactions on Microwave Theory and Techniques* MTT-31(3): 249–254.

Marcuvitz, N. (1951). *Waveguide handbook*, McGraw-Hill, New York, USA.

Marsland, T. and Evans, S. (1987). Dielectric measurements with an open-ended coaxial probe, *IEE Proceedings H: Microwaves, Antennas and Propagation* 134(4): 341–349.

Misra, D., Chabbra, M., Epstein, B. R., Mirotznik, M. and Foster, K. R. (1990). Noninvasive electrical characterization of materials at microwave frequencies using an open-ended coaxial line: Test of an improved calibration technique, *IEEE Transactions on Microwave Theory and Techniques* 38(1): 8–14.

Murata, K., Hanawa, A. and Nozaki, R. (2005). Broadband complex permittivity measurement techniques of materials with thin configuration at microwave frequencies, *Journal of Applied Physics* 98(8): 084107-1–084107-8.

Nicolson, A. M. and Ross, G. F. (1970). Measurement of the intrinsic properties of materials by time-domain techniques, *IEEE Transactions on Instrumentation and Measurement* IM-19(4): 377 –382.

Otto, G. and Chew, W. (1991). Improved calibration of a large open-ended coaxial probe for dielectric measurements, *IEEE Transactions on Instrumentation and Measurement* 40(4): 742–746.

Queffelec, P., Gelin, P., Gieraltowski, J. and Loaec, J. (1994). Microstrip device for the broad band simultaneous measurement of complex permeability and permittivity, *IEEE Transactions on Magnetics* 30(2 pt 1): 224–231.

Sadiku, M. N. O. (2007). *Elements of Electromagnetics*, 4th edn, Oxford University Press, Inc., New York, USA.

Stuchly, S. S., Sibbald, C. L. and Anderson, J. M. (1994). A new aperture admittance model for open-ended waveguides, *IEEE Transactions on Microwave Theory and Techniques* 42(2): 192–198.

Weir, W. B. (1974). Automatic measurement of complex dielectric constant and permeability at microwave frequencies, *Proceedings of the IEEE* 62(1): 33–36.

Wolfram Research, Inc. (2008). *Mathematica v. 7.0*, Wolfram Research, Inc., Champaign, Illinois, USA.

Xu, D., Liu, L. and Jiang, Z. (1987). Measurement of the dielectric properties of biological substances using an improved open-ended coaxial line resonator method, *IEEE Transactions on Microwave Theory and Techniques* MTT-35(12): 1424–1428.

Xu, Y., Bosisio, R. G. and Bose, T. K. (1991). Some calculation methods and universal diagrams for measurement of dielectric constants using open-ended coaxial probes, *IEE Proceedings–H: Microwaves, Antennas and Propagation* 138(4): 356–360.

Yoshikawa, H. and Nakayama, A. (2008). Measurements of complex permittivity at millimeter-wave frequencies with an end-loaded cavity resonator, *IEEE Transactions on Microwave Theory and Techniques* 56(8): 2001–2007.

2

Features of Electromagnetic Waves Scattering by Surface Fractal Structures

O. Yu. Semchuk[1] and M.Willander[2]
[1]*Chuiko Institute of Surface Chemistry National Academy of Sciences of Ukraine,*
[2]*Department of Science and Technology Linkoping University*
[1]*Ukraine*
[2]*Sweden*

1. Introduction

Accurate measurement of surface roughness of machined workpieces is of fundamental importance particularly in the precision engineering and manufacturing industry. This is mainly due to the more stringent demand on material quality as well as the miniaturization of product components in these industries [1-3]. For instance, in the disk drive industry, to maintain the quality of the electrical components mounted on an optical disk, the surface roughness of the disk must be accurately measured and controlled. Hence, the surface finish, normally expressed in terms of surface roughness, is a critical parameter used for the acceptance or rejection of a product.

Surface roughness is usually determined by a mechanical stylus profilometer. However, the stylus technique has certain limitations: the mechanical contact between the stylus and the object can cause deformations or damage on the specimen surface and it is a pointwise measurement method and is time consuming. Hence a noncontact and more speed optical method would be attractive. Different optical noncontact methods for measuring surface roughness have been developed mainly based on reflected light detection, focus error detection, laser scattering, speckle and the interference method [4-10]. Some of these have good resolutions and are being applied in some sectors where mechanical measuring methods previously enjoyed clear predominance. Among these methods, the light scattering method [11] which is a noncontact area-averaging technique, is potentially more speedy for surface inspection than other profiling techniques particularly the traditional stylus technique. Other commercially available products such as the scanning tunneling microscope (STM), the atomic force microscope (AFM) and subwavelength photoresist gratings [12-15], which are pointwise techniques, are used mainly for optically smooth surfaces with roughnesses in the nanometer range.

In this chapter in the frame of the Kirchhoff method (scalar model) the average coefficient of light scattering by surface fractal structures was calculated. A normalized band-limited Weierstrass function is presented for modeling 2D fractal rough surfaces. On the basis of numerical calculation of average scattering coefficient the scattering indicatrises diagrams for various surfaces and falling angles were calculated. The analysis of the diagrams results in the following conclusions: the scattering is symmetrically concerning a plane of fall; with increase the degree of a surface calibration the picture becomes complicated; the greatest intensity of a scattering wave is observed in a mirror direction; there are other direction in which the bursts of intensity are observed.

2. Fractal model for two-dimensional rough surfaces

At theoretical research of processes of electromagnetic waves scattering selfsimilar heterogeneous objects (by rough surfaces) is a necessity to use the mathematical models of dispersive objects. As a basic dispersive object we will choose a rough surface. As is generally known, she is described by the function $z(x,y)$ of rejections z of points of M of surface from a supporting plane (x,y) (fig.1) and requires the direct task of relief to the surface.

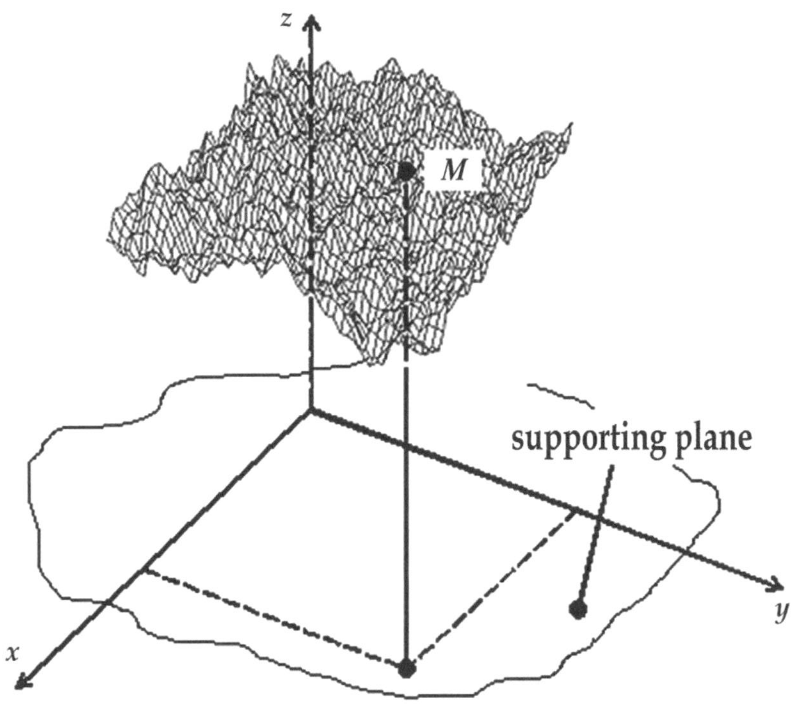

Fig. 1. Schematic image of rough surface

There are different modifications of Weierstrass-Mandelbrot function in the modern models of rough surface are used. For a design a rough surface we is used the Weierstrass limited to the stripe function [3,4]

$$z(x,y) = c_w \sum_{n=0}^{N-1} \sum_{m=1}^{M} q^{(D-3)n} \sin\left\{ Kq^n \left[x\cos\frac{2\pi m}{M} + y\sin\frac{2\pi m}{M} \right] + \varphi_{nm} \right\}, \tag{1}$$

where c_w is a constant which ensures that $W(x, y)$ has a unit perturbation amplitude; $q(q > 1)$ is the fundamental spatial frequency; D $(2 < D < 3)$ is the fractal dimension; K is the

fundamental wave number; N and M are number of tones, and φ_{nm} is a phase term that has a uniform distribution over the interval $[-\pi,\pi]$.
The above function is a combination of both deterministic periodic and random structures. This function is anisotropic in the two directions if M and N are not too large. It has a large derivative and is self similar. It is a multi-scale surface that has same roughness down to some fine scales. Since natural surfaces are generally neither purely random nor purely periodic and often anisotropic, the above proposed function is a good candidate for modeling natural surfaces.
The phases φ_{nm} can be chosen determinedly or casually, receiving accordingly determine or stochastic function $z(x,y)$. We further shall consider φ_{nm} as casual values, which in regular distributed on a piece $[-\pi;\pi]$. With each particular choice of numerical meanings all $N \times M$ phases φ_{nm} (for example, with the help of the generator of random numbers) we receive particular (with the beforehand chosen meanings of parameters c_w, q,K,D,N,M) realization of function $z(x,y)$. The every possible realizations of function $z(x,y)$ form ensemble of surfaces.
A deviation of points of a rough surface from a basic plane proportional c_w, therefore this parameter is connected to height of inequalities of a structure of a surface. Further it is found to set a rough surface, specifying root-mean-square height of its structure σ, which is determined by such grade:

$$\sigma \equiv \sqrt{\langle h^2 \rangle}, \tag{2}$$

where $h = z(x,y)$, $\langle ... \rangle = \prod_{n=0}^{N-1} \prod_{m=1}^{M} \int_{-\pi}^{\pi} \frac{d\varphi_{nm}}{2\pi}(...)$ - averaging on ensemble of surfaces.
The connection between c_w and σ can be established, directly calculating integrals:

$$\sigma = \left[\prod_{n=0}^{N-1} \prod_{m=1}^{M} \int_{-\pi}^{\pi} \frac{d\varphi_{nm}}{2\pi} z^2(x,y) \right]^{\frac{1}{2}} = c_w \left[\frac{M\left(1-q^{2N(D-3)}\right)}{2\left(1-q^{2(D-3)}\right)} \right]^{\frac{1}{2}}. \tag{3}$$

So, the rough surface in our model is described by function from six parameters: c_w (or), q,K,D,N,M. The influence of different parameters on a kind of a surface can be investigated analytically, and also studying structures of surfaces constructed by results of numerical accounts of Weierstrass function. Analysis of the surface profiles built by us on results of numeral calculations (fig. 2) due to the next conclusions:

- the wave number K sets length of a wave of the basic harmonic of a surface;
- the numbers N, M, D and q determine a degree of a surface calibration at the expense of imposing on the basic wave from additional harmonics, and N and M determine the number of harmonics, which are imposed;
- D determines amplitude of harmonics;
- q - both amplitude, and frequency of harmonics.

Let's notice that with increase N,M,D and q the spatial uniformity of a surface on a large scale is increased also.

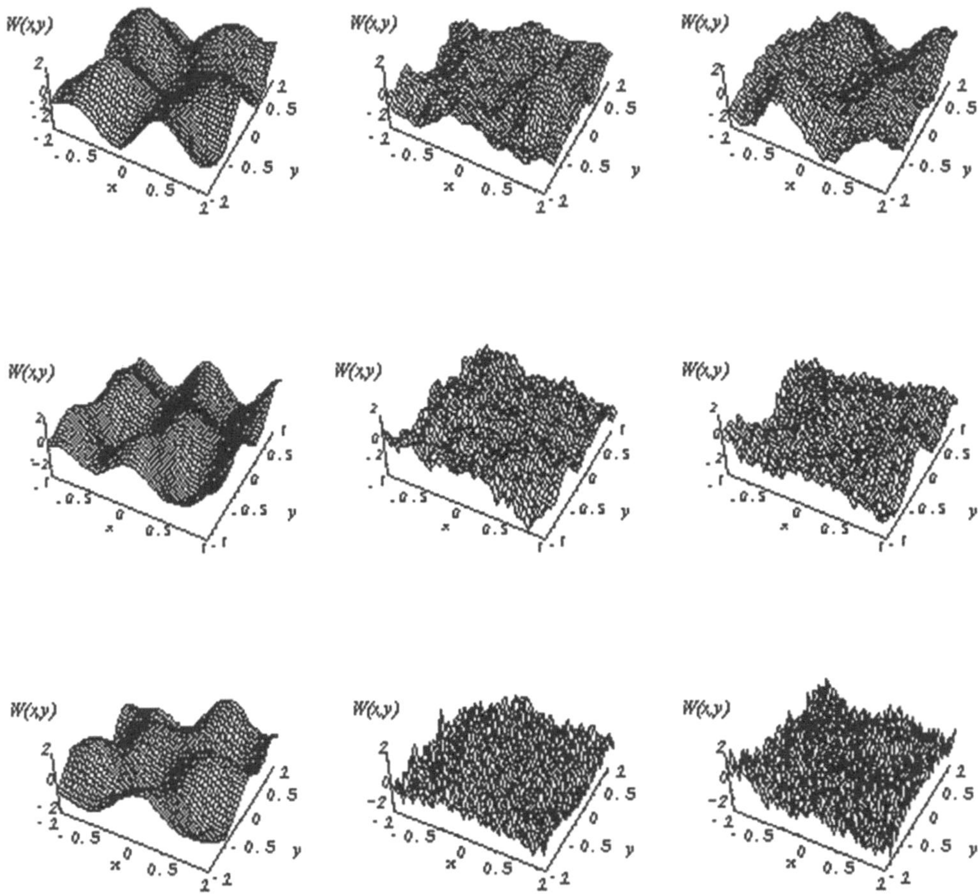

Fig. 2. Examples of rough surface by the Weierstrass function $K = 2\pi$; $N = M = 5$; $\sigma = 1$. $D = 2{,}1$; $D = 2{,}5$; $D = 2{,}9$ (from above to the bottom) $q = 1{,}1$; $q = 3$; $q = 7$ (from left to right)

By means of the original program worked out by us in the environment of Mathematika 5.1 there was the created base of these various types of fractal dispersive surfaces on the basis of Weierstrass function.

Influence each of parameters q, K, D, N, M on character of profile of surface it appears difficult enough and determined by values all other parameters. So, for example, at a value $D = 2{,}1$, what near to minimum ($D = 2$), the increase of size q does not almost change the type of surface (see the first column on fig.2). With the increase of size D the profile of surface becomes more sensible to the value q (see the second and third columns on fig.2). Will notice that with an increase N, M, D and q increases and spatial homogeneity of surface on grand the scale: large-scale "hills" disappear, and finely scale heterogeneities remind a more mesentery on a flat surface.

3. Electromagnetic wave scattering on surface fractal structures

At falling of electromagnetic wave there is her dispersion on the area of rough surface - the removed wave scattering not only in direction of floppy, and, in general speaking, in different directions. Intensity of the radiation dissipated in that or other direction is determined by both parameters actually surfaces (by a reflectivity, in high, by a form and character of location of inequalities) and parameters of falling wave (frequency, polarization) and parameters of geometry of experiment (corner of falling). The task of this subdivision is establishing a connection between intensity of the light dissipated by a fractal surface in that or other direction, and parameters of surface.

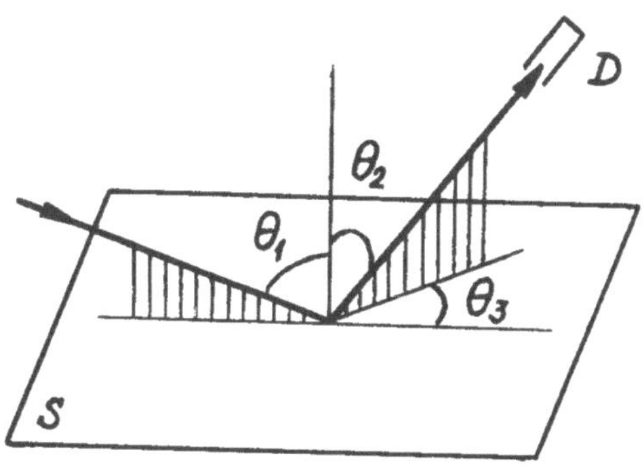

Fig. 3. The scheme of experiment on light scattering by fractal surface: S is a scattering surface; D-detector, θ_1 is a falling angle; θ_2 is a polar angle; θ_3 is an azimuthally angle

The initial light wave falls on a rough surface S under a angle θ_1 and scattering in all directions. The scattering wave is observed by means of the detector D in a direction which is characterized by a polar angle θ_2 and an azimuthally angle θ_3. The measured size is intensity of light I_s scattered at a direction (θ_2,θ_3). Our purpose is construction scattering indicatrise of an electromagnetic wave by a fractal surface (1).
As $I_s = \vec{E}_s \cdot \vec{E}_s^*$ (where $\vec{E}_s$ is an electric field of the scattering wave in complex representation) that the problem of a finding I_s is reduced to a finding of the scattered field $\vec{E}_s$.
The scattered field we shall find behind Kirchhoff method [16], and considering complexity of a problem, we shall take advantage of more simple scalar variant of the theory according to which the electromagnetic field is described by scalar size. Thus we lose an opportunity to analyze polarizing effects
The base formula of a Kirchhoff method allows to find the scattered field under such conditions:

- the falling wave is monochromatic and plane;
- a scattered surface rough inside of some rectangular ($-X < x_0 < X$, $-Y < y_0 < Y$) and corpulent outside of its borders;
- the size of a rough site much greater for length of a falling wave;
- all points of a surface have the ended gradient;
- the reflection coefficient identical to all points of a surface;

- the scattering field is observed in a wave zone, i.e. is far enough from a scattering surface.

Under these conditions the scattered field is given by

$$E_s(\vec{r}) = -ikrF(\theta_1,\theta_2,\theta_3)\frac{\exp(ikr)}{2\pi r}\int_{S_0}\exp[ik\varphi(x_0,y_0)]dx_0dy_0 + E_e(\vec{r}), \tag{4}$$

Where k is the wave number of falling wave; $F(\theta_1,\theta_2,\theta_3) = -\frac{R}{2C}(A^2+B^2+C^2)$ is a angle factor; R - scattering coefficient; $\varphi(x_0,y_0) = Ax_0 + By_0 + Ch(x_0,y_0)$ is the phase function; $h(x_0,y_0) = z(x_0,y_0)$; $A = \sin\theta_1 - \sin\theta_2\cos\theta_3$; $B = -\sin\theta_2\sin\theta_3$; $C = -\cos\theta_1 - \cos\theta_2$; $E_e(\vec{r}) = -\frac{R}{C}\cdot\frac{\exp(ikr)}{4\pi r}(AI_1 + BI_2)$,

$$I_1 = \int_{-Y}^{Y}\left[e^{ik\varphi(X,y_0)} - e^{ik\varphi(-X,y_0)}\right]dy_0,$$

$$I_2 = \int_{-X}^{X}\left[e^{ik\varphi(x_0,Y)} - e^{ik\varphi(x_0,-Y)}\right]dx_0. \tag{5}$$

After calculation of integrals (4) and (5) by means of the formula

$$e^{iz\sin\phi} = \sum_{l=-\infty}^{\infty} I_l(z)e^{il\phi},$$

(where $I_l(z)$ is the Bessel function of the whole order), we receive

$$E_s(\vec{r}) = -2ikFXY\frac{\exp(ikr)}{\pi r}\sum_{l\{rs\}}\left\{\left[\prod_{uv} I_{l_{uv}}(\xi_u)\right]e^{i\sum_{nm} l_{nm}\varphi_{nm}}\right\}\operatorname{sinc}(k_cX)\operatorname{sinc}(k_sY) + E_e(\vec{r}). \tag{6}$$

where

$$F = F(\theta_1,\theta_2,\theta_3),\ \sum_{l\{rs\}} \equiv \sum_{l_{0,1}=-\infty}^{\infty}\sum_{l_{0,2}=-\infty}^{\infty}\cdots\sum_{l_{(N-1),M}=-\infty}^{\infty},\ \prod_{uv} \equiv \prod_{u=1}^{N-1}\prod_{v=0}^{M},\ \sum_{nm} \equiv \sum_{n=1}^{N-1}\sum_{m=0}^{M},$$

$$\xi_u \equiv kc_wCq^{(D-3)u},\ \operatorname{sinc} x \equiv \frac{\sin x}{x},\quad k_c \equiv kA + K\sum_{nm} q^n l_{nm}\cos\frac{2\pi m}{M},\quad k_s \equiv kB + K\sum_{nm} q^n l_{nm}\sin\frac{2\pi m}{M},$$

$$E_e(\vec{r}) = -ikXY\frac{R}{C}(A^2+B^2)\frac{e^{ikr}}{\pi r}\operatorname{sinc}(kAX)\operatorname{sinc}(kBY).$$

Thus, expression (6) gives the decision of a problem about finding a field scattering by a fractal surface, within the limits of Kirchhoff method.

Now under the formula (4) it is possible to calculate intensity of scattered waves if to set parameters of a disseminating surface c_w (or) $\sigma\ D,q,K,N,M,X,Y,\phi_{nm}$, parameter k (or $\lambda = \frac{2\pi}{k}$) a falling wave and parameters $\theta_1,\theta_2,\theta_3$ of geometry of experiment. This intensity will be to characterize scattering on concrete realization of a surface $z(x,y)$ (with a concrete set of casual phases ϕ_{nm}). For comparison of calculations with experimental data it is necessary to operate with average on ensemble of surfaces intensity $\langle I_s \rangle = \langle \vec{E}_s \vec{E}_s^* \rangle$. Such intensity has appeared proportional intensity $I_0 = \left(\frac{2kXY\cos\Theta_1}{\pi r} \right)^2$ of the wave reflected from the corresponding smooth basic surface, therefore for the theoretical analysis of results it is more convenient to use average scattering coefficient

$$\langle \rho_s \rangle = \frac{\langle I_s \rangle}{I_0}.$$

After calculation $\langle I_s \rangle$ and leaving from (6), we shall receive exact expression

$$\langle \rho_s \rangle = \left[\frac{F(\theta_1,\theta_2,\theta_3)}{\cos\Theta_1} \right]^2 \sum_{l\{rs\}} \{ \prod_{uv} I_{l_{uv}}^2(\xi_u)\, \mathrm{sinc}^2(k_c X)\, \mathrm{sinc}^2(k_s Y) \} + \\ + \left[\frac{R\left(A^2+B^2\right)}{2C\cos\theta_1} \right]^2 \mathrm{sinc}^2(kAX)\, \mathrm{sinc}^2(kBY) \qquad (7)$$

As expression (7) consist the infinite sum to use it for numerical calculations inconveniently. Essential simplification is reached in case of $\xi_n < 1$. Using thus decomposition function in a line

$$I_\nu(z) = \left(\frac{3}{2}\right)^\nu \sum_{k=0}^{\infty} \frac{\left(-z^2/4\right)^k}{k!\,\Gamma(\nu+k+1)},$$

that rejecting members of orders, greater than ξ_n^2. We shall receive the approached expression for average scattering coefficient

$$\langle \rho_s \rangle \approx \left[\frac{F(\theta_1,\theta_2,\theta_3)}{\cos\theta_1} \right]^2 \{ \left[1-(k\sigma C)^2 \right] \mathrm{sinc}^2(kAX)\, \mathrm{sinc}^2(kBY) + \\ + \frac{1}{2} c_f^2 \sum_{nm} q^{2(D-3)n}\, \mathrm{sinc}^2\left[\left(kA + Kq^n \cos\frac{2\pi m}{M} \right) X \right] \mathrm{sinc}^2\left[\left(kB + Kq^n \sin\frac{2\pi m}{M} \right) Y \right] \} + \\ + \left[\frac{R}{2C\cos\theta_1} \left(A^2+B^2\right) \right]^2 \mathrm{sinc}^2(kAX)\, \mathrm{sinc}^2(kBY), \qquad (8)$$

where

$$c_f \equiv kc_wC = k\sigma C\left[\frac{2}{M}\cdot\frac{1-q^{2(D-3)}}{1-q^{2N(D-3)}}\right]^{\frac{1}{2}}.$$

4. Results of numerical calculations

On the basis of numerical calculations of average factor of dispersion under the formula (8) we had been constructed the average scattering coefficient $\langle\rho_s\rangle$ from θ_2 and θ_3. (scattering indicatrix diagrams) for different types of scattering surfaces. At the calculations we have supposed $R=1$, and consequently did not consider real dependence of reflection coefficient R from the length of a falling wave λ and a falling angle θ_1. The received results are presented on fig. 4.

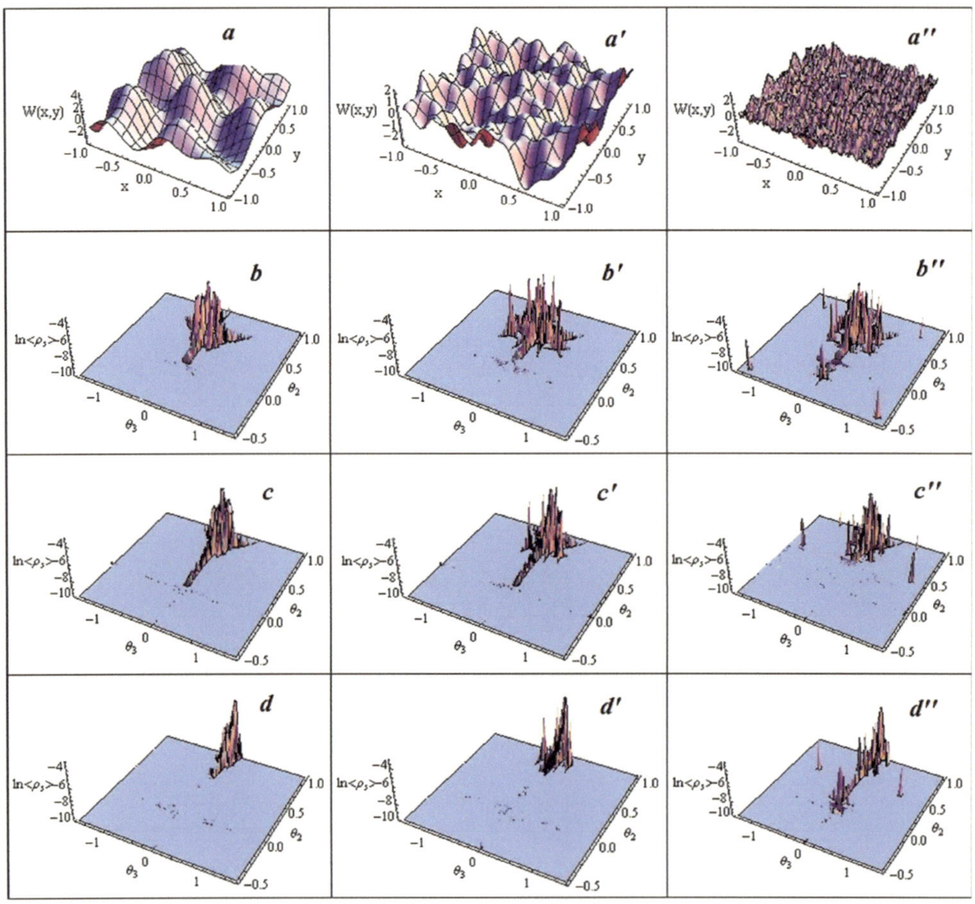

Fig. 4. Dependencies of the $\log\langle\rho_s\rangle$ from the angles θ_2 and θ_3 for the various type of fractal surfaces: *a, a', a''* - the samples of rough surfaces, which the calculation of dispersion indexes was produced; from top to bottom the change of scattering index is rotined for three angles of incidence $\theta_1 = 30, 40, 60^0$ (*a-d, a'-d', a''-d''*) at N=5, M=10, D=2.9, q=1.1; n=2, M=3, D=2.5, q=3; N=5, M=10, D.2.5, q=3 accordingly

The analysis of schedules leads to such results:

- Scattering is symmetric concerning of a falling plane;
- The greatest intensity of the scattering wave is observed in a direction of mirror reflection;
- There are other directions in which splashes in intensity are observed;
- With increase in a calibration degree of surfaces (or with growth of its large-scale heterogeneity) the picture of scattering becomes complicated. Independence of the type of scattering surface there is dependence of the scattering coefficient from the incidence angle of light wave. As far as an increase of the incidence angle from 30^0 to 60^0 amounts of additional peaks diminishes. Is their most number observed at $\theta_1 = 30^0$. It is related to influence on the scattering process of the height of heterogeneity of the surface. At the increase of the angle of incidence of the falling light begins as though not to "notice" the height of non heterogeneity and deposit from them diminishes.

The noted features of dispersion are investigation of combination of chaoticness and self-similarity relief of scattering surface.

5. Conclusion

In this chapter in the frame of the Kirchhoff method the average coefficient of light scattering by surface fractal structures was calculated. A normalized band-limited Weierstrass function is presented for modeling 2D fractal rough surfaces. On the basis of numerical calculation of average scattering coefficient the scattering indicatrises diagrams for various surfaces and falling angles were calculated. The analysis of the diagrams results in the following conclusions: the scattering is symmetrically concerning a plane of fall; with increase the degree of a surface calibration the picture becomes complicated; the greatest intensity of a scattering wave is observed in a mirror direction; there are other direction in which the bursts of intensity are observed.

6. References

[1] Bifano, T. G. Fawcett, H. E. & Bierden, P. A. (1997). Precision manufacture of optical disc master stampers. *Precis. Eng.* 20(1), 53-62.

[2] Wilkinson, P. et al. (1997). Surface finish parameters as diagnostics of tool wear in face milling, *Wear* 25(1), 47-54.

[3] Sherrington, I.' & Smith, E. H. (1986). The significance of surface topography in engineering. *Precis. Eng.* 8(2). 79-87.

[4] Kaneami, J. & Hatazawa, T. (1989). Measurement of surface profiles bv the focus method. *Wear* 134, 221-229.

[5] Mitsui, K. (1986). In-process sensors for surface roughness and their applications. *Precis. Eng.* 8(40), 212-220.

[6] Baumgart , J. W. & Truckenbrodt H. (1998). Scatterometrv of honed surfaces. *Opt. Eng.* 37(5), 1435-1441.

[7] Tay, C. J. Toh, L S., Shang, H. M. & Zhang, J. B. (1995). Whole-field determination of surface roughness bv speckle correlation. *Appl. Opt.* 34(13). 2324-2335.

[8] Peiponen, K. E. & Tsuboi, T. (1990). Metal surface roughness and optical reflectance *Opt. Laser Technol.* 22(2). 127-130.

[9] Whitley, J. Q., Kusy, R. P., Mayhew, M. J. & Buckthat, J. E. (1987). Surface roughness of stainless steel and electroformed nickle standards using HeNe laser. *Opt. Laser Technol* 19(4), 189-196.

[10] Mitsui, M., Sakai, A. & Kizuka, O. (1988'). Development of a high resolution sensor for surface roughness, *Opt. Eng.* 27(^6). 498-502.

[11] Vorburger, T. V., Marx, E. & Lettieri, T. R. (1993). Regimes of surface roughness measurable with light scattering. *Appl Opt.* 32(19!. 3401-3408.

[12] Raymond, C. J., Murnane, M. R., H. Naqvi, S. S. & Mcneil J. R. (1995). Metrology of subwavelength photoresist gratings usine optical scat-terometry. *J. Vac. Sci. Technol B* 13(4), 1484-1495.

[13] Whitehouse, D. J. (1991). Nanotechnologv instrumentation. *Meas. Control* 24(3), 37-46.

[14] Madsen, L. L, J. Srgensen,J. F., Carneiro, K. & Nielsen, H. S. (1993-1994). Roughness of smooth surfaces: STM versus profilometers. *Metro-logia* 30. 513-516.

[15] Stedman, M. (1992). The performance and limits of scanning probe microscopes. In *Proc. Int. Congr. X-ray Optics and Microanalysis,* pp. 347-352, Manchester. IOP Publishing Ltd.

[16] Berry, M.V. & Levis Z.V. (1980). On the Weirstrass-Mandelbrot fractal function. Proc.Royal Soc. London A, v.370, p.459.

3

Electromagnetic Wave Scattering from Material Objects Using Hybrid Methods

Adam Kusiek, Rafal Lech and Jerzy Mazur
Gdansk University of Technology, Faculty of Electronics, Telecommunications and Informatics
Poland

1. Introduction

Recent progress in wireless communication systems requires the development of fast and accurate techniques for designing and optimizing microwave components. Among such components we focus on the structures where a set of metallic and dielectric objects is applied. The investigation of such structures can be divided into two areas of interest. The first approach includes open problems, i.e. the electromagnetic wave scattering by posts arbitrarily placed in free space and illuminated by plane wave or Gaussian beam. In these problems the scattered field patterns of the investigated structures in near and far zones are calculated. Such structures are applied to the reduction of strut radiation of reflector antennas Kildal et al. (1996), novel PBG and EBG structures realized as periodical arrays Toyama & Yasumoto (2005); Yasumoto et al. (2004) and polarizers Gimeno et al. (1994). The second approach concerns closed problems, e.g. the electromagnetic wave scattering by posts located in different type of waveguide junctions or cavities. The main parameters describing these structures are the frequency responses or resonant and cut-off frequencies. The aforementioned waveguide discontinuities, as well as cylindrical and rectangular resonators, play important role in the design of many microwave components and systems. Rectangular waveguide junctions and circular cavities consisting of single or multiple posts are applied to filters Alessandri et al. (2003), resonators Shen et al. (2000), phase shifters Dittloff et al. (1988), polarizers Elsherbeni et al. (1993), multiplexers and power dividers Sabbagh & Zaki (2001).

One group of the developed techniques used to analyze scattering phenomena is a group of hybrid methods which combine those of functional analysis with the discrete ones Aiello et al. (2003); Arndt et al. (2004); Mrozowski (1994); Mrozowski et al. (1996); Sharkawy et al. (2006); Xu & Hong (2004). The advantage of this approach is that the complexity of the problem can be reduced, and time and memory efficiency algorithms can be achieved. The aforementioned methods are focused on objects located in free space Sharkawy et al. (2006); Xu & Hong (2004) or in waveguide junctions Aiello et al. (2003); Arndt et al. (2004); Esteban et al. (2002). Here the objects are enclosed in a finite region where the solution is obtained with the use of discrete methods such as finite element method (FEM) Aiello et al. (2003), finite-difference time-domain (FDTD) Xu & Hong (2004) or frequency-domain (FDFD) Sharkawy et al. (2006) methods and method of moments (MoM) Arndt et al. (2004); Xu & Hong (2004). In open problems Sharkawy et al. (2006); Xu & Hong (2004) the relation between the fields in the inner and outer regions is found by calculating the currents on the interface between the regions.

The total scattered field from a configuration of objects is obtained from the time domain analysis where the steady state is calculated Xu & Hong (2004) or from the iterative scattering procedure in the case of frequency domain solution Sharkawy et al. (2006). In closed problems Aiello et al. (2003); Arndt et al. (2004) the boundary Dirichlet conditions Aiello et al. (2003) or general scattering matrix (GSM) approach Arndt et al. (2004) are used to combine both of the investigated regions.
In this chapter we would like to describe a hybrid MM/MoM/FDFD/ISP method of analysis of scattering phenomena. In comparison to alternative methods Aiello et al. (2003); Arndt et al. (2004); Aza et al. (1998); Rogier (1998); Roy et al. (1996); Rubio et al. (1999); Sharkawy et al. (2006); Xu & Hong (2004) the presented approach allows one to analyze scattering from arbitrary set of objects which can be located both in free space or in waveguide junctions. In the presented method an equivalent cylindrical or spherical object, enclosing a single object or a set of objects, is introduced. At its surface the total incident and scattered fields are defined and used to determine the transmission matrix representation of the object Waterman (1971). Since the transmission matrix contains the information about the geometry and the boundary conditions of the structure, instead of analyzing the object or group of objects with arbitrary geometry the effective cylinders or spheres described by their transmission matrices are used. In this approach the transmission matrix representation of each single scatterer with simple or complex geometry is calculated with the use of analytical MM and MoM techniques or discrete FDFD technique, respectively. Utilizing the iterative scattering procedure (ISP) Elsherbeni et al. (1993); Hamid et al. (1991); Polewski & Mazur (2002) to analyze a set of scatterers allows to obtain the total transmission matrix defined on a cylindrical or spherical contour surrounding the considered configuration of objects. As the total transmission matrix does not depend on the external excitation, it is possible to utilize the presented approach to analyze a variety of both closed and open problems.
The presented here considerations are limited to the analysis of sets of objects homogenous in one dimension. The validity and accuracy of the approach is verified by comparing the results with those obtained from own measurements, derived from the analytical approach (defined for simple structures) and the commercial finite-difference time-domain (FDTD) simulator *Quick Wave 3D (QWED)* (n.d.).

2. Formulation of the problem

It is assumed that the arbitrary configuration of objects is illuminated by a known incident field (see Fig. 1(a)). The aim of the analysis is to determine the scattered field which is a result of this illumination. In the approach we assume the existence of an artificial cylindrical or spherical surface (surface $\mathcal{S}$) that encloses the analyzed set of objects. With this assumption we divide the structure into two regions of investigation: inner region and outer region. On the surface $\mathcal{S}$, that separates the regions, the incident and scattered field can be related by an aggregated transmission matrix $\overline{\overline{\mathbf{T}}}$ ($\overline{\overline{\mathbf{T}}}$-matrix) Waterman (1971) (see Fig. 1(b)). The values of the $\overline{\overline{\mathbf{T}}}$-matrix terms depend on the material and geometry properties (e.g. shape of the posts, location and orientation in space) and do not depend on the excitation. Therefore, the investigated configuration can be placed in any external region, and the outer fields can be combined with fields coming from the inner region. In particular $\overline{\overline{\mathbf{T}}}$-matrix approach can be easily applied to the analysis of open problems e.g. beam-forming structures or periodic structures. Moreover, it can be utilized in the analysis of closed structures e.g. waveguide filters or resonators.

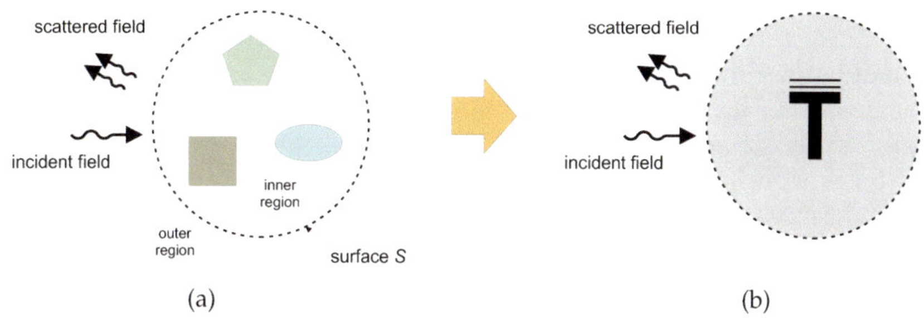

Fig. 1. Schematic representation of the analyzed problem: (a) multiple object configuration and (b) aggregated $\overline{\overline{\mathbf{T}}}$-matrix representation

In order to analyze the configuration of multiple objects placed arbitrarily in the inner region we utilize the analytical iterative scattering procedure (ISP) Elsherbeni et al. (1993); Hamid et al. (1991); Polewski & Mazur (2002). This method is based on the interaction of individual posts and allows to find a total scattered field on surface $\mathcal{S}$ from all the obstacles. This technique can be easily applied in orthogonal coordinate systems where the analytical solution of wave equation can be derived, e.g. cylindrical, elliptical or spherical coordinates. The ISP technique is thoroughly described in literature and previously was applied to the analysis of arbitrary sets of inhomogeneous height parallel cylinders Polewski & Mazur (2002) or arbitrary sets of spheres Hamid et al. (1991) (see Fig. 2). A more detailed description of the

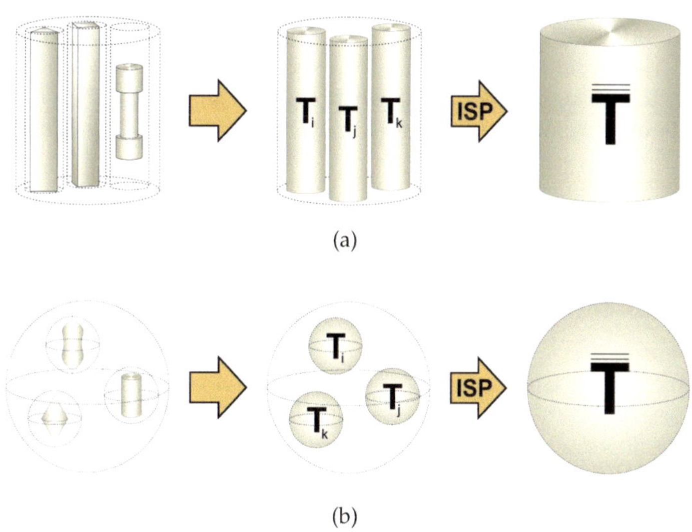

Fig. 2. Analytical iterative scattering procedure (ISP) defined in (a) cylindrical coordinates and (b) spherical coordinates

ISP will be presented in section 2.1

In order to generalize ISP to the configurations of objects with arbitrary geometry we employ different numerical techniques, depending on the post geometry. The basic concept of this approach is to enclose the analyzed object with irregular shape by the artificial homogeneous cylinder or sphere (see Fig. 2). This allows us to utilize ISP formulated in cylindrical or spherical coordinates to determine aggregated transmission matrix $\overline{\overline{\mathbf{T}}}$ of the investigated configuration of posts with irregular shape.

We focus here on two groups of objects. One group includes cylinders with arbitrary cross-section and homogeneous along height and the other group includes axially symmetrical posts with irregular shape. The geometry properties of these objects allows one to simplify the three-dimensional (3D) problem to two-and-a-half-dimensional (2.5D) one which is more numerically efficient and less time-consuming. In the case of objects with

(a) (b) (c)

Fig. 3. Analyzed objects: (a) homogenous posts, (b) segments of cylinders and cylinders with conducting strips and (c) posts with irregular shape

simple geometry as presented in Fig. 3(a) the analytical mode-matching (MM) technique is utilized. In the case of objects presented in Fig. 3(b), e.g. metallized cylinders, fragments of metallic cylinders or corrugated posts the method of moments (MoM) is used. Finally, in the analysis of objects with irregular shape such as cylinders with arbitrary cross-section and axially-symmetrical posts shown in Fig. 3(c) the hybrid finite-difference frequency-domain/mode-matching (FDFD-MM) technique is applied. The aim of the single object analysis is to determine its own isolated transmission matrix $\mathbf{T}$ ($\mathbf{T}$-matrix). All the mentioned techniques and $\mathbf{T}$-matrix expressions for chosen types of posts will be presented in section 2.2.

2.1 Iterative Scattering Procedure

The ISP method is based on the interaction of individual posts and assumes that the incident field on a single post in one iteration is derived from the scattered field from the remaining posts in the previous iteration. In order to describe the ISP we assume that the analyzed configuration is composed of set of K objects located arbitrarily in global coordinate system xyz. Each analyzed object is represented by its transmission matrix $\mathbf{T}_i$ (where $i = 1, \ldots, K$). In the homogeneous region around the investigated post configuration we define the artificial cylindrical or spherical surface $\mathcal{S}$. The aim of analysis is to determine the relation between incident and scattered fields in the outer region on the surface $\mathcal{S}$.

In the first step we assume that objects are illuminated by an unknown incident field $F^{inc(0)}$ defined in global coordinate system. Depending on the formulation of the method these fields are defined as infinite series of cylindrical or spherical eigenfunctions with unknown coefficients. As the excitation wave illuminates all the posts inside the inner region, it has to be transformed from global coordinates of the inner region to the local coordinates of each object (see Fig. 4). As a result of this excitation a zero order scattered field $F_i^{scat(0)}$ from each post is created (see Fig. 5). The scattered field $F_i^{scat(0)}$ is defined in local coordinates $x_i y_i z_i$ and can be derived for desired excitation with the use of transmission matrix $\mathbf{T}_i$. Next the

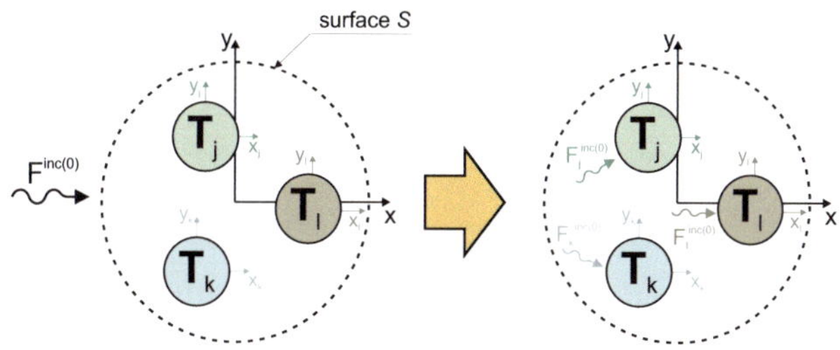

Fig. 4. Incident field $F^{inc(0)}$ transformation from global coordinates xyz to local coordinates $x_i y_i z_i$.

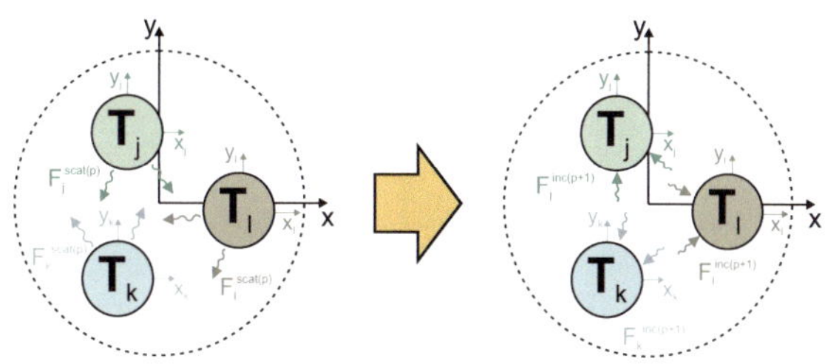

Fig. 5. Determination of new incident field in $(p+1)$th iteration based on the scattered field in pth iteration.

scattered field from the previous iteration coming from $K-1$ objects is assumed to be a new incident field $F_i^{inc(1)}$ on ith object in first iteration (see Fig. 5) and is defined as follows:

$$F_i^{inc(1)} = \sum_{\substack{j=1 \\ j \neq i}}^{K} F_{ij}^{inc(0)}, \tag{1}$$

where: $F_{ij}^{inc(0)}$ is a scattered field from jth object in zero iteration transformed to local coordinates $x_i y_i z_i$. Once again we derive the scattered field $F_i^{scat(1)}$ for each ith object in the configuration.

During the iteration process the scattered field from the previous iteration (from K-1 posts) is utilized as a new incident field on the remaining post and the coefficients of the pth iteration depend only on the coefficients of the $(p-1)$th iteration.

Using this method, after a sufficient number of iterations P, the scattered electric and magnetic fields from the ith post F_{Ti}^{scat} in its local coordinates are obtained as a superposition of the scattered fields from each iteration (see Fig. 6)

$$F_{Ti}^{scat} = \sum_{p=1}^{P} F_i^{scat(p)}. \tag{2}$$

Having transformed the scattered fields from the local coordinates of each post to the global coordinate system, one can define the total scattered field on the surface S as a superposition of the scattered fields from all K posts (see Fig. 6)

$$F_T^{scat} = \sum_{i=1}^{K} F_{T(i)}^{scat}. \tag{3}$$

Bearing in mind that the scattered field obtained during the iteration process depends on the

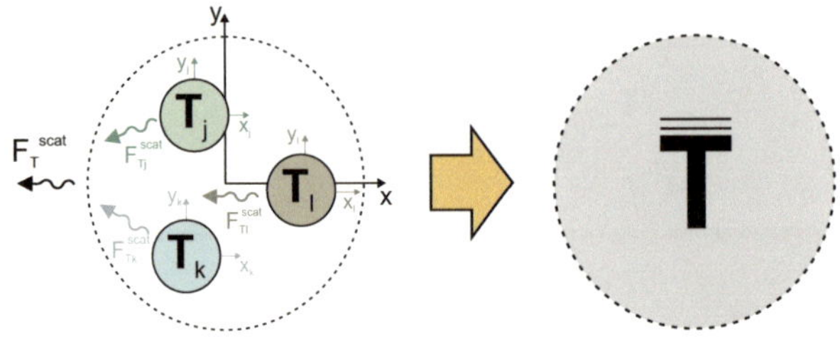

Fig. 6. Aggregated transmission matrix $\overline{\overline{\mathbf{T}}}$ of investigated objects configuration

unknown coefficients of zero order incident field, the investigated configuration of posts can be described by aggregated $\overline{\overline{\mathbf{T}}}$-matrix (see Fig. 6) defined in global coordinates xyz.
It should be emphasized that the described above analytical ISP allows one for fast and numerically efficient calculations of scattering parameters of arbitrary configuration of objects. In this approach the field transformation (interaction) matrices are evaluated only once for desired frequency and used in all iterations. Moreover, in each iteration the scattered fields from each post are obtained by simple multiplication and summation of matrices that makes this procedure numerically very efficient.

2.2 Single object analysis

In the single object analysis we try to describe the analyzed object by its isolated transmission matrix **T** defined on the surface S of the artificial cylinder or sphere which surrounds the object (see Figs. 7(a) and 7(b)). The introduction of this artificial surface $\mathcal{S}$ allows us to divide the computation domain into two regions: region I - inside the surface $\mathcal{S}$ and region II - outside the surface $\mathcal{S}$. The general solution of Helmholtz equation in cylindrical or spherical coordinates

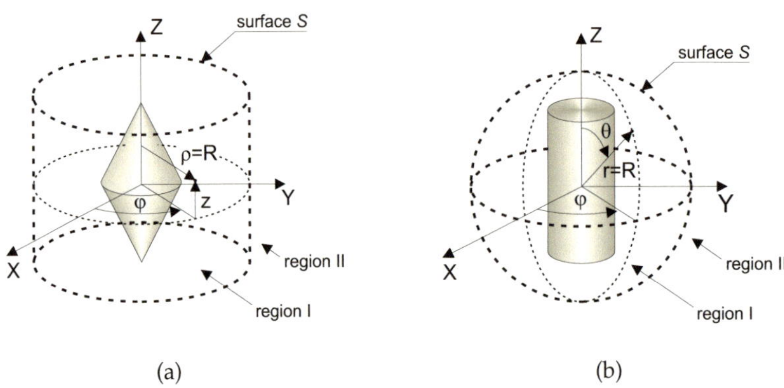

Fig. 7. Formulation of the problem for single object with irregular shape in: (a) cylindrical coordinates and (b) spherical coordinates.

in region II takes the following form:

$$E_t(\alpha, \beta, \gamma) = \sum_{k=1}^{2} \sum_{n=-\infty}^{\infty} \sum_{m=-\infty}^{\infty} A^E_{knm} f^{\alpha}_{knm}(\alpha) f^{\beta}_{m}(\beta) f^{\gamma}_{n}(\gamma), \tag{4}$$

$$H_t(\alpha,\beta,\gamma) = \sum_{k=1}^{2}\sum_{n=-\infty}^{\infty}\sum_{m=-\infty}^{\infty} A^H_{knm} f^{\alpha}_{knm}(\alpha) f^{\beta}_{m}(\beta) f^{\gamma}_{n}(\gamma), \quad (5)$$

where $f^{\alpha}_{k}(\alpha)$ determines the variation of the fields in direction normal to the surface $\mathcal{S}$ and $f^{\beta}_{m}(\beta)f^{\gamma}_{n}$ determines the variation of the fields in directions tangential to the surface $\mathcal{S}$. Variables α, β, γ denotes ρ, φ, z in cylindrical coordinates and r, φ, θ in spherical coordinates. In above equations field expansion coefficients A_{1nm} and A_{2nm} can be related with the use of isolated transmission matrix $\mathbf{T}$ as follows:

$$\mathbf{A}_2 = \mathbf{T}\mathbf{A}_1, \quad (6)$$

where $\mathbf{A}_1$ and $\mathbf{A}_2$ are the column vectors of A^E_{1nm}, A^H_{1nm} and A^E_{2nm}, A^H_{2nm}, respectively. In order to determine the transmission matrix of artificial cylinder or sphere containing analyzed post with irregular shape different techniques can be used as described in sections 2.2.1-2.2.2.

2.2.1 Analytical techniques

Transmission matrix $\mathbf{T}$ of homogeneous material cylinder can be simply evaluated using mode-matching technique. In this approach in each considered region the fields are expressed as series of eigenfunctions with unknown expansion coefficients. In order to eliminate the unknown coefficients in region 1 of the structure and to determine the relation between coefficients in the region 2 we need to satisfy the boundary conditions at the surface of the object. As a result we are obtaining the desired transmission matrix of the object. In the case of some material cylinders we present the form of transmission matrix $\mathbf{T}$ below. For the sake of brevity our considerations are limited to TM case.

- **metallic cylinder of radius r**

  The metalic cylinders find the application in structures where the high reflection coefficient is needed, e.g. microwave filters, power dividers and polarizers. The transmission matrix of such posts is defined as follows:

$$\mathbf{T} = \mathrm{diag}\left(-\frac{J_m(k_0 r)}{H^{(2)}_m(k_0 r)}\right)_{m=-M}^{m=M}, \quad (7)$$

  where $J_m(x)$ and $H^{(2)}_m(x)$ are Bessel and the second kind Hankel functions, respectively, of order m, and $k_0 = \omega\sqrt{\mu_0\varepsilon_0}$.

- **dielectric cylinder of radius r and relative permittivity ε_r**

  The dielectric cylinders are commonly utilized as different types of resonators in microwave structures. Their transmission matrix is defined as follows:

$$\mathbf{T} = \mathrm{diag}\left(\frac{k_0 J_m(kr)J'_m(k_0 r) - k_0 J'_m(kr)J_m(k_0 r)}{kJ'_m(kr)H^{(2)}_m(k_0 r) - k_0 J_m(kr)H'^{(2)}_m(k_0 r)}\right)_{m=-M}^{m=M}, \quad (8)$$

  where $k = \omega\sqrt{\mu_0\varepsilon_0\varepsilon_r}$ and prime denotes a first derivative of the function with respect of the argument.

- **ferrite cylinder of radius r and tensor permeability $\boldsymbol{\mu}$**

  The ferrite cylinders are used in many microwave nonreciprocal devices such as circulators, isolators and phase shifters. The nonreciprocal properties can be controlled by the direction and value of bias magnetization field. For the ferrite cylinder case the constitutive equations are expressed as follows:

$$\mathbf{D} = \varepsilon_0 \varepsilon_f \mathbf{E}, \tag{9}$$

$$\mathbf{B} = \mu_0 \boldsymbol{\mu} \mathbf{E}, \tag{10}$$

We assume that tensor $\boldsymbol{\mu}$ has the following dyadic form:

$$\boldsymbol{\mu} = \mu(\mathbf{i}_\rho \mathbf{i}_\rho + \mathbf{i}_\varphi \mathbf{i}_\varphi) + j\mu_a(\mathbf{i}_\rho \mathbf{i}_\varphi - \mathbf{i}_\varphi \mathbf{i}_\rho) + 1\mathbf{i}_z \mathbf{i}_z \tag{11}$$

with $\mu = 1 + \frac{p\delta}{\delta^2 - 1}$, $\mu_a = \frac{p}{\delta^2-1}$, $\delta = \frac{\gamma H_i}{f}$, $p = \frac{\gamma M_s}{f}$, H_i denotes internal bias magnetic field intensity, M_s saturated magnetization and γ gyromagnetic coefficient of the ferrite. For lossy ferrite $\delta = \delta_0 + j\alpha$ where $\delta_0 = \frac{\gamma H_i}{f}$, $\alpha = \frac{\gamma \Delta H}{f}$ and ΔH denotes resonance linewidth of ferrite.

Therefore, the **T**-matrix for the ferrite cylinder is defined as:

$$\mathbf{T} = \operatorname{diag}\left(\frac{k_0 J_m(kr) J_m'(k_0 r) - J_m(k_0 r)\left[\frac{k}{\mu_{eff}} J_m'(kr) - \frac{m}{r}\frac{\mu_a}{\mu_{eff}} J_m(kr)\right]}{H_m^{(2)}(k_0 r)\left[\frac{k}{\mu_{eff}} J_m'(kr) - \frac{m}{r}\frac{\mu_a}{\mu_{eff}} J_m(kr)\right] - k_0 J_m(kr) H_m'^{(2)}(k_0 r)} \right)_{m=-M}^{m=M}. \tag{12}$$

where ε_f and $\mu_{eff} = \frac{\mu^2 - \mu_a^2}{\mu}$ denotes relative ferrite permittivity and effective ferrite permeability.

- **pseudo-chiral cylinders**
  Pseudo-chiral medium cylinders in spite of its isotropic nature allows to control the field distribution by changing the sign of pseudo-chiral admittance (direction of Ω particles in the post).

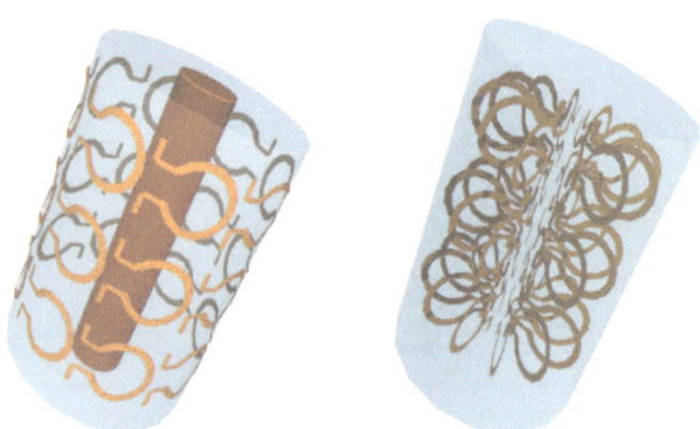

(a) (b)

Fig. 8. 3D view of a psuedo-chiral cylinders:
(a) type '1' and (b) type '2'

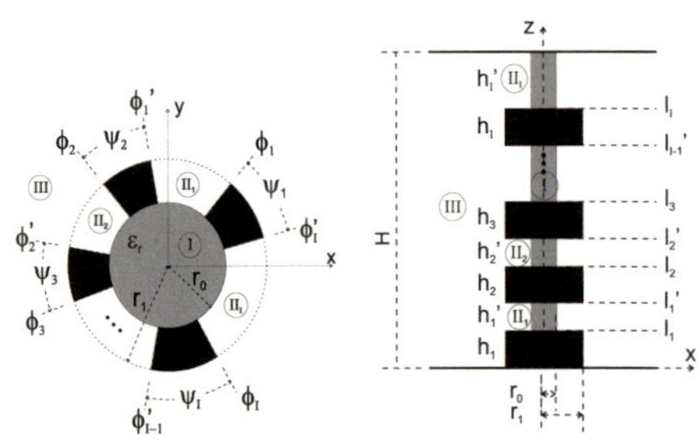

(a) (b)

Fig. 9. General configuration of the posts: (a) metallized dielectric, (b) corrugated cylinder

  - **type '1'**
    For this case the Ω particles are arranged in the cylinder as shown in Fig. 8(a). Assuming the TM^z excitation and homogeneity of the field along z the constitutive equations are of the form:

$$\mathbf{D} = \varepsilon_0 \varepsilon \mathbf{E} + j\boldsymbol{\Omega}_{\rho z} \mathbf{B}_\rho, \tag{13}$$

$$\mathbf{B} = \mu_0 \boldsymbol{\mu} \mathbf{H} - j\mu_0 \mu \boldsymbol{\Omega}_{\rho z} \mathbf{E}_z, \tag{14}$$

where $\boldsymbol{\varepsilon}$, $\boldsymbol{\mu}$ and $\boldsymbol{\Omega}$ are given in a dyadic form as:

$$\boldsymbol{\varepsilon} = \varepsilon \mathbf{i}_\rho \mathbf{i}_\rho + \varepsilon \mathbf{i}_\varphi \mathbf{i}_\varphi + \varepsilon_z \mathbf{i}_z \mathbf{i}_z, \tag{15}$$

$$\boldsymbol{\mu} = \mu_\rho \mathbf{i}_\rho \mathbf{i}_\rho + \mu \mathbf{i}_\varphi \mathbf{i}_\varphi + \mu \mathbf{i}_z \mathbf{i}_z \tag{16}$$

and $\boldsymbol{\Omega}_{z\rho} = \Omega \mathbf{i}_z \mathbf{i}_\rho$, $\boldsymbol{\Omega}_{\rho z} = \Omega \mathbf{i}_\rho \mathbf{i}_z$, Ω denotes pseudo-chiral admittance, $\varepsilon_z > \varepsilon$, $\mu_\rho > \mu$ and ε, μ are the parameters of a host medium where ε_z, μ_ρ depend on Ω. For $\Omega = 0$, $\varepsilon_z \longrightarrow \varepsilon$ and $\mu_\rho \longrightarrow \mu$.

For this case the **T**-matrix is defined as follows:

$$\mathbf{T} = \text{diag}\left(\frac{J_m(k_0 r^{(2)}) X_v' - k_0 J_m'(k_0 r^{(2)}) X_v}{k_0 H_m'^{(2)}(k_0 r^{(2)}) X_v - H_m^{(2)}(k_0 r^{(2)}) X_v'} \right)_{m=-M}^{m=M}, \tag{17}$$

where: $X_v = J_v(k^{(2)} r^{(2)}) + A_v Y_v(k^{(2)} r^{(2)})$, $X_v = J_v(k^{(2)} r^{(2)}) + A_v Y_v(k^{(2)} r^{(2)})$ and A_v has the following form for dielectric and metallic inner core, respectively:

$$A_v = \frac{k^{(1)} J_v(k^{(2)} r^{(1)}) J_m'(k^{(1)} r^{(1)}) - k^{(2)} J_v'(k^{(2)} r^{(1)}) J_m(k^{(1)} r^{(1)})}{k^{(2)} Y_v'(k^{(2)} r^{(1)}) J_m(k^{(1)} r^{(1)}) - k^{(1)} Y_v(k^{(2)} r^{(1)}) J_m'(k^{(1)} r^{(1)})}, \tag{18}$$

$$A_v = -\frac{J_v(k^{(2)} r^{(1)})}{Y_v(k^{(2)} r^{(1)})}. \tag{19}$$

The prime symbol denotes the derivative with respect to argument.

- **type '2'**

For this case the Ω particles are arranged in the cylinder as shown in Fig. 8(b). Assuming the TM^z excitation and homogeneity of the field along z the constitutive equations are of the form:

$$\mathbf{D} = \varepsilon_0 \boldsymbol{\varepsilon_c} \mathbf{E} + j \boldsymbol{\Omega}_{z\phi} \mathbf{B}, \tag{20}$$

$$\mathbf{B} = \mu_0 \boldsymbol{\mu_c} \mathbf{H} - j \mu_0 \boldsymbol{\mu_c} \boldsymbol{\Omega}_{\phi z} \mathbf{E}. \tag{21}$$

The relative electric permittivity and magnetic permeability have dyadic form:

$$\boldsymbol{\varepsilon_c} = \varepsilon(\mathbf{i}_\rho \mathbf{i}_\rho + \mathbf{i}_\phi \mathbf{i}_\phi) + \varepsilon_z \mathbf{i}_z \mathbf{i}_z, \tag{22}$$

$$\boldsymbol{\mu_c} = \mu(\mathbf{i}_\rho \mathbf{i}_\rho + \mathbf{i}_z \mathbf{i}_z) + \mu_\phi \mathbf{i}_\phi \mathbf{i}_\phi. \tag{23}$$

The coupling dyadics are defined as $\boldsymbol{\Omega}_{z\phi} = \Omega_c \mathbf{i}_z \mathbf{i}_\phi$ and $\boldsymbol{\Omega}_{\phi z} = \Omega_c \mathbf{i}_\phi \mathbf{i}_z$.
T-matrix for this type of object has the following form:

$$\mathbf{T} = \text{diag}\left(\frac{P_\nu'^{(a)}(k_\rho r) J_m(k_0 r) - k_0 \eta_0 \mu_\phi \Omega_c P_\nu^{(a)}(k_\rho r) J_m(k_0 r) - \mu_\phi P_\nu^{(a)}(k_\rho r) J_m'(k_0 r)}{\mu_\phi P_\nu^{(a)}(k_\rho r) H_m^{(2)\prime}(k_0 r) - P_\nu'^{(a)}(k_\rho r) H_m^{(2)}(k_0 r) + k_0 \eta_0 \mu_\phi \Omega_c P_\nu^{(a)}(k_\rho r) H_m^{(2)}(k_0 r)} \right)_{m=-M}^{m=M} \tag{24}$$

where $P_\nu^{(a)}(\xi)$ is of the form:

$$P_\nu^{(a)}(\xi) = \sum_{n=0}^{\infty} c_n \xi^{n+\nu}. \tag{25}$$

where c_0 is an arbitrary constant, $c_1 = \dfrac{a c_0}{2\nu + 1}$ and $c_n = \dfrac{a c_{n-1} - c_{n-2}}{n(2\nu + n)}$ for $n \geq 2$.

Selecting $c_0 = \dfrac{1}{2^\nu \Gamma(\nu + 1)}$, $P_\nu^{(a)}(\xi)$, amounts to Bessel function of the first kind when $\Omega_c = 0$ $(a = 0)$.

- **metallized or corrugated cylinder**

  The object with nonhomogeneous cross-section allows to modify the scattering parameters of the structures by the means of its simple rotation or vertical displacement. This property can be utilized in such structures as tunable filters, resonators or antenna beam-forming structures. This group includes the objects presented in Fig. 9. In order to find **T**-matrix of these structures we use the procedure that departs from standard mode-matching technique, and instead assumes the tangential component of electric field at the interface between regions of the structure by an unknown function $U(\phi)$ or $W(z)$. The functions $U(\cdot)$ and $W(\cdot)$ are then expanded in a series of basis functions with unknown coefficients. With this assumption, an additional degree of freedom is introduced in the problem. In other words, the introduction of the functions $U(\cdot)$ and $W(\cdot)$ allows to include in the formulation whatever information we have on the tangential electric field at the interface. The basis functions should contain as much of the a priori information we have on the behaviour of the tangential electric field at the interface as possible, its conditions at the sharp metallic edges of the ridge especially. A set of basis functions satisfying this local requirement is given as follows:

$$U_k^{(i)}(\phi) = \frac{f\left(\dfrac{k\pi(\phi-\phi_i)}{2(\pi-\theta_i)}\right)}{\sqrt[3]{(\phi-\phi_i)\left[2(\pi-\theta_i)-(\phi-\phi_i)\right]}} \qquad i = 1,2\ldots I \tag{26}$$

$$W_k^{(i)}(z) = \frac{g\left(\dfrac{k\pi(z-l_i)}{h'_i}\right)}{\sqrt[3]{(z-l_i)(l'_i-z)}} \qquad i = 1,2\ldots I-1 \tag{27}$$

$$W_k^{(I)}(z) = \frac{g\left(\dfrac{k\pi(z-l_I)}{2h'_I}\right)}{\sqrt[3]{(z-l_I)(2(H-l_I)-(z-l_I))}} \tag{28}$$

  where $f(\cdot)$ and $g(\cdot)$ are $\sin(\cdot)$ or $\cos(\cdot)$ functions and $2\theta_i = 2\pi - (\phi'_i - \phi_i)$.

  The inclusion of the information about the edge conditions at each of the metallic edges of the ridge allows for an efficient and accurate analysis of the spectrum of the system. It also eliminates the phenomenon of relative convergence which is introduced by the truncation of the field expansions.

More detailed descriptions of presented analytical techniques can be found in the following papers: Polewski et al. (2004) which presents formulation for metallic, dielectric and ferrite cylinders, Lech & Mazur (2007) considering formulation for segments of cylinders and cylinders with conducting strips and Lech et al. (2006); Polewski et al. (2006) presenting formulation for pseudo-chiral cylinders.

2.2.2 Hybrid technique

In hybrid approach discrete FDFD and analytical solutions of Maxwell equations are used in region I and II, respectively. The tangential components of electric and magnetic fields defined on surface S in region I can be expressed as follows:

$$E_t^I(\alpha = R, \beta, \gamma) = \sum_{n=-\infty}^{\infty} \sum_{m=-\infty}^{\infty} C_{nm} f_m^{\beta}(\beta) f_n^{\gamma}(\gamma), \tag{29}$$

$$H_t^I(\alpha = R, \beta, \gamma) = \sum_{n=-\infty}^{\infty} \sum_{m=-\infty}^{\infty} D_{nm} f_m^{\beta}(\beta) f_n^{\gamma}(\gamma), \tag{30}$$

where C_{mn} and D_{mn} are the unknown expansion field coefficients of electric and magnetic fields. In above equations $f_m^{\beta}(\beta)$ and $f_n^{\gamma}(\gamma)$ are the eigenfunctions which determines the variation of the field in the tangential to the surface $\mathcal{S}$ directions.

First stage of the analysis is to determine an impedance matrix **Z** which relates unknown electric (29) and magnetic (30) field coefficients and is defined as follows:

$$\mathbf{C} = \mathbf{Z}\mathbf{D}, \tag{31}$$

where **C** and **D** are column vectors of C_{mn} and D_{mn} coefficients. In order to determine the **Z**-matrix the discrete finite-difference frequency-domain (FDFD) technique is used. In our considerations we focused on two group of objects presented in Fig. 10. The first group,

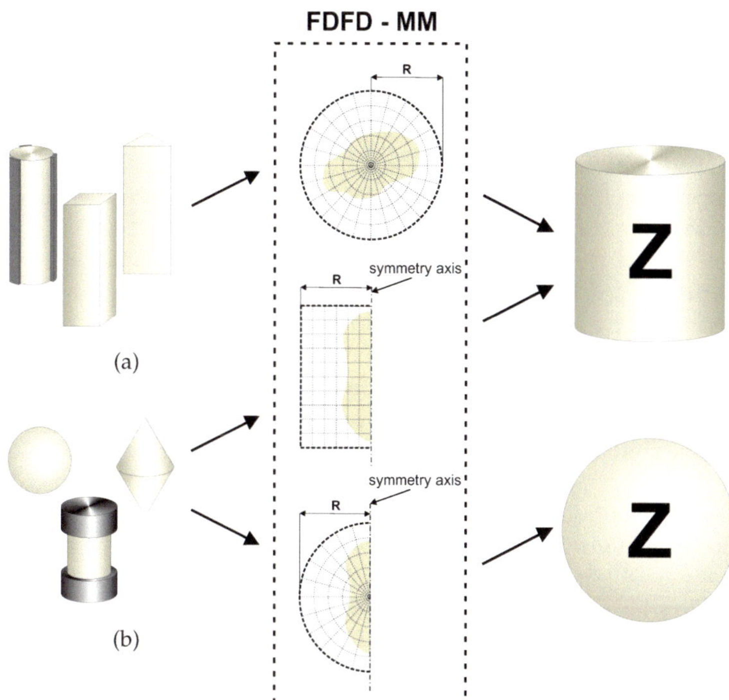

Fig. 10. Determination of Z-matrix with the use of hybrid FDFD-MM technique

presented in Fig. 10(a), includes cylinders homogenous along their height with arbitrary cross-section. The second group includes the axially-symmetrical posts with irregular shape (see Fig. 10(b)). The geometry properties of this objects allow one to simplify the three-dimensional (3D) problem to two-and-a-half dimensional (2.5D) one. As a result the investigated post is discretized only in the $\alpha - \beta$ or $\alpha - \gamma$ plane, while in the γ or β direction we assume analytical form of the fields which is determined by the series of eigenfunctions $f_n^{\gamma}(\gamma)$ or $f_m^{\beta}(\beta)$, respectively.

When the **Z**-matrix of the object is calculated we can treat the considered post with irregular shape as a homogenous cylinder or sphere with the boundary conditions defined by its impedance matrix. Now, imposing the boundary continuity conditions between tangential components of fields in the outer (defined by equations (4) and (5)) and inner region (defined by equations (29) and (30)) we are obtaining **T**-matrix of analyzed object. More detailed description of the hybrid FDFD-MM method formulated in cylindrical and spherical coordinates is presented in Kusiek & Mazur (2009; 2010; 2011).

3. Applications

The investigation of electromagnetic wave scattering in inner region led us to the description of the analyzed set of objects by an aggregated $\overline{\overline{\mathbf{T}}}$-matrix defined on the surface $\mathcal{S}$ which surrounds the whole set. As the surface $\mathcal{S}$ is either cylindrical with circular cross-section or spherical the investigated structure is seen from a point of view of an incoming field in outer region as a simple cylinder or sphere with the boundary condition included in $\overline{\overline{\mathbf{T}}}$-matrix.
The outer region can be assumed as closed or open space allowing to analyze a wide group of application such as a rectangular waveguide junction, an open space with arbitrary incident plane wave, a circular waveguide or resonator. The aim of the analysis is to match the fields coming from the outer and the inner regions and formulate the generalized scattering matrix in closed structures and the scattering coefficients for the open structures.
In order to verify the obtained mathematical models of electromagnetic wave scattering a few configurations of the investigated structures in closed and open structures have been investigated. The results have been compared with those obtained from commercial simulators, found in literature and the author's experiment.

3.1 Waveguide resonators, filters and periodic structures

For the closed structures the outer region is assumed as the waveguide junction or resonator. The chosen applications to be investigated concern placing the circular inner region with a set of analyzed posts in a rectangular waveguide junction or limiting the circular inner region by electric walls forming a circular resonator. As a result of the closed structure investigation a multimode scattering matrix of a junction or a set of resonance frequencies of a cavity are obtained.
The first investigated configuration is a circular cavity resonator loaded with square dielectric cylinder. In the analysis an additional boundary conditions needed to be applied for the side wall of the cavity to obtain resonance frequencies. In the resonator case the Sommerfeld's radiation condition does not need to be satisfied, thus, for the numerical efficiency it is more convenient to replace the Hankel function of the second kind in the ρ-dependent field eigenfunctions in region outside the post with Bessel function of the second kind. This will ensure the values of resultant resonance frequencies of the cavity are real. For the chosen example the resonant frequencies calculated for different values of post displacement are presented in Table 1 and are compared with measurement. The obtained results show that by utilizing proposed method, a very good agreement with the measurements can be obtained.
By cascading single sections it is possible to utilize the presented method for filter or periodic structure design. In order to test the validity of the method the filter structure proposed in Alessandri et al. (2003) was investigated. The performance of the filter is predicted by cascading S-matrices of the separated sections. The results show again that utilizing this method a good agreement with calculations of the alternative numerical method was obtained. The possibility of applying this method to the investigation of other structures follows from the results.

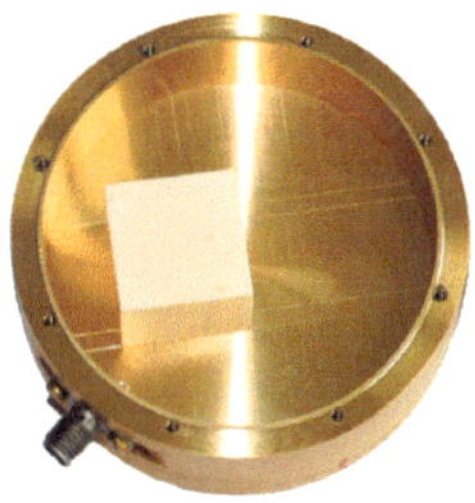

Table 1. Resonance frequencies (in GHz) of circular cavity resonator loaded with single square dielectric post: ($a = 20$mm, $\varepsilon_r = 11.25$) for different values of post shift d mm.

Mode	n	FDFD-MM			Measured	δ [%]
		$d = 15$	$d = 10$	$d = 0$	$d = 0$	
TM^1	0	1.7475	1.6299	1.5553	1.5568	0.10
TM^2	0	3.2373	3.1275	3.0646	3.0644	0.01
TM^3	0	3.2969	3.1375			
hybrid^1	1	4.0254	4.0305	4.0302	4.0056	0.61
hybrid^2	1	4.0556	4.0343			
TM^4	0	4.6021	4.4715	4.4229	4.4181	0.11
TM^5	0	4.7434	4.7523	4.7494	4.75	0.01

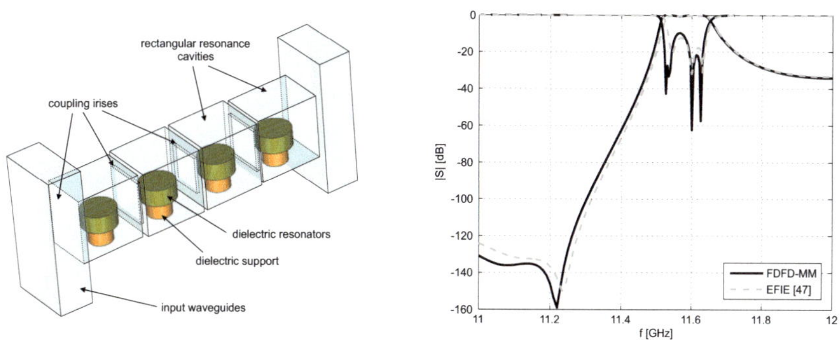

(a) (b)

Fig. 11. Waveguide filter presented in Alessandri et al. (2003): (a) schematic view of the structure and (b) frequency response (input waveguides: 19.05×9.52mm, resonance cavities: $6.91 \times 9 \times 9$mm, $7.93 \times 9 \times 9$mm, coupling irises: $6.91 \times 9 \times 0.5$mm, $5.93 \times 5.86 \times 0.5$mm, $4.85 \times 5.25 \times 0.5$mm, dielectric resonators: $r = 2.53$mm, $h = 2.3$mm, $\varepsilon_r = 30$, dielectric support: $r = 1.75$mm, $h = 2.31$mm, $\varepsilon_r = 9$).

It was seen from the numerical results the post with arbitrary cross-section, as distinct from common cylinders, enables to vary the resonant frequency by a simple rotation of the object. In cascade filters with several posts, their rotation and shift influence the coupling between the filter resonators, which enables tuning of the circuit to the demanded frequency. The changes of a single post position affect the shift of the resonance frequency more than the rotation of the post. Changes of the post positions in the cascade affect the coupling values between the posts more and thereby introduce more perturbation to the resultant frequency response characteristic of the filter. The influence which the rotation of the post has on the resonance frequency can be used in cascade filter structures only to introduce slight adjustments to the filter frequency response. This effect permits us to compensate for material defects and improper dimensions or other mechanical inaccuracies of the structure which have an effect on the length of the cavities. The other advantage of using the nonhomogeneous cross-section of the resonators in filter structures is that there is no need to introduce additional tuning elements which would require some design modifications. The example of such filter is presented in Fig. 12.

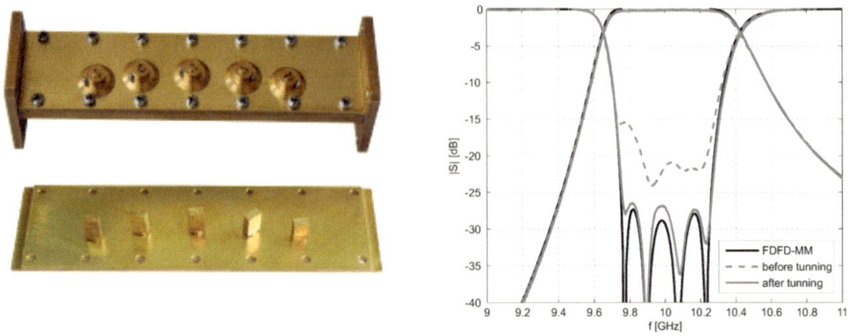

(a) (b)

Fig. 12. Waveguide direct coupled filter with rectangular posts: (a) photo of the fabricated structure and (b) simulated and measured frequency response (posts dimensions: 3×6mm and location: $l_1 = 18.31$mm, $l_2 = 21.31$mm, $d_1 = 5.76$mm, $d_2 = 3$mm, $d_3 = 2.37$mm). Waveguide dimensions: 22.86×10.16mm.

As can be seen the preliminary fabricated model with the post orientation identical to the designed filter did not give satisfactory agreement with the designed parameters. However, only slight corrections made by rotation of the post were needed to obtain the demanded shape of the frequency response characteristics. Despite the small discrepancies between the designed and measured patterns, the obtained results are more then satisfactory and, thanks to tuning ability provided by the applied rectangular posts, the design goal was accomplished proving the correctness of the approach.

The last analyzed structure presented in Fig. 13 is a waveguide filled with dielectric material ($\varepsilon_r = 2$) and loaded periodically with metallic circular cylinders of finite height. At first the

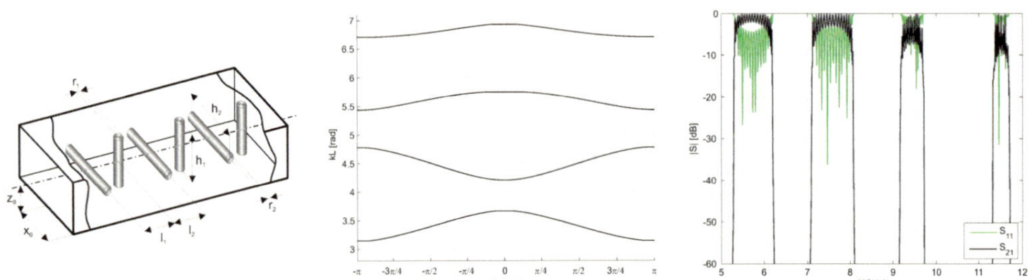

(a) (b) (c)

Fig. 13. Rectangular waveguide loaded periodically with metallic finite height circular cylinders: (a) view of the structure, (b) $k - \beta$ diagram and scattering parameters of pseudo-periodic structures composed of (c) twenty sections. Parameters of the structure: waveguide: 22.86×10.16mm, waveguide filling: $\varepsilon_r = 2$, $\mu_r = 1$, metallic cylinders: $r_1 = r_2 = 1$mm, $h_1 = h_2 = 8$mm, $x_0 = 11.43$mm, $z_0 = 5.08$mm, $l_1 = l_2 = 20$mm, single section length: $L = l_1 + l_2$.

dispersion diagram was determined for the investigated periodic structure (see Fig. 13(b)). This figure represents the plot of propagation coefficient βL as a function of kL. The results presented in $kL - \beta L$ diagrams clearly show passbands and stopbands formed in the periodic structure.

The scattering parameters for the finite periodic structures containing twenty sections of cylindrical posts were calculated and presented in Fig. 13(c). It is worth noticing that the bands shown in dispersion diagram for the infinite periodic structure are formed and are visible even for a finite periodic structure with small number of sections.

3.2 Antenna beam-forming structures

The method has been used to investigate open structures. Assuming the plane wave illumination it is possible to calculate the scattering coefficients and thus obtain scattered field pattern in near and far zones.

The rotation of the nonhomogeneous posts located in free space and illuminated by a plane wave triggers the possibility of shaping the scattered patterns of the post arrays and allows for some adjustments to the characteristics, such as reduction of the back and side lobes.

The first example concerns an array of five dielectric circular cylinders loaded with metallic rectangular cylinders and illuminated by TM plane wave. The results for two different angles φ_0 of plane wave illumination are presented in Fig. 14. From the presented results it can be noticed that the far field pattern is modified by changing the plane wave illumination angle. When the plane wave angle of incidence is changed to $\varphi_0 = 45°$ the four main lobe appears in scattered field characteristic. The results well agree with those obtained form the alternative method. The second investigated example concerns a configuration of linear array of three

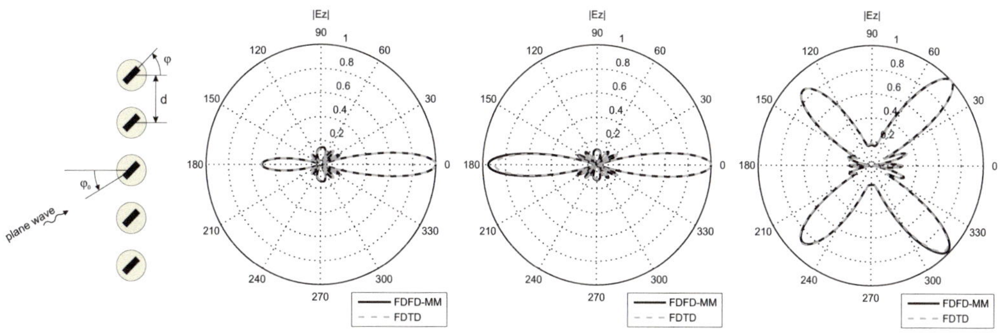

(a) (b) (c) (d)

Fig. 14. Normalized amplitude of scattered z component of electric field for configuration of five dielectric circular cylinders ($\varepsilon_{ri} = 3$) loaded with rectangle metallic cylinders ($a_i = 0.2\lambda_0$, $b_i = 0.04\lambda_0$ for $i = 1, \ldots, 5$) for three different sets: (a) investigated structure and normalized scattered field patterns for: (b) $\varphi = 0$, $\varphi_0 = 0$, (c) $\varphi = 90°$, $\varphi_0 = 0$, (d) $\varphi = 90°$, $\varphi_0 = 45°$.

Ω cylinders type '2'. The results are illustrated in Fig. 15. It is seen that the change of the sign of pseudochiral admittance Ω_c in the presented cases causes the reverse of the scattered field direction. Therefor, even thou the medium is isotropic, with the change of the sign of pseudochiral admittance Ω_c it is possible to obtain analogous effects as in the ferrite medium with the reverse of magnetization field.

The next investigated structure presented in Fig. 16 is a configuration of four dielectric cylinders illuminated by plane wave. The normalized scattered electric field pattern is presented in Figs. 16(b)-16(d). The results of proposed method well agree with the ones obtained from commercial software QuickWave 3D (FDTD). From the presented results it can be also noticed that the usage of the investigated dielectric posts allows to obtain the

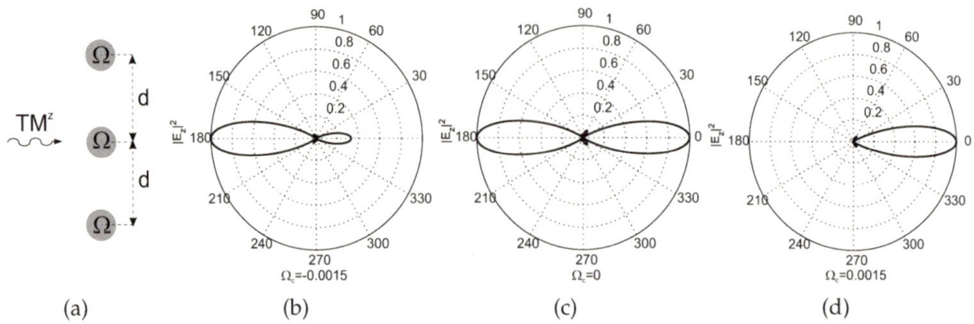

Fig. 15. Tipple post configuration ($f = 15GHz$, $r = 3mm$, $d = 10mm$, $\varepsilon_z = 4$, $\mu_\phi = 3.1$, $\mu = 1$): (a) investigated structure and normalized scattered field patterns for (b) $\Omega_c = -0.0015$ (c) $\Omega_c = 0$ and (d) $\Omega_c = 0.0015$

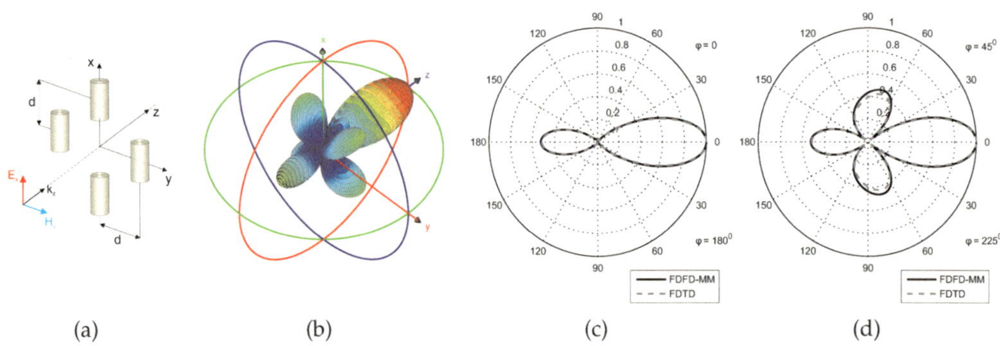

Fig. 16. Plane wave scattering on configuration of four dielectric cylinders: (a) investigated structure (r = 3mm, h = 10mm, d = 12mm, $\varepsilon_r = 5$, f_0 = 14GHz)), (b) normalized scattered electric field pattern and its cross-sections for (c) $\varphi = \{0, 180°\}$ and (d) $\varphi = \{45°, 225°\}$ planes.

directional scattered field pattern with the low level of side lobes. This property of dielectric objects can be applied in antenna beam focusing systems.

The last structure is an array of sixteen metallic cylinders presented in Fig. 17(a). As in the previous case the analyzed set of objects is illuminated by plane wave. The normalized energy characteristics of scattered field pattern for two different frequencies of analysis are presented in Figs. 17(b) and 17(c). It can be noticed that the periodically situated posts allows to transmit or reflect almost all the power of the incident wave for different frequencies. This effect can be used in periodic structures to obtain the frequency selective surfaces which has a wide range of applications in novel microwave systems.

3.3 Multilayered periodic structures - tunneling and electromagnetic curtain effects

In this section the scattering and tunneling properties of multilayered periodic arrangement composed of frequency selective surfaces (FSS) will be presented.

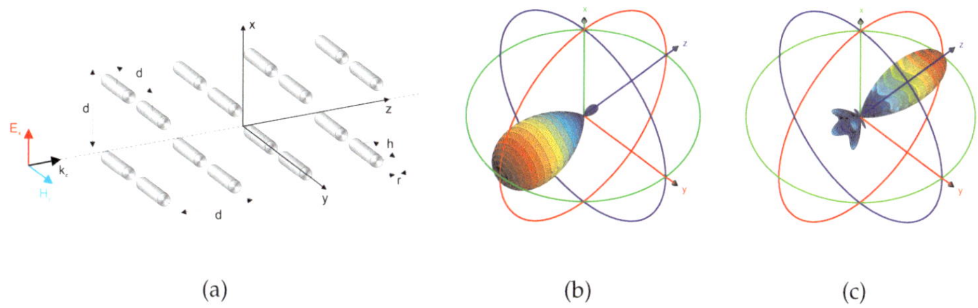

Fig. 17. Plane wave scattering by an array of sixteen metallic cylinders ($r = 2$mm, $h = 18$mm, $d = 16$mm): (a) investigated structure and normalized scattered power pattern and its cross-sections for (b) $f_0 = 5$GHz, and (c) $f_0 = 6$GHz

3.3.1 Description of the structure and methods of analysis

The structure under investigation is composed of a multilayered arrays of uniformly spaced identical sections situated in a free space and illuminated by a plane wave (see Fig. 18).

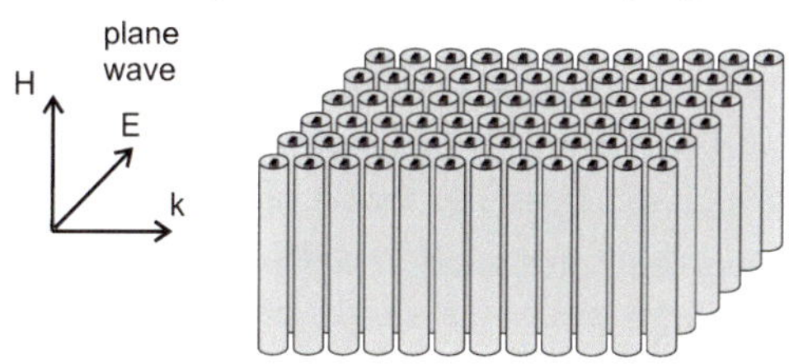

Fig. 18. Multilayered periodic structure composed of cylindrical objects

Such structures possess a stop band in their transmission characteristics and therefore are called electromagnetic band gap (EBG) structures in the microwave wavelength range. The constructive elements of these structures are comparable in size to the operation wavelength and may be composed of different media.

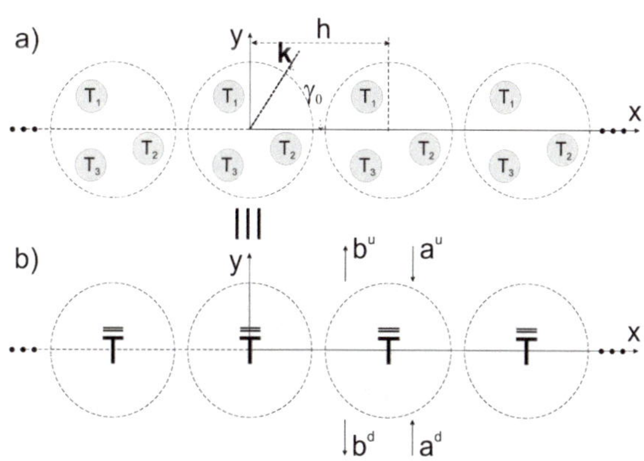

Fig. 19. Linear array of of uniformly spaced identical sections containing cylindrical objects described by their (a) **T**-matrix and (b) aggregated $\overline{\overline{\mathbf{T}}}$-matrix equivalent

The schematic representation of a single periodic array of cylindrical unit cells is presented in Fig. 19a. The unit cell contains cylindrical metallo-dielectric object or a set of objects each of which is described by its isolated **T**-matrix. In case of multiple objects in a unit cell we use the ISP procedure to calculate the aggregated $\overline{\overline{\mathbf{T}}}$-matrix which then represents the whole unit cell as shown schematically in Fig. 19b. The aggregated $\overline{\overline{\mathbf{T}}}$-matrix is substantially changed by the geometric parameters of the posts such as dimension and location. For the single post located in the center of a unit cell the isolated **T**-matrix is utilized in the further analysis as a **T**-matrix of a whole unit cell. Thus, in both cases the investigation boils down to plane wave scattering from a periodic array of cylinders (unit cells), each of which is characterized by the aggregated $\overline{\overline{\mathbf{T}}}$-matrix.

It is assumed in the analysis that the investigated periodic array of cylinders is situated in a free space and the cylinders are infinitely long, parallel to each other and spaced with distance h along x axis on the plane $y = 0$. The direction of electromagnetic plane wave incidence is assumed to be normal to the cylinder axis. The electric or magnetic field vector of an incident plane wave may be slanted with respect to the post axis, but with the above assumptions the problem is reduced to a two-dimensional one and therefore the incident polarization may be decomposed into the two fundamental polarizations: TM and TE relative to the z axis.

In order to calculate a multimodal scattering matrix of a periodic array of cylindrical scatterers we utilize an efficient numerical technique described in Kushta & Yasumoto (2000) which is based on the transmission matrix approach and uses the lattice sums technique Yasumoto & Yoshitomi (1999). In this method the scattering matrix which relates the incident space-harmonics to the scattered, both reflected and transmitted ones, is defined for a periodic array. The scattering matrix is expressed in terms of lattice sums characterizing a periodic arrangement of scatterers and the aggregated $\overline{\overline{\mathbf{T}}}$-matrix for posts located within a unit cell.

Having obtained the scattering matrix of a single periodic array of cylindrical objects one can investigate the N-layered periodic array as shown in Fig. 20. The cylindrical sections in different layers need not be identical, i.e. the dimensions, the material parameters and the arrangement of the posts in the unit cell can be arbitrary for different layers. These differences will affect the aggregated $\overline{\overline{\mathbf{T}}}$-matrix of the unit cell. Additionally, the period between the cells in different layers can also be arbitrary, which will in turn require calculating different lattice sums for each layer. For the sake of generalization, the spacing between layers can also be chosen arbitrarily.

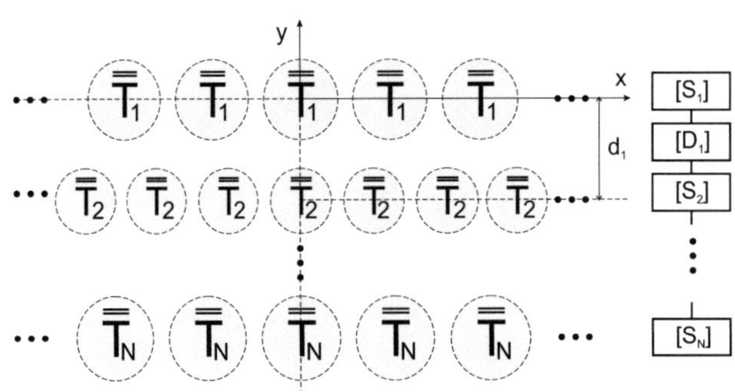

Fig. 20. Cross section of a square lattice periodic array of cylindrical sections described by aggregated $\overline{\overline{\mathbf{T}}}$-matrix

In our approach, instead of utilizing a popular recursive formula Yasumoto (2000) for analyzing multilayered periodic arrays, we calculate the scattering matrix of N-layered structure by cascading the scattering matrices of each periodic array and the scattering matrices of free space, defined utilizing the decaying space harmonic components, which include the spacing between the layers (see Fig. 20). To calculate the scattering matrix of the entire structure we utilize the formula for cascading discontinuities described by their **S**-matrices presented in Uher et al. (1993).

3.3.2 Results

A few examples of multilayered periodic structures have been analyzed and the reflection characteristics of these arrays have been calculated. The results are shown for the wavelength range $h/\lambda_0 < 1$ under normal incidence, as only for such a situation the use of periodic arrays as frequency and polarization selective components fulfils its function. Only the fundamental Floquet mode $\nu = 0$ is propagating in this range, thus the characteristics are plotted for R_0 denoting the reflection coefficient of the fundamental space harmonic.

Although the fundamental mode is the only one propagating in the investigated range, the analysis of multilayered systems demands utilizing the multimodal scattering matrix of a single periodic array in the cascading process. The importance of this fact is depicted in the first example proposed in Yasumoto et al. (2004). The power reflectance R_0 for the structure composed of two identical arrays of dielectric cylinders is calculated utilizing the cascading procedure with the evanescent space harmonics neglected in the analysis, and the results are presented in Fig. 21a. The scattering matrices of both periodic arrays were of the dimensions 2×2 and contain only the reflection and transmission parameters of the fundamental mode. As can be seen from the results, a single peak appeared in the wavelength response of the double-layered system for both TM and TE polarizations. However, calculating the same system as a one-layered periodic array of cylindrical sections containing two dielectric cylinders per section, one obtains the wavelength response in reflection coefficient as depicted in Fig. 21b. In this case, the aggregate $\overline{\overline{\mathbf{T}}}$-matrix of a single section in the array was calculated utilizing the ISP. Comparing the results, one can see considerable differences in the wavelength responses. Calculating the system again as two-layered periodic structure of circular dielectric cylinders, but taking into account the higher evanescent space harmonics, one obtains the wavelength responses depicted in Fig. 21c. The results coincide with those from Fig. 21b, which shows that the evanescent space harmonics play an important role in wavelength response of multilayered arrays and the presented simplifications lead to erroneous results in the scattering characteristics of the system. The results are with an excellent agreement with those presented in Yasumoto et al. (2004), where the recursive algorithm for the generalized reflection matrix was used for calculation of the periodic system.

3.3.3 Electromagnetic curtain effect

In a layered system, the multiple interaction of space harmonics scattered from each of the array layers modifies the frequency response, and bandgaps or stopbands in which any electromagnetic wave propagation is forbidden are formed. The power reflection coefficients for one-hundred-layered square lattice periodic arrays with one strip-metallized post embedded in a dielectric cylinder per unit cell versus frequency are illustrated in Figs. 23 for normal incidence of TM and TE waves. Each figure consists of three characteristics for different rotation angles of the posts. For both polarizations a single stop band is chosen and depicted in the figure. It is noticeable that for rotation angle $\xi = 90°$ stop bands for both polarizations are formed in the same frequency range. The characteristic plotted for TE wave is insensitive on the posts rotation angle, however the stop band width for TM

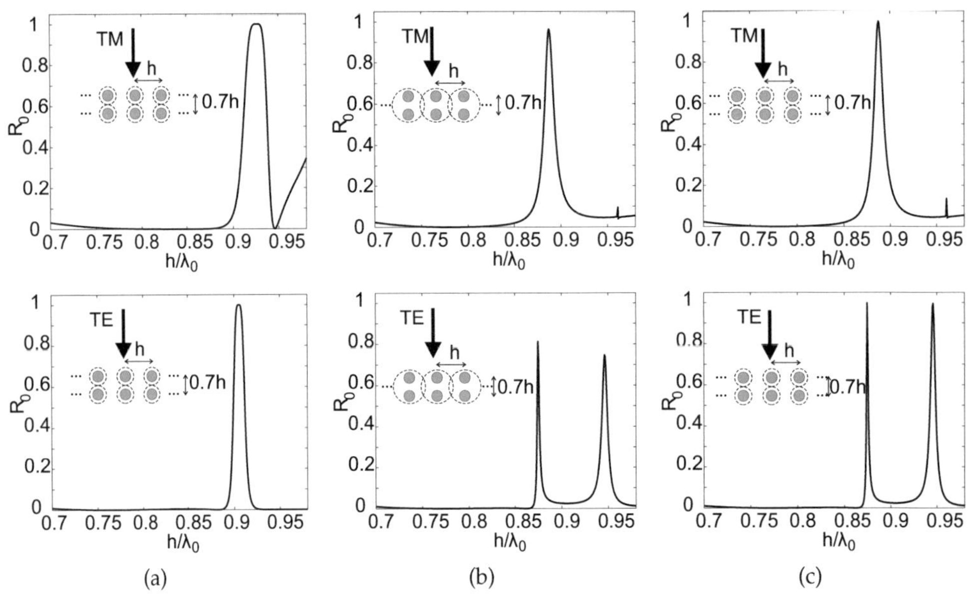

Fig. 21. Power reflection coefficients of the fundamental space harmonics versus non-dimensional frequency h/λ_0 for normal incidence of TM and TE waves into two-layered periodic structure of circular dielectric cylinders. Parameters of the post: $r = 0.3h$, $\varepsilon_r = 2$. (a) results obtained without the inclusion of evanescent space-harmonics, (b) results calculated for single periodic array of double cylinder configuration (ISP procedure) and (c) results obtained with the inclusion of evanescent space-harmonics

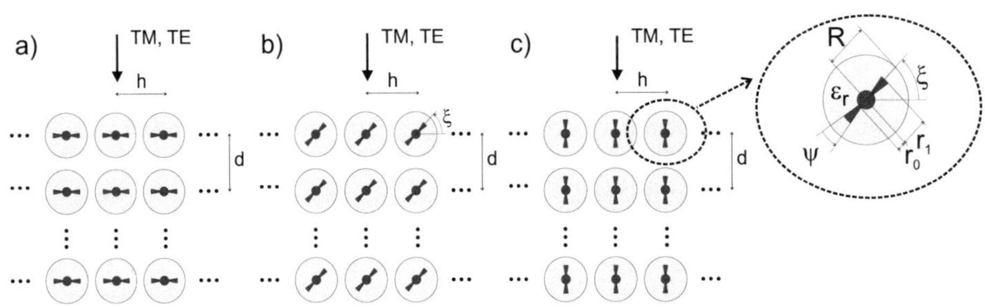

Fig. 22. Cross section of a square lattice periodic array of metallized cylindrical objects for different ξ angles: a) $\xi = 0°$, b) $\xi = 45°$, c) $\xi = 90°$

wave is gradually narrowing with the posts rotating from $\xi = 90°$ to $\xi = 0°$ and almost vanishes for the latter. The effect described above can be called an electromagnetic curtain effect, which allows to control the transmission of TM polarized wave though the periodic structure, while TE wave is blocked. A similar effect was obtained and presented in Toyama & Yasumoto (2005) in the case of a periodic array of composite circular dielectric cylinders with two cylindrical inclusions, only in that case stop bands for both TE and TM wave polarizations were narrowing and did not vanish in any case.

Another interesting case of band shifting can be observed for the structures with the whole layers rotated with respect to each other. In the example presented in Fig. 24 both EBG

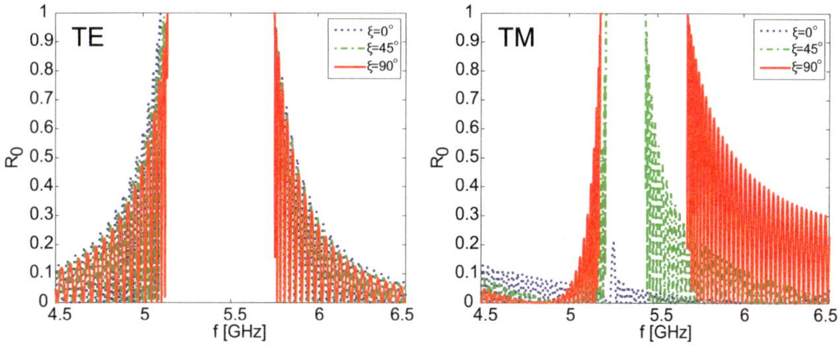

Fig. 23. Power reflection coefficients of the fundamental space harmonics versus frequency for one-hundred-layered square lattice periodic arrays of metallic posts embedded in dielectric cylinders from Fig 22. Parameters of the structure: $h = 19.5$mm, $d = 25$mm, $r_0 = 0.01h$, $r_1 = 0.09h$, $R = 0.19h$, $\psi_{1,2} = 10^\circ$, $\varepsilon_r = 20$

structures are identical but rotated by 90° with respect to each other. For the same plane wave illuminating both configurations, they produce stop bands which only slightly overlap. When half of the structure (i.e. 10 last or first arrays) are being rotated with respect to the other half one obtains the effect of stop band shifting. The stop bands, which are almost identical in width, can be shifted from one bandwidth to another. The case of 90° rotation of stacks is presented in Fig. 24(b). When only every other periodic array are being rotated the produced stop band is widening and in the case of 90° rotation it embraces both stop bands as can be seen in Fig. 24(c).

3.3.4 Tunneling effect

An interesting effect of wave tunneling can be obtained in the structure under investigation. This effect, along with the "growing evanescent envelope" for field distributions, was previously observed in metamaterial medium (negative value of real permittivity and permeability) and a structure composed of a pair of only-epsilon-negative and only-mu-negative layers Alu & Engheta (2003). This effect was also discussed in Alu & Engheta (2005) for periodically layered stacks of frequency selective surfaces (FSS). It was shown in Alu & Engheta (2005) that a complete electromagnetic wave tunneling may be achieved through a pair of different stacked FSSs which are characterized by dual behaviors, even though each stack is completely alone opaque (operates in its stop band).

Similar effect can be obtained for the structure composed of a pair of identical stacks of periodic arrays of cylindrical posts rotated by 90° with respect to each other. This effect can also be controlled by introducing a gap d between the stacks (see Fig. 25). The calculation of a total scattering matrix for a pair of such stacks boils down to cascading the scattering matrix of a stack calculated for TE wave excitation with the scattering matrix calculated for TM wave excitation.

The tunneling effect has been obtained for the periodic structure described in Fig. 25. The stop bands are formed in the same frequency range for both TE and TM waves. Therefore, we obtain a pair of stacks with dual behavior both of which operate in their stop bands. In the equivalent circuit analogy one stack is represented by a periodical line loaded with capacitors, while the other one is loaded with inductances.

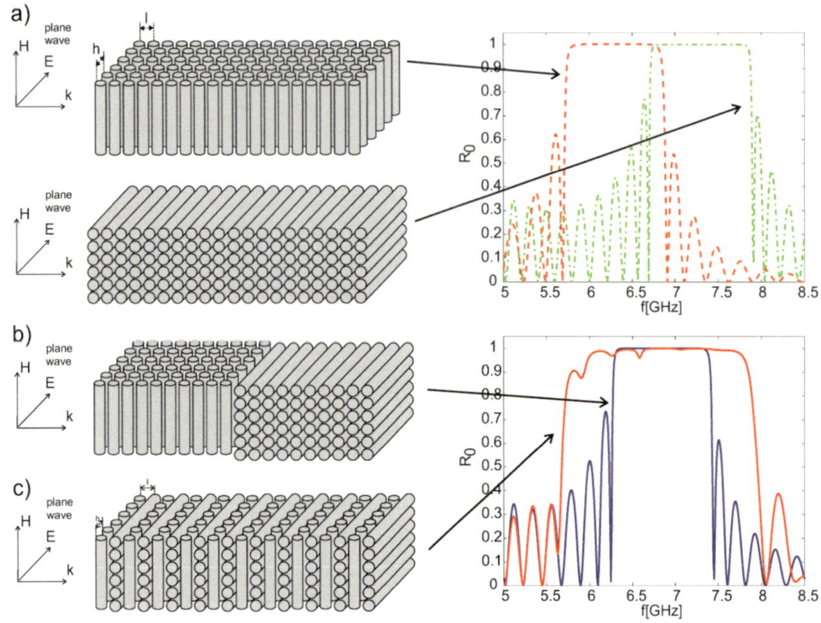

Fig. 24. Power reflection coefficients of the fundamental space harmonics versus frequency for normal incidence of TE wave on a periodic structures; Parameters of the structures $h = 20\text{mm}$, $l = h$, $r = 0.06h$, $R = 0.35h$, $\varepsilon_{r1} = 3$, $\varepsilon_{r2} = 2.5$, number of sections 20.

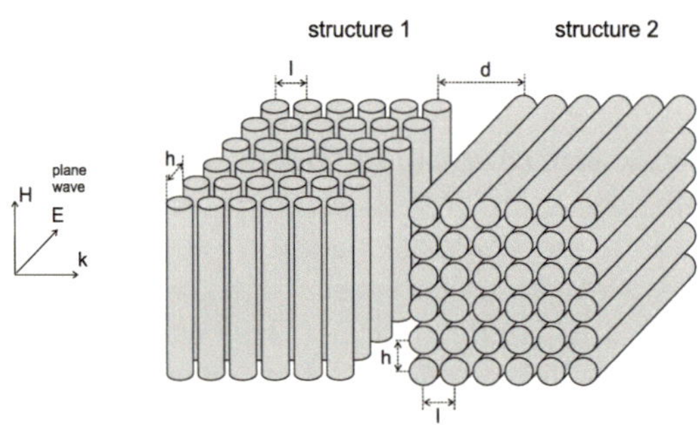

Fig. 25. Schematic 3-D representation of a periodic structure under investigation

Fig. 26 illustrates the power reflection coefficients for the normal incidence of a TE polarized plane wave on stacks of periodic structures of dielectric cylinders with double dielectric inclusions. The characteristics for scattering from structure 1, 2 and 3 are illustrated. The results show that stacks 1 and 2 are completely opaque in presented frequency ranges. However, when half of these configurations are rotated by 90° with respect to the other half, forming the structure 3, the tunneling effect can be observed. The obtained configurations enable the signal from a very narrow frequency range to tunnel through the structure.
This tunneling effect can be controlled by adjusting the distance d between stacks (see Fig. 25). Fig. 27 shows the characteristics of the power reflection coefficients for the normal incidence of a TE polarized plane wave on structure 3 for different values of distance d. It can be clearly seen that this value is directly connected to the frequency of the tunneled wave.

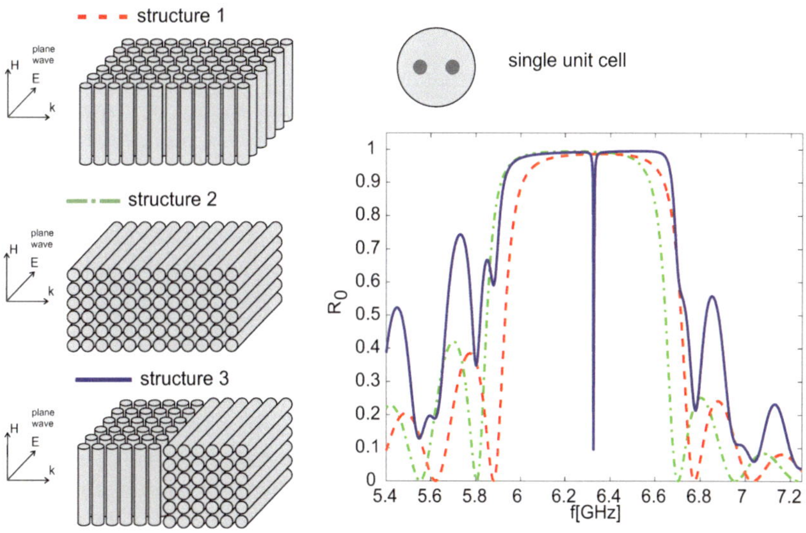

Fig. 26. Power reflection coefficients of the fundamental space harmonics versus frequency for normal incidence of TE wave on a periodic structures; Parameters of the structures: $h = 20\text{mm}$, $l = 20\text{mm}$, $d = l$, $R = 0.48h$, $\varepsilon_r = 1.5$, inclusion - two dielectric cylinders $r = 0.16h$, $\varepsilon_{rc} = 2$, displacement from the center $.24h$ number of sections 20;

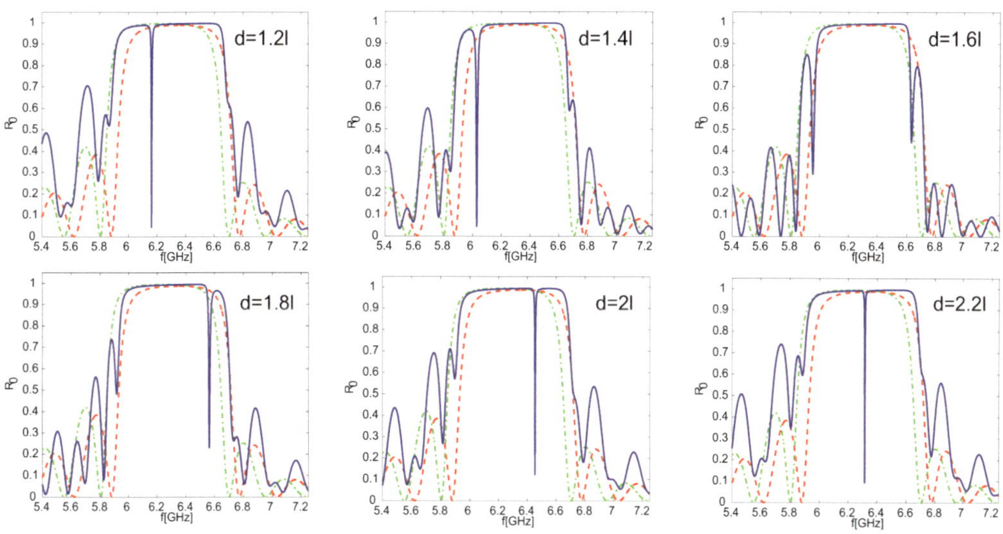

Fig. 27. Power reflection coefficients of the fundamental space harmonics versus frequency for normal incidence of TE wave on a periodic structures described in Fig. 26 for different values of distance d between stacks; dashed line (red) - structure 1, dash-dot line (green) - structure 2; solid line (blue) - structure 3

4. Conclusion

In the chapter a hybrid method of electromagnetic wave scattering from structures containing complex cylindrical or spherical objects is presented. Depending of the investigated post geometry different numerical techniques were utilized such as mode-matching technique, method of moments and finite difference method defined in the frequency domain. The proposed approach enables to determine the scattering parameters of open and closed structures containing the configuration of cylindrical objects of arbitrary cross-section and axially symmetrical posts. The proposed technique rests on defining the collateral cylindrical or spherical object containing the investigated element and then utilizing the analytical iterative model for determining scattering parameters of arbitrary configuration of objects. The obtained solution can be combined with the arbitrary external excitation which allows analyzing the variety of open and closed microwave structures. The convergence of the method have been analyzed during the numerical studies. Additionally, in order to verify the correctness of the developed method the research of a number of open and closed microwave structures such as beam shaping configurations, resonators, filters and periodic structures have been conducted. The obtained numerical results have been verified by comparing them with the ones obtained form alternative numerical methods or own measurements. A good agreement between obtained results was achieved.

5. Acknowledgement

This work was supported in part by the Polish Ministry of Science and Higher Education under Contract N515 501740, decision No 5017/B/T02/2011/40 and in part from sources for science in the years 2010-2012 under COST Action IC0803, decision No 618/N-COST/09/2010/0.

6. References

Aiello, G., Alfonzetti, S. & Dilettoso, E. (2003). Finite-element solution of eddy-current problems in unbounded domains by means of the hybrid FEM-DBCI method, *MAGN* 39(3): 1409 – 1412.

Alessandri, F., Giordano, M., Guglielmi, M., Martirano, G. & Vitulli, F. (2003). A new multiple-tuned six-port Riblet-type directional coupler in rectangular waveguide, *IEEE Trans. Microw. Theory Tech.* 51(5): 1441 – 1448.

Alu, A. & Engheta, N. (2003). Pairing an epsilon-negative slab with a mu-negative slab: Resonance, tunneling and transparency, *IEEE Trans. Antennas Propag.* 51(10): 2558–2571.

Alu, A. & Engheta, N. (2005). Evanescent growth and tunneling through stacks of frequency-selective surfaces, *IEEE Antennas Wireless Propag. Letters* 4: 417–420.

Arndt, F., Catina, V. & Brandt, J. (2004). Efficient hybrid MM/MoM technique for the CAD of circular combline filters with resonators of more general shape, *IEEE Microwave Symposium Digest* 3: 1407 – 1410.

Aza, G., Encinar, J. A., Zapata, J. & Lambea, M. (1998). Full-wave analysis of cavity-backed and probe-fed microstrip patch arrays by a hybrid mode-matching generalized scattering matrix and finite-element method, *IEEE Trans. Antennas Propag.* 46(2): 234 – 242.

Dittloff, J., Arndt, F. & Grauerholz, D. (1988). Optimum design of waveguide E-plane stub-loaded phase shifters, *IEEE Trans. Microw. Theory Tech.* 36(3): 582–587.

Elsherbeni, A. Z., Hamid, M. & Tian, G. (1993). Iterative scattering of a Gaussian beam by an array of circular conducting and dielectric cylinders, *J. of Electromag. Waves App.* 7(10): 1323–1342.

Esteban, H., Cogollos, S., Boria, V., Blas, A. S. & Ferrando, M. (2002). A new hybrid mode-matching/numerical method for the analysis of arbitrarily shaped inductive obstacles and discontinuities in rectangular waveguides, *IEEE Trans. Microw. Theory Tech.* 50(4): 1219–1224.

Gimeno, B., Cruz, J. L., Navarro, E. A. & Such, V. (1994). A polarizer rotator system for three-dimensional oblique incidence, *IEEE Trans. Antennas Propag.* 42(7): 912–919.

Hamid, A.-K., Ciric, I. & Hamid, M. (1991). Iterative solution of the scattering by an arbitrary configuration of conducting or dielectric spheres, *IEE Proc. H Microw., Antennas Propag.* 138(6): 565 – 572.

Kildal, P.-S., Kishk, A. A. & Tengs, A. (1996). Reduction of forward scattering from cylindrical objects using hard surfaces, *IEEE Trans. Antennas Propag.* 44: 1509–1520.

Kushta, T. & Yasumoto, K. (2000). Electromagnetic scattering from periodic arrays of two circular cylinders per unit cell, *Prog. Electromag. Research* 29: 69–85.

Kusiek, A. & Mazur, J. (2009). Analysis of scattering from arbitrary configuration of cylindrical objects using hybrid finite-difference mode-matching method, *Prog. Electromag. Research* 97: 105 – 127.

Kusiek, A. & Mazur, J. (2010). Hybrid finite-difference/mode-matching method for analysis of scattering from arbitrary configuration of rotationally-symmetrical posts, *Prog. Electromag. Research* 110: 23 – 42.

Kusiek, A. & Mazur, J. (2011). Application of hybrid finite-difference mode-matching method to analysis of structures loaded with axially-symmetrical posts, *Microw. Opt. Tech. Lett.* 53(1): 189–194.

Lech, R. & Mazur, J. (2007). Analysis of circular cavity with cylindrical objects, *IEEE Trans. Microw. Theory Tech.* 55(10): 2115 – 2123.

Lech, R., Polewski, M. & Mazur, J. (2006). Electromagnetic wave scattering by a set of axially symmetric ω-medium cylinder, *Microw. Opt. Tech. Lett.* 48(9): 1880 – 1884.

Mrozowski, M. (1994). A hybrid PEE-FDTD algorithm for accelerated time domain analysis of electromagnetic waves, *IEEE Microw. Guided Wave Lett.* 4: 323 – 325.

Mrozowski, M., Okoniewski, M. & Stuchly, M. (1996). A hybrid PEE-FDTD method for efficient field modeling in cyllindrical coordinates, *Elect. Lett.* 32: 194 – 195.

Polewski, M., Lech, R. & Mazur, J. (2004). Rigorous modal analysis of structures containing inhomogeneous dielectric cylinders, *IEEE Trans. Microw. Theory Tech.* 52(5): 1508–1516.

Polewski, M., Lech, R. & Mazur, J. (2006). Radiation phenomena from pseudochiral cylinders at plane wave incidence, *Microw. Opt. Tech. Lett.* 48(9): 1880 – 1884.

Polewski, M. & Mazur, J. (2002). Scattering by an array of conducting lossy dielectric, ferrite and pseudochiral cylinders, *Prog. Electromag. Research* pp. 283–310.

Quick Wave 3D (QWED) (n.d.). http://www.qwed.com.pl/.

Rogier, H. (1998). A new hybrid FDTD-BIE approach to model electromagnetic scattering problems, *IEEE Microw. Guided Wave Lett.* 8(3): 138 – 140.

Roy, T., Sarkar, T. K., DjordjeviC, A. R. & Salazar-Palma, M. (1996). A hybrid method solution of scattering by conducting cylinders (TM case), *IEEE Trans. Microw. Theory Tech.* 44(12): 2145 – 2151.

Rubio, J., Arroyo, J. & Zapata, J. (1999). Analysis of passive microwave circuits by using a hybrid 2-D and 3-D finite-element mode-matching method, *IEEE Trans. Microw. Theory Tech.* 47(9): 1746 – 1749.

Sabbagh, M. E. & Zaki, K. (2001). Modeling of rectangular waveguide junctions containing cylindrical posts, *Prog. Electromag. Research* 33: 299–331.

Sharkawy, M. A., Demir, V. & Elsherbeni, A. Z. (2006). Plane wave scattering from three dimensional multiple objects using the iterative multiregion technique based on the FDFD method, *IEEE Trans. Antennas Propag.* 54(2): 666–673.

Shen, T., Zaki, K. A. & Wang, C. (2000). Tunable dielectric resonators with dielectric tuning disks, *IEEE Trans. Microw. Theory Tech.* 48: 2439Ű2445.

Toyama, H. & Yasumoto, K. (2005). Electromagnetic scattering from periodic arrays of composite circular cylinder with internal cylindrical scatterers, *Prog. Electromag. Research* 52: 355–358.

Uher, J., Bornemann, J. & Rosenberg, U. (1993). *Waveguide Components for Antenna Feed Systems: Theory and CAD*, Artech House Antennas and Propagation Library, Norwood.

Waterman, P. C. (1971). Symmetry, unitarity and geometry in electromagnetic scattering, *Physical Review D* 3: 825–839.

Xu, F. & Hong, W. (2004). Analysis of two dimensions sparse multicylinder scattering problem using DD-FDTD method, *IEEE Trans. Antennas Propag.* 52(10): 2612 – 2617.

Yasumoto, K. (2000). Generalized method for electromagnetic scattering by two-dimensional periodic discrete composites using lattice sums, *Proc. Int. Conf. Microwave and Millimeter Wave Technology* (9): 29–34.

Yasumoto, K., Toyama, H. & Kushta, T. (2004). Accurate analysis of two-dimensional electromagnetic scattering from multilayered periodic arrays of circular cylinders using lattice sums technique, *IEEE Trans. Antennas Propag.* 52: 2603–2611.

Yasumoto, K. & Yoshitomi, K. (1999). Efficient calculation of lattice sums for freespace periodic green's function, *IEEE Trans. Antennas Propag.* 47(6): 1050–1055.

4

The Eigen Theory of Electromagnetic Waves in Complex Media

Shaohua Guo
Zhejiang University of Science and Technology
P. R. China

1. Introduction

Since J. C. Maxwell presented the electromagnetic field equations in 1873, the existence of electromagnetic waves has been verified in various medium (Kong, 1986; Monk, 2003). But except for Helmholtz's equation of electromagnetic waves in isotropic media, the laws of propagation of electromagnetic waves in anisotropic media are not clear to us yet. For example, how many electromagnetic waves are there in anisotropic media? How fast can these electromagnetic waves propagate? Where are propagation direction and polarization direction of the electromagnetic waves? What are the space patterns of these waves? Although many research works were made in trying to deduce the equations of electromagnetic waves in anisotropic media based on the Maxwell's equation (Yakhno, 2005, 2006; Cohen, 2002; Haba, 2004), the explicit equations of electromagnetic waves in anisotropic media could not be obtained because the dielectric permittivity matrix and magnetic permeability matrix were all included in these equations, so that only local behaviour of electromagnetic waves, for example, in a certain plane or along a certain direction, can be studied.

On the other hand, it is a natural fact that electric and magnetic fields interact with each other in classical electromagnetics. Therefore, even if most of material studies deal with the properties due to dielectric polarisation, magneitc materials are also capable of producing quite interesting electro-magnetic effects (Lindellm et al., 1994). From the bi-anisotropic point of view, magnetic materials can be treated as a subclass of magnetoelectric materials. The linear constitutive relations linking the electric and magnetic fields to the electric and magnetic displacements contain four dyadics, three of which have direct magnetic contents. The magnetoelectric coupling has both theoretical and practical significance in solid state physics and materials science. Though first predicted by Pierre Curie, magnetoeletric coupling was originally through to be forbidden because it violates time-reversal symmetry, until Laudau and Lifshitz (Laudau & Lifshitz, 1960) pointed out that time reversal is not a symmetry operation in some magnetic crystal. Based on this argument, Dzyaloshinskii (Dzyaloshinskii, 1960) predicted that magnetoelectric effect should occur in antiferromagnetic crystal Cr_2O_3, which was verified experimentally by Astrov (Astrov, 1960). Since then the magnetoelectric coupling has been observed in single-phase materials where simultaneous electric and magnetic ordering coexists, and in two-phase composites where the participating phase are pizoelectric and piezomagnetic (Bracke & Van Vliet,1981; Van Run et al., 1974) . Agyei and Birman (Agyei & Birman, 1990) carried out a detailed analysis of the linear magnetoelectric effect, which showed that the effect should occur not only in some magnetic but also in some electric crystals. Pradhan (Pradhan, 1993) showed that an electric charge placed in a magnetoelectric medium becomes a source of induced magnetic field with non-zero divergence of volume integral. Magnetoelectric effect in two-phase composites has been analyzed by Harshe et al. (Harshe et al., 1993), Nan (Nan, 1994) and Benveniste (Benveniste, 1995). Broadband transducers based on magnetoelectric effect have also been developed (Bracke & Van Vliet, 1981). Although the development mentioned above, no great progress in the theories of electromagnetic waves in bi-anisotropic media because of the difficulties in deal with the bi-coupling in electric field and magnetic one of the Maxell's equation and the bi-anisotropic constitutive equation by classical electromagnetic theory.

Recently there is a growing interest modeling and analysis of Maxwell's equations (Lee & Madsen, 1990; Monk, 1992; Jin et al., 1999). However, most work is restricted to simple medium such as air in the free space. On the other hand, we notice that lossy and dispersive media are ubiquitous, for example human tissue, water, soil, snow, ice, plasma, optical fibers and radar-absorbing materials. Hence the study of how electromagnetic wave interacts with dispersive media becomes very important. Some concrete applications include geophysical probing and subsurface studied of the moon and other planets (Bui et al., 1991), High power and ultra-wide-band radar systems, in which it is necessary to model ultra-wide-band electromagnetic pulse propagation through plasmas (Dvorak & Dudley, 1995), ground penetrating radar detection of buried objects in soil media (liu & Fan, 1999). The Debye medium plays an important role in electromagnetic wave interactions with biological and water-based substances (Gandhi & Furse, 1997). Until 1990, some paper on modeling of wave propagation in dispersive media started making their appearance in computational electromagnetics community. However, the published papers on modeling of dispersive media are exclusively restricted to the finite-difference time-domain methods and the finite element methods (Li & Chen, 2006; Lu et al., 2004). To our best knowledge, there exist only few works in the literature, which studied the theoretical model for the Maxwell's equation in the complex anisotropic dispersive media, and no explicit equations of electromagnetic waves in anisotropic dispersive media can be obtained due to the limitations of classical electromagnetic theory.

Chiral materials have been recently an interesting subject. In a chiral medium, an electric or magnetic excitation will produce simultaneously both electric and magnetic polarizations. On the other hand, the chiral medium is an object that cannot be brought into congruence with its mirror image by translation and rotation. Chirality is common in a variety of naturally occurring and man-made objects. From an operation point of view, chirality is introduced into the classical Maxwell equations by the Drude-Born-Fedorov relative constitutive relations in which the electric and magnetic fields are coupled via a new materials parameter (Lakhtakia, 1994; Lindell et al., 1994), the chirality parameter. These constitutive relations are chosen because they are symmetric under time reversality and duality transformations. In a homogeneous isotropic chiral medium the electromagnetic fields are composed of left-circularly polarized (LCP) and right- circularly polarized (RCP) components (Jaggard et al., 1979; Athanasiadis & Giotopoulos, 2003), which have different wave numbers and independent directions of propagation. Whenever an electromagnetic wave (LCP, RCP or a linear combination of them) is incident upon a chiral scatterer, then the scattered field is composed of both LCP and RCP components and therefore both LCP and RCP far-field patterns are derived. Hence, in the vector problem we need to specify two directions of propagation and two polarizations. In recent years, chiral materials have been increasingly studied and there is a growing literature covering both their applications and the theoretical investigation of their properties. It will be noticed that the works dealing with wave phenomena in chiral materials have been mainly concerned with the study of time-harmonic waves which lead to frequency domain studies (Lakhtakia et al., 1989; Athanasiadis et al., 2003).

In this chapter, the idea of standard spaces is used to deal with the Maxwell's electromagnetic equation (Guo, 2009, 2009, 2010, 2010, 2010). By this method, the classical Maxwell's equation under the geometric presentation can be transformed into the eigen Maxwell's equation under the physical presentation. The former is in the form of vector and the latter is in the form of scalar. Through inducing the modal constitutive equations of complex media, such as anisotropic media, bi-anisotropic media, lossy media, dissipative media, and chiral media, a set of modal equations of electromagnetic waves for all of those media are obtained, each of which shows the existence of electromagnetic sub-waves, meanwhile its propagation velocity, propagation direction, polarization direction and space pattern can be completely determined by the modal equations.This chapter will make introductions of the eigen theory to reader in details. Several novel theoretical results were discussed in the different parts of this chapter.

2. Standard spaces of electromagnetic media

In anisotropic electromagnetic media, the dielectric permittivity and magnetic permeability are tensors instead of scalars. The constitutive relations are expressed as follows

$$\boldsymbol{D}=\boldsymbol{\varepsilon}\cdot\boldsymbol{E},\quad \boldsymbol{B}=\boldsymbol{\mu}\cdot\boldsymbol{H} \tag{1}$$

Rewriting Eq.(1) in form of scalar, we have

$$D_i=\varepsilon_{ij}E_j,\quad B_i=\mu_{ij}H_j \tag{2}$$

where the dielectric permittivity matrix $\boldsymbol{\varepsilon}$ and the magnetic permeability matrix $\boldsymbol{\mu}$ are usually symmetric ones, and the elements of the matrixes have a close relationship with the selection of reference coordinate. Suppose that if the reference coordinates is selected along principal axis of electrically or magnetically anisotropic media, the elements at non-diagonal of these matrixes turn to be zero. Therefore, equations (1) and (2) are called the constitutive equations of electromagnetic media under the geometric presentation. Now we intend to get rid of effects of geometric coordinate on the constitutive equations, and establish a set of coordinate-independent constitutive equations of electromagnetic media under physical presentation. For this purpose, we solve the following problems of eigen-value of matrixes.

$$(\boldsymbol{\varepsilon}-\lambda\boldsymbol{I})\boldsymbol{\phi}=\boldsymbol{0},\quad (\boldsymbol{\mu}-\gamma\boldsymbol{I})\boldsymbol{\varphi}=\boldsymbol{0} \tag{3}$$

where $\lambda_i\,(i=1,2,3)$ and $\gamma_i\,(i=1,2,3)$ are respectively eigen dielectric permittivity and eigen magnetic permeability, which are constants of coordinate-independent. $\boldsymbol{\phi}_i\,(i=1,2,3)$ and $\boldsymbol{\varphi}_i\,(i=1,2,3)$ are respectively eigen electric vector and eigen magnetic vector, which show the electrically principal direction and magnetically principal direction of anisotropic media, and are all coordinate-dependent. We call these vectors as standard spaces. Thus, the matrix of dielectric permittivity and magnetic permeability can be spectrally decomposed as follows

$$\boldsymbol{\varepsilon} = \boldsymbol{\Phi}\boldsymbol{\Lambda}\boldsymbol{\Phi}^{\mathrm{T}}, \quad \boldsymbol{\mu} = \boldsymbol{\Psi}\boldsymbol{\Pi}\boldsymbol{\Psi}^{\mathrm{T}} \tag{4}$$

where $\boldsymbol{\Lambda} = diag[\lambda_1, \lambda_2, \lambda_3]$ and $\boldsymbol{\Pi} = diag[\gamma_1, \gamma_2, \gamma_3]$ are the matrix of eigen dielectric permittivity and eigen magnetic permeability, respectively. $\boldsymbol{\Phi} = \{\boldsymbol{\phi}_1, \boldsymbol{\phi}_2, \boldsymbol{\phi}_3\}$ and $\boldsymbol{\Psi} = \{\boldsymbol{\varphi}_1, \boldsymbol{\varphi}_2, \boldsymbol{\varphi}_3\}$ are respectively the modal matrix of electric media and magnetic media, which are both orthogonal and positive definite matrixes, and satisfy $\boldsymbol{\Phi}^{\mathrm{T}}\boldsymbol{\Phi} = \boldsymbol{I}$, $\boldsymbol{\Psi}^{\mathrm{T}}\boldsymbol{\Psi} = \boldsymbol{I}$.
Projecting the electromagnetic physical qualities of the geometric presentation, such as the electric field intensity vector $\boldsymbol{E}$, magnetic field intensity vector $\boldsymbol{H}$, magnetic flux density vector $\boldsymbol{B}$ and electric displacement vector $\boldsymbol{D}$ into the standard spaces of the physical presentation, we get

$$\boldsymbol{D}^* = \boldsymbol{\Phi}^{\mathrm{T}} \cdot \boldsymbol{D}, \quad \boldsymbol{E}^* = \boldsymbol{\Phi}^{\mathrm{T}} \cdot \boldsymbol{E} \tag{5}$$

$$\boldsymbol{B}^* = \boldsymbol{\Psi}^{\mathrm{T}} \cdot \boldsymbol{B}, \quad \boldsymbol{H}^* = \boldsymbol{\Psi}^{\mathrm{T}} \cdot \boldsymbol{H} \tag{6}$$

Rewriting Eqs.(5) and (6) in the form of scalar, we have

$$D_i^* = \boldsymbol{\phi}_i^{\mathrm{T}} \cdot \boldsymbol{D} \quad i = 1,2,3, \quad E_i^* = \boldsymbol{\phi}_i^{\mathrm{T}} \cdot \boldsymbol{E} \quad i = 1,2,3 \tag{7}$$

$$B_i^* = \boldsymbol{\varphi}_i^{\mathrm{T}} \cdot \boldsymbol{B} \quad i = 1,2,3, \quad H_i^* = \boldsymbol{\varphi}_i^{\mathrm{T}} \cdot \boldsymbol{H} \quad i = 1,2,3 \tag{8}$$

These are the electromagnetic physical qualities under the physical presentation.
Substituting Eq. (4) into Eq. (1) respectively, and using Eqs.(5) and (6) yield

$$D_i^* = \lambda_i E_i^* \quad i = 1,2,3 \tag{9}$$

$$B_i^* = \gamma_i H_i^* \quad i = 1,2,3 \tag{10}$$

The above equations are just the modal constitutive equations in the form of scalar.

3. Eigen expression of Maxwell's equation

The classical Maxwell's equations in passive region can be written as

$$\boldsymbol{\nabla} \times \boldsymbol{H} = \nabla_t \boldsymbol{D}, \quad \boldsymbol{\nabla} \times \boldsymbol{E} = -\nabla_t \boldsymbol{B} \tag{11}$$

Now we rewrite the equations in the form of matrix as follows

$$\begin{bmatrix} 0 & -\partial_z & \partial_y \\ \partial_z & 0 & -\partial_x \\ -\partial_y & \partial_x & 0 \end{bmatrix} \begin{Bmatrix} H_1 \\ H_2 \\ H_3 \end{Bmatrix} = \nabla_t \begin{Bmatrix} D_1 \\ D_2 \\ D_3 \end{Bmatrix} \tag{12}$$

or

$$[\Delta]\{H\} = \nabla_t \{D\} \tag{13}$$

$$\begin{bmatrix} 0 & -\partial_z & \partial_y \\ \partial_z & 0 & -\partial_x \\ -\partial_y & \partial_x & 0 \end{bmatrix} \begin{Bmatrix} E_1 \\ E_2 \\ E_3 \end{Bmatrix} = -\nabla_t \begin{Bmatrix} B_1 \\ B_2 \\ B_3 \end{Bmatrix} \tag{14}$$

or

$$[\Delta]\{E\} = -\nabla_t \{B\} \tag{15}$$

where $[\Delta]$ is defined as the matrix of electric and magnetic operators.
Substituting Eq. (1) into Eqs. (13) and (15) respectively, we have

$$[\Delta]\{H\} = \nabla_t [\varepsilon]\{E\} \tag{16}$$

$$[\Delta]\{E\} = -\nabla_t [\mu]\{H\} \tag{17}$$

Substituting Eq. (16) into (17) or Eq. (17) into (16), yield

$$[\square]\{H\} = -\nabla_t^2 [\mu][\varepsilon]\{H\} \tag{18}$$

$$[\square]\{E\} = -\nabla_t^2 [\mu][\varepsilon]\{E\} \tag{19}$$

where $[\square] = [\Delta][\Delta]$ is defined as the matrix of electromagnetic operators as follows

$$[\square] = \begin{bmatrix} -\left(\partial_z^2 + \partial_y^2\right) & \partial_{xy}^2 & \partial_{xz}^2 \\ \partial_{yx}^2 & -\left(\partial_x^2 + \partial_z^2\right) & \partial_{yz}^2 \\ \partial_{zx}^2 & \partial_{zy}^2 & -\left(\partial_x^2 + \partial_y^2\right) \end{bmatrix} \tag{20}$$

In another way, substituting Eqs. (5) and (6) into Eqs. (13) and (15), respectively, we have

$$[\Delta][\Phi]\{E^*\} = -\nabla_t [\Psi]\{B^*\} \tag{21}$$

$$[\Delta][\Psi]\{H^*\} = \nabla_t [\Phi]\{D^*\} \tag{22}$$

Rewriting the above in indicial notation, we get

$$\{\Delta_i^*\} E_i^* = -\nabla_t \{\varphi_i\} B_i^* \qquad i = 1,2,3 \tag{23}$$

$$\{\Delta_i^*\} H_i^* = \nabla_t \{\phi_i\} D_i^* \qquad i = 1,2,3 \tag{24}$$

where, Δ_i^* is the electromagnetic intensity operator, and i th row of $\left[\Delta^*\right] = [\Delta][\Phi]$.

4. Electromagnetic waves in anisotropic media

4.1 Electrically anisotropic media

In anisotropic dielectrics, the dielectric permittivity is a tensor, while the magnetic permeability is a scalar. So Eqs. (18) and (19) can be written as follows

$$[\square]\{H\} = -\nabla_t^2 \mu_0 [\varepsilon]\{H\} \tag{25}$$

$$[\square]\{E\} = -\nabla_t^2 \mu_0 [\varepsilon]\{E\} \tag{26}$$

Substituting Eqs. (4) - (6) into Eqs. (25) and (26), we have

$$[\square^*]\{H^*\} = -\nabla_t^2 \mu_0 [\Lambda]\{H^*\} \tag{27}$$

$$[\square^*]\{E^*\} = -\nabla_t^2 \mu_0 [\Lambda]\{E^*\} \tag{28}$$

where $[\square^*] = [\Phi]^{\mathrm{T}} [\square][\Phi]$ is defined as the eigen matrix of electromagnetic operators under the standard spaces. We can note from Appendix A that it is a diagonal matrix. Thus Eqs. (27) and (28) can be uncoupled in the form of scalar

$$\square_i^* H_i^* + \mu_0 \lambda_i \nabla_t^2 H_i^* = 0 \qquad i = 1,2,3 \tag{29}$$

$$\square_i^* E_i^* + \mu_0 \lambda_i \nabla_t^2 E_i^* = 0 \qquad i = 1,2,3 \tag{30}$$

Eqs.(29) and (30) are the modal equations of electromagnetic waves in anisotropic dielectrics.

4.2 Magnetically anisotropic media

In anisotropic magnetics, the magnetic permeability is a tensor, while the dielectric permittivity is a scalar. So Eqs. (18) and (19) can be written as follows

$$[\square]\{H\} = -\nabla_t^2 \varepsilon_0 [\mu]\{H\} \tag{31}$$

$$[\square]\{E\} = -\nabla_t^2 \varepsilon_0 [\mu]\{E\} \tag{32}$$

Substituting Eqs. (4) - (6) into Eqs. (31) and (32), we have

$$[\square^*]\{H^*\} = -\nabla_t^2 \varepsilon_0 [\Pi]\{H^*\} \tag{33}$$

$$[\square^*]\{E^*\} = -\nabla_t^2 \varepsilon_0 [\Pi]\{E^*\} \tag{34}$$

where $[\square^*] = [\Psi]^{\mathrm{T}} [\square][\Psi]$ is defined as the eigen matrix of electromagnetic operators under the standard spaces. We can also note from Appendix A that it is a diagonal matrix. Thus Eqs. (33) and (34) can be uncoupled in the form of scalar

$$\square_i^* H_i^* + \varepsilon_0 \gamma_i \nabla_t^2 H_i^* = 0 \qquad i = 1,2,3 \tag{35}$$

$$\square_i^* E_i^* + \varepsilon_0 \gamma_i \nabla_t^2 E_i^* = 0 \qquad i = 1,2,3 \tag{36}$$

Eqs.(35) and (36) are the modal equations of electromagnetic waves in anisotropic magnetics.

5. Electromagnetic waves in bi-anisotropic media

5.1 Bi-anisotropic constitutive equations

The constitutive equations of bi-anisotropic media are the following (Lindellm & Sihvola, 1994; Laudau & Lifshitz, 1960)

$$\boldsymbol{D} = \boldsymbol{\varepsilon} \cdot \boldsymbol{E} + \boldsymbol{\xi} \cdot \boldsymbol{H} \tag{37}$$

$$\boldsymbol{B} = \boldsymbol{\xi} \cdot \boldsymbol{E} + \boldsymbol{\mu} \cdot \boldsymbol{H} \tag{38}$$

where $\boldsymbol{\xi}$ is the matrix of magneto-electric parameter, and a symmetric one.

Substituting Eqs. (5) and (6) into Eqs. (37) and (38), respectively, and multiplying them with the transpose of modal matrix in the left, we have

$$\boldsymbol{\Phi}^{\mathrm{T}} \boldsymbol{D} = \boldsymbol{\Phi}^{\mathrm{T}} \boldsymbol{\varepsilon} \boldsymbol{\Phi} \boldsymbol{E}^{*} + \boldsymbol{\Phi}^{\mathrm{T}} \boldsymbol{\xi} \boldsymbol{\Psi} \boldsymbol{H}^{*} \tag{39}$$

$$\boldsymbol{\Psi}^{\mathrm{T}} \boldsymbol{B} = \boldsymbol{\Psi}^{\mathrm{T}} \boldsymbol{\xi} \boldsymbol{\Phi} \boldsymbol{E}^{*} + \boldsymbol{\Psi}^{\mathrm{T}} \boldsymbol{\mu} \boldsymbol{\Psi} \boldsymbol{H}^{*} \tag{40}$$

Let $\boldsymbol{G} = \boldsymbol{\Phi}^{\mathrm{T}} \boldsymbol{\xi} \boldsymbol{\Psi} = \boldsymbol{\Psi}^{\mathrm{T}} \boldsymbol{\xi} \boldsymbol{\Phi}$, that is a coupled magneto-electric matrix, and using Eq. (4), we have

$$\boldsymbol{D}^{*} = \boldsymbol{\Lambda} \boldsymbol{E}^{*} + \boldsymbol{G} \boldsymbol{H}^{*} \tag{41}$$

$$\boldsymbol{B}^{*} = \boldsymbol{G} \boldsymbol{E}^{*} + \boldsymbol{\Pi} \boldsymbol{H}^{*} \tag{42}$$

Rewriting the above in indicial notation, we get

$$D_i^* = \lambda_i E_i^* + g_{ij} H_j^* \qquad i = 1,2,3 \qquad j = 1,2,3 \tag{43}$$

$$B_i^* = \gamma_i H_i^* + g_{ij} E_j^* \qquad i = 1,2,3 \qquad j = 1,2,3 \tag{44}$$

Eqs. (43) and (44) are just the modal constitutive equations for bi-anisotropic media.

5.2 Eigen equations of electromagnetic waves in bi-anisotropic media

Substituting Eqs. (43) and (44) into Eqs. (23) and (24), respectively, we have

$$\{\Delta_i^*\} E_i^* = -\nabla_t \{\varphi_i\} \left(\gamma_i H_i^* + g_{ij} E_j^* \right) \tag{45}$$

$$\{\Delta_i^*\} H_i^* = \nabla_t \{\phi_i\} \left(\lambda_i E_i^* + g_{ij} H_j^* \right) \tag{46}$$

From them, we can get

$$\left(\{\Delta_i^*\} - \nabla_t \{\phi_i\} g_{ij} \delta_{ij} \right)^T \left(\{\Delta_i^*\} + \nabla_t \{\varphi_i\} g_{ij} \delta_{ij} \right) E_i^* = -\nabla_t^2 \{\varphi_i\} \{\phi_i\}^T \lambda_i \gamma_i E_i^* \tag{47}$$

$$\left(\{\Delta_i^*\} + \nabla_t \{\varphi_i\} g_{ij} \delta_{ij} \right)^T \left(\{\Delta_i^*\} - \nabla_t \{\phi_i\} g_{ij} \delta_{ij} \right) H_i^* = -\nabla_t^2 \{\phi_i\} \{\varphi_i\}^T \gamma_i \lambda_i H_i^* \tag{48}$$

The above can also be written as the standard form of waves

$$\square_i^* E_i^* + \nabla_t \{\Delta_i^*\}^T \cdot (\{\varphi\} - \{\phi\}) g_{ii} E_i^* + \nabla_t^2 \{\phi\}^T \cdot \{\varphi\}(\lambda_i \gamma_i - g_{ii}^2) E_i^* = 0 \quad i = 1,2,3 \tag{49}$$

$$\square_i^* H_i^* + \nabla_t \{\Delta_i^*\}^T \cdot (\{\varphi\} - \{\phi\}) g_{ii} H_i^* + \nabla_t^2 \{\phi\}^T \cdot \{\varphi\}(\lambda_i \gamma_i - g_{ii}^2) H_i^* = 0 \quad i = 1,2,3 \tag{50}$$

where, $\square_i^* = \{\Delta_i^*\}^T \cdot \{\Delta_i^*\}$ is the electromagnetic operator. Eqs.(49) and (50) are just equations of electric field and magnetic field for bi-anisotropic media.

5.3 Applications

5.3.1 Bi-isotropic media

The constitutive equations of bi-isotropic media are the following

$$\boldsymbol{D} = \begin{bmatrix} \varepsilon & 0 & 0 \\ 0 & \varepsilon & 0 \\ 0 & 0 & \varepsilon \end{bmatrix} \cdot \boldsymbol{E} + \begin{bmatrix} \xi & 0 & 0 \\ 0 & \xi & 0 \\ 0 & 0 & \xi \end{bmatrix} \cdot \boldsymbol{H} \tag{51}$$

$$\boldsymbol{B} = \begin{bmatrix} \xi & 0 & 0 \\ 0 & \xi & 0 \\ 0 & 0 & \xi \end{bmatrix} \cdot \boldsymbol{E} + \begin{bmatrix} \mu & 0 & 0 \\ 0 & \mu & 0 \\ 0 & 0 & \mu \end{bmatrix} \cdot \boldsymbol{H} \tag{52}$$

The eigen values and eigen vectors of those matrix are the following

$$\boldsymbol{\Lambda} = diag[\varepsilon, \varepsilon, \varepsilon] \;, \quad \boldsymbol{\Pi} = diag[\mu, \mu, \mu] \tag{53}$$

$$\boldsymbol{\Phi} = \boldsymbol{\Psi} = \begin{bmatrix} 1 & 0 & 0 \\ 0 & 1 & 0 \\ 0 & 0 & 1 \end{bmatrix} \tag{54}$$

We can see from the above equations that there is only one eigen-space in isotropic medium, which is a triple-degenerate one, and the space structure is the following

$$\boldsymbol{W} = W_1^{(3)} [\phi_1, \phi_2, \phi_3] \tag{55}$$

$$\boldsymbol{\phi}_1^* = \boldsymbol{\varphi}_1^* = \frac{\sqrt{3}}{3} [1,1,1]^T \tag{56}$$

Then the eigen-qualities and eigen-operators of bi-isotropic medium are respectively shown as belows

$$E_1^* = \boldsymbol{\phi}_1^{*\mathrm{T}} \cdot \boldsymbol{E} = \frac{\sqrt{3}}{3} (E_1 + E_2 + E_3) \tag{57}$$

$$\square_1^* = -(\partial_x^2 + \partial_y^2 + \partial_z^2) \,, \; g_{11} = \xi \tag{58}$$

So, the equation of electromagnetic wave in bi-isotropic medium becomes

$$\left(\partial_x^2+\partial_y^2+\partial_z^2\right)E_1^*=\left(\mu\varepsilon-\xi^2\right)\partial_t^2 E_1^* \tag{59}$$

the velocity of electromagnetic wave is

$$c^{(1)}=\frac{1}{\sqrt{\mu\varepsilon-\xi^2}} \tag{60}$$

5.3.2 Dzyaloshinskii's bi-anisotropic media

Dzyaloshinskii's constitutive equations of bi-anisotropic media are the following

$$\boldsymbol{D}=\begin{bmatrix}\varepsilon & 0 & 0\\ 0 & \varepsilon & 0\\ 0 & 0 & \varepsilon_z\end{bmatrix}\cdot\boldsymbol{E}+\begin{bmatrix}\xi & 0 & 0\\ 0 & \xi & 0\\ 0 & 0 & \xi_z\end{bmatrix}\cdot\boldsymbol{H} \tag{61}$$

$$\boldsymbol{B}=\begin{bmatrix}\xi & 0 & 0\\ 0 & \xi & 0\\ 0 & 0 & \xi_z\end{bmatrix}\cdot\boldsymbol{E}+\begin{bmatrix}\mu & 0 & 0\\ 0 & \mu & 0\\ 0 & 0 & \mu_z\end{bmatrix}\cdot\boldsymbol{H} \tag{62}$$

The eigen values and eigen vectors of those matrix are the following

$$\boldsymbol{\Lambda}=diag\left[\varepsilon,\varepsilon,\varepsilon_z\right]\ ,\ \ \boldsymbol{\Pi}=diag\left[\mu,\mu,\mu_z\right] \tag{63}$$

$$\boldsymbol{\Phi}=\boldsymbol{\Psi}=\begin{bmatrix}1 & 0 & 0\\ 0 & 1 & 0\\ 0 & 0 & 1\end{bmatrix} \tag{64}$$

We can see from the above equations that there are two eigen-spaces in Dzyaloshinskii's bi-anisotropic medium, in which one is a binary-degenerate one, the space structure is the following

$$\boldsymbol{W}=W_1^{(2)}[\phi_1,\phi_2]\oplus W_2^1[\phi_3] \tag{65}$$

Then the eigen-qualities and eigen-operators of Dzyaloshinskii's bi-anisotropic medium are respectively shown as belows

$$E_2^*=\boldsymbol{\phi}_2^{\mathrm{T}}\cdot\boldsymbol{E}=E_3\ ,\ \ \left|E_1^*\right|=\sqrt{\left(\boldsymbol{E}-\boldsymbol{\phi}_2^{\mathrm{T}}E_2^*\right)^{\mathrm{T}}\left(\boldsymbol{E}-\boldsymbol{\phi}_2^{\mathrm{T}}E_2^*\right)}=\sqrt{E_1^2+E_2^2} \tag{66}$$

$$\square_1^*=-\left(\partial_x^2+\partial_y^2+2\partial_z^2-2\partial_{xy}^2\right),\ \square_2^*=-\left(\partial_x^2+\partial_y^2\right),\ g_{11}=\xi,\ \ g_{22}=\xi_z \tag{67}$$

So, the equations of electromagnetic wave in Dzyaloshinskii's bi-anisotropic medium become

$$\left(\partial_x^2+\partial_y^2+2\partial_z^2-2\partial_{xy}^2\right)\sqrt{E_1^2+E_2^2}=\left(\mu\varepsilon-\xi^2\right)\partial_t^2\sqrt{E_1^2+E_2^2} \tag{68}$$

$$\left(\partial_x^2+\partial_y^2\right)E_3=\left(\mu_z\varepsilon_z-\xi_z^2\right)\partial_t^2 E_3 \tag{69}$$

the velocities of electromagnetic wave are

$$c^{(1)} = \frac{1}{\sqrt{\mu\varepsilon - \xi^2}} \tag{70}$$

$$c^{(2)} = \frac{1}{\sqrt{\mu_z\varepsilon_z - \xi_z^2}} \tag{71}$$

It is seen both from bi-isotropic media and Dzyaloshinskii's bi-anisotropic medium that the electromagnetic waves in bi-anisotropic medium will go faster duo to the bi-coupling between electric field and magnetic one.

6. Electromagnetic waves in lossy media

6.1 The constitutive equation of lossy media

The constitutive equation of lossy media is the following

$$\boldsymbol{D} = \boldsymbol{\varepsilon} \cdot \boldsymbol{E} + \int_t \sigma \cdot \frac{d\boldsymbol{E}}{d\tau} d\tau \tag{72}$$

It is equivalent to the following differential constitutive equation

$$\dot{\boldsymbol{D}} = \boldsymbol{\varepsilon} \cdot \dot{\boldsymbol{E}} + \sigma \cdot \boldsymbol{E} \tag{73}$$

Let

$$\boldsymbol{D}^{\mathrm{e}} = \boldsymbol{\varepsilon} \cdot \boldsymbol{E} \ , \ \dot{\boldsymbol{D}}^{\mathrm{d}} = \sigma \cdot \boldsymbol{E} \tag{74}$$

Eq.(73) can be written as

$$\dot{\boldsymbol{D}} = \dot{\boldsymbol{D}}^{\mathrm{e}} + \dot{\boldsymbol{D}}^{\mathrm{d}} \tag{75}$$

or

$$\nabla_t \{D\} = ([\varepsilon]\nabla_t + [\sigma])\{E\} \tag{76}$$

Using Eq.(5), the above becomes

$$\nabla_t \{D^*\} = \left([\Phi]^T [\varepsilon][\Phi]\nabla_t + [\Phi]^T [\sigma][\Phi]\right)\{E^*\} \tag{77}$$

According to Appendix B and Eq.(77), we have

$$\nabla_t \{D^*\} = ([\Lambda]\nabla_t + [\Gamma])\{E^*\} \tag{78}$$

Rewriting the above in indicial notation, we get

$$\nabla_t D_i^* = (\lambda_i \nabla_t + \eta_i) E_i^* \tag{79}$$

Eq.(79) is just the modal constitutive equations for lossy media.

6.2 Eigen equations of electromagnetic waves in lossy media

Substituting Eqs. (10) and (79) into Eqs. (23) and (24), respectively, we have

$$\{\Delta_i^*\}E_i^* = -\nabla_t\{\varphi_i\}\gamma_i H_i^* \quad i=1,2,3 \tag{80}$$

$$\{\Delta_i^*\}H_i^* = \{\phi_i\}(\lambda_i\nabla_t + \eta_i)E_i^* \quad i=1,2,3 \tag{81}$$

From them, we can get

$$\square_i^* E_i^* + \nabla_{tt}\xi_i\gamma_i\lambda_i E_i^* + \nabla_t\xi_i\gamma_i\eta_i E_i^* = 0 \quad i=1,2,3 \tag{82}$$

$$\square_i^* H_i^* + \nabla_{tt}\xi_i\gamma_i\lambda_i H_i^* + \nabla_t\xi_i\gamma_i\eta_i H_i^* = 0 \quad i=1,2,3 \tag{83}$$

where $\xi_i = \{\phi_i^*\}^T \cdot \{\varphi_i^*\}$. Eqs.(82) and (83) are just equations of electric field and magnetic field for bi-anisotropic media.

6.3 Applications

In this section, we discuss the propagation laws of electromagnetic waves in an isotropic lossy medium. The material tensors in Eqs.(1) and (72) are represented by the following matrices

$$\boldsymbol{\varepsilon} = \begin{bmatrix} \varepsilon_{11} & 0 & 0 \\ 0 & \varepsilon_{11} & 0 \\ 0 & 0 & \varepsilon_{11} \end{bmatrix}, \ \boldsymbol{\mu} = \begin{bmatrix} \mu_{11} & 0 & 0 \\ 0 & \mu_{11} & 0 \\ 0 & 0 & \mu_{11} \end{bmatrix}, \ \boldsymbol{\sigma} = \begin{bmatrix} \sigma_{11} & 0 & 0 \\ 0 & \sigma_{11} & 0 \\ 0 & 0 & \sigma_{11} \end{bmatrix} \tag{84}$$

The eigen values and eigen vectors of those matrix are the following

$$\boldsymbol{\Lambda} = diag[\varepsilon_{11}, \varepsilon_{11}, \varepsilon_{11}], \ \boldsymbol{\Pi} = diag[\mu_{11}, \mu_{11}, \mu_{11}], \ \boldsymbol{\Gamma} = diag[\sigma_{11}, \sigma_{11}, \sigma_{11}] \tag{85}$$

$$\boldsymbol{\Phi} = \boldsymbol{\Psi} = \boldsymbol{\Theta} = \begin{bmatrix} 1 & 0 & 0 \\ 0 & 1 & 0 \\ 0 & 0 & 1 \end{bmatrix} \tag{86}$$

We can see from the above equations that there is only one eigen-space in an isotropic lossy medium, which is a triple-degenerate one, and the space structure is the following.

$$\boldsymbol{W}_{mag} = W_1^{(3)}[\boldsymbol{\phi}_1, \boldsymbol{\phi}_2, \boldsymbol{\phi}_3], \ \boldsymbol{W}_{ele} = W_1^{(3)}[\boldsymbol{\varphi}_1, \boldsymbol{\varphi}_2, \boldsymbol{\varphi}_3] \tag{87}$$

where, $\boldsymbol{\phi}_1^* = \frac{\sqrt{3}}{3}\{1,1,1\}^T$, $\boldsymbol{\varphi}_1^* = \frac{\sqrt{3}}{3}\{1,1,1\}^T$, $\xi_1 = 1$.

Then the eigen-qualities and eigen-operators of an isotropic lossy media are respectively shown as follows

$$E_1^* = \frac{\sqrt{3}}{3}(E_1 + E_2 + E_3) \tag{88}$$

$$H_1^* = \frac{\sqrt{3}}{3}(H_1 + H_2 + H_3) \tag{89}$$

$$\Box_1^* = \frac{1}{3}\left[-\left(\partial_x^2 + \partial_y^2 + \partial_z^2\right)\right] \tag{90}$$

So, the equation of electromagnetic wave in lossy media becomes

$$\left(\partial_x^2 + \partial_y^2 + \partial_z^2\right)E_1^* = \frac{1}{c^2}\partial_t^2 E_1^* + \frac{1}{\tau^2}\partial_t E_1^* \tag{91}$$

Rewriting it in the component form, we have

$$\left(\partial_x^2 + \partial_y^2 + \partial_z^2\right)E_1 = \frac{1}{c^2}\partial_t^2 E_1 + \frac{1}{\tau^2}\partial_t E_1 \tag{92}$$

$$\left(\partial_x^2 + \partial_y^2 + \partial_z^2\right)E_2 = \frac{1}{c^2}\partial_t^2 E_2 + \frac{1}{\tau^2}\partial_t E_2 \tag{93}$$

$$\left(\partial_x^2 + \partial_y^2 + \partial_z^2\right)E_3 = \frac{1}{c^2}\partial_t^2 E_3 + \frac{1}{\tau^2}\partial_t E_3 \tag{94}$$

where, c is the velocity of electromagnetic wave, τ is the lossy coefficient of electromagnetic wave

$$c = \frac{1}{\sqrt{\mu_{11}\varepsilon_{11}}}, \quad \tau = \frac{1}{\sqrt{\mu_{11}\sigma_{11}}} \tag{95}$$

Now, we discuss the the propagation laws of a plane electromagnetic wave in x-axis. In this time, Eq. (92) becomes

$$\frac{\partial}{\partial x^2}E_1 = \frac{1}{c^2}\partial_t^2 E_1 + \frac{1}{\tau^2}\partial_t E_1 \tag{96}$$

Let the solution of Eq. (96) is as follows

$$E_1 = A\exp\left[i(kx - \omega t)\right] \tag{97}$$

Substituting the above into Eq. (96), we have

$$k^2 = \frac{\omega^2}{c^2} + i\frac{\omega}{\tau^2} \tag{98}$$

From Eq.(96), we can get

$$\overline{k} = k_1 + ik_2 \tag{99}$$

where

$$k_1 = \frac{\omega}{c}\left[\frac{1+\left(1+\frac{c^4}{\omega^2\tau^4}\right)^{\frac{1}{2}}}{2}\right]^{\frac{1}{2}}, \quad k_2 = \frac{\omega}{c}\left[\frac{-1+\left(1+\frac{c^4}{\omega^2\tau^4}\right)^{\frac{1}{2}}}{2}\right]^{\frac{1}{2}}.$$

Then, the solutions of electromagnetic waves are the following

$$E_1 = Ae^{-k_2 x} \cdot e^{i(k_1 x - \omega t)} = \overline{A} \cdot e^{i(k_1 x - \omega t)} \tag{100}$$

It is an attenuated sub-waves.

7. Electromagnetic waves in dispersive media

7.1 The constitutive equation of dispersive media

The general constitutive equations of dispersive media are the following

$$\boldsymbol{D} = \boldsymbol{\varepsilon} \cdot \boldsymbol{E} + \boldsymbol{\varepsilon}_1 \cdot \dot{\boldsymbol{E}} + \boldsymbol{\varepsilon}_2 \cdot \ddot{\boldsymbol{E}} + \cdots \tag{101}$$

$$\boldsymbol{B} = \boldsymbol{\mu} \cdot \boldsymbol{H} + \boldsymbol{\mu}_1 \cdot \dot{\boldsymbol{H}} + \boldsymbol{\mu}_2 \cdot \ddot{\boldsymbol{H}} + \cdots \tag{102}$$

where $\boldsymbol{\varepsilon}_i, (i = 1,2,\cdots)$ and $\boldsymbol{\mu}_i, (i = 1,2,\cdots)$ are the higher order dielectric permittivity matrix and the magnetic permeability matrix respectively, and all symmtric ones.
Substituting Eqs. (5) and (6) into Eqs. (101) and (102), respectively, and multiplying them with the transpose of modal matrix in the left, we have

$$\boldsymbol{D}^* = \boldsymbol{\Phi}^T \boldsymbol{\varepsilon} \boldsymbol{\Phi} \boldsymbol{E}^* + \boldsymbol{\Phi}^T \boldsymbol{\varepsilon}_1 \boldsymbol{\Phi} \dot{\boldsymbol{E}} + \boldsymbol{\Phi}^T \boldsymbol{\varepsilon}_2 \boldsymbol{\Phi} \ddot{\boldsymbol{E}} + \cdots \tag{103}$$

$$\boldsymbol{B}^* = \boldsymbol{\Psi}^T \boldsymbol{\mu} \boldsymbol{\Psi} \boldsymbol{H} + \boldsymbol{\Psi}^T \boldsymbol{\mu}_1 \boldsymbol{\Psi} \dot{\boldsymbol{H}} + \boldsymbol{\Psi}^T \boldsymbol{\mu}_2 \boldsymbol{\Psi} \ddot{\boldsymbol{H}} + \cdots \tag{104}$$

It can be proved that there exist same standard spaces for various order electric and magnetic fields in the condition close to the thermodynamic equilibrium. Then, we have

$$D_i^* = \lambda_i E_i^* + \lambda_i^{(1)} \dot{E}_i^* + \lambda_i^{(2)} \ddot{E}_i^* + \cdots \tag{105}$$

$$B_i^* = \gamma_i H_i^* + \gamma_i^{(1)} \dot{H}_i^* + \gamma_i^{(2)} \ddot{H}_i^* + \cdots \tag{106}$$

Eqs. (105) and (106) are just the modal constitutive equations for the general dispersive media.

7.2 Eigen equations of electromagnetic waves in dispersive media

Substituting Eqs. (105) and (106) into Eqs. (23) and (24), respectively, we have

$$\{\Delta_i^*\} E_i^* = -\nabla_t \{\varphi_i\} \left(\gamma_i H_i^* + \gamma_i^{(1)} \dot{H}_i^* + \gamma_i^{(2)} \ddot{H}_i^* + \cdots\right) \tag{107}$$

$$\{\Delta_i^*\}H_i^* = \nabla_t\{\phi_i\}\left(\lambda_i E_i^* + \lambda_i^{(1)}\dot{E}_i^* + \lambda_i^{(2)}\ddot{E}_i^* + \cdots\right) \tag{108}$$

From them, we can get

$$\frac{1}{\xi_i}\square_i^* E_i^* + \gamma_i\lambda_i\nabla_{tt}E_i^* + \left(\gamma_i\lambda_i^{(1)} + \gamma_i^{(1)}\lambda_i\right)\nabla_{ttt}E_i^* + \left(\gamma_i\lambda_i^{(2)} + \gamma_i^{(1)}\lambda_i^{(1)} + \gamma_i^{(2)}\lambda_i\right)\nabla_{tttt}E_i^* + \cdots = 0 \tag{109}$$

$$\frac{1}{\xi_i}\square_i^* H_i^* + \gamma_i\lambda_i\nabla_{tt}H_i^* + \left(\gamma_i\lambda_i^{(1)} + \gamma_i^{(1)}\lambda_i\right)\nabla_{ttt}H_i^* + \left(\gamma_i\lambda_i^{(2)} + \gamma_i^{(1)}\lambda_i^{(1)} + \gamma_i^{(2)}\lambda_i\right)\nabla_{tttt}H_i^* + \cdots = 0 \tag{110}$$

Eqs.(109) and (110) are just equations of electric field and magnetic field for general dispersive media.

7.3 Applications

In this section, we discuss the propagation laws of electromagnetic waves in an one-order dispersive medium. The material tensors in Eqs.(101) and (102) are represented by the following matrices

$$\boldsymbol{\varepsilon} = \begin{bmatrix} \varepsilon_{11} & 0 & 0 \\ 0 & \varepsilon_{11} & 0 \\ 0 & 0 & \varepsilon_{11} \end{bmatrix},\ \boldsymbol{\mu} = \begin{bmatrix} \mu_{11} & 0 & 0 \\ 0 & \mu_{11} & 0 \\ 0 & 0 & \mu_{11} \end{bmatrix},\ \boldsymbol{\varepsilon}_1 = \begin{bmatrix} \varepsilon'_{11} & 0 & 0 \\ 0 & \varepsilon'_{11} & 0 \\ 0 & 0 & \varepsilon'_{11} \end{bmatrix},\ \boldsymbol{\mu}_1 = \begin{bmatrix} \mu'_{11} & 0 & 0 \\ 0 & \mu'_{11} & 0 \\ 0 & 0 & \mu'_{11} \end{bmatrix} \tag{111}$$

The eigen values and eigen vectors of those matrix are the following

$$\boldsymbol{\Lambda} = diag[\varepsilon_{11}, \varepsilon_{11}, \varepsilon_{11}],\ \boldsymbol{\Lambda}_1 = diag[\varepsilon'_{11}, \varepsilon'_{11}, \varepsilon'_{11}] \tag{112}$$

$$\boldsymbol{\Pi} = diag[\mu_{11}, \mu_{11}, \mu_{11}],\ \boldsymbol{\Pi}_1 = diag[\mu'_{11}, \mu'_{11}, \mu'_{11}] \tag{113}$$

$$\boldsymbol{\Phi} = \boldsymbol{\Psi} = \begin{bmatrix} 1 & 0 & 0 \\ 0 & 1 & 0 \\ 0 & 0 & 1 \end{bmatrix} \tag{114}$$

We can see from the above equations that there is only one eigen-space in isotropic one-order dispersive medium, which is a triple-degenerate one, and the space structure is the following

$$\boldsymbol{W}_{mag} = W_1^{(3)}[\boldsymbol{\phi}_1, \boldsymbol{\phi}_2, \boldsymbol{\phi}_3],\ \boldsymbol{W}_{ele} = W_1^{(3)}[\boldsymbol{\varphi}_1, \boldsymbol{\varphi}_2, \boldsymbol{\varphi}_3] \tag{115}$$

where, $\boldsymbol{\phi}_1^* = \frac{\sqrt{3}}{3}\{1,1,1\}^T$, $\boldsymbol{\varphi}_1^* = \frac{\sqrt{3}}{3}\{1,1,1\}^T$, $\xi_1 = 1$. Thus the eigen-qualities and eigen-operators of isotropic one-order dispersive medium are known as same as Eqs. (88) – (90). The equations of electromagnetic wave in one-order dispersive medium become

$$\left(\partial_x^2 + \partial_y^2 + \partial_z^2\right)E_1^* = \frac{1}{c^2}\nabla_{tt}E_1^* + \left(\mu_{11}\varepsilon_{11}^{(1)} + \mu_{11}^{(1)}\varepsilon_{11}\right)\nabla_{ttt}E_1^* \tag{116}$$

$$\left(\partial_x^2+\partial_y^2+\partial_z^2\right)H_1^*=\frac{1}{c^2}\nabla_{tt}H_1^*+\left(\mu_{11}\varepsilon_{11}^{(1)}+\mu_{11}^{(1)}\varepsilon_{11}\right)\nabla_{ttt}H_1^* \tag{117}$$

in which c is the velocity of electromagnetic wave

$$c=\frac{1}{\sqrt{\mu_{11}\varepsilon_{11}}} \tag{118}$$

Now, we discuss the propagation laws of a plane electromagnetic wave in x-axis. In this time, Eq.(116) becomes

$$\frac{\partial}{\partial x^2}E_1=\frac{1}{c^2}\nabla_{tt}E_1+\left(\mu_{11}\varepsilon_{11}^{(1)}+\mu_{11}^{(1)}\varepsilon_{11}\right)\nabla_{ttt}E_1 \tag{119}$$

Let

$$E_1=A\exp\left[i(kx-\omega t)\right] \tag{120}$$

Substituting the above into Eq.(119), we have

$$k^2=\frac{\omega^2}{c^2}-i\omega^3\left(\mu_{11}\varepsilon_{11}^{(1)}+\mu_{11}^{(1)}\varepsilon_{11}\right) \tag{121}$$

From the above, we can get

$$\bar{k}=k_1+ik_2 \tag{122}$$

where

$$k_1=\frac{\omega}{c}\left[\frac{1+\left(1+c^4\left(\mu_{11}\varepsilon_{11}^{(1)}+\mu_{11}^{(1)}\varepsilon_{11}\right)^2\omega^2\right)^{\frac{1}{2}}}{2}\right]^{\frac{1}{2}},\quad k_2=\frac{\omega}{c}\left[\frac{-1+\left(1+c^4\left(\mu_{11}\varepsilon_{11}^{(1)}+\mu_{11}^{(1)}\varepsilon_{11}\right)^2\omega^2\right)^{\frac{1}{2}}}{2}\right]^{\frac{1}{2}}.$$

Then, the solutions of electromagnetic waves are

$$E_1=Ae^{-k_2x}\cdot e^{i(k_1x-\omega t)}=\bar{A}\cdot e^{i(k_1x-\omega t)} \tag{123}$$

It is an attenuated sub-waves.

8. Electromagnetic waves in chiral media

8.1 The constitutive equation of chiral media

The constitutive equations of chiral media are the following

$$\boldsymbol{D}=\boldsymbol{\varepsilon}\cdot\boldsymbol{E}-\boldsymbol{\chi}\cdot\nabla_t\boldsymbol{H} \tag{124}$$

$$\boldsymbol{B}=\boldsymbol{\chi}\cdot\nabla_t\boldsymbol{E}+\boldsymbol{\mu}\cdot\boldsymbol{H} \tag{125}$$

where $\boldsymbol{\chi}$ is the matrix of chirality parameter, and a symmtric one.
Substituting Eqs. (5) and (6) into Eqs. (124) and (125), respectively, and multiplying them with the transpose of modal matrix in the left, we have

$$\boldsymbol{D}^{*} = \boldsymbol{\Phi}^{T}\boldsymbol{\varepsilon}\boldsymbol{\Phi}\boldsymbol{E}^{*} - \boldsymbol{\Phi}^{T}\boldsymbol{\chi}\boldsymbol{\Psi}\nabla_{t}\boldsymbol{H}^{*} \tag{126}$$

$$\boldsymbol{B}^{*} = \boldsymbol{\Psi}^{T}\boldsymbol{\chi}\boldsymbol{\Phi}\nabla_{t}\boldsymbol{E}^{*} + \boldsymbol{\Psi}^{T}\boldsymbol{\mu}\boldsymbol{\Psi}\boldsymbol{H}^{*} \tag{127}$$

Let $\boldsymbol{\Gamma} = \boldsymbol{\Psi}^{\mathrm{T}}\boldsymbol{\chi}\boldsymbol{\Phi}$, that is a coupled chiral matrix, and using Eq. (4) , we have

$$\boldsymbol{D}^{*} = \boldsymbol{\Lambda}\boldsymbol{E}^{*} - \boldsymbol{\Gamma}^{T}\nabla_{t}\boldsymbol{H}^{*} \tag{128}$$

$$\boldsymbol{B}^{*} = \boldsymbol{\Gamma}^{T}\nabla_{t}\boldsymbol{E}^{*} + \boldsymbol{\Pi}\boldsymbol{H}^{*} \tag{129}$$

For most chiral, $\boldsymbol{\Gamma} = diag[\varsigma_1, \varsigma_2, \varsigma_3]$. Then we have

$$D_i^{*} = \lambda_i E_i^{*} - \varsigma_i \nabla_t H_i^{*} \tag{130}$$

$$B_i^{*} = \varsigma_i \nabla_t E_i^{*} + \gamma_i H_i^{*} \tag{131}$$

Eqs.(130) and (131) are just the modal constitutive equations for anisotropic chiral media.

8.2 Eigen equations of electromagnetic waves in chiral media

Substituting Eqs. (130) and (131) into Eqs. (23) and (24), respectively, we have

$$\{\Delta_i^{*}\}E_i^{*} = -\nabla_t\{\varphi_i\}\left(\varsigma_i\nabla_t E_i^{*} + \gamma_i H_i^{*}\right) \qquad i = 1,2,3 \tag{132}$$

$$\{\Delta_i^{*}\}H_i^{*} = \nabla_t\{\phi_i\}\left(\lambda_i E_i^{*} - \varsigma_i\nabla_t H_i^{*}\right) \qquad i = 1,2,3 \tag{133}$$

From them, we can get

$$\square_i^{*} E_i^{*} + \xi_i\varsigma_i^{2}\nabla_{tttt}E_i^{*} + \left(2\varsigma_i\partial_i^{*} + \xi_i\lambda_i\gamma_i\right)\nabla_{tt}E_i^{*} = 0 \qquad i = 1,2,3 \tag{134}$$

$$\square_i^{*} H_i^{*} + \xi_i\varsigma_i^{2}\nabla_{tttt}H_i^{*} + \left(2\varsigma_i\partial_i^{*} + \xi_i\lambda_i\gamma_i\right)\nabla_{tt}H_i^{*} = 0 \qquad i = 1,2,3 \tag{135}$$

where $\xi_i = \{\phi_i\}^{T}\cdot\{\varphi_i\} = 1$, $\partial_i = \{\Delta_i^{*}\}^{T}\cdot\{\varphi_i\}$. Eqs.(134) and (135) are just equations of electric field and magnetic field for chiral media.

8.3 Applications

In this section, we discuss the propagation laws of electromagnetic waves in an isotropic chiral medium. The material tensors in Eqs.(124) and (125) are represented by the following matrices

$$\boldsymbol{\varepsilon} = \begin{bmatrix} \varepsilon_{11} & 0 & 0 \\ 0 & \varepsilon_{11} & 0 \\ 0 & 0 & \varepsilon_{11} \end{bmatrix}, \ \boldsymbol{\mu} = \begin{bmatrix} \mu_{11} & 0 & 0 \\ 0 & \mu_{11} & 0 \\ 0 & 0 & \mu_{11} \end{bmatrix}, \ \boldsymbol{\chi} = \begin{bmatrix} \chi_{11} & 0 & 0 \\ 0 & \chi_{11} & 0 \\ 0 & 0 & \chi_{11} \end{bmatrix} \tag{136}$$

The eigen values and eigen vectors of those matrix are the following

$$\boldsymbol{\Lambda} = diag\left[\varepsilon_{11}, \varepsilon_{11}, \varepsilon_{11}\right], \; \boldsymbol{\Pi} = diag\left[\mu_{11}, \mu_{11}, \mu_{11}\right], \; \boldsymbol{\Gamma} = diag\left[\chi_{11}, \chi_{11}, \chi_{11}\right] \tag{137}$$

$$\boldsymbol{\Phi} = \boldsymbol{\Psi} = \begin{bmatrix} 1 & 0 & 0 \\ 0 & 1 & 0 \\ 0 & 0 & 1 \end{bmatrix} \tag{138}$$

We can see from the above equations that there is only one eigen-space in isotropic medium, which is a triple-degenerate one, and the space structure is the following

$$\boldsymbol{W}_{mag} = W_1^{(3)}\left[\boldsymbol{\phi}_1, \boldsymbol{\phi}_2, \boldsymbol{\phi}_3\right], \; \boldsymbol{W}_{ele} = W_1^{(3)}\left[\boldsymbol{\varphi}_1, \boldsymbol{\varphi}_2, \boldsymbol{\varphi}_3\right] \tag{139}$$

where, $\boldsymbol{\phi}_1^* = \frac{\sqrt{3}}{3}\{1,1,1\}^T$, $\boldsymbol{\varphi}_1^* = \frac{\sqrt{3}}{3}\{1,1,1\}^T$, $\xi_1 = 1$.

Then the eigen-qualities and eigen-operators of isotropic chiral medium are respectively shown as follows

$$E_1^* = \frac{\sqrt{3}}{3}\left(E_1 + E_2 + E_3\right), \; H_1^* = \frac{\sqrt{3}}{3}\left(H_1 + H_2 + H_3\right) \tag{140}$$

$$\Box_1^* = \frac{1}{3}\left[-\left(\partial_x^2 + \partial_y^2 + \partial_z^2\right)\right], \; \partial_1^* = \frac{\sqrt{3}}{3}\left(\partial_x + \partial_y + \partial_z\right) \tag{141}$$

So, the equations of electromagnetic wave in isotropic chiral medium become

$$\left(\partial_x^2 + \partial_y^2 + \partial_z^2\right)E_1^* = \chi_{11}^2 \nabla_{tttt} E_1^* + \left[2\sqrt{3}\chi_{11}\left(\partial_x + \partial_y + \partial_z\right) + \frac{1}{c^2}\right]\nabla_{tt} E_1^* \tag{142}$$

$$\left(\partial_x^2 + \partial_y^2 + \partial_z^2\right)H_1^* = \chi_{11}^2 \nabla_{tttt} H_1^* + \left[2\sqrt{3}\chi_{11}\left(\partial_x + \partial_y + \partial_z\right) + \frac{1}{c^2}\right]\nabla_{tt} H_1^* \tag{143}$$

where, c is the velocity of electromagnetic wave

$$c = \frac{1}{\sqrt{\mu_{11}\varepsilon_{11}}} \tag{144}$$

Now, we discuss the propagation laws of a plane electromagnetic wave in x-axis. In this time, Eq.(142) becomes

$$\frac{\partial}{\partial x^2}E_1 = \chi_{11}^2 \nabla_{tttt} E_1 + \left[2\sqrt{3}\chi_{11}\frac{\partial}{\partial x} + \frac{1}{c^2}\right]\nabla_{tt} E_1 \tag{145}$$

Let

$$E_1 = A\exp\left[i\left(kx - \omega t\right)\right] \tag{146}$$

Substituting the above into Eq.(145), we have

$$k^2 = \left(\frac{1}{c^2} + i2\sqrt{3}\chi_{11}k\right)\omega^2 - \chi_{11}^2\omega^4 \tag{147}$$

or

$$k^2 - i2\sqrt{3}\chi_{11}\omega^2 k + \left(\chi_{11}^2\omega^2 - \mu_{11}\varepsilon_{11}\right)\omega^2 = 0 \tag{148}$$

1. when $\omega^2 - \dfrac{\mu_{11}\varepsilon_{11}}{\chi_{11}^2} < 0$

By Eq.(148), we have

$$k_1 = k_1' + ik_1'' = i\chi_{11}\omega^2\left[\sqrt{3} + (x + iy)\right] \tag{149}$$

$$k_2 = k_2' + ik_2'' = i\chi_{11}\omega^2\left[\sqrt{3} - (x + iy)\right] \tag{150}$$

From them, we can get

$$k_1' = -\chi_{11}\omega^2 y\ ,\ k_1'' = \left(\sqrt{3} + x\right)\chi_{11}\omega^2 \tag{151}$$

$$k_2' = \chi_{11}\omega^2 y\ ,\ k_2'' = \left(\sqrt{3} - x\right)\chi_{11}\omega^2 \tag{152}$$

where, $x = \dfrac{\sqrt{2}}{2}\sqrt{3 + \left[9 + \dfrac{\left(\dfrac{\mu_{11}\varepsilon_{11}}{\chi_{11}^2} - \omega^2\right)^2}{\omega^4}\right]^{\frac{1}{2}}}$, $y = \dfrac{\sqrt{2}}{2}\sqrt{-3 + \left[9 + \dfrac{\left(\dfrac{\mu_{11}\varepsilon_{11}}{\chi_{11}^2} - \omega^2\right)^2}{\omega^4}\right]^{\frac{1}{2}}}$.

Then, the solution of electromagnetic waves is the following

$$E_1 = A_1 e^{-k_1''x} \cdot e^{i(k_1'x - \omega t)} + A_2 e^{-k_2''x} \cdot e^{i(k_2'x - \omega t)} \tag{153}$$

It is composed of two attenuated sub-waves.

2. when $\omega^2 - \dfrac{\mu_{11}\varepsilon_{11}}{\chi_{11}^2} > 0$

By Eq.(148), we have

$$k_1 = k_1' + ik_1'' = i\chi_{11}\omega^2\left[\sqrt{3} + \sqrt{3 + \frac{1}{\omega^2}\left(\omega^2 - \frac{\mu_{11}\varepsilon_{11}}{\chi_{11}^2}\right)}\right] \tag{154}$$

$$k_2 = k_2' + ik_2'' = i\chi_{11}\omega^2\left[\sqrt{3} - \sqrt{3 + \frac{1}{\omega^2}\left(\omega^2 - \frac{\mu_{11}\varepsilon_{11}}{\chi_{11}^2}\right)}\right] \tag{155}$$

where

$$k_1' = 0\,,\; k_1'' = \chi_{11}\omega^2\left[\sqrt{3} + \sqrt{3 + \frac{1}{\omega^2}\left(\omega^2 - \frac{\mu_{11}\varepsilon_{11}}{\chi_{11}^2}\right)}\right] \tag{156}$$

$$k_2' = -\chi_{11}\omega^2\left[\sqrt{3 + \frac{1}{\omega^2}\left(\omega^2 - \frac{\mu_{11}\varepsilon_{11}}{\chi_{11}^2}\right)} - \sqrt{3}\right],\; k_2'' = 0 \tag{157}$$

Then, the solution of electromagnetic waves is the following

$$E_1 = A_1 e^{-k_1'' x} e^{-i\omega t} + A_2 e^{i(k_2' x - \omega t)} \tag{158}$$

It is seen that there only exists an electromagnetic sub-wave in opposite direction.

3. when $\omega^2 - \dfrac{\mu_{11}\varepsilon_{11}}{\chi_{11}^2} = 0$

By Eq.(148), we have

$$k = k' + ik'' = i2\sqrt{3}\chi_{11}\omega^2 \tag{159}$$

where

$$k' = 0\,,\; k'' = 2\sqrt{3}\chi_{11}\omega^2 \tag{160}$$

Then, the solution of electromagnetic waves is the following

$$E = Ae^{-k''x} \cdot e^{-i\omega t} \tag{161}$$

No electromagnetic sub-waves exist now.

9. Conclusion

In this chapter, we construct the standard spaces under the physical presentation by solving the eigen-value problem of the matrixes of dielectric permittivity and magnetic permeability, in which we get the eigen dielectric permittivity and eigen magnetic permeability, and the corresponding eigen vectors. The former are coordinate-independent and the latter are coordinate-dependent. Because the eigen vectors show the principal directions of electromagnetic media, they can be used as the standard spaces. Based on the spaces, we get the modal equations of electromagnetic waves for anisotropic media, bi-anisotropic media, dispersive medium and chiral medium, respectively, by converting the classical Maxwell's vector equation to the eigen Maxwell's scalar equation, each of which shows the existence of an electromagnetic sub-wave, and its propagation velocity, propagation direction, polarization direction and space pattern are completely determined in the equations. Several novel results are obtained for anisotropic media. For example, there is only one kind of electromagnetic wave in isotropic crystal, which is identical with the classical result; there are two kinds of electromagnetic waves in uniaxial crystal; three kinds of electromagnetic waves in biaxial crystal and three kinds of distorted electromagnetic waves in monoclinic crystal. Also for bi-anisotropic media, there exist two electromagnetic waves in Dzyaloshinskii's bi-anisotropic media, and the electromagnetic waves in bi-anisotropic medium will go faster duo to the bi-coupling between electric field and magnetic one. For isotropic dispersive medium, the electromagnetic wave is an attenuated sub-waves. And for chiral medium, there exist different propagating states of electromagnetivc waves in different frequency band, for example, in low frequency band, the electromagnetic waves are composed of two attenuated sub-waves, in high frequency band, there only exists an electromagnetic sub-wave in opposite direction, and in the critical point, no electromagnetic can propagate. All of these new theoretical results need to be proved by experiments in the future.

10. Appendix A: Proof of the eigenmode of electromagnetic operator matrix

The Maxwell's equation of anisotropic dielectrics is the following

$$[\square]\{H\} = -\nabla_t^2 \mu_0 [\varepsilon]\{H\} \tag{A1}$$

Using the representation transform relationship Eq. (6), we have

$$[\square][\Phi]\{H^*\} = -\nabla_t^2 \mu_0 [\varepsilon][\Phi]\{H^*\} \tag{A2}$$

Substituting the spectral decomposition matrix of dielectric permittivity Eq. (4) into above, we have

$$[\square][\Phi]\{H^*\} = -\nabla_t^2 \mu_0 [\Phi][\Lambda]\{H^*\} \tag{A3}$$

Comparing the both sides of above equation, we can get

$$[\square][\Phi] = -\nabla_t^2 \mu_0 [\Phi][\Lambda] \tag{A4}$$

Multiplying the both sides of above with the transpose of modal matrix in the left, we have

$$[\Phi]^{\mathrm{T}} [\square][\Phi] = -\nabla_t^2 \mu_0 [\Lambda] \tag{A5}$$

It is seen that the right side above is a diagonal matrix, which shows that the electromagnetic operators matrix can also be spectrally decomposed in standard spaces, then we get

$$[\square^*] = -\nabla_t^2 \mu_0 [\Lambda] \tag{A6}$$

Rewriting above in the form of scalar, we have

$$\square_i^* = -\nabla_t^2 \mu_0 \lambda_i \tag{A7}$$

11. Appendix B: Spectrally decomposition of lossy matrix

The Helmholtz's free energy of electromagnetic system with lossy property is the following

$$\psi=\psi\left(\boldsymbol{D},\boldsymbol{B},\boldsymbol{D}^{d}\right)=\frac{1}{2}\left(\boldsymbol{D}-\boldsymbol{D}^{d}\right)\boldsymbol{\varepsilon}^{-1}\left(\boldsymbol{D}-\boldsymbol{D}^{d}\right)+\frac{1}{2}\boldsymbol{B}\boldsymbol{\mu}^{-1}\boldsymbol{B} \tag{B1}$$

Differentiating the above with lossy variable, and using Eq.(1), we have

$$\boldsymbol{R}=-\frac{\partial\psi}{\partial\boldsymbol{D}^{d}}=\left(\boldsymbol{D}-\boldsymbol{D}^{d}\right)\boldsymbol{\varepsilon}^{-1}=\boldsymbol{E} \tag{B2}$$

According to the Onsager's principle, for the process of closing to equilibrium, the rate of lossy variable is proportion to the driving force, that is

$$\frac{\partial\psi}{\partial D_i^{d}}+\beta_{ij}\frac{dD_j^{d}}{dt}=0 \tag{B3}$$

where, $\beta_{ij}=\beta_{ji}$ is the general friction coefficient. Rewritting the above in matrix form, we have

$$\{R\}=[B]\{\dot{D}^{d}\} \tag{B4}$$

Projecting the lossy electric displacement vector $\boldsymbol{D}$ into the standard spaces of the physical presentation, we get

$$\{D^{d}\}=a\{\varphi\} \tag{B5}$$

Using Eq.(B2), we have

$$\{R\}=\omega\dot{a}\{\varphi\} \tag{B6}$$

Substituting Eqs. (B5) and (B6) into Eq. (B4), the condition of non-zero solution to α is the following

$$([B]-\omega[I])\{\varphi\}=0 \tag{B7}$$

It is seen that the general friction coefficient matrix can also be spectrally decomposed in standard spaces, so we have

$$[B]=[\Phi][\Omega][\Phi]^{T} \tag{B8}$$

Comparing Eq. (B4) with Eq. (74), it is known that we can also spectrally decompose the lossy matrix in standard spaces

$$[\sigma]=[\Phi]^{T}[\Gamma][\Phi] \tag{B9}$$

where, $\boldsymbol{\Gamma}=\boldsymbol{\Omega}^{-1}=diag[\eta_1,\eta_2,\eta_3]$.

12. References

[1] Agyei, A. & Birman, J. (1990). On the linear magnetoelectric effect. *Journal of Physics: Condensed Matter,* Vol. 2, No. 13, (April 1990), pp. 3007-3020, ISSN 0953-8984

[2] Astrov, D. (1960). The magnetoelectric effect in antiferromagnetics. *Soviet Physics, Journal of Experimental and Theoretical Physics*, Vol. 38, No. 11, (May 1960), pp. 984-985, ISSN 0038-5646
[3] Athanasiadis, C. & Giotopoulos, S. (2003). The Atkinson-Wilcox expansion theorem for electromagnetic chiral waves. *Applied Mathematics Letters*, Vol. 16, No. 5, (July 2003), pp. 675-681, ISSN 0893-9659
[4] Athanasiadis, C.; Roach, G. & Stratis, I. (2003). A time domain analysis of wave motions in chiral materials. *Mathematische Nachrichten*, Vol. 250, No. 1, (February 2003), pp. 3-16, ISSN 1522-2616
[5] Benveniste, Y. (1995). Magnetoelectric effect in fibrous composites with piezoelectric and piezomagnetic phases. *Physical Review B*, Vol. 51, No. 9, (September 1995), pp. 424-427, ISSN 1098-0121
[6] Bracke, L. & Van Vliet, M. (1981). A broadband magneto-electric transducer using a composite material. *International Journal of Electronics*, Vol. 51, No. 2, (February 1981), pp. 255-262, ISSN 0020-7217
[7] Bui, M.; Stuchly, S. & Costache, G. (1991). Propagation of transients in dispersive dielectric media. *IEEE Transactions on Microwave Theory and Techniques*, Vol. 39, No. 7, (July 1991) , pp. 1165-1171, ISSN 0018-9480
[8] Cohen, G. (2002). *Higher-order numerical methods for transient wave equations*, Springer, ISBN 3-540-41598-x, Berlin
[9] Dzyaloshinskii, I. (1959). On the magneto-electrical effect in antiferromagnets. *Soviet Physics, Journal of Experimental and Theoretical Physics*, Vol. 10, No. 3, (April 1959), pp. 628-669, ISSN 0038-5646
[10]Dvorak, S. & Dudley, D. (1995). Propagation of ultra-wide-band electromagnetic pulses through dispersive media. *IEEE Transactions on Electromagnetic Compatibility*, Vol. 37, No. 1, (February 1995), pp. 192-200, ISSN 0018-9375
[11] Gandhi, O. & Furse, C. (1997). Currents induced in the human body for exposure to ultrawideband electromagnetic pulses. *IEEE Transactions on Electromagnetic Compatibility*, Vol. 39, No. 2, (February 1997), pp. 174-180, ISSN 0018-9375
[12] Guo, S. (2009). An eigen theory of electromagnetic waves based on the standard spaces. *International Journal of Engineering Science*, Vol. 47, No. 3, (March 2009), pp. 405-412, ISSN 0020-7225
[13] Guo, S. (2009). An Eigen theory of Static Electromagnetic Field for Anisotropic Media. *Applied Mathematics and Mechanics*, Vol. 30, No. 5, (May 2009), pp. 643-652, ISSN 0253-4827
[14] Guo, S. (2010). A second-order eigen theory for static electromagnetic fields. *Journalof Modern Physics*, Vol. 1, No. 2, (June 2010), pp. 100-107, ISSN 2153-1196
[15] Guo, S. (2010). An eigen theory of electro-magnetic acoustic waves in magnetoelectroelastic media. *Acta Mechanica*, Vol. 211, No. 2, (April 2010), pp. 173-180, ISSN 0001-5970
[16] Guo, S. (2010). An fully dynamical theory of piezoelectromagnetic waves. *Acta Mech.*, Vol. 215, No. 3, (December 2010), pp. 335-344, ISSN 0001-5970
[17] Haba, Z. (2004). Green functions and propagation of waves in strongly inhomogeneous media. *Journal of Physis A: Mathematical and General*, Vol. 37, No. 9, (September 2004), pp. 9295-9302, ISSN 0305-4470

[18] Harshe, G.; Dougherty, J. & Newnham, R. (1993). Theoretical modelling of multilayer magnetoelectric composites. *International Journal of Applied Electromagnetic Matter*, Vol. 4, No. 2, (February 1993), pp. 145-159, ISSN 1383-5416
[19] Jaggard, D.; Mickelson, A. & Papas, C. (1979). On electromagnetic waves in chiral media. *Applied Physics A: Materials Science and Processing*, Vol. 18, No. 2, (February 1979), pp. 211-216, ISSN 0947-8396
[20] Jin, J,; Zunoubi, M.; Donepudi, K. & Chew, W. (1999). Frequency-domain and time-domain finite element solution of Maxwell's equations using spectral Lanczos decomposition method. *Computer Methods in Applied Mechanics and Engineering*, Vol. 169, No. 2, (February 1999), pp. 279-296, ISSN 0045-7825
[21] Kong, J. (1986). *Electromagnetic wave theory*, John Wiley & Sons, ISBN 978-0966814392 , New York
[22] Lakhtakia, A. (1994). *Beltrami fields in chiral media*. World Scientific, ISBN 9810214030 , Singapore
[23] Lakhtakia, A.; Varadan, V. K. & Varadan, V. V. (1989). *Time-harmonic electromagnetic fields in chiral media*. Springer-Verlag, ISBN 3-540-51317-5, Berlin
[24]Landau, L. & Lifshitz, I. (1960). *Electrodynamics of Continuous Media*. Pergamon Press, ISBN 0-444-86165-3, Oxford
[25] Lee, R. & Madsen, N. (1990). A mixed finite element formulation for Maxwell's equations in the time domain. *Journal Computational Physics*, Vol. 88, No. 2, (March 1990), pp. 284-304, ISSN 0021-9991
[26] Lindell, I.; Sihvola, A.; Tretyakov, S. & Viitanen, A. (1994). *Electromagnetic Waves in Chiral and Bi-isotropic Media*. Artech House, ISBN , Boston
[27] Lindell, I.; Sihvola, A.; Tretyakov, S. & Viitanen, A. (1994). *Electromagnetic Waves in Chiral and Bi-isotropic Media.*, Artech House, ISBN 1598293206 , Boston
[28] Li, J, & Chen, Y. (2006). Analysis of a time-domain finite element method for 3D Maxwell's equations in dispersive media. *Computer Methods in Applied Mechanics and Engineering*, Vol. 195, No. 11, (December 2006), pp. 4220-4229, ISSN 0045-7825
[29] Liu, Q. & Fan, G. (1999). Simulation of GPR in dispersive media using a frequency-dependent PSTD algorithm. *IEEE Transactions on Geoscience and Remote Sensing*, Vol. 37, No. 5, (June 1999), pp. 2317-2324, ISSN 0196-2892
[30] Lu, T.; Zhang, P. & Cai, W. (2004). Discontinuous Galerkin methods for dispersive and lossy Maxwell's equations and PML boundaryconditions. *Journal Computational Physics*, Vol. 200, No. 4, (May 2004), pp. 549-580, ISSN 0021-9991
[31] Monk, P. (1992). A comparison of three mixed methods for time-dependent Maxwell's equations. *SIAM Journal of Science Statistics Computer*, Vol. 13, No. 10, (December 1992), pp. 1097-1122, ISSN 0960-3174
[32] Monk, P. (2003). *Finite Element methods for Maxwell's equations*, Clarendon Press, ISBN 0198508883, Oxford
[33] Pradhan, T. (1993). Magnetic charge induced by electric charge in a magneto-electric medium. *Europhysics Letters*, Vol. 21, No. 7, (July 1993), pp. 971-973, ISSN 0295-5075
[34] Van Run, A.; Terrell, D. & Scholing, J. (1974). An in situ grown eutectic magnetoelectric composite material. *Journal of Materials Science*, Vol. 9, No. 11, (December 1974), pp. 1710-1714, ISSN 0022-2461

[35] Yakhno, V.; Yakhno, T. & Kasap, M. (2006). A novel approach for modeling and simulation of electromagnetic waves in anisotropic dielectrics. *Internatioal Journal of Solids and Structures*, Vol. 43, No. 12, (December 2006), pp. 6261-6276, ISSN 0020-7683

5

Electromagnetic Waves in Cavity Design

Hyoung Suk Kim
Kyungpook National University
Korea

1. Introduction

Understanding electromagnetic wave phenomena is very important to be able to design RF cavities such as for atmospheric microwave plasma torch, microwave vacuum oscillator/amplifier, and charged-particle accelerator. This chapter deals with some electromagnetic wave equations to show applications to develop the analytic design formula for the cavity. For the initial and crude design parameter, equivalent circuit approximation of radial line cavity has been used. The properties of resonator, resonant frequency, quality factor, and the parallel-electrodes gap distance have been considered as design parameters. The rectangular cavity is introduced for atmospheric microwave plasma torch as a rectangular example, which has uniform electromagnetic wave distribution to produce wide area plasma in atmospheric pressure environment. The annular cavity for klystrode is introduced for a microwave vacuum oscillator as a circular example, which adapted the grid structure and the electron beam as an annular shape which gives high efficiency compared with conventional klystrode. Some simulation result using the commercial software such as HFSS and MAGIC is also introduced for the comparison with the analytical results.

2. Equivalent circuit approximation of radial-line cavity

Microwave circuits are built of resonators connected by waveguides and coaxial lines rather than of coils and condensers. Radiation losses are eliminated by the use of such closed elements and ohmic loss is reduced because of the large surface areas that are provided for the surface currents. Radio-frequency energy is stored in the resonator fields. The linear dimensions of the usual resonator are of the order of magnitude of the free-space wavelength corresponding to the frequency of excitation. A simple cavity completely enclosed by metallic walls can oscillate in any one of an infinite number of field configurations. The free oscillations are characterized by an infinite number of resonant frequencies corresponding to specific field patterns of modes of oscillation. Among these frequencies there is a smallest one,

$$f_0 = c/\lambda_0 \quad (1)$$

, where the free-space wavelength is of the order of magnitude of the linear dimensions of the cavity, and the field pattern is unusually simple; for instance, there are no internal nodes in the electric field and only one surface node in the magnetic field.

The oscillations of such a cavity are damped by energy lost to the walls in the form of heat. This heat comes from the currents circulating in the walls and is due to the finite conductivity of the metal of the walls. The total energy of the oscillations is the integral over the volume of the cavity of the energy density,

$$W = \frac{1}{2}\int_v \left(\varepsilon_0 E^2 + \mu_0 H^2\right) dv \tag{2}$$

$$\mu_0 = 4\pi \times 10^{-7}\, H/m \text{ and } \varepsilon_0 = \frac{1}{36\pi} \times 10^{-9}\, F/m \tag{3}$$

, where E and H are the electric and magnetic field vectors, in volts/meter and ampere-turns/meter, respectively. The cavity has been assumed to be empty. The total energy W in a particular mode decreases exponentially in time according to the expression,

$$W = W_0 e^{-\frac{\omega_0 t}{Q}} \tag{4}$$

, where $\omega_0 = 2\pi f_0$ and Q is a quality factor of the mode which is defined by

$$Q = \frac{2\pi(\text{energy stored in the cavity})}{(\text{energy lost in one cycle})}. \tag{5}$$

The fields and currents decrease in time with the factor $e^{-\frac{\omega_0 t}{Q}}$.

Most klystrons and klystrodes are built with cavities of radial-line types. Several types of reentrant cavities are shown in Fig. 1. It is possible to give for the type of cylindrical reentrant cavities a crude but instructive mathematical description in terms of approximate solutions of Maxwell's equations.

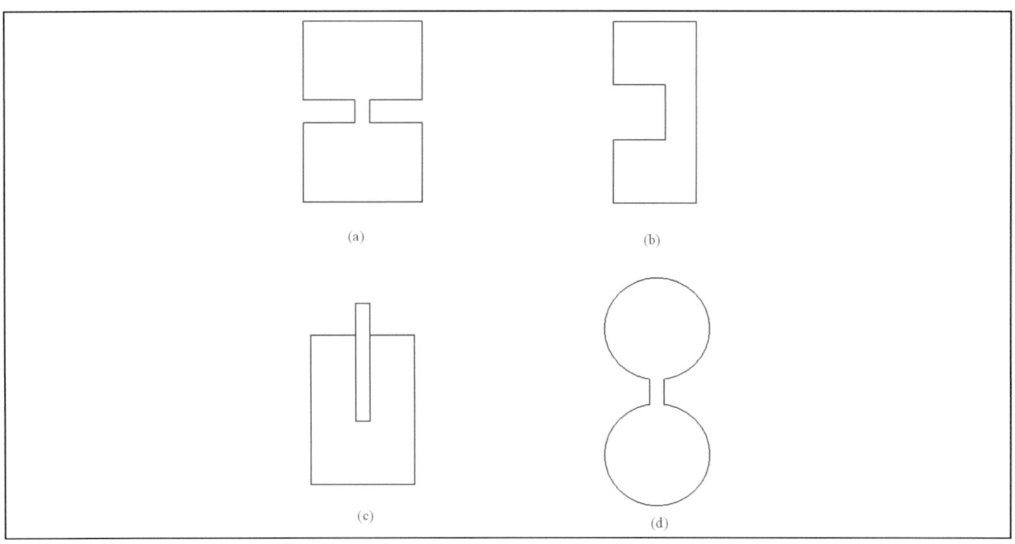

Fig. 1. Resonant cavities; (a) Coaxial cavity, (b) Radial cavity, (c) Tunable cavity, (d) Toroidal cavity

The principle, or fundamental, mode of oscillation of such cavity, and the one with the longest free-space wavelength , has electric and magnetic fields that do not depend on the azimuthal angle defining the half plane though both the axis and the point at which the fields are being considered. In addition, the electric field is zero only at wall farthest apart from the gap and the magnetic field is zero only at the gap. In this mode the magnetic field is everywhere perpendicular to the plane passing through the axis the electric field lies in that plane. Lines of magnetic flux form circles about the axis and lines of electric flux pass from the inner to the outer surfaces.

In the principle mode of radial-line cavity only E_z , and H_z are different from zero and these quantities are independent of ϕ . (see Fig. 2 for cylindrical coordinates and dimensions of the cavity). The magnetic field automatically satisfies the conditions of having no normal component at the walls.

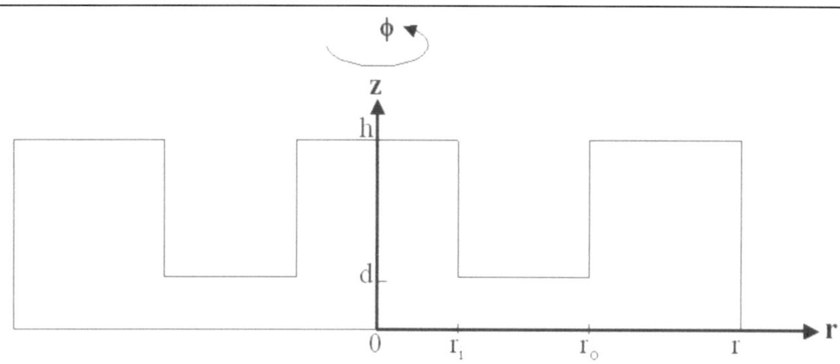

Fig. 2. Cylindrical coordinates and dimensions of the radial cavity

The cavities in which RF interaction phenomena happens with charged particles almost always have a narrow gap, that is, the depth of the gap d (see Fig. 2), is small compared with the radius r_0 of the post ($d\langle\langle r_0$ in Fig. 2).. If the radius of the post is much less than one-quarter of the wavelength, and if the rest of the cavity is not small, the electric field in the gap is relatively strong and approximately uniform over the gap. It is directed parallel to the axis and falls off only slightly as the edge of the gap is approached. On the other hand, the magnetic field increases from zero at the center of the gap in such a manner that it is nearly linear with the radius.

In a radial-line cavity the electric field outside the gap tends to remain parallel to the axis, aside from some distortion of the field that is caused by fringing near the gap; it is weaker than in the gap and tends to become zero as the outer circular wall is approached. The magnetic field, on the other hand, increases from its value at the edge of the gap and has its maximum value at the outer circular wall.

It is seen that, whereas the gap is a region of very large electric field and small magnetic field, the reentrant portion of the cavity is a region of large magnetic field and small electric field. The gap is the capacitive region of the circuit, and the reenetrant portion is the inductive region. Charge flows from the inner to the outer conducting surface of the gap by passing along the inner wall, across the outer end. The current links the magnetic flux and the magnetic flux links the current, as required by the laws of Faraday, Biot and Savart.

2.1 Capacitance in cylindrical cavity

If the gap is narrow, the electric field in the gap is practically space-constant. Thus the electric field E_z in the gap of the circular cavity (see Fig. 2) comes from Gauss' law,

$$E_z = \frac{\sigma}{\varepsilon_0} = \frac{Q}{\varepsilon_0 \pi (r_0^2 - r_i^2)}. \tag{6}$$

At the end of the cavity near to the gap both E_z and E_r exist and the field equations are more complicated. If d is small compared with h and r, it can be assumed that the fields in the gap are given approximately by the preceding equation.
Therefore,

$$C = \frac{Q}{V} = \frac{Q}{E_z d} = \frac{\varepsilon_0 \pi (r_0^2 - r_i^2)}{d}. \tag{7}$$

2.2 Inductance in cylindrical cavity

The magnetic field H_φ comes from Ampere's law,

$$\oint H r d\theta = I. \tag{8}$$

Therefore,

$$B = \frac{\mu_0 I}{2\pi r}. \tag{9}$$

The total magnetic flux in the cylindrical cavity,

$$\Phi = \int_{r_0}^{r} B h dr = \frac{\mu_0 I h}{2\pi} \int_{r_0}^{r} \frac{dr}{r} = \frac{\mu_0 I h}{2\pi} \ln \frac{r}{r_0}. \tag{10}$$

Comparing this with inductance definition, $\Phi = LI$, we get the followings;

$$L = \frac{\mu_0}{2\pi} h \ln \frac{r}{r_0}. \tag{11}$$

2.3 Resonance frequency in cylindrical cavity

The resonant wavelength of a particular mode is found from a proper solution of Maxwell's equation, that is, one that satisfies the boundary conditions imposed by the cavity. When the walls of the cavity conduct perfectly, these conditions are that the electric field must be perpendicular to the walls and the magnetic field parallel to the walls over the entire surface, where these fields are not zero.
The resonant frequency f_0 could be calculated for the principal mode of the simple reentrant cavity. The resonant cavity is modeled by parallel LC circuits as can be seen in Fig. 3. In fact, cavities are modeled as parallel resonant LC circuits in order to facilitate discussions or analyses. The resonant frequency is inversely proportional to the square root of inductance and capacitance;

$$\omega = \frac{1}{\sqrt{LC}} = \frac{1}{\sqrt{\frac{\varepsilon_0\mu_0}{2}\frac{r_0^2 - r_i^2}{d} h \ln\frac{r}{r_0}}}. \tag{12}$$

2.4 Unloaded Q in cylindrical cavity

In the cavity undergoing free oscillations, the fields and surface currents all vary linearly with the degree of excitation, that is, a change in one quantity is accompanied by a proportional change in the others. The stored energy and the energy losses to the walls vary quadratically with the degree of excitation.

Since the quality factor Q of the resonator is the ratio of the stored energy and the energy losses per cycle to the walls, it is independent of the degree of excitation.

$$Q = \frac{2\pi f(\textit{energy stored in the cavity})}{(\textit{power lost})} = \omega_0 \frac{U}{P_{loss}} = \omega \frac{\frac{1}{2}LI^2}{\frac{1}{2}RI^2} = \omega \frac{L}{R}. \tag{13}$$

The resonator losses per second, besides being proportional to the degree of excitation, are inversely proportional to the product of the effective depth of penetration of the fields and currents into the walls, the skin depth, and the conductivity of the metal of the walls. Since the skin depth is itself inversely proportional to the square root of the conductivity, the losses are inversely proportional to the square root of the conductivity. The losses are also roughly proportional to the total internal surface area of the cavity; and this area is proportional to the square of the resonant wavelength for geometrically similar resonators. The skin depth is proportional to the square root of the wavelength, and hence the losses per second are proportional to the three-halves power of the resonant wavelength.

The loss per cycle, which is the quantity that enters in Q, is proportional to the five-halves power of the resonant wavelength. Since the energy stored is roughly proportional to the volume, or the cube of the wavelength, the Q varies as the square root of the wavelength for geometrically similar cavities, a relationship that is exact if the mode is unchanged because the field patterns are the same. In general, large cavities, which have large resonant wavelengths in the principal mode, have large values of Q. Cavities that have a surface area that is unusually high in proportion to the volume, such as reentrant cavities, have Q's that are lower than those of cavities having a simpler geometry.

The surface current, J, is equal in magnitude to H_ϕ at the wall. The power lost is the surface integral over the interior walls of the cavity.

$$\begin{aligned} P_{loss} &= \frac{R_s}{2}\int J^2 ds = \frac{R_s}{2}\int |H_\phi|^2 ds \\ &= \frac{R_s}{2}\left[\left(\frac{I}{2\pi r_o}\right)^2 2\pi r_0(h-d) + 2\int_{r_0}^{r}\left(\frac{I}{2\pi r_0}\right)^2 2\pi r dr + \left(\frac{I}{2\pi r_o}\right)^2 2\pi r h\right] \\ &= \frac{R_s I^2}{2\pi}\left[\frac{h-d}{r_0} + 2\ln\frac{r}{r_0} + \frac{h}{r}\right] \equiv \frac{1}{2}RI^2, \end{aligned} \tag{14}$$

where the shunt resistance is

$$R \equiv \frac{R_s}{2\pi}\left[\frac{h-d}{r_0} + 2\ln\frac{r}{r_0} + \frac{h}{r}\right]. \tag{15}$$

The surface current can be considered concentrated in a layer of resistive material of thickness. Surface resistance is that

$$R_s = \frac{1}{\sigma\delta} = \frac{1}{\sigma\sqrt{\frac{1}{\pi f \mu\sigma}}} = \sqrt{\frac{\pi f \mu}{\sigma}}. \tag{16}$$

As an example, the conductivity of copper is $\sigma_{copper} = 5.8\times10^7 / \Omega m$ and since copper is nonmagnetic $\mu_{copper} = \mu_0 = 4\pi\times10^{-7} H/m$, hence, in case of that cavity material is copper, for f=6GHz,

$$\begin{aligned} &\delta = 0.85\mu m \\ &R_s = 2.02\times10^{-2}\Omega \\ &R = 2.66\times10^{-3}\Omega \\ &L = 1.01\times10^{-10} H \\ &Q_0 = \omega\frac{L}{R} = 1431. \end{aligned} \tag{17}$$

The shunt conductance G is, as given by the expression,

$$G = \frac{(energy\ lost\ per\ second)}{V(t)^2} \tag{18}$$

is defined only when the voltage $V(t)$ is specified. In a reentrant cavity the potential across the gap varies only slightly over the gap if the gap is narrow and the rest of the cavity is not small. A unique definition is obtained for G by using for $V(t)$ the potential across the center of the gap. The gap voltage is proportional to the degree of excitation, and hence the shunt conductance is independent of the degree of excitation. For geometrically similar cavities the shunt conductance varies inversely as the square root of the resonant wavelength for the same mode of excitation. This relationship exists because for the same excitation $V(t)^2$ is proportional to the square of the wavelength and the loss per second to the three-halves power of the wavelength.

2.5 Lumped-constant circuit representation

The main value of the analogy between resonators and lumped-constant circuits lies not in the extension of characteristic parameters to other geometries, in which the analogy is not very reliable, but in the fact that the equations for the forced excitation of resonators and lumped-constant circuits are of the same general form.

If, for example, it is assumed that the current i(t) passes into the shunt combination of L, C and conductance G, by Kirchhoff's laws, (see Fig. 3)

$$i(t) = C\frac{dV(t)}{dt} + \frac{1}{L}\int V(t)dt + GV(t). \quad (19)$$

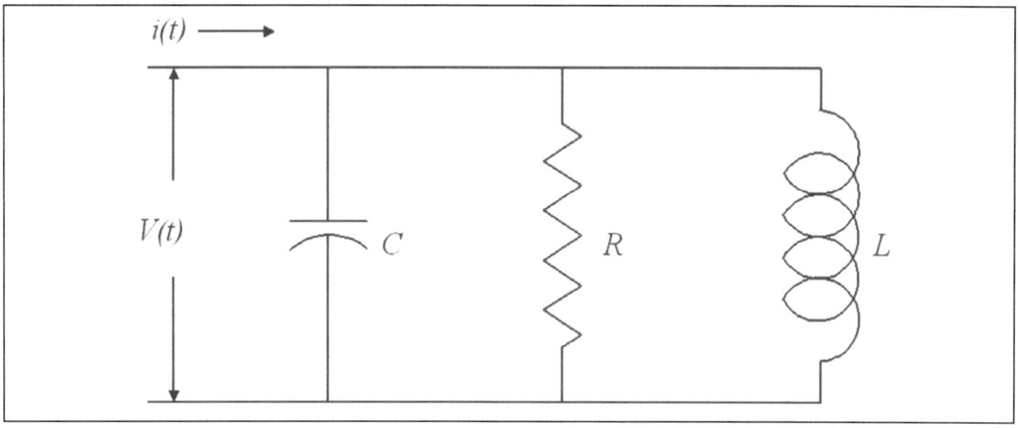

Fig. 3. Limped-constant circuit

On taking the derivative and eliminating L,

$$\frac{di(t)}{dt} = C\left[\frac{d^2V(t)}{dt^2} + \omega_0^2 V(t)\right] + G\frac{V(t)}{dt}. \quad (20)$$

In other word,

$$\frac{\omega}{C}\frac{di(t)}{d\theta} = \omega^2\frac{d^2V(t)}{d\theta^2} + 2\gamma\omega\frac{dV(t)}{d\theta} + \omega_0^2 V(t), \quad (21)$$

where $\gamma = G/2C$, $\omega_0 = 1/\sqrt{LC}$ and $\theta = \omega t$, which are used to calculate numerically the initial beam effect in the last chapter.
For a forced oscillation with the frequency ω,

$$i_\omega = \left[G + j\omega_0 C\left(\frac{\omega}{\omega_0} - \frac{\omega_0}{\omega}\right)\right]V_\omega. \quad (22)$$

Thus, there is defined circuit admittance

$$Y = G + j\omega_0 C\left(\frac{\omega}{\omega_0} - \frac{\omega_0}{\omega}\right). \quad (23)$$

These equations describe the excitation of the lumped-constant circuit.

3. Numerical analysis for the high frequency oscillator system with cylindrical cavity

In this section, we will meet an circular cavity example of a klystrode as a high frequency oscillator system with the knowledge which is described in previous sections.

Conventional klystrodes and klystrons often have toroidal resonators, i.e., reentrant cavity with a loop or rod output coupler for power extraction. These resonators commonly use solid-electron-beam which could limit the output power. One way to get away this limitation is to use the annular beam as was commonly done in TWTs. The main reason using reentrant cavities in most microwave tubes with circular cross sections is that the gap region should produce high electric field and thus high interaction impedance of the electron beam when the cavity is excited. In our design we assume a short cavity length, d, along the longitudinal direction parallel to the electron motion. In the meantime the width of electron beam tunnel, $r_0 - r_i$, is much larger, i.e. $d \langle\langle r_0 - r_i$ as shown in Fig 4. And thus the efficiency of beam and RF interaction in this klystrode cavity depends sensitively upon the cavity shape at the beam entrance of the RF cavity in the beam tunnel. A simple trade-off study suggests to put to use of gridded plane, so-called a cavity grid (anode), so that the eigenmode of the reentrant cavity is maintained. With the gridded plane removed and left open, the TM01-mode has many competing modes and the interaction efficiency disappears. The use of thin cavity grid in the beam tunnel, however, can slightly reduce the electron beam transmission, which will not pose a much of problem when the same type of grid is used in between the cathode and anodic cavity grid. In the simulations with the MAGIC and HFSS codes, the anodic cavity grid could be assumed to be a smooth conducting surface, and pre-bunched electrons were launched from those surfaces of cavity grid. This kind of concept can provide a compact microwave source of low cost and high efficiency that is of strong interest for industrial, home electronics and communications applications.

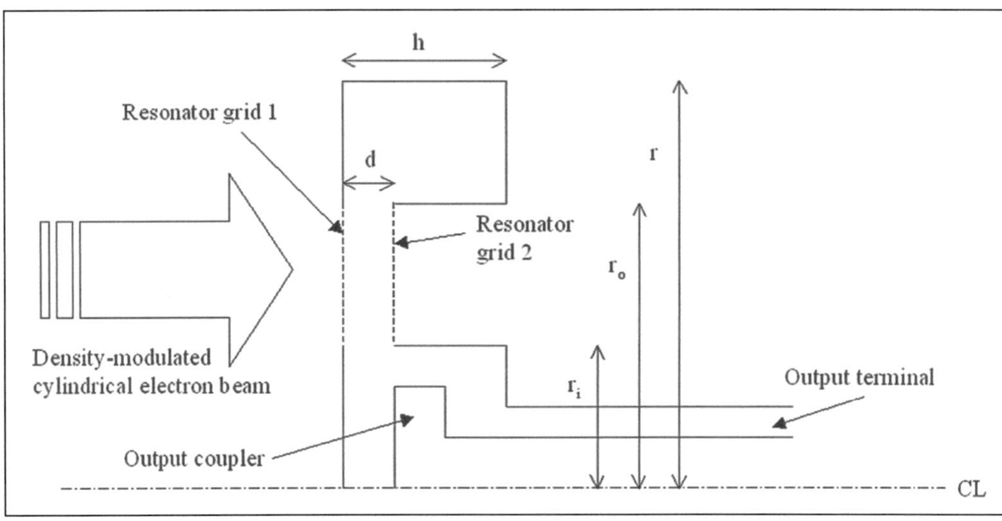

Fig. 4. Schematics of the annular beam klystrode with the resonator grids for the high electric field and high interaction efficiency in the gap region. This cavity structure allows easier power extraction through the center coax coupler

The klystrodes consist of the gated triode electron gun, the resonator and the collector. The gated electron gun provides with the pre-modulated electron bunches at the fundamental frequency of the input resonator, where the voltage on the grid electrode is controlled by an external oscillator or feedback system. The other possible type of gated electron guns could be the field-emitter-array gun, RF gun, and photocathode. The electron bunches arrive at the output gap with constant kinetic energy but with the density pre-modulated. Here, we assumed the electron beam is operated on class B operation, that is, electron bunch length is equal to one half of the RF period. Through the interaction between electron beam and RF field, the kinetic energy is extracted from the pre-modulated electrons and converted into RF energy.

Figure 4 shows the schematics of the circular gridded resonator with center coupling mechanism for the easy and efficient power extraction. In this section, we will describe the design of annular beam klystrode in C-band.

3.1 RF interaction cavity design

As we have seen in previous section, using the lumped-circuit approach, the resonant frequency of this protuberance cavity with the annular beam is expressed as

$$\omega = \frac{1}{LC} = \frac{1}{\sqrt{\frac{\varepsilon_0 \mu_0}{2} \frac{r_0^2 - r_i^2}{d} h \ln \frac{r}{r_0}}}. \tag{24}$$

Since this expression is an approximation which gives the tendency of frequency variation when we are adjusting design parameters, we can perform parameter tuning exercise using design tools such as HFSS. Fig. 5 shows an example of the detailed design using HFSS where the emission was introduced at the gap region between inner radial distances of 5.7 and 9.4mm. In the figure, the electric field is enhanced and fairly uniform due to the presence of resonator grid1 and resonator grid2. The grid structure in beam inlet and beam outlet make the electric field maintain fairly high intensity in the gap region through which the electron beam passes to interact with RF. Figure 6 also show scattering parameter plots where resonator grids of the klystrode are considered closed metal wall and the cavity has only output terminal as one port system. The bold line is the real value of S and the thin line is the imaginary one. The resonator frequency is 5.78 GHz in the absence of finite conductivity of cavity and electron beam.

The detailed tuning of beam parameters for efficient klystrode could be investigated using PIC code such as MAGIC. As an example, the current is assumed density-modulated in the input cavity and cut-off sinusoidal,

$$\begin{aligned} I(z=0,t) &= I_0 MAX\left(\sin(\omega t), 0\right) \\ &= I_0 \left[\frac{1}{\pi} + \frac{1}{2}\sin(\omega t) - \sum_{n-0}^{\infty} \frac{2}{\pi(4n^2-1)} \cos(2n\omega t) \right] \end{aligned} \tag{25}$$

whose peak current, I_{peak}, is 3 amperes.

The tube is supposed of being operated in class B as shown in Fig. 7. A class B amplifier is one in which the grid bias is approximately equal to the cut-off value of the tube, so that the plate current is approximately zero when no exciting grid potential is applied, and such that plate current flows for approximately one-half of each cycle when an AC grid voltage is applied.

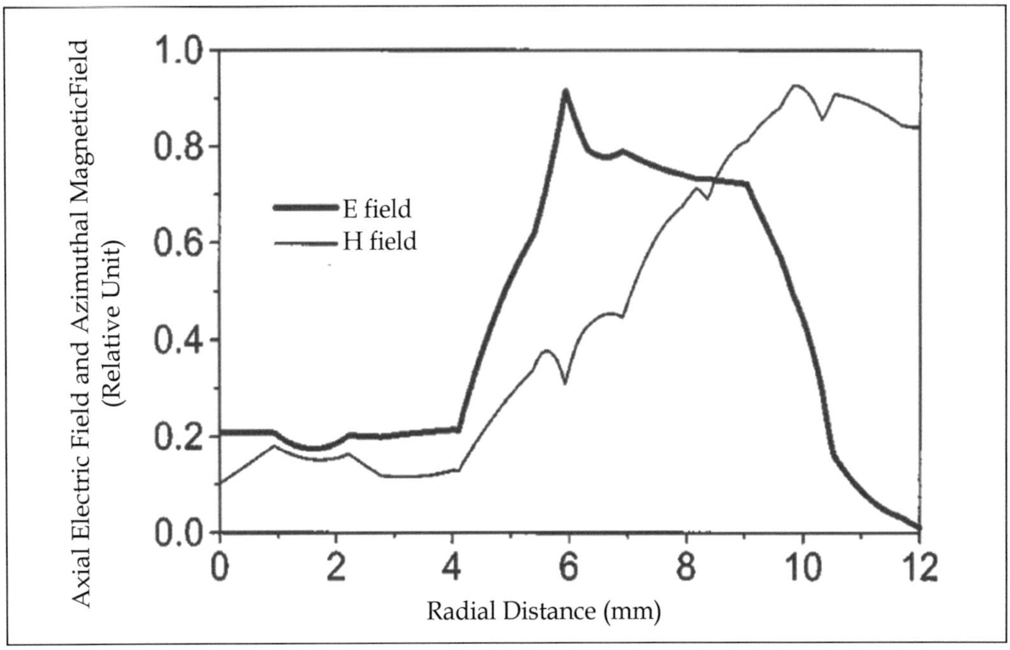

Fig. 5. Magnitude of axial electric field and azimuthal magnetic field (in relative unit) along the radial distance on the mid-plane between resonator grid 1 and grid 2 in the cavity. Emission surface is between the radial distances of 5.7 and 9.4 mm

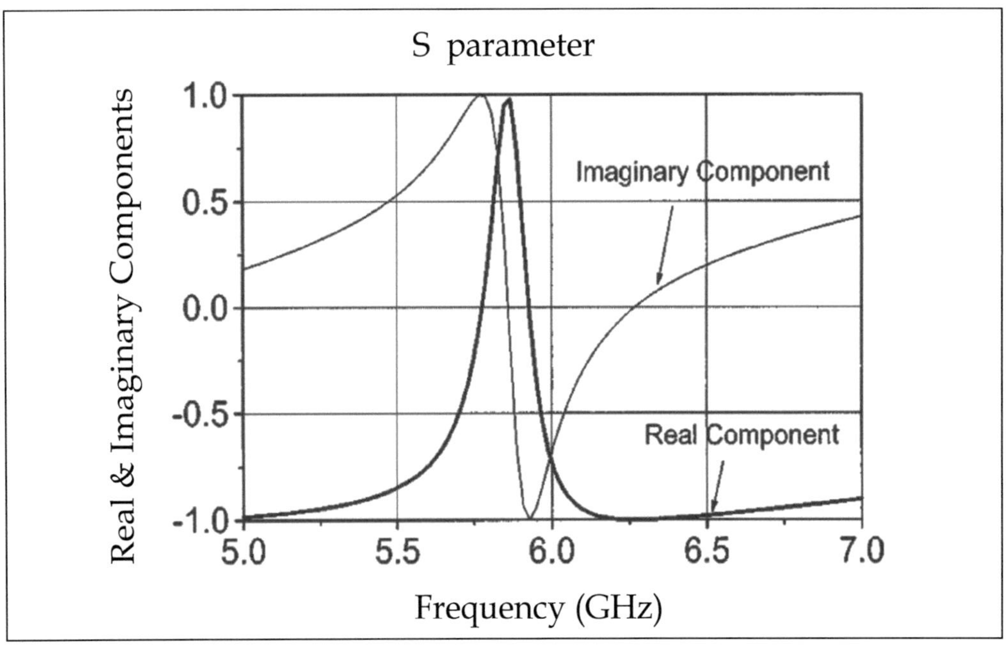

Fig. 6. Scattering parameter plots. The resonator frequency is 5.78 GHz in the absence of finite conductivity of cavity and electron beam

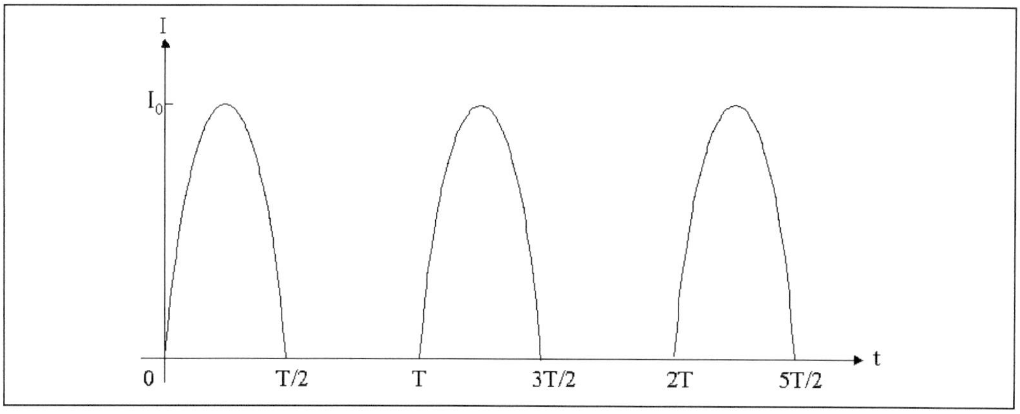

Fig. 7. Pre-modulated electron beam in current vs. time; cut-off sinusoidal current which is used in class B operation, $I = I_0 MAX(\sin(\omega t), 0)$

The fundamental mode (TM_{01}-mode) to be interacted with longitudinal traversing electron beam was adapted to our annular beam resonators for the high efficiency device.
Electron transit angle between electrodes gives limitation in the application of the conventional tubes at microwave frequencies. The electron transit angle is defined as

$$\beta = \omega\tau_g = \omega d / v_0 , \tag{26}$$

where $\tau_g = d/v_0$ is the transit time across the gap, d is separation between cathode and grid, $v_0 = \sqrt{2eV_0/m} = 0.593\times10^6\sqrt{V_0}$ is the velocity of the electron, and V_0 is DC voltage.

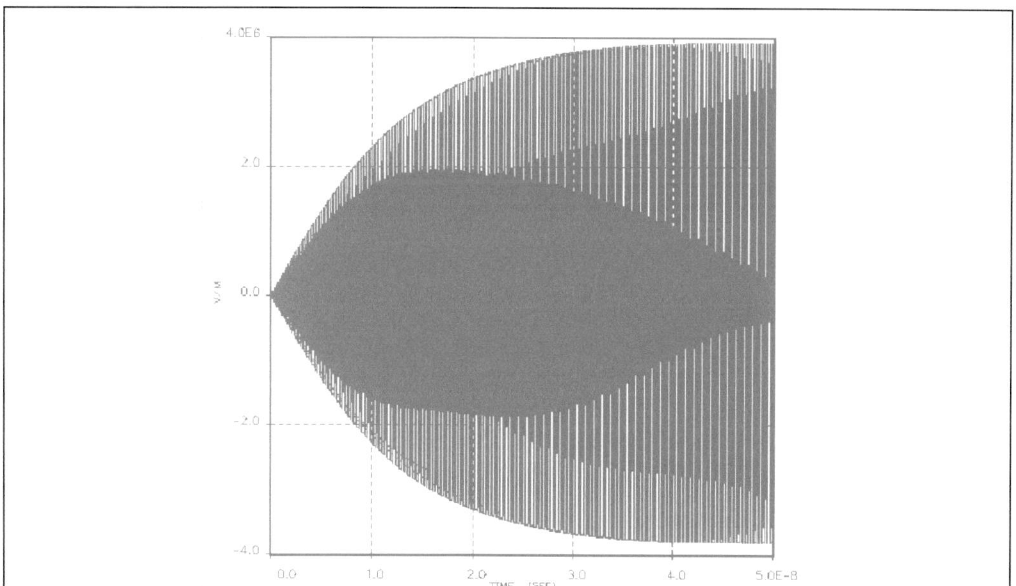

Fig. 8. Electric field in the gap region across the anode electrode 1(grid1) and electrode 2(grid2); The field reaches 4,000,000 V/m

The transit angle was chosen to give that the transit time is much smaller than the period of oscillation for the efficient interaction between RF and electron beam, so that, the beam coupling coefficient, M, is 0.987. The resonant frequency is 5.78 GHz in cold cavity and 6.0 GHz in hot cavity. Although the frequency shift may be greater than the value of normal case, this would be come from the fact that this annular beam covers much more area with electron beam than the conventional solid beam in a given geometry. As we can see in Fig. 8 and Fig. 9, this resonant cavity is filled and saturated with the RF power in 50 ns, and reveals high efficiency of about 67%. The output power is 1250 W so that the efficiency of this annular beam klystrode reveals 67 % at 6.004 GHz.

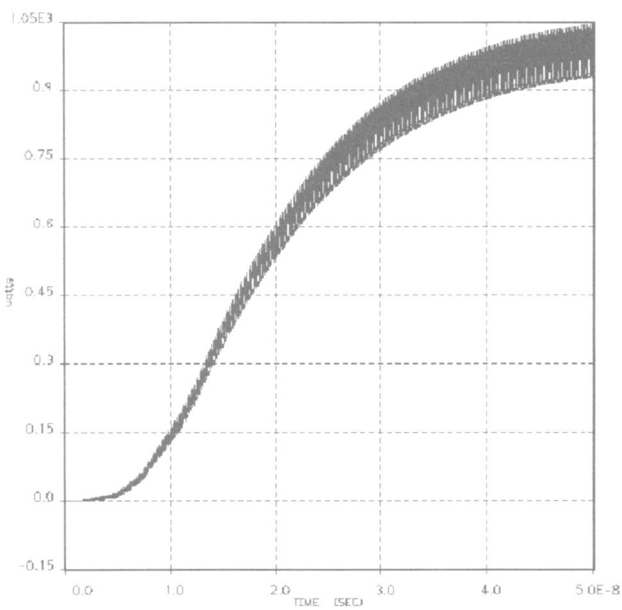

Fig. 9. Output power going through the output port vs. time where driving frequency is 6GHz. It goes to about 1.25kW

3.2 RF interaction efficiency calculation

There are some computational design codes for the klystrode. But in this section, 1-dimensional but realistic electron beam and electric field shape are introduced to develop analytical calculations for the klystrode design, which results in easy formulas for the efficiency and electric field in the gap region of the klystrode in steady state.

Maxwell's equations for electron beams are followings,

$$\nabla \cdot E = -\frac{\rho}{\varepsilon}, \tag{27}$$

$$\nabla \times E = -\frac{\partial B}{\partial t} \tag{28}$$

and

$$\nabla \times H = J + \frac{\partial D}{\partial t}. \tag{29}$$

As an approximation, the electron dynamics is in 1-D space. Then, the Maxwell's equations are simplified as the followings.

$$\frac{\partial E}{\partial z} = -\frac{\rho}{\varepsilon}, \tag{30}$$

$$\frac{\partial B}{\partial t} = 0 \tag{31}$$

and

$$-\rho v + \varepsilon \frac{\partial B}{\partial t} = \varepsilon \frac{\partial E}{\partial z} v + \varepsilon \frac{\partial E}{\partial t} = \varepsilon \frac{dE}{dt} = 0. \tag{32}$$

Therefore, this means that E remains constant for each electrons moving with velocity, v. From the Lorentz force equation,

$$\frac{dv}{dt} = -\frac{e}{m} E = (const.) \tag{33}$$

for each particle with velocity v.
Define the snapshot time be τ such that,

$$\tau = t_x + t_y, \tag{34}$$

where t_x is transit time for the moving particle from resonator grid1 to the transit distance, z, and t_y is leaving time for moving particle from the resonator grid1. Its definition is shown in schematic representation for the transit time, departure time, snapshot time, and transit distance in Fig. 10.
Therefore, we can say that the variables of electrons are denoted by

$$v = v_0 - \frac{e}{m} E(t_y) t_x, \tag{35}$$

$$z = v_0 t_x - \frac{e}{2m} E(t_y) t_x^2 \tag{36}$$

and

$$\rho = \frac{J_{peak}}{v} Max\left(\sin(\omega t_y), 0\right). \tag{37}$$

Let's assume that $E(t_y) = E_0 \sin(\omega t_y)$ which is synchronous to current modulation for the maximum interaction between electron beam and RF-field.

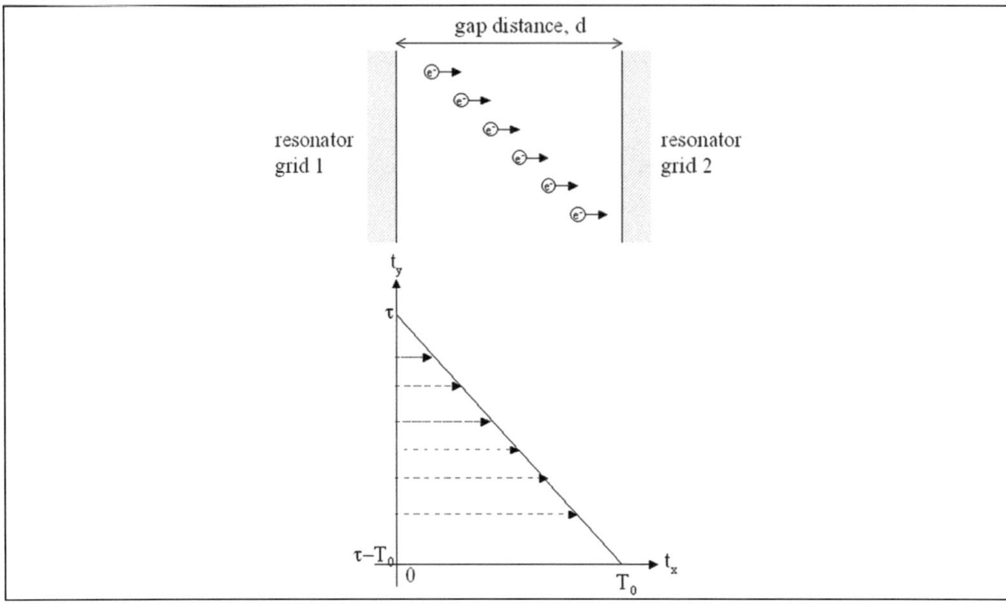

Fig. 10. Schematic representation for the definition of snapshot time (τ), transit time (t_x) to z , departure time (t_y)

Then, we have

$$v = v_0 - \frac{e}{m} E_0 \sin(\omega t_y) t_x \tag{38}$$

and

$$z = v_0 t_x - \frac{e}{2m} E \sin(\omega t_y) t_x^2. \tag{39}$$

By the way, from the above equation,

$$z = \frac{v_0 + v_z}{2} t_x \tag{40}$$

and

$$T = \frac{2d}{v_0 + v_d} \tag{41}$$

so that for $(period) = 167ps(\lambda = 5cm)$

$$T_0 \in [15ps, 30ps], \ \tau \in [0ps, 167ps], \tag{42}$$

and

$$t_x \in [0ps, 30ps]. \tag{43}$$

Figure 11 shows a typical case that electrons are decelerated due to the interaction between RF and electron beam. Extreme case would be 100 % donation of its kinetic energy to RF, which makes its velocity be zero at the resonator grid2. In that case electrons are delayed by 30 ps to the phase of RF field, where the half period of RF is 84 ps (6 GHz).

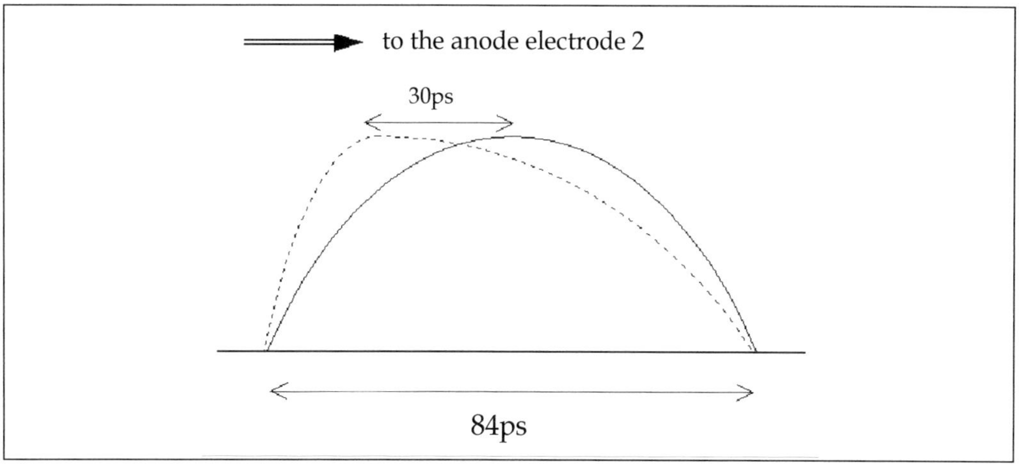

Fig. 11. Electrons are decelerated due to the interaction between RF and electron beam. Extreme case would be 100% donation of its kinetic energy to RF, which makes its velocity be zero at the resonator grid2. In that case electrons are delayed by 30ps to the phase of RF field, where the half period of RF is 84ps (6GHz)

The resonator field theory is described by the equation

$$\frac{dU_E}{d\tau} = -\frac{\omega}{Q}U_E + P_{out}, \tag{44}$$

where P_{out} is the energy output from the bunch and U_E is the electric energy in the cavity. The value of P_{out} depends on the behavior of the individual electrons as they move across the gap which in turn depends on the gap voltage and field profile. The axial electric field is only assumed by sinusoidal shape as $E(t_y) = E_0 \sin(\omega t_y)$.
Define

$$\langle f(t_y)\rangle \equiv \frac{\omega}{2\pi}\int_0^{\frac{2\pi}{\omega}} f(t_y)dt_y. \tag{45}$$

Then, the time-averaged output power becomes

$$\begin{aligned}\langle P_{out}\rangle &= \left\langle \int_0^d \rho E v dz \right\rangle = \left\langle \int_0^d J_{peak} E_0 Max\left(\sin^2(\omega t_y), 0\right) dz \right\rangle \\ &= \frac{\omega}{2\pi}\int_0^{\frac{2\pi}{\omega}}\int_0^d J_{peak} E_0 \sin^2(\omega t_y)\, dz dt_y = \frac{J_{peak} E_0 d}{4}.\end{aligned} \tag{46}$$

Because $E = 0$ when there are non electron charges, from the Maxwell's equation set,

$$\left\langle \frac{\omega}{Q} U_E \right\rangle = \frac{\varepsilon\omega}{2Q} \left\langle \int_0^d E_0^2 Max\left(\sin\left(\omega t_y\right),0\right)^2 dz \right\rangle \cong \frac{\varepsilon\omega}{2Q} \frac{E_0^2}{2} \left\langle \int_0^d \sin^2\left(\omega t_y\right) dz \right\rangle$$
$$= \frac{\varepsilon\omega E_0^2}{4Q} \left\langle \int_0^d \sin^2\left(\omega t_y\right) dz \right\rangle = \frac{\varepsilon\omega E_0^2 d}{4Q} \left\langle \sin^2\left(\omega t_y\right) \right\rangle = \frac{\varepsilon\omega E_0^2 d}{8Q}. \tag{47}$$

Since

$$\langle P_{out} \rangle = \left\langle \frac{\omega}{Q} U_E \right\rangle \tag{48}$$

at steady state, we have

$$E_0 = \frac{2QJ_{peak}}{\varepsilon\omega}. \tag{49}$$

Therefore, the efficiency becomes that

$$\eta = \frac{\langle P_{out} \rangle}{\langle J \rangle V} = \frac{\frac{J_{peak} E_0 d}{4}}{\frac{J_{peak}}{\pi} V} = \frac{\pi E_{0d}}{4V} = \frac{\pi Q J_{peak} d}{2\varepsilon\omega V}. \tag{50}$$

As an example,

$$\begin{aligned} &\varepsilon = 8.85 \times 10^{-12} F/m, \\ &\omega = 2\pi \times 10^9 / \sec, \\ &d = 0.4 \times 10^{-3} m, \\ &Q = 37 \end{aligned} \tag{51}$$

give us the followings

$$J_{peak} = \frac{3}{\pi\{(9.4 \times 10^{-3})^2 - (5.7 \times 10^{-3})^2\}} = 1.71[A/cm^2], \tag{52}$$

where $V = 2000 volts$, $e = 1.6 \times 10^{-19} C$, and $m = 9.1 \times 10^{-31} kg$.
Therefore,

$$\eta = \frac{\pi Q J_{peak} d}{2\varepsilon\omega V} = 0.60 \tag{53}$$

and

$$E_0 = \frac{2QJ_{peak}}{\varepsilon\omega} = 3.8 \times 10^6 V/m. \tag{54}$$

In the previous example, we have seen that the tube reveals 67% efficiency and the electric field in the gap region was $4 \times 10^9 V/m$. This result of the numerical analysis is well matched with the above theoretical calculations.

On the other hand, we can compare MAGIC analysis with analytical calculation as can be seen Fig. 12 which shows efficiency vs. peak current density by MAGIC PIC simulation and analytical calculation. The slope of the line fitted with the results from MAGIC analysis shows 0.266.
From the analytical equation of efficiency, we can get the slope of the line, m, as the followings,

$$m = \frac{\eta}{J_{peak}} = \frac{\pi Q d}{2\varepsilon\omega V} = 0.348 cm^2 / A. \tag{55}$$

Analytical calculation says the slope of the line be 0.348. MAGIC and analytical equation reveal that as we increase the current density, the efficiency of the tube also increases.

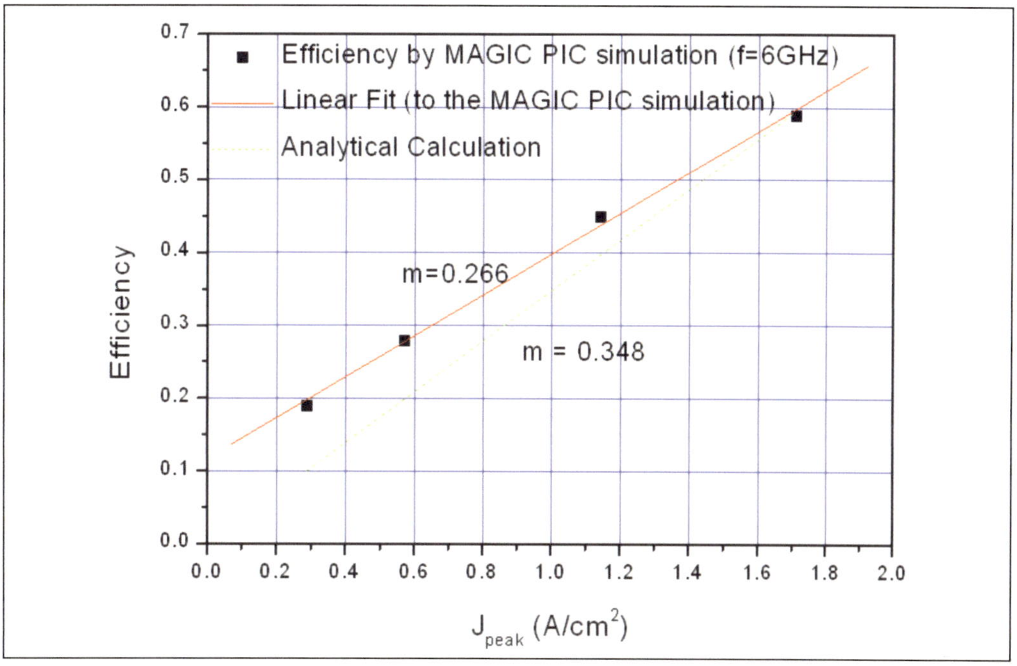

Fig. 12. Efficiency vs. peak current density. Square dots are results from MAGIC simulation. Solid line is line fit with the results. Dotted line is analytical calculation. The slope of the solid line, line fit with the MAGIC results, is 0.266 and that of dotted line from analytical calculation is 0.348

4. Cavity design for the uniform atmospheric microwave plasma source

The atmospheric-pressure microwave-sustained plasma has aroused considerable interest for many application areas. The advantages of this plasma are electrode-less operation and efficient microwave-to-plasma coupling. It has the characteristics of high plasma density ($\approx 10^{13} / cm^{-3}$) and efficient energy conversion from microwaves to the discharged plasma which is high-efficient plasma discharge (~80%). An atmospheric-pressure microwave-sustained plasma can be formed in a rectangular resonant cavity, a waveguide, or a surface-effect system. This plasma has been widely used in the laboratory spectroscopic analysis, continuous emissions monitoring in the field, commercial processing, and other environmental applications.

Atmospheric-pressure microwave plasma sources consist of a magnetron as a microwave source, an isolator to isolate the magnetron from the harmful reflected microwave, a directional coupler to monitor the reflected microwave power, a 3-stub tuner to match impedance from the magnetron to that from the plasma generator, and a plasma generator through which gases pass and are discharged by the injected microwave. The plasma in the discharge generator is sustained in a fused quartz tube which penetrates perpendicularly through the wide walls of a tapered and shorted WR-284 waveguide.

The plasma's cross-sectional area of the atmospheric-pressure microwave plasma source is limited by the quartz tube's diameter, which is also limited by the maximum intensity of the electric field profile sustainable by the injected microwave and the shorted waveguide structure, as shown in the Fig. 13. The diameter of the quartz tube is set around 30 mm to maintain a dischargeable electric field intensity in the discharge area. If the diameter of the quartz tube becomes much larger than that, no discharge occurs. This is the main reason that the cross-sectional area cannot be extended more widely in the atmospheric-pressure microwave-sustained plasma source.

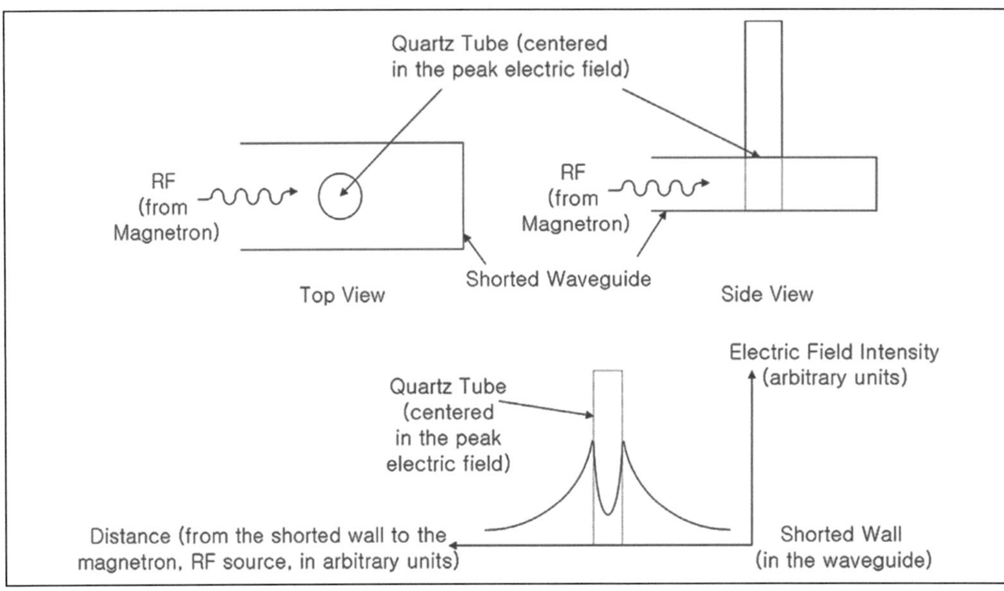

Fig. 13. Schematic illustrations of the atmospheric pressure microwave plasma source using a quartz tube located a quarter wavelength from the end of the shorted waveguide to maintain a high electric field intensity to discharge the passing- gases to generate a plasma efficiently

Here is an example of a rectangular reentrant resonator, a single ridge cavity, to overcome this size-limited plasma discharge in an atmospheric microwave-sustained plasma source. The box-type 915 MHz reentrant resonator has gridded walls on both sides of the gap so that the plasma gases pass thorough the grid holes and are discharged by the induced electric gap field due to the injected microwave energy. As we have studied in last section, it could be calculated analytically the cavity design formula by using lumped-circuit approximation, then, we performed a RF simulation by using the HFSS (High Frequency Structure Simulator) code. The designed plasma cross-sectional area is 810 mm x5 mm. The cavity has a fundamental TE_{10}-like mode to discharge gases uniformly between the gridded walls and has a 915 MHz resonant frequency. This large cross-sectional area plasma torch can be effectively used in commercial processing and other environmental applications.

4.1 Numerical analysis of the plasma source rectangular cavity

A reentrant cavity in cylindrical symmetry has been widely used in vacuum tubes such as klystrodes and klystrons. However, in order to keep the resonant frequency constant whatever the cavity lengths for atmospheric microwave plasma torch, it would be better to consider a rectangular reentrant cavity, as shown in Fig. 14. The resonant frequency of the cavity for the atmospheric microwave plasma source has a fundamental TE_{10}-like mode.

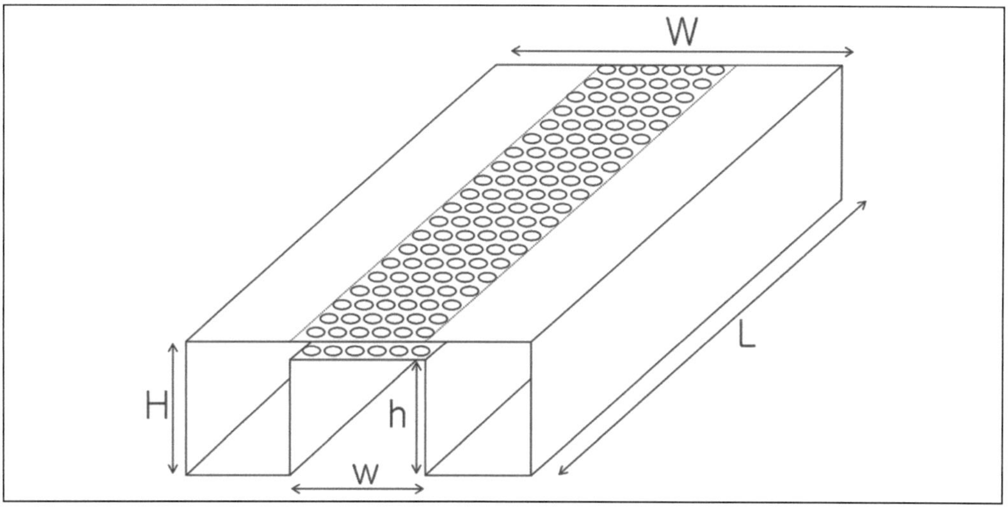

Fig. 14. Schematic of the box-type resonator cavity as a atmospheric pressure microwave plasma source with grids to maintain a high electric field intensity to discharge the passing gases to generate plasma efficiently in the gap region

The gap region between the two gridded walls in the reentrant cavity sustains a high electric field intensity and, thus, easily discharges the gases to the plasma state when it is excited by the microwave energy. With the grid planes surrounding the gap region, the TE_{10}-like mode is the dominant one, and 915 MHz is the cavity's resonant frequency. In the simulation process using the HFSS code, the cavity grids were assumed to be smooth conducting surfaces to design the box-type reentrant cavity.

In this example, the cavity for the microwave plasma source was set to be resonated at 915 MHz, rather than 2.45 GHz, so as to have a much larger cross-sectional plasma area. Its design, setting a cavity to 915 MHz, is based on the design formula obtained by using a theoretical calculation based on the lumped-circuit approximation for the cavity. Then, we used a simulation tool, the HFSS code, to check the cavity frequency.

From Gauss' law, the equivalent capacitance value of the reentrant cavity for a TE_{10}-like mode is

$$C = \frac{Q}{V} = \frac{Q}{E_{\perp}(H-h)} = \frac{\varepsilon_0 L w}{H-h}, \tag{56}$$

where Q, V, and $E_{\perp}$ are the stored electric charges on the grid surfaces, the electric voltage difference, and the perpendicular electric field intensity between the opposite grid surfaces respectively, and w, H, and h are designated dimensional parameters, as shown in Fig. 14. From Ampere's law, the equivalent inductance value of the reentrant cavity for a TE_{10}-like mode can be expressed as follows. Since the magnetic field flux density in the cavity is $B = \mu_0 I / 2L$, and since the total magnetic flux in the box-type reentrant cavity is $\Phi = BH(W-w)$, from the inductance definition,

$$L = \frac{\mu_0 H(W-w)}{4L} \quad . \tag{57}$$

Therefore, the cavity resonant frequency is given by

$$f = \frac{c}{\pi}\sqrt{\frac{H-h}{Hw(W-w)}} \quad . \tag{58}$$

This resonant frequency formula shows that the resonant frequency is independent of the cavity length L, which means that if we extrude the plasma through this grid planes, the plasma's cross-sectional area can be uniformly extended as much as we want without any influence on the cavity resonant frequency. However, because we use an available microwave source, which has a limited RF power, to discharge the gases, the cavity length cannot be extended to an unlimited extent. We will discuss this relationship between the consumption RF power and the cavity length in the next paragraph.

Since we have a theoretically-approximated expression for the design parameters, we should use this formula to pick up the initial design parameters to get detailed parameters for the resonator reentrant cavity as a plasma source. Then we should investigate the RF characteristics using the design tool, HFSS.

The energy stored in the gap region can be expressed as

$$U = \frac{1}{2} C V^2 \quad , \tag{59}$$

where C and V are the equivalent capacitance value of the cavity and the induced gap peak voltage between the grid planes, respectively. If we assume that all of this stored energy can be delivered to the gas discharge reaction process to generate the plasma, the microwave power consumption can be expressed as

$$P = \frac{1}{2} f C V^2 = \frac{L V^2}{2\pi}\sqrt{\frac{\varepsilon_0 \omega}{\mu_0 H(H-h)(W-w)}} \quad . \tag{60}$$

In other words, the cavity length for the atmospheric-pressure microwave-sustained plasma source is

$$L = \frac{2000\pi P}{V^2}\sqrt{\frac{\mu_0 H(H-h)(W-w)}{w}}, \tag{61}$$

where V is in kV, P is in kW, and the dimensional parameters, L, H, h, W, and w, are in mm units in this formula.
Fig. 15 and Fig. 16. show a rectangular cavity design example using one of the available magnetrons, 915 MHz, 60 kW, as a RF source, and the geometrical parameters to $H = 29mm$, $H-h = 1mm$, and $W = 80mm$. Furthermore, if the gap voltage is the upper-limit RF air breakdown voltage, that is, the DC breakdown voltage for air at the level of $33kV/cm$ at atmospheric pressure, the cavity length is limited by $810mm$ for self discharge. In this case, the plasma's cross-sectional area is uniformly distributed within $810mm \times 5mm$, i.e., $w = 5mm$.
This uniformly distributed plasma area is $810mm \times 5mm$, which is much larger than the areas of the conventional atmospheric-pressure microwave-sustained plasma sources. This large-sized microwave plasma at atmospheric pressure can be used in commercial processing and other environmental applications area.

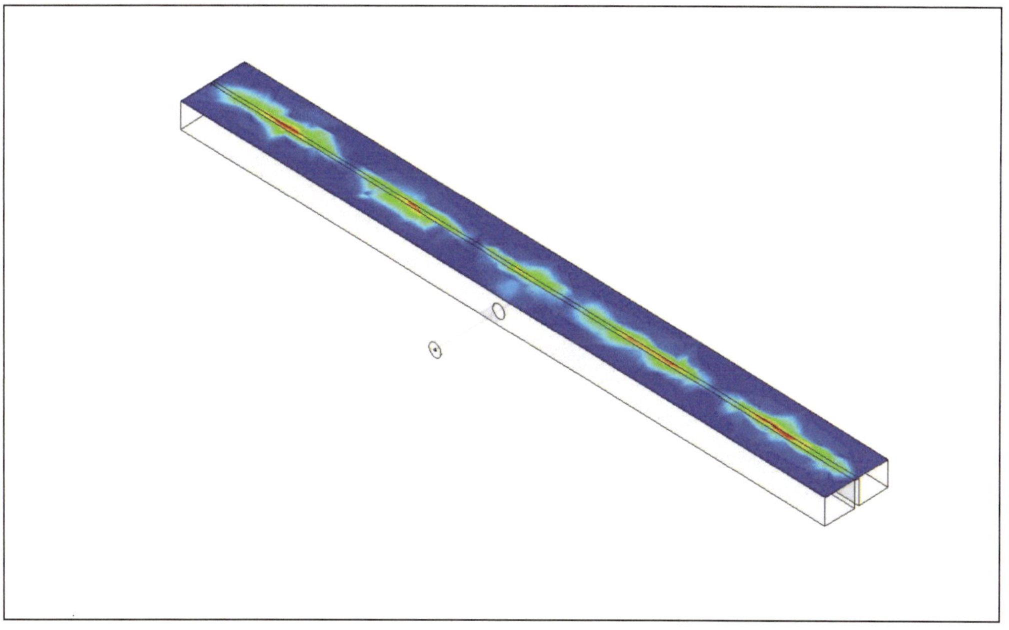

Fig. 15. Electric field intensity profile of the TE10-like mode from the HFSS simulation on the mid-plane between the grid surfaces in the gap region. Both ends are shorted walls, and the electric field profiles are distributed within $810mm \times 5mm$ in the gap region. Microwave energy is feeding through the coaxial coupler to the cavity. If we use a pair of these cavities in parallel, the crest of one will compensate for the other's trough electric field intensity areas, which could result in a uniform microwave plasma source

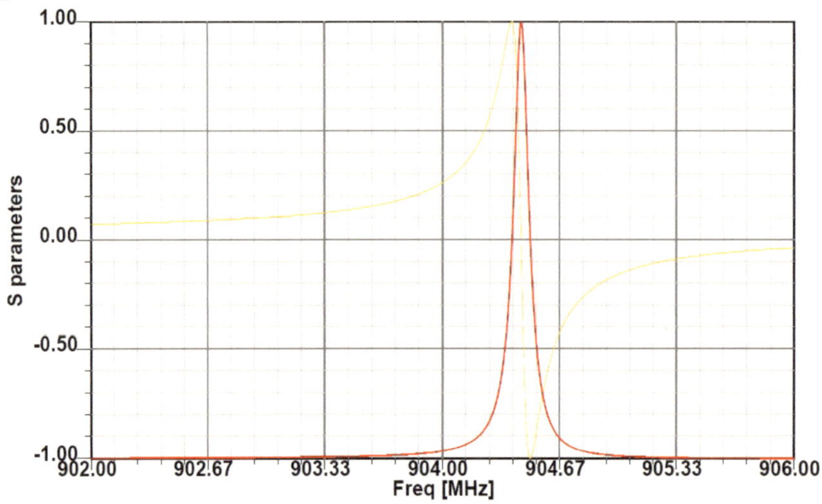

Fig. 16. Scattering parameters of the box-type reentrant cavity. The solid red and dashed blue lines are the real and imaginary values of S11, respectively. This shows that the cavity has a resonant frequency of 904 MHz and a high Q value of 7400

5. Conclusion

The annular beam cavity design was investigated analytically and simulated using the HFSS and MAGIC PIC codes to find the fine-tuned design parameters and optimum efficiency of the TM_{01}-mode operation in the klystrode with the reentrant interaction cavity. We also studied how to induce the governing efficiency formular for the microwave vacuum tube, klystrode. The efficiency of this exampled annular beam klystrode reveals 67 % at 6.004 GHz. The theoretical calculation anticipated that the efficiency would be 60% and the electric field intensity 4000V/m which is well matched with the results of numerical analysis. This concept could yield an economic sized device comparable to commercial magnetron devices.

A single ridge cavity was designed to show how to design uniform atmospheric microwave plasma source which does not need ignitors for the initial discharge. The single ridge cavity in the shape of resonator was built in the grid structures for the wide cavity torch of atmospheric pressure microwave plasma. This cavity design process has been studied by analytical calculation using lumped circuit approximation and simulation using commercial 3D HFSS code. A self-dischargeable study also has been investigated for the breakdown feasibility in the cavity gap region. These cavities show the uniform electromagnetic wave distribution between the grids through which large-sized microwave plasma could be generated.

6. References

Ansoft Co. (1999). High Frequency Structure Simulator, version 6.0, *Ansoft Co.*, PA, USA, 1999

Burton, A. J.; & Miller, G. F. (1971). The application of integral equation methods to the numerical solution of some exterior boundary-value problems, *Proc. Roy. Soc. Lond.*, A. 323, 201 1971

Curnow, H. J. (1965). A General Equivalent Circuit for Coupled-Cavity Slowe-Wave Structures, *IEEE Transactions on Microwave Theory and Techniques*, Vol. MTT-13, No. 5, 671, Sep. 1965

Dai, F. & Omar, A. S. (1993). Field-analysis model for predicting dispersion property of coupled-cavity circuits, *IEEE MTT-S Digest*, 901 1993

Fallgatter, K.; Svoboda, V. & Winefordner, J.D. (1971). Physical and analytical aspects of a microwave excited plasma, *Applied Spectrocopy*, 25, no.3, 347 1971

Fujisawa, K. Fujisawa (1951). Theory of slotted cylindrical cavities with transverse electric field, *Tech. Reps. Osaka Univ.*, Vol. 1, 69, 1951

Fujisawa,. K. (1958). General treatment of klystron resonant cavities, *IRE Trans.*, MTT-6, 344, Oct. 1958

Gilmour, Jr.,A.S. (1986) Microwave Tube, *Artech House*, Chap. 15 1986

Goplen, B.; Ludeking, L.; Nguyen, K. & Warren, G. (1990). Design of an 850MHz klystrode, *International Electron Devices Meeting*, 889 1990

Goplen, B.; Ludeking, L.; Smithe, D. & Warren, G. (1994). MAGIC User's Manual, Rep. MRC/WDC-R-326, *Mission Res. Corp., Newington, VA*, 1994

Hamilton, D. R.; Knipp, J. K. & Kuper, J. B. Horner (1966). Klystrons and Microwave Triodes, *Dover Pub., New York*, 1966

Hansen, W. W. (1938). A type of electrical resonator, *Journal of Applied Physics*, Vol. 9, 654, Oct. 1938

HeBe, Z. (1997). The cold quality Qcold for magnicon on resonator in the rotating TMn10 mode, *International Journal of Infrared and Millimeter Waves*, Vol. 18, No. 2, 437 1997

Kim, Hyoung S. & Uhm, Han S. (2001). Analytical Calculations and Comparison with Numerical Data for Annular Klystrode, *IEEE Transactions on Plasma Science*, Vol. 29, Issue 6, pp. 875-880, Dec. 2001

Kim, Hyoung Suk & Ahn, Saeyoung (2000). Numerical Analysis of C-band Klystrode with Annular Electron Beam, *International Journal of Infrared and Millimeter Waves*, Vol. 21, No. 1, pp. 11-20, Jan. 2000

Kleinman, R. E. & Roach, G. F. (1974). Boundary Integral equations for the three-dimensional Helmholtz equation, *SIAM Review*, Vol. 16, No. 2, 214, April 1974

Lee, S. W.; Kang, H. J.; Kim, H. S. & Hum, H. S. (2009). Numerical Design of the Cavity for the Uniform Atmospheric Microwave Plasma Source, *Journal of Korean Physical Society*, Vol. 54, No. 6, pp. 2297-2301, June 2009

Liao, Samuel Y. (1990). Microwave Devices & Circuits, *Prentice Hall*, 1990

Marks, R. B. (1986). Application of the singular function expansion to an integral equation for scattering, *IEEE Transactions on Antennas and Propagation*, Vol. AP-34, No. 5, 725, May 1986

Mcdonald, S. W.; Finn, J. M.; Read, M. E. & Manheimer, W. M. (1986). Boundary integral method for computing eigenfunctions in slotted gyrotron cavities of arbitrary cross-sections, *Int. J. Electronics*, Vol. 61, No. 6, 795, 1986

Pearson, L. W. Pearson, (1984). A Note on the representation of scattered fields as a singularity expansion, *IEEE Transactions on Antennas and Propagation*, Vol. AP-32, No. 5, 520, May 1984

Ramm, A. G. (1982). Mathematical foundations of the singularity and eigenmode expansion methods, *Journal of Mathematical Analysis and Applications*, 86, 562 1982

Riddell, R. J. Jr. (1979). Boundary-Distribution Solution of The Helmholtz Equation for a Region with Corners, *Journal of Computational Physics*, 31, 21 1979

Riekmann, C.; Jostingmeier, A. & Omar, A. S. (1996). Application of the eigenmode transformation technique for the analysis of planar transmission lines, *IEEE MTT-S Digest*, 1023 1996

Samuel Seely, (1950). Electron-Tube Circuits, *McGraw Hill Book Com. Inc., New York*, 1950

Stevenson, A. F. (1948). Theory of slots in rectangular wave-guides, *Journal of Applied Physics*, Vol. 19, 24, Jan. 1948

Tan, J. & Pan, G. (1996). A general functional analysis to dispersive structures, *IEEE MTT-S Digest*, 1027 1996

Tobocman, W. (1984). Calculation of acoustic wave scattering by means of the Helmholtz integral equation. I, *J. Acoust. Soc. Am.*, 76 (2), 599, Aug. 1984

Zander, A.T. & Hieftje, G.M. (1981). Microwave-supported discharges, *Applied Spectroscopy*, 35, no. 4, 357 1981

6

Wide-band Rock and Ore Samples Complex Permittivity Measurement

Sixin Liu[1], Junjun Wu[1], Lili Zhang[2] and Hang Dong[3]
[1]*Jilin University,*
[2]*Shenyang Aerospace University,*
[3]*Northeast Institute of Geography and Agroecology, CAS*
China

1. Introduction

Ground penetrating radar (GPR) is based on high-frequency electromagnetic wave propagation and its detecting targets are below the ground surface (Daniels, 2004; Jol, 2009). Velocity and attenuation are two important factors describing the electromagnetic wave in the media composed of rocks or soils. Velocity is inverse proportional to the root of the permittivity while other parameters are fixed generally. In order to understand the performance of GPR, permittivity testing and analysis are critical. In addition, during the metal ore exploration by borehole radar which is an operating mode of GPR, the permittivity difference between ore body and surrounding rock is the foundation for exploration. As the sampling site, geological environment, and seasons are changing, the permittivity are different even for the same rock. Therefore, permittivity measurement is very important.

Currently, measurement methods are basically indirect methods which are based on transmission line theory, characteristic impedance, and propagation constant. These variables have intrinsic relationship with permittivity which can be inverted from the measured data by certain calculation procedure. At the RF frequency band, common measurement methods include short-circuited wave-guide measurement, coaxial line transmission/reflection method, open-ended coaxial probe, resonant cavity method, free-space transmission technique, parallel-plate capacitance method, etc.

Roberts and von Hippel (1946) developed the short-circuited wave-guide measurement, sample is inserted at the end of the wave-guide or coaxial line, the standing wave is formed as the incident wave and the reflected wave coexist in the wave-guide. The sanding wave ratios (SWR's) were required to measure in the case with and without sample. Permittivity can be determined by the change in the widths of nodes, sample length, and the waveguide dimension.

Resonant cavity method is a perturbation technique, which is frequently used for measuring permittivity because of its simplicity, accuracy, and high temperature capability (Venkatesh & Raghavan, 2005). This technique is based on the resonant frequency shift, and the change in absorption characteristics due to the insertion of sample material. The measurement is made by placing a sample completely in the center of a waveguide. The size of the cavity is designed for special frequency. Calibration is required before measurement, once the calibration is finished, the measurement is fast. The samples preparation is little difficult.

Free-space transmission technique is carried out by placing the sample between the transmitting and the receiving antennas (Kraszewski, 1980). The electrical parameters can be calculated by measuring the attenuation and the phase shift in the media. Free-space measurement can measure permittivity in a wide frequency band. It is a no-destructive and contact-less method. The frequency can be as high as 30GHz. The usual assumption for this technique is that a uniform plane wave incident upon the flat surface of a homogeneous material, and the planar sample has infinite extent laterally (Venkatesh & Raghavan, 2005).

Parallel disk capacitor technique (Shi & Shen, 1989; Shen, Marouni, Zhang, & shi, 1987) is based on the theory that the capacitance is dependant on the medium permittivity filling the capacitor. As a capacitor is filled with medium having permittivity of ε_r , its capacitance C is ε_r times of vacuum capacitance C_0. If we measure the capacitance before and after the capacitor is filled with the medium, we can obtain the permittivity. This method is mainly fitted for low-frequency range between 20MHz-200MHz.

Coaxial reflection/transmission method was introduced by Nicolson, Rose, and Wire, in 1970s, therefore, it is also called NRW method (Nicolson &Ross, 1970; Weir, 1974). The material under test (MUT) is inserted into coaxial line (waveguide). As the electromagnetic wave is traveling in the line, it meets the MUT, a part of the wave is transmitted, and the other is reflected. Attenuation and phase shift occur at the same time. Vector network analyzer (VNA) can measure the reflection and transmission coefficients at a wide frequency range. The permittivity can be inverted by electromagnetic theory from these measured data. This technique was first used for rock measurement (Shen, 1985), and the water-bearing, oil-bearing, or gas-bearing rocks can be distinguished from the measured data. Coutanceau-Monteil and Jacquin improved this technique in 1993. The rock samples are coated with a low melting point alloy and this improvement enables measurements to be made on larger samples and avoids air-gap correction at the solid-cell interface. However, the sample preparation is still difficult.

Open-ended coaxial probe technique (Zhen &Smith, 1991; Blackham&Pollard, 1997; Hoshina, Kanai, Miyakawa, 2001) is a technique of which the open-end of a coaxial is touched on the MUT, and the other end is connected to a VNA. The measured reflection coefficient is used to invert the permittivity. A flange is connected to the open-end to improve the accuracy. The circular waveguide probe, the rectangular waveguide probe, and the coaxial probe are often used. The coaxial probe can be used at frequency-domain, time-domain, point frequency, and it is suitable for kinds of media with different electric characteristic (low to high permittivity, lossy or lossless, magnetic lossy materials). This method is no-destructive, no-intrusive, and also characterized with wide frequency band, simple and open structure. It is fitted for on-line, in-vivo, and in-situ testing, it is receiving the attention from many scientists and engineers.

Open-ended probe electromagnetic radiation dates back to 1950s or earlier. Open-ended measurement technique was improved rapidly in recent 20 years (Stuchly M.A. & Stutch, S.S, 1980). It was first introduced and applied for biological tissue measurement at RF frequency, and was used for dielectric material testing at RF band shortly, and the lumped parameter model is used at the time for data processing. Later, as the testing frequency expanded to X and Ku band, quisi-static analysis model (Misra, 1987; Fans et al., 1952; Nyshadham et al., 1992; Wei, & Sridhar, 1991; Blackham & Pollard, 1997; Nelson& Bartley,

1998) and full-wave analysis model (Li & Chen, 1995) appeared, and the testing material also expanded to high permittivity from low permittivity. But the sample thickness was infinite. Until the early stage of 1990s, spectral-domain quisi-static analysis model appeared, and it was suitable for multi-layered material with finite thickness. Then, many researchers analyzed the high modes in the coaxial line, and reached the spectral-domain full-wave analysis model which is a complete mathematic model.

Several rules must be considered before we choose the techniques for measurement, such as frequency range preferred, sample preparation easiness, the equipments accessible. We choose open-ended coaxial probe technique backed by metal plate because the model is easy to be analyzed, the samples making is relatively easy, and the VNA is accessible for us.

We used open-ended coaxial probe to measure 342 samples within 31 categories from 6 mine sites, including: (1) granite, marble, hybrid diorite, altered hornblende pyroxenite, pyroxene peridotite, low-grade ore, medium-grade ore, high-grade ore from the Changren nickel-cupper mine, Jilin province, China; (2) tuffaceous fine-grained sandstone, tuffaceous breccia, dacitoid crystal tuff, dacite, crystal tuff, low-grade ore, medium-grade ore, high-grade ore, and pyrite from the Huanghuagou lead-zinc mine, Chifeng, Inner Mongolian, China; (3) altered K-feldspar granite, high-grade ore, low-grade ore from the Nianzigou Molybdenum mine, Chifeng, Inner Mongolian, China; (4) albite rhyolite porphyry, breccia porphyry, quartz albitophyre, albitophyre copper, malachite copper oxide ore from the Qunji copper mine, Xinjiang, China; (5) vesicular amygdaloidal andesite, massive diorite, andesitic copper ore from the Musi copper mine, Xinjiang, China; (6) glutenite, copper ore, lead-zinc ore from the Zengnan copper mine, Xinjiang, China. It is found that high-grade ore can be distinguished from surrounding rocks by permittivity, low-grade ore shows similar permittivity as other veins. We first measured and analyzed so many samples including different rocks and ores at wide frequency range according to current references world widely.

2. Open-ended coaxial technique

Firstly, the theory of open-coaxial probe technique is introduced. Then, the single-layered short-circuit spectral domain quasi-static model is established. Finally, the permittivity is calculated through the optimized inversion procedure.

2.1 The open-coaxial probe technique

The process of open-ended coaxial probe measuring medium includes three steps.

The first step is experimental testing (hardware operation). Reflection coefficient, which contains electromagnetic parameters information of the material, is measured by an open-coaxial probe with a flange-plane connected to reflection equipment which may be vector reflectometry (vector network analyzer or six ports) or scalar reflectometry. The vector network analyzer is used here (Agilent E5071B). The second step is the theoretical modeling. The physics model for the measurement is built based on electromagnetic knowledge. Namely, the relationship between the reflection coefficient of open-coaxial probe and the electromagnetic parameters of the sample measured is usually complicated and nonlinear one. And this modeling process is also called forward process. The third step is about the rebuilding electromagnetic parameters. In this step, the electromagnetic parameters corresponding with reflection coefficient of the material measured are obtained. This process is often called inverse process. Because the equations established are often nonlinear, we often use numerical solution.

Both theoretical modeling and numerical inversion are important steps in electromagnetic parameters measurement technology. Their accuracy decides the reliability of the measured results. For open-coaxial probe technique, modeling process is a mathematics analysis process, while the inversion is a process of appropriate optimization iteration.

2.2 Spectral-domain quasi-static reflection coefficient model

At present, popular analysis models used are full wave model and quasi-static model. The former is a theory model solved according to electromagnetic field. But the latter ignores the influences of high-order mode to reflection coefficient near the aperture of a coaxial probe. In theory, full wave model can accurately measure dielectric properties of materials. In fact, some theories and test indicate that the model error introduced by ignoring high-order mode is equal to the measuring error induced by the hardware and the operation. So, we can use quasi-static model (Wu et al., 2000), if the measurement error cannot be ignored and the rapid data processing is wanted.

We use spectral-domain quasi-static method here. The flow chart for the measurement is as shown in Fig. 1. Firstly, the reflection coefficient of a MUT is measured by a VNA. Then the measured reflection coefficient is calibrated (or corrected) to achieve real reflection coefficient. Finally the permittivity is calculated through inversion from the reflection coefficient.

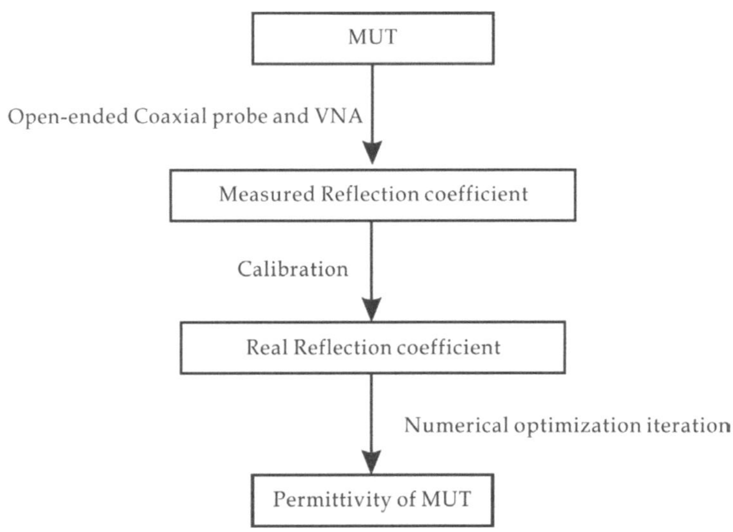

Fig. 1. Measurement flow chart

We used a single-layered short-circuit model here. Fig. 2 shows a schematic diagram of open-ended coaxial probe. "a" is the radius of inner conductor of coaxial probe. "b" is the radius of outer conductor. "d" is the thickness of sample measured. " ε_1 "and " μ_1 " are relative permittivity and relative magnetic conductivity of the medium filled in coaxial probe. " ε_s " and " μ_s " are permittivity and relative magnetic conductivity of the sample measured.

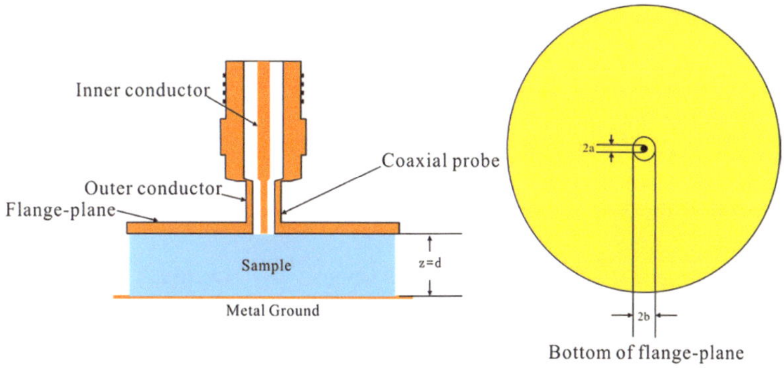

Fig. 2. Open-ended coaxial probe schematic diagram

It is assumed that only the *TEM* mode wave is traveling in coaxial probe and the time-harmonic factor $e^{j\omega t}$ is ignored. The complex electric field and complex magnetic field of forward wave and backward wave in coaxial probe can be expressed respectively as,

$$E_{r1} = \frac{A}{r}[\exp(-jk_1 z) + \Gamma \exp(jk_1 z)] \tag{1}$$

$$H_{\varphi 1} = \frac{A}{\eta_1 r}[\exp(-jk_1 z) - \Gamma \exp(jk_1 z)] \tag{2}$$

where, r, φ and z are the coordinate variable of cylindrical coordinate system in the equations. $k_1 = \omega\sqrt{\mu_0\varepsilon_0\mu_1\varepsilon_1}$, $\eta_1 = \sqrt{\mu_0\mu_1/(\varepsilon_0\varepsilon_1)}$. ε_0, μ_0 are the permittivity and magnetic conductivity in the vacuum. ω is the angular frequency . Γ is the reflection coefficient. A is the electric field amplitude of the forward wave on probe terminal surface.
The electromagnetic field E_{rs}, $H_{\varphi s}$ in the material measured can be expressed as the integral sum of all plane waves (including high-order mode) in spectral domain.

$$E_{rs} = \int_0^\infty B(k_c)[\exp(-\gamma z) + \Gamma_b(k_c)\exp(\gamma z)]J_1(k_c r)k_c dk_c \tag{3}$$

$$H_{\varphi s} = \int_0^\infty B(k_c)Y(k_c)[\exp(-\gamma z) - \Gamma_b(k_c)\exp(\gamma z)]J_1(k_c r)k_c dk_c \tag{4}$$

In the equations, $\gamma = \sqrt{k_c^2 - k_s^2}$, and $\mathrm{Re}(\gamma) \geq 0$, $k_s = \omega \cdot \sqrt{\mu_0\varepsilon_0\mu_s\varepsilon_s}$, $Y(k_c) = j\omega\varepsilon_0\varepsilon_s / \gamma$, $\Gamma_b(k_c) = -\exp(-2\gamma d)$, $J_1(x)$ is first-order first-kind Bessel function, $B(k_c)$ is the field amplitude expressed in the spectral domain. k_c is the continuous eigenvalue.
As the boundary conditions at the $z = 0$ plane are satisfied, the tangent components of the field are equal; following equations can be derived,

$$\int_0^\infty B(k_c)\left[1 + \Gamma_b(k_c)\right]J_1(k_c r)k_c dk_c = \begin{cases} \dfrac{A(1+\Gamma)}{r} & a \leq r \leq b \\ 0 & r \leq a, r \geq b \end{cases} \tag{5}$$

$$\int_0^\infty B(k_c)Y(k_c)\left[1 - \Gamma_b(k_c)\right]J_1(k_c r)k_c dk_c = \frac{A(1-\Gamma)}{\eta_1 r} \qquad a \leq r \leq b \tag{6}$$

Multiply $J_1(k_c'r)r$ to two sides of equation (5), and then take an integral with the form $\int_0^\infty dr$, we get,

$$\int_0^\infty \int_0^\infty B(k_c)\left[1+\Gamma_b(k_c)\right]J_1(k_c r)k_c dk_c \cdot J_1(k_c' r)rdr = \int_a^b \frac{A(1+\Gamma)}{r}\cdot J_1(k_c' r)rdr \tag{7}$$

The integral order of the left of equation (7) is rearranged as,

$$\text{left} = \int_0^\infty B(k_c)\left[1+\Gamma_b(k_c)\right]k_c\left[\int_0^\infty J_1(k_c r)J_1(k_c' r)rdr\right]dk_c \tag{8}$$

According to the orthogonality of Bessel functions, we get

$$\int_0^\infty J_n(k_c r)J_n(k_c' r)rdr = \frac{1}{k_c}\delta(k_c - k_c') \tag{9}$$

According to the sifting properties of δ function:

$$\int_{-\infty}^{\infty} \delta(t-t_0)f(t)dt = f(t_0) \tag{10}$$

Then the equation (8) can be modified:

$$\text{left} = \int_0^\infty B(k_c)\left[1+\Gamma_b(k_c)\right]k_c \frac{1}{k_c}\delta(k_c - k_c')dk_c = B(k_c')\left[1+\Gamma_b(k_c')\right] \tag{11}$$

Since $\frac{d}{dx}J_0(x) = -J_1(x)$, the right of the equation (7) can be calculated:

$$\text{right} = \int_a^b \frac{A(1+\Gamma)}{r}\cdot J_1(k_c' r)rdr = A(1+\Gamma)\int_a^b J_1(k_c' r)dr = A(1+\Gamma)\left[\frac{J_0(k_c' a) - J_0(k_c' b)}{k_c'}\right] \tag{12}$$

Then the equation (7) can be rearranged:

$$B(k_c')\left[1+\Gamma_b(k_c')\right] = A(1+\Gamma)\left[\frac{J_0(k_c' a) - J_0(k_c' b)}{k_c'}\right] \tag{13}$$

Let $k_c' = k_c$, the equation (13) becomes,

$$B(k_c) = \frac{A(1+\Gamma)}{1+\Gamma_b(k_c)}\left[\frac{J_0(k_c a) - J_0(k_c b)}{k_c}\right] \tag{14}$$

The equation (14) is substituted to equation (6) and takes an integral with $\int_a^b dr$ with the equation (6):

$$\int_a^b \int_0^\infty \frac{A(1+\Gamma)}{1+\Gamma_b(k_c)}\left[\frac{J_0(k_c a) - J_0(k_c b)}{k_c}\right]Y(k_c)\left[1-\Gamma_b(k_c)\right]J_1(k_c r)dk_c dr = \int_a^b \frac{A(1-\Gamma)}{\eta_1 r}dr \tag{15}$$

Eliminating A and rearranging, (15) becomes,

$$\int_0^{\infty}\left[\int_a^b J_1(k_c r)\,dr\right]Y(k_c)\frac{1-\Gamma_b(k_c)}{1+\Gamma_b(k_c)}\left[\frac{J_0(k_c a)-J_0(k_c b)}{k_c}\right]dk_c=\frac{(1-\Gamma)}{\eta_1(1+\Gamma)}\int_a^b\frac{1}{r}dr \qquad (16)$$

The result is:

$$\frac{1-\Gamma}{1+\Gamma}=\frac{\eta_1}{\ln(b/a)}\int_0^{\infty}Y(k_c)\frac{1-\Gamma_b(k_c)}{1+\Gamma_b(k_c)}\cdot\frac{\left[J_0(k_c a)-J_0(k_c b)\right]^2}{k_c}dk_c \qquad (17)$$

$J_0(x)$ is a Zero-order Bessel function of the first kind. It becomes the following:

$$\frac{1-\Gamma}{1+\Gamma}=\frac{\eta_1}{\ln(b/a)}\cdot\int_0^{\infty}\frac{\left[J_0(k_c a)-J_0(k_c b)\right]^2\left[1+2\exp\left(-\sqrt{k_c^2-\omega^2\varepsilon_0\varepsilon_s\mu_0\mu_s}\times d\right)\right]\varepsilon_s i\omega\varepsilon_0}{\left[1-2\exp\left(-\sqrt{k_c^2-\omega^2\varepsilon_0\varepsilon_s\mu_0\mu_s}\times d\right)\right]k_c\sqrt{k_c^2-\omega^2\varepsilon_0\varepsilon_s\mu_0\mu_s}}dk_c \qquad (18)$$

Because the other parameters are all known. The equation (18) is a nonlinear equation about independent variable ε_s and variable Γ. The question of solving permittivity becomes a process of solving this nonlinear equation.

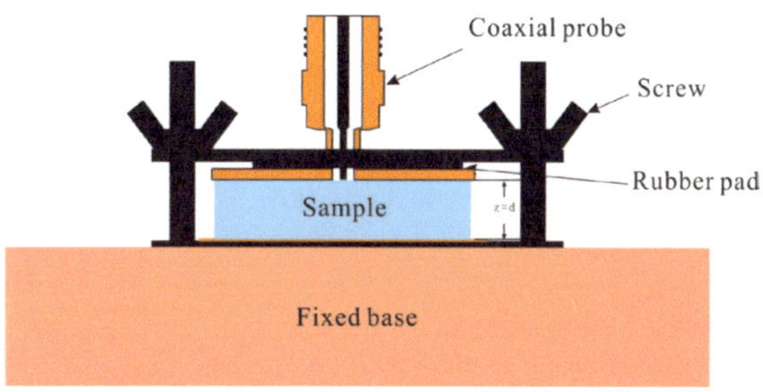

Fig. 3. Schematic diagram of rock sample holder

Fig. 4. VNA and the rock sample holder

To avoid the influence of air-gap to the testing result, a rock sample holder is make as shown in fig. 3.
The practical testing equipment including a VNA is shown in Fig. 4.
In equation (17) and (18), because the Bessel function has oscillating property, the main difficulty focuses on the Bessel function with integral variable. Obviously these nonlinear equations have no analytical solution. So we uses numerical solution here. A big number (12500) is used for the positive infinite of the upper limit during the numerical intergal. The value of the big number is determined by different testing of many conditions.

2.3 Calibration

The calibration in this measurement includes two steps, one is the transmission line calibration, and the other is the probe calibration.

2.3.1 Transmission line calibration

The VNA is a exact device and is connected to coaxial probe through a coaxial cable. According to the operation requirement of the network analyzer, the coaxial line is calibrated using calibrating kits. The detailed operation process as follows.

1. VNA parameters setting. The frequency range is between 1MHz-1GHz in this test according to our request. Power can be selected by our need, for example, 0dBm. Large power is believed to sense large sample volume. We choose 1000 sampling points here.
2. Calibration process. We use single port calibration here because we only measure S11. The calibration kits which including the device SHORT, LOAD, and OPEN are used to calibrate the VNA. After this process, the reference plane for VNA is at the end of the coaxial cable. However, because the reflection surface is on the flange surface, not the end of the cable, further calibration is still needed.

2.3.2 Probe calibration

Probe calibration is an indirect method. We use short-circuit, the air, and the de-ionized water to calibrate the probe.
If Γ_m is the reflection coefficient obtained through measuring and Γ_a is the practical reflection coefficient of probe terminal, Γ_m can be expressed as (Blackham & Pollard, 1997):

$$\Gamma_m = e_d + \frac{e_r \Gamma_a}{1 - e_s \Gamma_a} \tag{19}$$

where, e_d is the limited directivity error; e_r is frequency response error; e_s is equivalent source matching error. The reflection coefficient of the material Γ_a can be calculated through equation (17) or (18). Through measuring the reflection coefficient of three kinds of materials Γ_m, the three equations about e_d, e_r, e_s can be obtained. There are three variables and three equations, the error coefficients e_d, e_r, e_s can be obtained.
Short-circuit, and air are ideal calibration materials. The third material must have known permittivity. The de-ionized water is selected as the third calibration material here. When it is of short-circuit, $\Gamma_a = -1$; when the calibration material is air, the reflection coefficient of every frequency can be calculated through equation (17) or (18), because the permittivity of air is 1. According to the same theory, the reflection coefficient of de-ionized water can be calculated. Here, the reflection coefficient of water is obtained through the Cole-Cole formula.

$$\varepsilon = \varepsilon' - j\varepsilon'' = \varepsilon_\infty + \frac{\varepsilon_s - \varepsilon_\infty}{1 + (j\omega / \omega_0)^{1-\alpha}} \tag{20}$$

where, ε_s is a direct current permittivity. ε_∞ is an optical frequency permittivity. ω_0 is a Debye relaxation angle frequency. α is a Cole-Cole factor.

By substituting the reflection coefficient of air film Γ_{air_a} and Γ_{air_m}, the reflection coefficient of de-ionized water Γ_{water_a} and Γ_{water_m} and the reflection coefficient of short-circuit $\Gamma_{short\_a} = -1$ and Γ_{short_m} into the equation (19) separately. We get,

$$e_s = \frac{C\Gamma_{air\_a} + C - \Gamma_{water\_a} + \Gamma_{air\_a}}{C\Gamma_{water\_a}\Gamma_{air\_a} + C\Gamma_{water\_a} + \Gamma_{water\_a} - \Gamma_{air\_a}} \tag{21}$$

where, $C = \dfrac{\Gamma_{water\_m} - A}{A - B}$, $A = \Gamma_{air\_m}$, $B = \Gamma_{short\_m}$.

We can also get,

$$e_r = \frac{\Gamma_{water\_m} - A}{\dfrac{\Gamma_{water\_a}}{1 - e_s\Gamma_{water\_a}} - \dfrac{\Gamma_{air\_a}}{1 - e_s\Gamma_{air\_a}}} \tag{22}$$

$$e_d = A - \frac{e_r\Gamma_{air\_a}}{1 - e_s\Gamma_{air\_a}} \tag{23}$$

It can be concluded based on the equation (19) that:

$$\Gamma_a = \frac{\Gamma_m - e_d}{e_s(\Gamma_m - e_d) + e_r} \tag{24}$$

Equation (24) determines the second step calibration. Fig. 7 shows the comparison among results before and after calibration for PTFE and de-ionized water, separately.

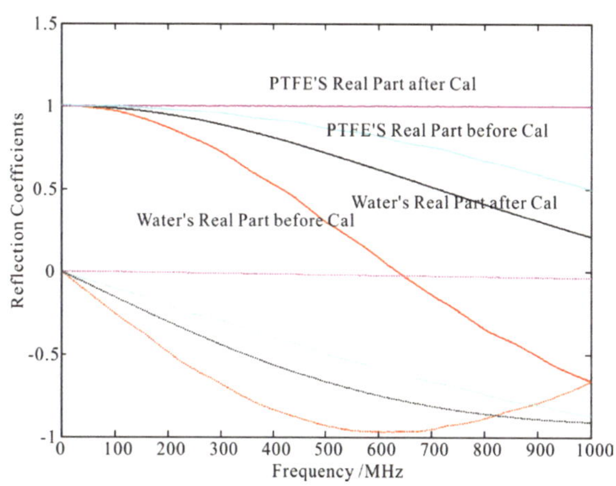

Fig. 5. Reflection coefficient before and after calibration

It can be seen that the real part of PTFE measured can be calibrated to around but different from 1.

2.4 Inversion calculation and error evaluation

If the permittivity of a material measured is known, the interface reflection coefficient (or admittance) can be calculated. This process is a forward one. The reverse process can be solved numerically. The following equation can be obtained from equation (18),

$$Y = \frac{1-\Gamma}{1+\Gamma} = \frac{\eta_1}{\ln(b/a)} \int_0^{\infty} Y(k_c) \frac{1-\Gamma_b(k_c)}{1+\Gamma_b(k_c)} \cdot \frac{\left[J_0(k_c a) - J_0(k_c b)\right]^2}{k_c} dk_c \tag{25}$$

The solution is:

$$\Gamma = \frac{1-Y}{1+Y} \tag{26}$$

Because it is a complex calculation, the objective function is defined as

$$f(\varepsilon) = a\left|\operatorname{Re}(\Gamma_m - \Gamma_c)\right|^2 + \left|\operatorname{Im}(\Gamma_m - \Gamma_c)\right|^2 \tag{27}$$

α is a weighting coefficient in this equation, Γ_m and Γ_c are measured and the calculated reflection coefficients. The real part and the imaginary part should be treated equally to avoid that the large part dominates over the small part too much. This is the typical optimization problem. Here, ε can be thought as a complex-single variable. But the most mathematical software optimization tool can not process complex variable optimization question. So the complex permittivity is divided to real part and imaginary part. The variable x is a vector array, where, $x_1 = \operatorname{Re}(\varepsilon)$, $x_2 = \operatorname{Im}(\varepsilon)$. The selection of weighting coefficient is based on the numerous tests.

We solve the optimization process using the simplex method. The value of $f(\varepsilon)$ after the optimization for every frequency is displayed in Fig. 6 for the material PTFE. It can be seen that the precision is very well. When the optimization stops, the objective function of minimum point satisfy the error requirement.

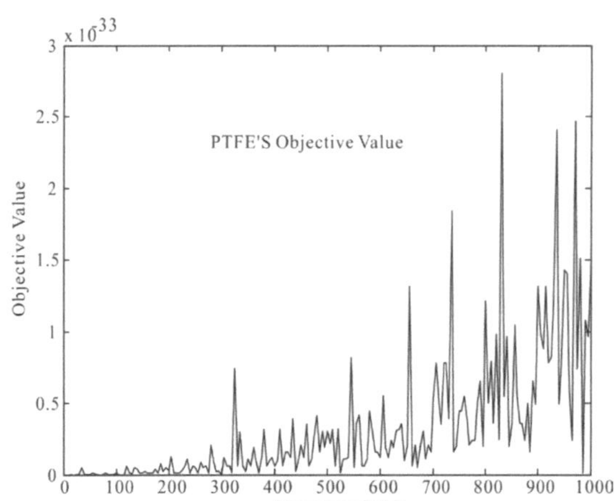

Fig. 6. The value of optimization objective function

We testify this technique using a standard material PTFE, air, and methanol.

We first test this technique with PTFE whose thickness is 10.50mm in this paper and has permittivity of 2.1-j0.0004 (Li & Chen, 1995) in microwave band. Because the imaginary part can not be measured exactly for lowly lossy medium (Wu et al., 2001) by this technique, we ignore the analysis for the imaginary part. The inverted permittivity is displayed in Fig. 7. The real part relative error at every frequency is displayed in Fg. 8.

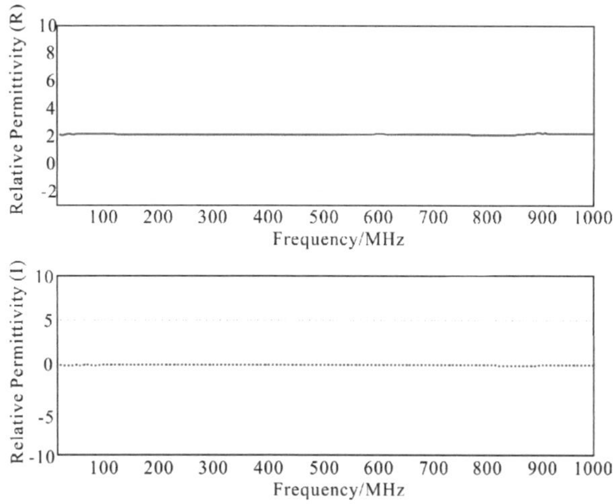

Fig. 7. Permittivity of PTFE sample

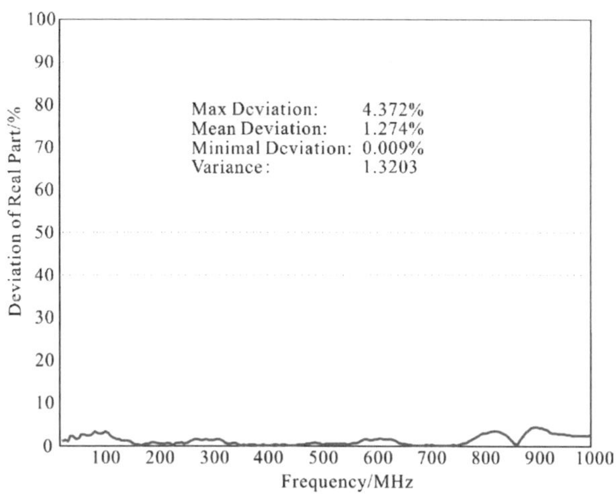

Fig. 8. Real part relative error of the permittivity of PTFE sample

We noticed that the arisen relative error is within 5% basically. The average relative error is 1.2749%. One of the many reasons leading to the error is the air gap between the flange and the sample. The main reasons of producing air gap are that the upper surface and down surface are not parallel and clean enough, and the upper surface and the down surface do not touch enough with coaxial probe flange-plane and short-circuit board, although we already tried our best.

The permittivity calculated by the air film is displayed in Fig. 9.

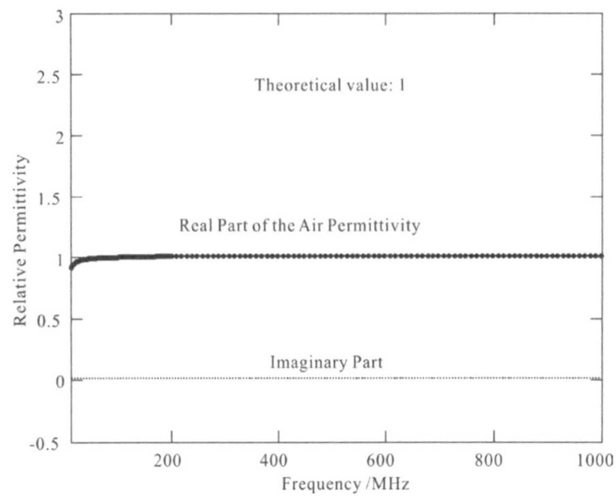

Fig. 9. Permittivity of the air

The relative error is 0.7692%. Because the air is a kind of calibration material, the permittivity of air calculated should be theoretical value 1.The relative error is below 0.8%. It proves the validity of inversion process.
The measured permittivity for methanol is displayed in Fig. 10.

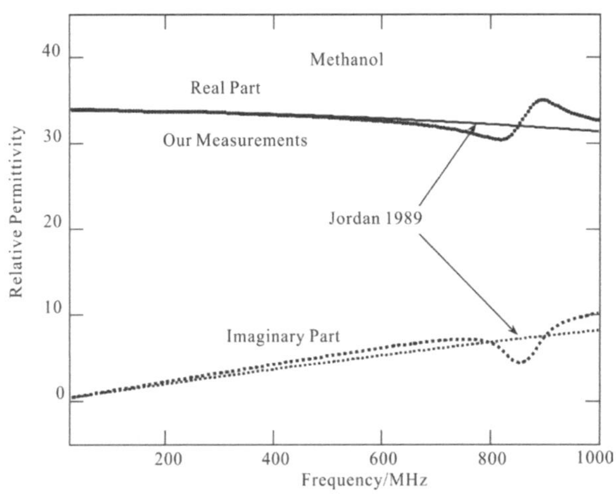

Fig. 10. Permittivity of methanol

The measured permittivity for methanol is compared with the theoritical values which is calculated by the debye equation or cole-cole equation (Jordan et al., 1978) as shown in Fig. 10. The measured data is accetable except that they have clear difference with the theotitical ones at high frequency range. The reducement of this error could be the future topic.

3. Measured results and analysis

342 rocks and ores sample within 31 categories from 6 mines are measured and analyzed in this part by using open-coaxial probe technique. The photos for these rocks and ores samples are shown in Fig. 11.

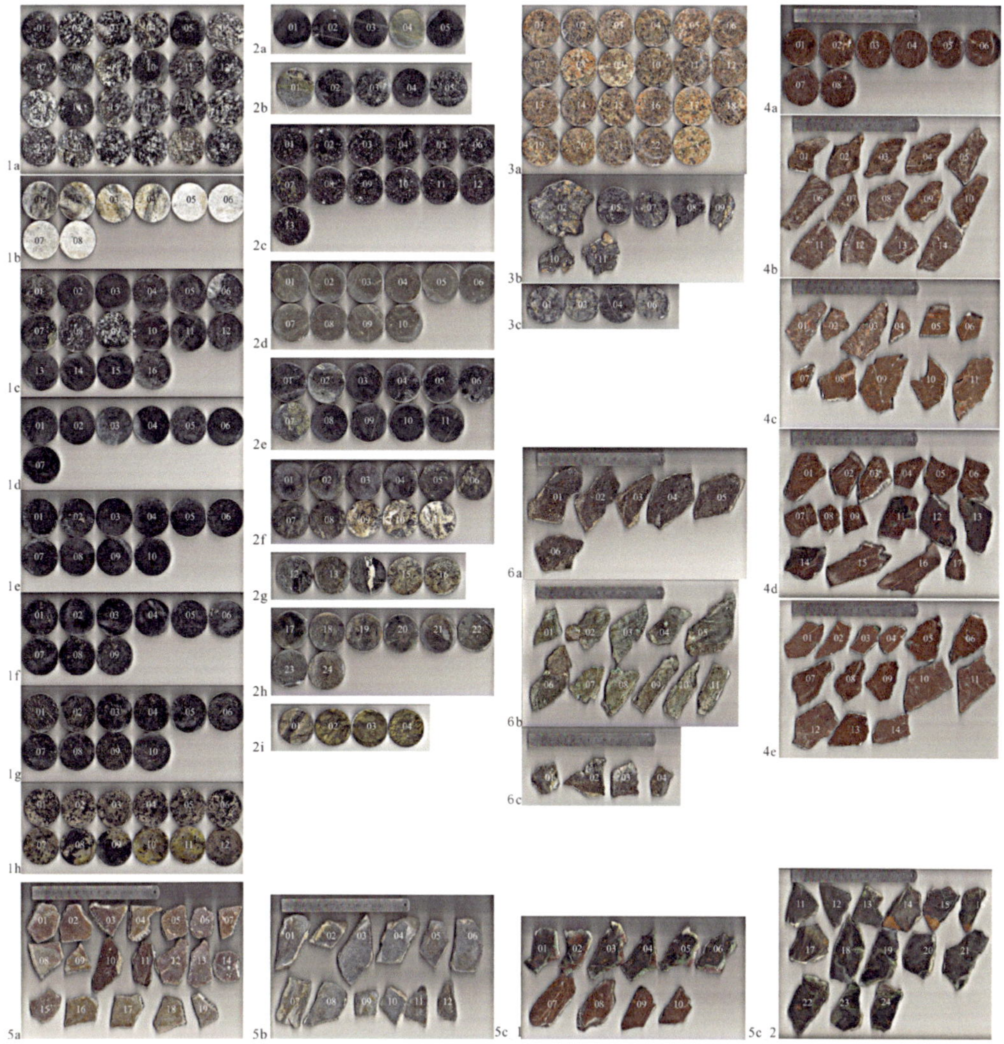

Fig. 11. Photographs of the rocks and ores samples from 6 metal mines

3.1 Samples from the Changren nickel-copper mine, Jilin, China

Table 1 shows the messages of rocks and ores from the Changren nickel-copper mine, Jilin, China.

Fig.12 shows marbles permittivities as an example, the solid and the dashed lines denote the real parts and the imagery parts. We find the values are diverse for the same rock. We think this kind of diversity is due to the fact of that the probe senses a small range and the samples are in-homogeneous. Therefore, we use the averaging value of these data to represent this sample, because the averaging could reflect the total characteristic.

Fig. 13 shows the average permittivities of all rocks and ores from the Changren nickel-cooper mine, China. We find high grade ore and medium grade ore have highest values, then the values range from high to low are the pyroxene peridotite, low grade ore, light alterative bornblende pyroxenite, marble, hybrid diorite, granitization granite.

Rocks	Rock or Ore names	Fig. no.	Measured permittivity	Sample number
1	granite	11(1a)	5-7.5	25
2	marble	11(1b)	5-10	8
3	hybrid diorite	11(1c)	5-10	16
4	altered hornblende pyroxenite	11(1d)	5-17	7
5	pyroxene peridotite	11(1f)	10-20	10
ores:				
6	low-grade ore	11(1g)	9-23	9
7	medium-grade ore	11(1f)	20-70	10
8	high-grade ore	11(1h)	5-95	12

Table 1. Rocks and ores from the Changren nickel-copper mine, Jilin, China

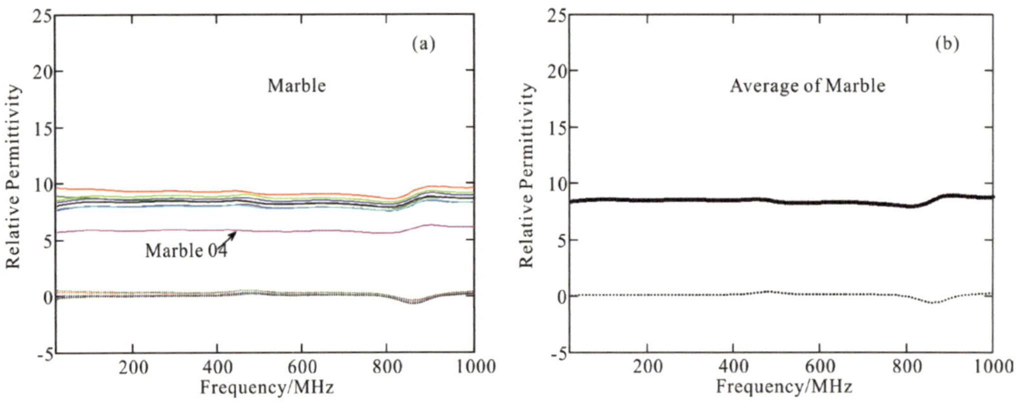

Fig. 12. Permittivity of marbles. (a) 8 Marble's samples permittivities; (a) average of mable samples' permittivities

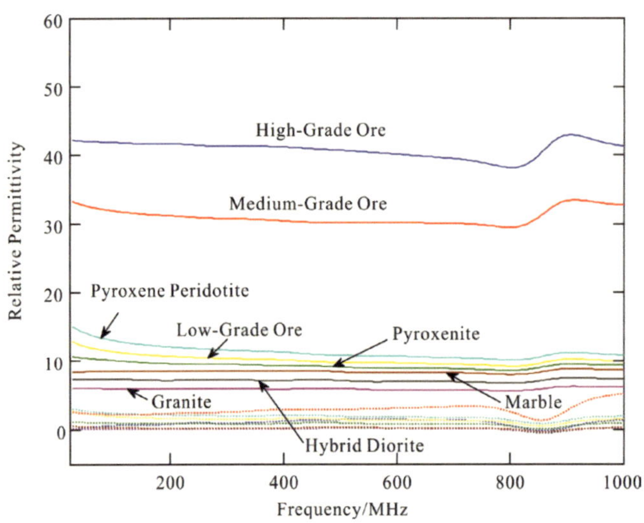

Fig. 13. Averaged relative permittivities of rocks and ores from the Changren nickel-cooper mine

Actually, the pyroxene peridotite, light alterative bornblende pyroxenite are basic rocks and ultra-basic rock which were ore carrier. When ore's grade is low, the permittivity represents the carrier rock's property. These basic rocks and ultra-basic rock come from tectonic emplacement. The granitized granite is the host rock which has distinguished lower values. These measured data show optimistic aspect for borehole radar detection for metal ore-body.

3.2 The samples from the Huanghuagou lead-zinc mine, Chifeng, Inner Mongolian, China

The table 2 shows the message of rocks and ores from the Huanghuagou Lead-Zinc mine Chifeng, China. Ores and rocks ranked by permittivity from high to low are high-grade ore, pyrite, medium-grade ore, dacitoid crystal tuff, low-grade ore, crystal tuff, tuffaceous breccia, tuffaceous sandstone, and dacite. The high-grade ore, pyrite, and the medium-grade ore are distinguishable from each other and the others.

Rocks	Rock or Ore names	Fig. no	permittivity	Samples number
1	tuffaceous fine-grained sandstone	11(2a)	5-7	5
2	tuffaceous breccia	11(2b)	5-6	5
3	dacitoid crystal tuff	11(2c)	5.5-8	13
4	dacite	11(2d)	5.5-6	10
5	crystal tuff	11(2e)	5-7.5	11
Ores				
6	high-grade ore	11(2f)	10-70	11
7	medium-grade ore	11(2g)	10-12	5
8	low-grade ore	11(2h)	5-10	8
9	pyrite	11(2i)	20-40	4

Table 2. Messages of the Huanghuagou lead-zinc mine, Chifeng, Inner Mongolian, China

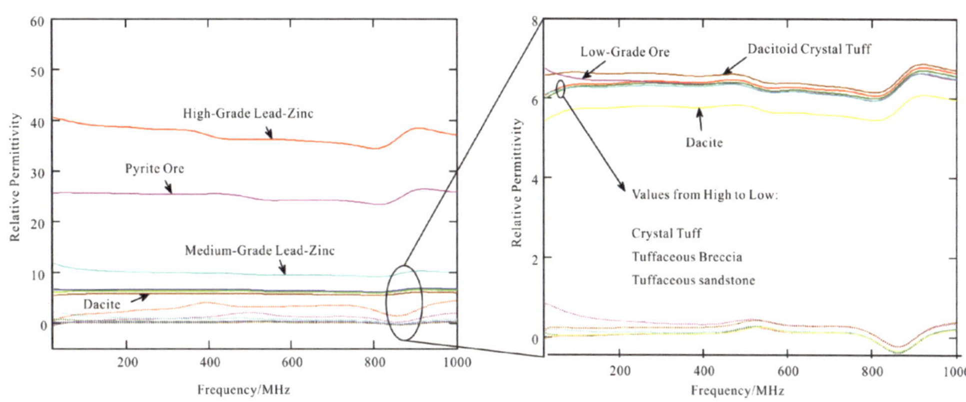

Fig. 14. Averaging permittivities of ores and rocks from the Huanghuagou lead-zinc mine, Chifeng, Inner Mongolian, China

3.3 Samples from the Nianzigou molybdenum mine, Chifeng, Inner Mongolian, China

The table 3 shows the messages of rocks and ores from the Nianzigou molybdenum mine, Chifeng, Inner Mongolian, China. Ores and rocks ranked by permittivity from high to low are high-grade ore, low-grade ore, and altered K-feldspar granite. The high-gride ore is distinguishable from other two, and the low-grade ore shows the nearly same permittivity as altered K-feldspar grinate.

Rocks	Rock or Ore names	Fig. no	permittivity	Samples number
1	altered K-feldspar granite	11(3a)	4.5-7.5	23 samples
Ores				
2	high-grade ore	11(3b)	5-15	7 (No: 02, 05, 07, 08, 09, 10, 11)
3	low-grade ore	11(3c)	4-10	4 (No: 01, 03, 04, 06)

Table 3. Messages of rocks and ores from the Nianzigou molybdenum mine, Chifeng, Inner Mongolian, China

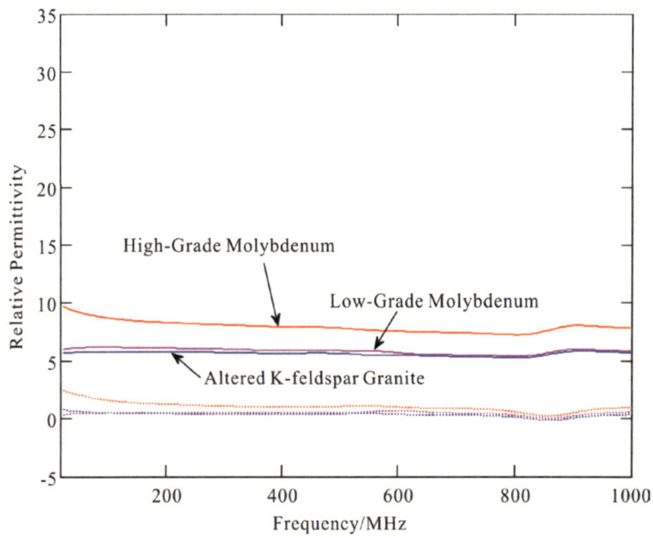

Fig. 15. Averaging permittivities of the rocks and ores from the Nianzigou molybdenum mine, Chifeng, Inner Mongolian, China

3.4 Samples from the Qunji copper mine, Xinjiang, China

The table 4 shows the messages of rocks and ores from the Qunji Copper mine, Xinjiang, China. Ores and rocks ranked by permittivity from high to low are albitophyre ore, quartz albitophyre, breccia porphyry, malachite copper oxide ore, and albite rhyolite porphyry. The albitophyre ore is clearly distinguishable from the others in the real part. Other rocks and ore are ambitious in permittivity.

Rocks	Rock or Ore names	Fig. no	Permittivity	Samples number
1	albite rhyolite porphyry (core)	11(4a)	5-5.5	8
2	breccia porphyry	11(4b)	5-5.5	14
3	quartz albitophyre	11(4c)	5-7.5	11
Ores				
4	albitophyre ore	11(4d)	5-10	16(No：01-09,11-17)
5	malachite oxide ore	11(4e)	5-5.5	14(No.：01-14)

Table 4. Messages of rocks and ores from the Qunji Copper mine, Xinjiang, China

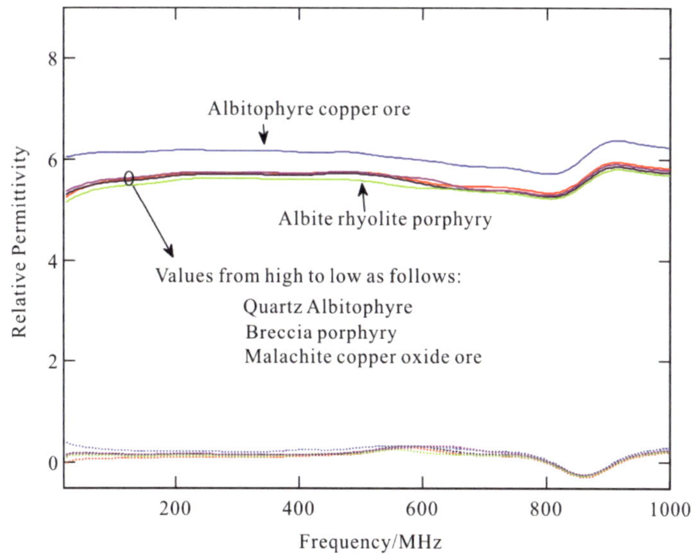

Fig. 16. Average permittivities of rocks and ores from the Qunji Copper mine, Xinjiang, China

3.5 Samples from the Musi copper mine, Xinjiang, China

The table 5 shows the messages of rocks and ores from the Musi copper mine, Xinjiang, China. Ores and rocks ranked by permittivity from high to low are vesicular amygdaloidal andesite, massive diorite, and andesitic copper ore. The andesitic copper ore is distinguishable from the others and shows low permittivity characteristic which is opposite to other mines.

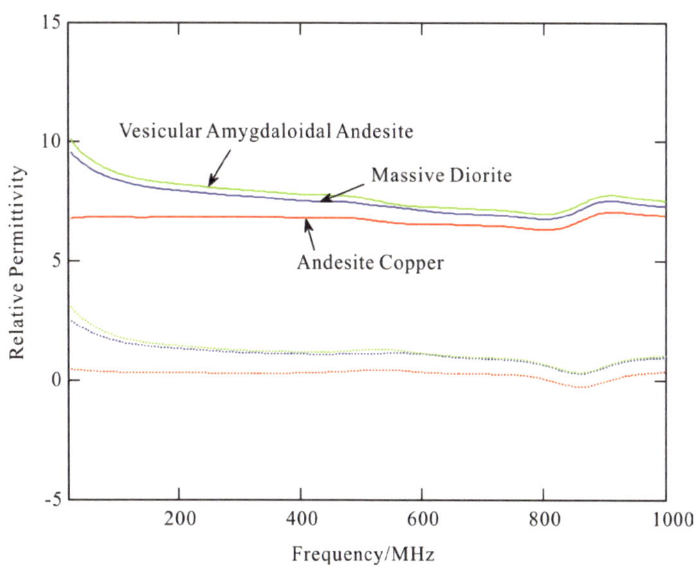

Fig. 17. Averaging permittivities of rocks and ores from the Musi copper mine, Xinjiang, China

Rocks	Rock or Ore names	Fig. no	Permittivity	Samples number
1	vesicular amygdaloidal andesite	11(5a)	5.5-12.5	19 samples
2	massive diorite	11(5b)	7.5-11	12 samples
Ores				Total:24
3	andesitic copper ore	11(5c_1; 5c_2)	5-10	24 samples

Table 5. Messages of rocks and ores from the Musi copper mine, Xinjiang, China

3.6 Samples from the Zengnan copper mine, Xinjiang, China

The table 6 shows the messages of rocks and ores from the Zengnan copper mine, Xinjiang, China. Ores and rocks ranked by permittivity from high to low are lead-zinc ore, copper ore, and glutenite. Three of them can be distinguished from each other.

Rocks	Rock or Ore names	Fig. no	Permittivity	Samples number
1	glutenite	11(6a)	8-14	6
Ores				
2	copper ore	11(6b)	8-40	11
3	lead-zinc ore	11(6c)	15-45	4

Table 6. Messages of rocks and ores from the Zengnan copper mine, Xinjiang, China

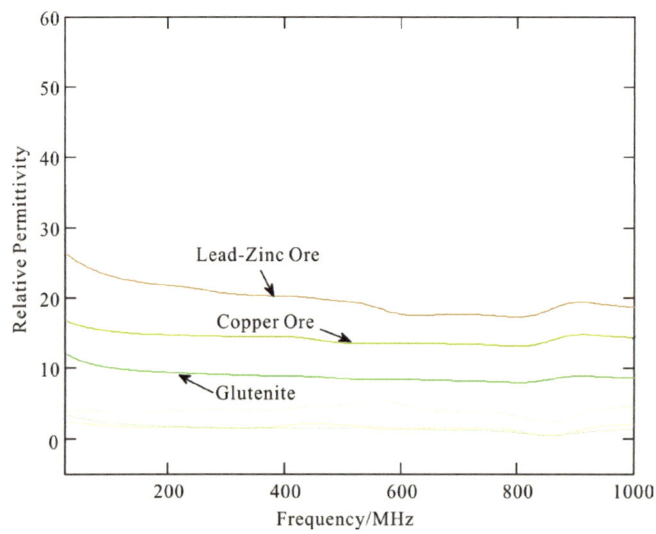

Fig. 18. Averaging permittivities of rocks and ores from the Zengnan copper mine, Xinjiang, China

4. Conclusion

Open-ended coaxial technique can measure the permittivity in wide frequency range quickly. The sample machining is relatively simple, and only the smooth surfaces of the sample sheets are required. Because the sensing range of the probe concentrates mainly at the center of the probe and the samples measured are no so homogeneous, we use averaging value from several samples of a rock or ore to reduce the random effect due to their in-

homogeneity. It is shown that permittivity of metal ore is higher than other rocks, and high-grade ore is distinguishable from surrounding rocks. These measurements provide insights into the wide-frequency permittivity of metal ores and rocks, and also provide basis for electromagnetic exploration by borehole radar.
There are still couple of problems with the current research. The sizes of the flange, the aperture of the probe, sheet sample thickness, are not optimized yet. The sensing area for the current probe is small for the inhomogeneous rocks and ores. These are all future works for us.

5. Acknowledgment

This research is supported by the National Natural Science Foundation of China (Grant No 40874073 and 41074076), and by the National High-Tech R&D Program 863 (Grant No 2008AA06Z103)

6. References

Blackham, D.V. & Pollard, R. D. (1952). An improved technique for permittivity measurements using a coaxial probe. *IEEE Transactions on Instrumentation and Measurement*, Vol. 46, No. 5, (Sept., 1997), pp.1093-1099, ISSN 0018-9456

Coutanceau-Monteil, N., & Jacquin, C. (1960). Improvements of the coaxial line technique for measuring complex dielectric permittivity of centimetric samples in the 20 to 1000 MHz range: application to sedimentary rocks. *Log Analyst*, Vol. 34, No. 5, (September-October, 1993), PP. 21-33, ISSN 1529-9074

Daniels, D. (2004). *Ground Penetrating Radar, 2nd Edition,* The Institution of Engineering and Technology, ISBN 0863413609, London, United Kingdom.

Fan S.; Staebellk K. & Misra D. (1952). Static analysis of an open-ended coaxial line terminated by layered media. *IEEE Transactions on Instrumentation and Measurement*, Vol. 39, No. 2, (March, 1990), pp.435-437, ISSN 0018-9456

Hoshina, S.; Kanai, Y. & Miyakawa, M. (1965). A numerical study on the measurement region of a coaxial probe used for complex permittivity measurement. *IEEE Trans. Magn*, Vol. 37, No. 5, (Sept., 2001), pp. 3311-3314, ISSN 0018-9464

Jol, H. M. (2009). *Ground Penetrating Radar Theory and Applications*, ELSEVIER, ISBN 0444533486, Amsterdam, the Netherlands.

Jordan, B P ; Sheppard, R J & Szarnowski, S. (1950). The dielectric properties of formamide, ethanediol and methanol. *J. Phys. D: Appl. Phys*, Vol. 11, No. 5, (April, 1978), pp. 695-701, ISSN 0022-3727

Kraszewski, A.W. (1965). Microwave aquametry - A review. *Journal of Microwave Power*, Vol. 15, No. 4, (1980), pp. 209-220, ISSN 0022-2739

Li, C. L. & Chen, K. M. (1952). Determination of electromagnetic properties of materials using flanged open-ended coaxial probe-full-wave analysis. *IEEE Transactions on Instrumentation and Measurement*, Vol. 44, No.1, (Jan., 1995), pp. 19-27, ISSN 0018-9456

Misra, D. K. (1963). A quasi-static analysis of open-ended coaxial lines. *IEEE Transactions on Microwave Theory and Techniques*, Vol. 35, No. 10, (1987), pp. 925-928, ISSN 0018-9480

Nelson, S.O & Bartley, P. G. (1952). Open-ended coaxial line permittivity measurements on pulverized materials. *IEEE Transactions on Instrumentation and Measurement*, Vol. 47, No.1, (Jan., 1998), pp. 133-137, ISSN 0018-9456

Nicolson, A. M. & Ross, G. (1952). Measurement of intrinsic properties of Materials by time domain techniques. *IEEE Trans on Instrument & Measurement*, Vol. IM-19, No. 4, (November 1970), pp. 377-382, ISSN 0018-9456

Nyshadham, A.; Sibbaldcl, C. L. & Stuchly, S. S. (1963). Permittivity measurements using open-ended sensors and reference liquid calibration--an uncertainty analysis. *IEEE Transactions on Microwave Theory and Techniques*, Vol. 40, No. 2, (Feb., 1992), pp. 305-314, ISSN 0018-9480

Roberts, S. R. & Hippel, A. Von. (1930). A new method for measuring dielectric constant and loss in the range of centimeter wave. *Journal of Applied Physics*, Vol. 17, No. 7, (July, 1946), pp. 610-616, ISSN 0021-8979

Shen, L. C. (1961). A laboratory technique for measuring dielectric properties of core samples at ultrahigh frequencies. *Society of Petroleum Engineers Journal*, Vol. 25 No. 4, (April, 1985), pp. 502-514, ISSN 0197-7520

Shi, X. D. & Shen, L. C. (1958). Dielectric properties of reservoir rocks at very-high frequencies. *ACTA GEOPHYSICA SINICA*, Vol. 32, No. 1, (Jan. 1989) pp. 99-110, ISSN 00015733

Shen, L.C.; Marouni, H.; Zhang, Y.X. & Shi, X.D. (1963). Analysis of the parallel disk sample holder for dielectric permittivity measurement, *IEEE transaction on geoscience and remote sensing*, Vol. GE-25, No.5, (Sept. 1987), pp. 534-540, ISSN 0196-2892

Stuchly, M. A. & Stuchly S. S. (1952). Coaxial line reflection methods for measuring dielectric properties of biological substances at radio and microwave frequencie-a review. *IEEE Trans. Instrum. Meas.*, Vol. 29, No.3, (Sept., 1980), pp.176-183, ISSN 0018-9456

Venkatesh, M. S. & Raghavan, G. S. V. (1949). An overview of dielectric properties measuring techniques. *CANADIAN BIOSYSTEMS ENGINEERING*, Vol. 47, No. 7, (2005), pp. 15-30, ISSN1492-9058

Weir, W. B.(1913). Automatic Measurement of Complex Dielectric Constant and Permeability at Microwave Frequency. *Proc. IEEE*, Vol. 62, No. 1, (Jan. 1974), pp.33-36, ISSN 0018-9219

Wei, W. & Sridhar, W. (1963). Radiation-corrected open-ended coax line technique for dielectric measurements of liquids up to 20 GHz. *IEEE Transactions on Microwave Theory and Techniques*, Vol. 39, No.3, (March, 1991), pp. 526-531, ISSN 0018-9480

Wu, M. Z.; Yao , X. & Zhang, L, Y. (2000). An improved coaxial probe technique for measuring microwave permittivity of thin dielectric materials. *Measurement Science and Technology*, Vol. 11, No. 11, (2000), pp. 1617-1622, ISSN 0957-0233

Zheng, H. & Smith, C. E. (1991). Permittivity measurement using a short open-ended coaxial line probe. *IEEE Microw. Guided Wave Lett*, Vol. 1, No.11, (November, 1991), pp. 337-339, ISSN 1051-8207

7

Detection of Delamination in Wall Paintings by Ground Penetrating Radar

Wanfu Wang
Dunhuang Academy
People's Republic of China

1. Introduction

Wall painting is an important part of cultural heritage. Generally speaking, painting on the wall of buildings or rocks, and those on the wall of caves are called wall paintings. But painting on the rock face is called rock painting. Wall painting on the building can be approximately classified into drawing murals, relief frescoes, mosaic murals and etcetera material paintings. Chinese ancient wall paintings can be generally distinguished according to different drawing site, such as palace paintings, temple paintings, grotto frescoes, coffin chamber murals, residential paintings and so on. Most of the paintings, including grotto frescoes, palace paintings or temple paintings, have several hundred years, or even several thousand years of history. During this time, under the influence of environmental factors (light, temperature, humidity, wind, sand and so on), biotic factors (micro-organism, insect), painting support walls and materials, architectural composition and human factor, wall paintings have undergone various kinds of diseases and damage. The most common painting diseases are delamination, flaking, disruption, smoking, pollution, deep-loss, paint-losses, cracks-hatch, mechanical-damage and so on.

Delamination is the loss of adhesion between layers in the support (wall, rock mass or others) and plaster stratigraphy, causing separation between plaster and suport. Delamination can occur between plaster layers, plaster and support. Generally, delamination causes painting surface crack and protrusion, even leads to painting losses because of gravity force from wall painting itself.

Literally speaking, Tibet is a region with abundance of cultural relics. According to an incomplete statistics, there are more than 2,000 ancient architectures all over the region, among which 3 are included in the world heritage list, 27 are national key preservation units, 55 are provincial level ones and 96 are city or county level ones. A primary survey shows such cardinal ancient architectures, just like Potala Palace, Norbulingka and Sagya Monastery, and the wall paintings are in severe need to be conserved. The architecture deterioration mainly occurs in the forms of structural distortion, roof leakage, rafter mildew, moth-eaten, rat-bitten beams, while the wall painting deterioration displays in delaminated plaster, pigment flaking, plaster and wall crevice, plaster disruption, soot and contaminant, among which the most serious damage, taking up more than 75% areas in total seems to be delamination [1]. In this sense, the great task in the conservation of Tibetan cultural relics proves to be the combat against the delamination in wall paintings.

Fig. 1. Delaminated wall paintings in Cave 329 of Mogao Grottoes

Fig. 2. Delaminated wall paintings in Eastern Audience Hall of Potala Palace

Wall paintings in Tibet Potala Palace, Norbulingka and Sagya Monastery were made as follows: firstly coarse red *Argar* earth was coated on the stone wall, rammed earth wall or light *Bianma* grass, secondly fine white *Argar* earth was coated on them, and then paintings were drawed, finally varnish or tung oil was spread on the wall painting surface.

The causes of wall painting delamination [2], [3] can be summed up in the following aspects: first of all, the construction material and crafts applied. The result of the survey discloses that the ancient Tibetan architectures are mixed constructures made of stone, earth and wood, which leaves the connection sections between the beams and wall paintings at the ceiling as well as the upper side of doors vulnerable to the delamination. The layer-structured wall paintings in those sections suffer distortion and breach under the pressure of vertical shearing stress, showing the unequal distribution of different interface stress upon different materials. The load of the building and roofing on beams and purlines transfers through those frameworks to walls. The wall painting plasters leaning against walls are directly connected to roofing. During the drying process, different materials displaying dissimilar contraction rates are easy to form gaps around the combining parts of those materials, which, in combination with the transmission of forces, contributes to the formation of the delamination.

Secondly, the cause comes to the layout of the architecture and the effects brought by both natural and human activity vibration. The structures of ancient Tibetan architecture mainly belong to pillar mixed load carrying members. In those architectures, the top of the architecture serves not only as the roofing of the floor but the platform for its upper floor. Besides used as passages and aisles, the Buddhist ceremony was also held here. Therefore, the vibration brought by the human activity is ranked among the causes for the formation of wall painting crevices and delamination. Each year the renovation of roofing is carried out regularly, during which a large number of people performing ramming generates strong vibration when they are ramming a new layer of *Argar*. This is also a potential threat to the supporting structure of roofing. In a word, the original layout of the structure leaves wall paintings open to deterioration, the deterioration of delamination in particular, while the human activity accelerates this process. The vibration produced by the human activity and the architecture weight itself are the direct cause of the mural delamination. In addition, the frequent earthquakes of different magnitudes also impose important effect upon the architecture, resulting in its distortion and damage.

Thirdly, the roof leakage is anther cause of delamination, which in turn is caused by the architecture distortion and the malfunction of the *Argar* layer.

Fourthly, the environment also contributes to the delamination. The surrounding environment of cultural relics is among the most important factors in their intact preservation. However, at the same time, it is also the prerequisite for the formation of deterioration. The environmental factors affecting the preservation of cultural relics mainly include temperature, moisture, illumination and ventilation, which in the case of the Tibetan palace wall paintings, are the indoor temperature and moisture plus region environment, such as air temperature, precipitation and air moisture. Researches show that the microenvironment of the Tibetan palace and temple are conducive to the preservation of wall paintings, whose annual mild changes to some degree avert the wall painting damage imposed by the freeze-thaw action.

In the conservation of wall paintings, it is quite a problem in technology to investigate the area and the degree of wall painting delamination. Traditionally, the diagnosis of delamination in grotto wall paintings and palace wall paintings is achieved by distinguishing the tone when tapping wall paintings by hand, such experience is useful in determining the area and degree of delamination, but it depends a lot on subjective sensation.

Non-destructive detection by ground penetrating radar (GPR) is the method of using high-frequency electromagnetic waves in the form of wide-band short pulse to transmit signal underground by the transmitting antenna of ground penetrating radar, which reflects back to the receiving antenna at the mismatching interfaces of electromagnetic impedence, and analysing the amplitude characteristics of received waves in time domain or frequency domain to distinguish abnormal body. Ground penetrating radar is widely used in archaeology, karst exploration, concrete pavement assessment, tunnel lining quality evaluation, subgrade stratification and so on. With the increase of central frequency of radar antenna and the using of ultra-wideband technology, ground penetrating radar is applied to the recognition of shallow target.

The depth of mural delamination is generally 2 ~ 5 cm, rarely more than 10 cm (Fig. 3). Therefore, ground penetrating radar can detect depth of 20 cm to meet the requirements. Based on physical modeling experiment in the laboratory, the author uses the RAMAC GPR made in Sweden to detect delamination of wall paintings in Tibetan lamaseries and Lashao temple. During the in-situ test, the ground penetrating radar is equipped with a shielded antenna at the nominal central frequency of 1.6 GHz, the antenna is gently attached upon a piece of transparent parchment paper that has been covered on the vanishing surface of wall paintings, the sampling parameters of time window is set at 4 *ns* and sampling frequency at 142 GHz, and the signal triggering mode is adopted as distance or time. Having been processed by the band-pass filter and the filter of subtracting mean trace, the scope of delamination disease is determined and the thickness of wall painting delamination is estimated in the radar profile.

Fig. 3. Typical plaster section of wall paintings in Potala Palace

2. Detection of delamination in replica plaster

Under ideal condition, the vertical resolution limit is up to 1/10 of electromagnetic wavelength, but under poor circumstance, the resolution is only 1/3 of characteristic wavelength. As to the geotechnical detection by ground penetrating radar, it is typically considered 1/4 to 1/2 of impulsed electromagnetic wavelength as the vertical resolution to select the appropriate radar antenna. When the characteristic wavelength of electromagnetic waves is close to the thickness of cavity or delamination, the relative strong echo from the top or the bottom of cavity in the radar image is easy to identify. Because of the application of ultra-wideband radio technology, such kind of ground penetrating radar has higher resolution [4]-[7].

Replica of Tibetan wall painting plaster is made, and regular voids at different depth and with varied size are set inside, then the forward modeling detection is carried out in order to get appropriate parameters for acquisition of radar data, and to find effective filters for signal processing.

2.1 Characteristic of transmitting impulse

RAMAC GPR, made by MALÅ GeoScience in Sweden, is used to carry out the physical modeling experiment. It is designed on the basis of general modular, and it consists of control unit, antenna and computer terminals (Fig. 4). When detecting the delamination of wall paintings in Tibet Potala Palace, the 1.6 GHz shielded antenna with the highest center frequency at that time was used. At present, the latest product, 2.3 GHz radar antenna, takes a higher central frequency. As for the impulse electromagnetic wave generated by the transmitting antenna, its time domain (Fig. 5) and frequency domain (Fig. 6) characteristics affect the performance of ground penetrating radar, especially the vertical resolution of ground penetrating radar.

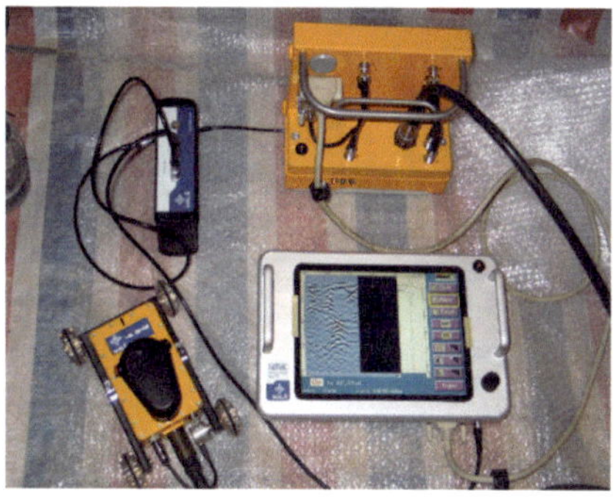

Fig. 4. RAMAC/GPR made by MALÅ GeoScience

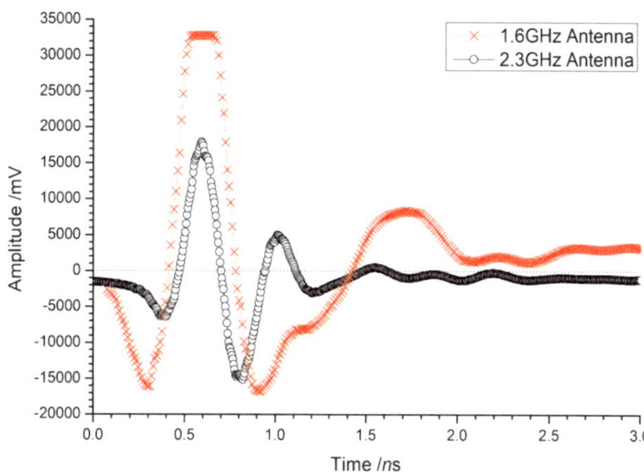

Fig. 5. Time domain waveform of carrier-free pulse emitted by GPR antenna

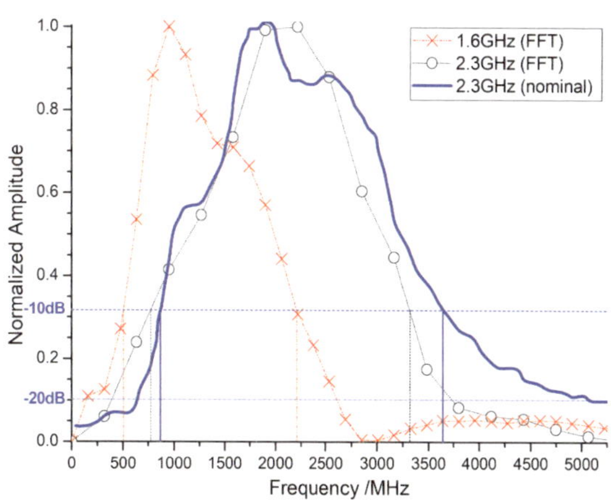

Fig. 6. Frequency domain spectrum of carrier-free pulse emitted by GPR antenna

According to the Federal Communications Commission (FCC), the band width of impulse signal of electromagnetic wave is defined as the range of frequencies in which the signal's spectral density **P**(f) is above -10 dB relative to its maximum:

$$P_{dB}(f) = 10 \cdot \log_{10}\left(\frac{|A(f)|^2}{|A_{max}(f_c)|^2}\right) dB \tag{1}$$

where: $\mathbf{P}_{dB}(f)$ is the normalized power when frequency is f and the measuring unit is dB; $A(f)$ is the amplitude when frequency is f and $A_{max}(f_c)$ is the peak amplitude at the central frequency of f_c.
When $\mathbf{P}_{dB}(f)$ is -10 dB, $A(f)=10^{-1/2}\cdot A_{max}(f)\approx 0.32\ A_{max}(f)$. As shown in Fig. 6, when the normalized amplitude is 0.32, its normalized power is equal to -10 dB.
In Fig. 6, the signals in time domain have been transformed into frequency domain by Fast Fourier Transform (FFT), the higher bound f_H and lower bound f_L of spectral band width of the electromagnetic wave transmitted by 1.6 GHz antenna is 502 MHz and 2,203 MHz respectively. As to 2.3 GHz antenna, the higher bound f_H and lower bound f_L is 772 MHz and 3,321 MHz respectively. The relative band width **B** is defined by the following equation:

$$B = \frac{f_H - f_L}{(f_H + f_L)/2} \times 100\% \tag{2}$$

where: **B** is the relative band width of electromagnetic wave in frequency spectrum and the measuring unit is %; f_H is the higher bound of the band width, f_L is the lower bound of the band width and both the measuring units are MHz.
According to equation (2), it can be figured out that the relative band width of the 1.6 GHz antenna is 126% and that of the 2.3 GHz antenna is 125%. So that, both of them belong to the type of ultra-wideband (UWB) antenna.

2.2 Vertical resolution

Having taken the technology of step frequency, RAMAC GPR expands the band width of impulse electromagnetic wave. The component of high frequency in the effective band width, $\mathbf{B}_{eff}$, possess higher resolution. The simplified equation of the vertical resolution, $\Delta\mathbf{R}$, of the ground penetrating radar can be worked out according to the Rayleigh criterion:

$$\Delta R = \frac{v}{2B_{eff}} = \frac{c/\sqrt{\varepsilon_r}}{2B_{eff}} \tag{3}$$

where: $\Delta\mathbf{R}$ is the vertical resolution of ground penetrating radar, also called as longitudinal resolution, its unit is m. $\boldsymbol{v}$ is the propagating velocity of impulse electromagnetic wave in the medium, with the unit of m/s. $\mathbf{B}_{eff}$ is the effective absolute band width in frequency spectrum of received signals and its unit is Hz. $\boldsymbol{c}$ is the traveling speed of electromagnetic wave in vacuum, its value is about 3.00×10^8 m/s. ε_r is the real part of the relative dielectric constant of the medium.
By the equipment of Agilent 8510C single terminal vector network analyzer (VNA), it is determined that the relative dielectric constant of the fine layer, namely white *Argar* earth, and the coarse layer, namely red *Argar* earth, in wall painting plaster is about 3.76 and 2.9

respectively in the frequency range of 0.2~3.0 GHz. As for 1.6 GHz antenna, its absolute wide band is 1.70×10^9 Hz, so that, according to equation (3), the vertical resolution is 0.051 m, that is 5.1 cm.

In equation (3), the half-wave length of the electromagnetic wave transmitting in the medium is regarded as the vertical resolution of ground penetrating radar. However, according to the Rayleigh criterion, 1/4 of the wave length is regarded as the limit of the vertical resolution. Under high signal to noise ratio, 1/8 of the wave length can be regarded as the limit of the theoretical vertical resolution. In fact, the replacement of effective band width by absolute band width to calculate the vertical resolution is a comprised method. Because the detection of delamination in wall paintings by ground penetrating radar belongs to the application of ultra shallow layer in the depth range of 10 cm, the two-way attenuation distance of electromagnetic wave in the dry plaster is relatively short. The component of high frequency with higher resolution can reflect back into the receiving antenna at the interface between plaster layer and cavity.

If the threshold of -20 dB spectral density, in equal to normalized amplitude of 0.1 in Fig. 6, is regarded as the recognition limit, the effective band width of 1.6 GHz antenna in frequency domain is 121~2,624 MHz. Therefore, the minimum wave length of the electromagnetic wave transmitting in the wall painting plaster is 6.62 cm. Then, the maximum theoretical vertical resolution of $\lambda/8$ is about 8 mm.

2.3 Physical modeling experiment

In order to determine the appropriate acquisition parameter of RAMAC GPR, and to obtain the method of digital signal processing (DSP), regular voids with different depth and thickness are made in the loam plaster (Fig. 7). The ground penetrating radar equipped with 1.6 GHz shielded antenna is used to carry out the lab test (Fig. 8, Fig. 9, Fig. 10, Fig.11).

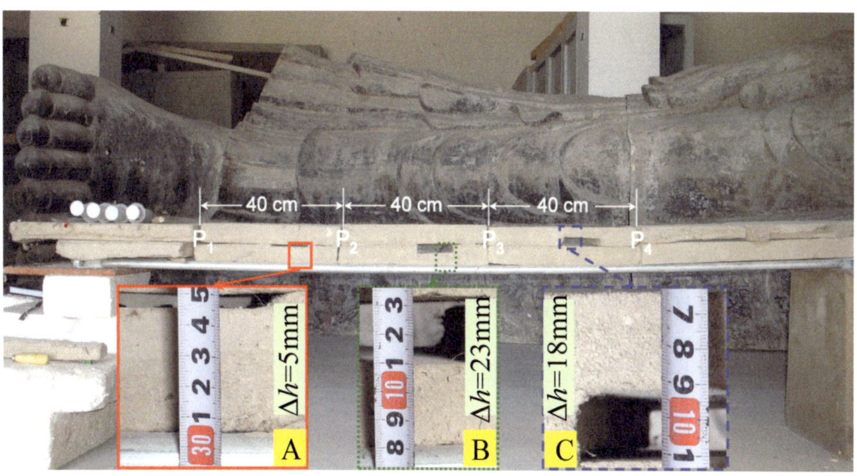

Fig. 7. Schematic layout of rectangular voids in plaster replica for detection by GPR

In Fig. 7, the length of the delamination parts A, B, and C is 100 mm. Their buried depth h and thickness Δh is 45 mm & 5 mm, 45 mm & 23 mm and 27 mm & mm respectively. The relative dielectical constant of the loam plaster is about 1.74, so that the propagation velecity of the electromagnetic wave in such medium is 2.27×10⁸ m/s, namely 0.227 m/*ns* or 227 m/*μ*s. It is faster than that of the electromagnetic wave in dry clay.

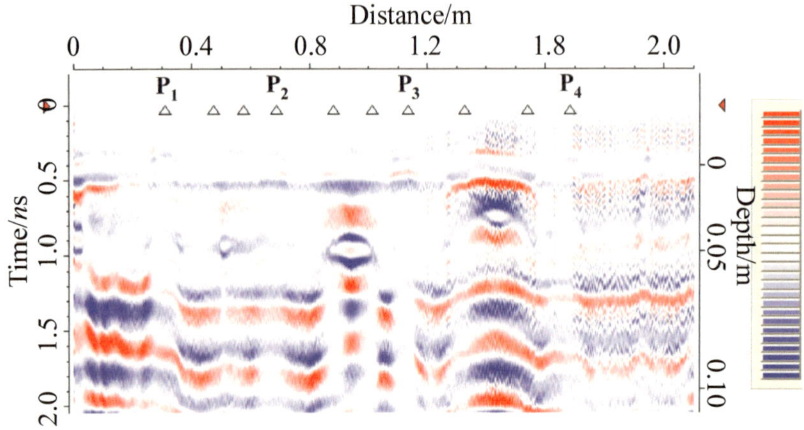

Fig. 8. Presentation of post-processed GPR profile in software of Ground Vision

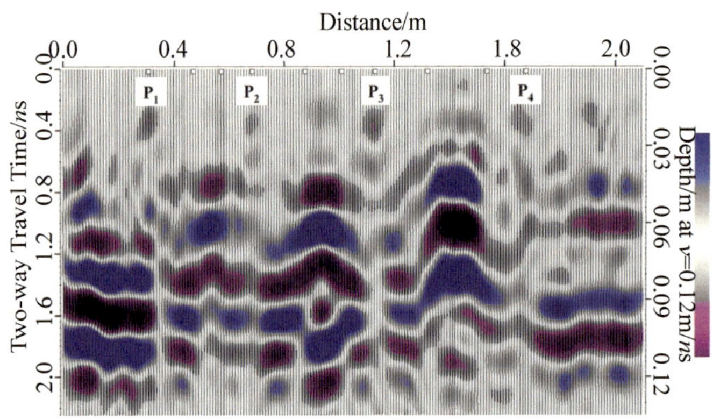

Fig. 9. Post-processed GPR profile in combination of wiggle mode and point mode

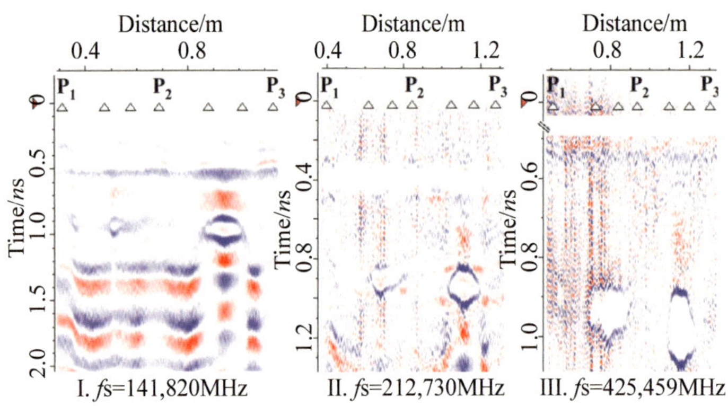

Fig. 10. FIR filtered GPR profiles at different sampling frequency

In Fig. 8, the length of the radar profile is about 2.1 m, the interval of the triggering time is 0.1 s, the total time spent is 62.1 s, and 621 traces of data have been collected. The average speed of the antenna is about 3.38 cm/s. The time window t of the profile is 2.26 *ns*, the sampling frequency, f_s, is 141.82 GHz. The number of samples, **N**, collected in each trace is 320, which is figured out by the following equation:

$$N=f_s \cdot t \tag{4}$$

What is shown in Fig. 8 is a radar profile, presented in the form of a matrix with 320 rows and 621 columns after loading the filter of finite impulse response (FIR). In the Ground Vision, which is a software affiliated to the ground penetrating radar equipment, through the processing of direct current (DC) removal, band pass filtering and subtract mean trace, the delamination in replica plaster can be distinguished clearly in the point mode of radar profile. Since the delamination A is not that thick, it is difficult to be located in Fig. 8. The delamination C is so shallow that the noise over the echo is strong. The delamination B is the most obvious and its thickness, Δh, can be calculated with the following equation:

$$\Delta h = c \cdot \frac{\Delta t}{2} = c \cdot \frac{(N_t - N_0) \cdot t}{2N} \tag{5}$$

where: Δh is the thickness of the delamination with the unit of m. c is the propagation velocity of electromagnetic wave in the delaminated area and the value is about 3.00×10^8 m/s. Δt is the two-way travel time when the electromagnetic wave propagates in the delaminated area and its unit is s. $\mathbf{N_t}$ is sample number of the lower surface of delamination in typical trace. $\mathbf{N_0}$ is sample number of the upper surface of delamination. ***t*** is time depth of the whole trace with the unit of s. **N** is sample number of the whole trace.

In equation (5), t is 2.26×10^{-9} s. According to the characteristic waveform of delamination B in time domain, N_t and N_0 is 179 and 155 respectively, and N is 320. The thickness of the delamination is figured out as 0.0254 m, namely 25.4 mm. It is very close to the actual thickness in replica plaster. By the same rule, N_t and N_0 for delamination C is 141 and 123 respectively, and its thickness is calculated as 19.1 mm.

Fig. 9 is the interpretation result of the same radar profile in software of REFLEX after the processing of subtract DC shift, band pass butterworth, background removal and F-K migration. It is the combination of presentation in point mode and wiggle mode. The delamination parts of A, B and C are obvious. Compared with the background, their common characteristics are the sudden increase in negtive amplitude and the phase inversion.

Fig. 10 is the interpretation result of delamination A and B at different sampling frequency. The higher the sampling frequency is, the more samples in illustration of delamination at the same time depth are. The bigger the difference of two way travel time between the upper and the lower surface of delamination is, the more serious the delamination is.

The results of modeling detection (Table 1) show that when the antenna couples well with the wall painting surface, the delamination in the radar profile is great clear. The higher the sampling frequency is, the more samples corresponding to the bound of delamination there are. This is good for the interpretation of radar profile, and the values of delamination thickness, figured out by equation (5), are close to each other.

Delamination	Replica dimension		Calculated thickness/mm		
	Depth/mm	Thickness/mm	Sampling frequency/GHz		
			142	213	425
A	45	5	16.5	19.2	15.6
B	45	23	25.4	26.6	26.3
C	27	18	19.1	N/A	N/A

Table 1. Interpreted thickness of delamination in comparison to nominal size

2.4 Analysis of typical traces

The typical wave forms (Fig. 11, Fig. 12) of delamination A with the thickness of 5 mm, delamination B of 23 mm thick and background are extracted from the radar profile. The comparison and analysis of transformed wave forms in time domain are presented in Fig. 13 and Fig. 14.

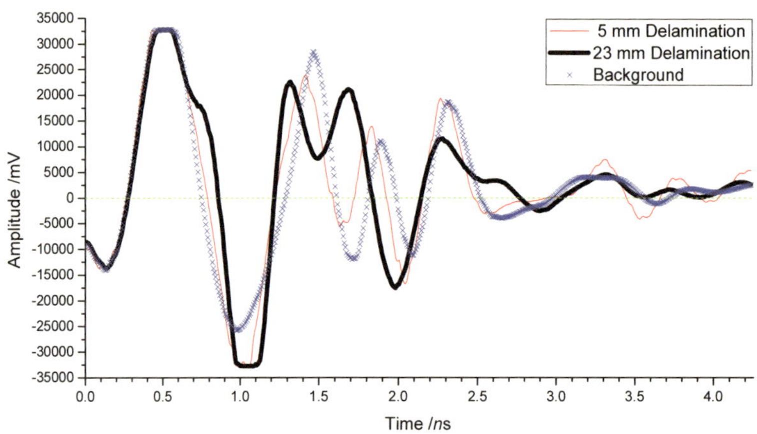

Fig. 11. Characteristic traces in time domain at sampling frequency of 142 GHz

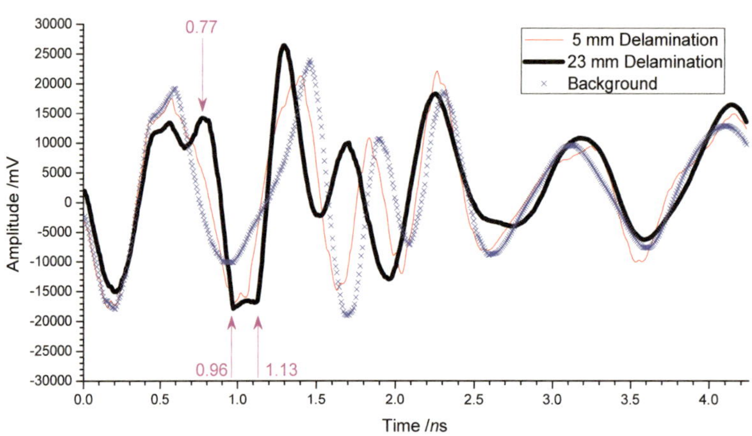

Fig. 12. Comparison of typical traces after high pass at 1.2 GHz in time domain

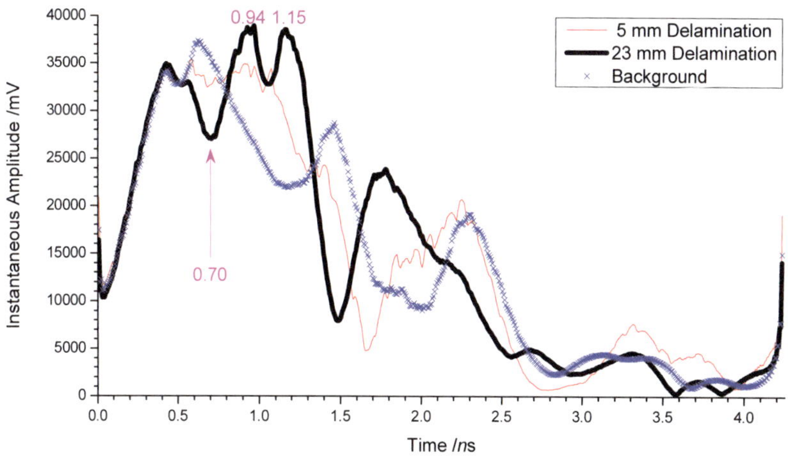

Fig. 13. Comparison of instantaneous amplitude after Hilbert transform

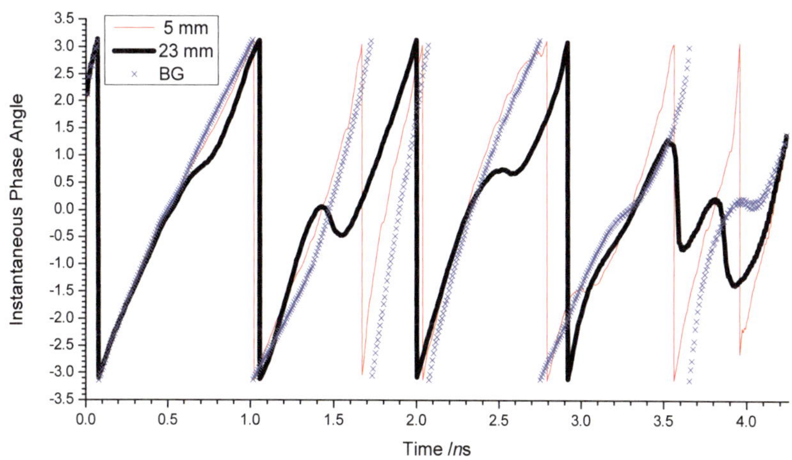

Fig. 14. Comparison of instantaneous phase after Hilbert transform

At the upper and lower interfaces of delamination, the instantaneous amplitude as well as the instantaneous phase of the delamination and the background in time domain go through alienation. The two way travel time is 0.7 *n*s and 0.9 *n*s respectively when the contrast is great, which is in consistent with the contrast of original waveforms in time domain.

Specifically to RAMAC/GPR and its accessory software of Ground Vision, it is suggested that the depth of time window should be about 3 *n*s and sampling frequency about 213 GHz. As a rule, the thickness of wall painting plaster in Tibetan lamaseries is less than 10 cm, and it depends on efficient removal of direct coupled waves in radar profile to detect delamination beneath wall painting plaster. As the GPR raw data is processed by applying filter of finite impulse response (FIR), delamination in wall paintings is characterized as sudden amplification of negative amplitude in waveform, and the extent of delamination is proportional to the time difference of two adjacent troughs, representing how serious the deterioration is.

3. In-situ detection of delamination in wall paintings

According to characteristic phenomena discovered and practical experience got in previous modeling experiment, GPR is applied to detect delamination in wall paintings in Potala Palace, Norbulingka and Sagya Monastery in Tibet. It is shown that strong negative amplitudes appear on certain GPR traces, which is in accordance with existence of serious delamination in wall paintings in these Tibetan lamaseries. Delamination in wall paintings in the Western Hall of Potala Palace is in great relation with wooden beam between plaster and blockstone wall. In the Mandala Hall of Sagya Monastery, delamination spreads from the top area of rammed wall, and it is proven to be more serious in the upper area.

3.1 Detection in Sagya Monastery

Visual inspection by means of video probe (Fig. 15) shows that delamination in wall paintings in the Sagya Monastery is very serious. The delamination is mainly distributed at the top of walls, usually in the form of wall painting plaster delaminating from the rammed plaster (Fig. 16).

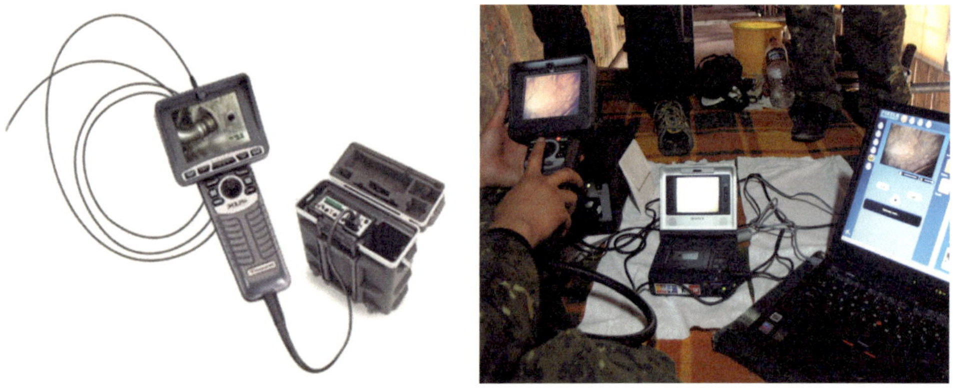

Fig. 15. XL PRO Video probe made by EverestVIT for remote visual test

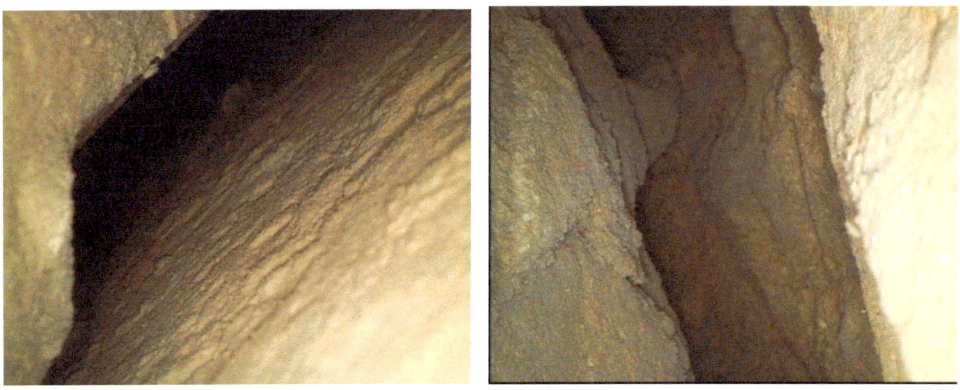

Fig. 16. Visualization of delamination in wall paintings by EverestVIT VideoProbe

It has been proven by video probe that the delamination in wall paintings around the probing holes is thick. The three holes (Fig. 17) are connected with each other in that when it is blown at one hole the dust come out from the other two holes. It is a pity that although the interior condition of the delamination can be visualized and imaged by probing test, the thickness of the delamination can not be determined without reference. Only the existence of the delamination can be ensured. The around area (Fig. 18) is selected as the site of field test for detection.

Fig. 17. Relative position of detection area to probing holes

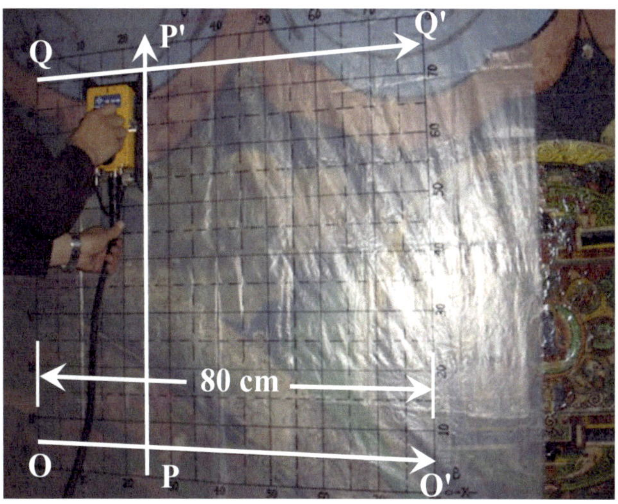

Fig. 18. In-situ detection of delamination in Mandala Hall of Sagya Monastery

The field test for detection of delamination in wall paintings was carried out on August 7, 2007. Test parameters were as follows: the sampling frequency is 60779.928037 MHz, interval of triggering distance is 0.002590 m and the depth of time window is 5.528141 *ns*. In the post-processing software of Easy 3D, as the radar profile is filtered by FIR, it is found that the delamination at the top is more serious than that at the bottom in the west wall (Fig. 19).

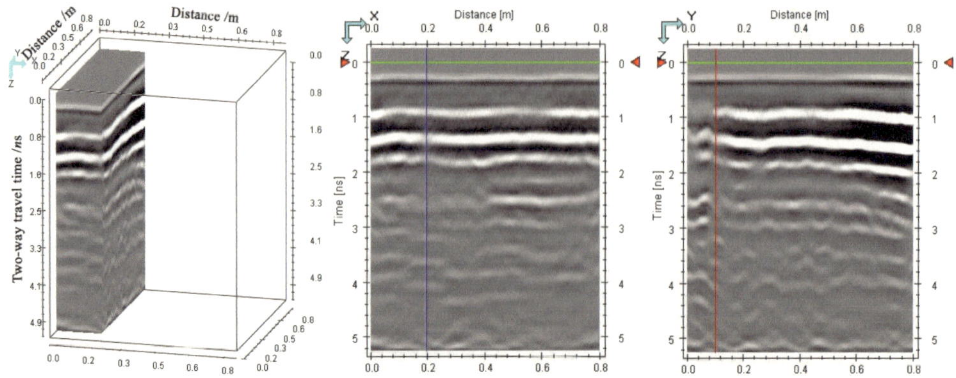

Fig. 19. Typical GPR profiles in illustration of delamination at Mandala Hall

3.2 Detection in Patala Palace

The south wall of Western Hall in Potala Palace is selected as the detection area (Fig. 20). From left to right, this area is divided into five square grids with a side length of 80 cm, so that each coverage is 0.64 m^2. The spacing distance of both transversal and vertical profiles in grid project is 10 cm (Fig. 21).

Fig. 20. Detection area in the Western Hall of Potala Palace

Fig. 21. In-situ orthogonal grid project

In-situ detection test in Potala Palace was carried out on July 10, 2006, the measurement configuration is as follows: sample number is 312, sampling frequency 28363.966797 MHz, triggering interval 0.003885 m, and time window 10.999872 *ns*. The sample number corresponding to the ground surface level is calibrated as 31.

Raw GPR profile data is digitally processed by the following filters in sequence: automatic gain control (scale: 500000, window: 21), direct current adjustment (start: 207, end: 311), delete mean trace (use entire date), finite impulse response (background: 15, lowpass: 5), and moving average (samples: 3). After the post processing as above, the wall made of stone adobe is distinctly illustrated in Fig. 22, since relatively more electromagnetic waves are reflected from the plaster-stone interface.

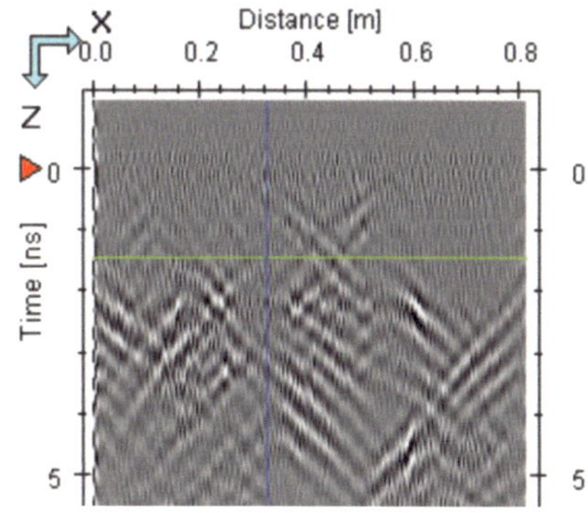

Fig. 22. Plaster-support interface illustrated in radar image

From Fig. 22, it can be seen that when the two-way travel time of electromagnetic wave is near 2 *ns*, the reflected signal is strong. Suppose that the electromagnetic wave travels at 0.1 m/*ns* in plaster, the thickness of the plaster layer is 10 cm. Besides, the signal of the wave appears unusual at the horizontal distance of 0.42 m, which may be the area of delamination.

If the data in Fig. 22 is only filtered by finite impulse response in Easy 3D, the length of the wooden beam, read as 0.8 m, can be seen clearly in the radar profile (Fig. 23). Furthermore, its width is about 0.28 m in the side view (Fig. 24).

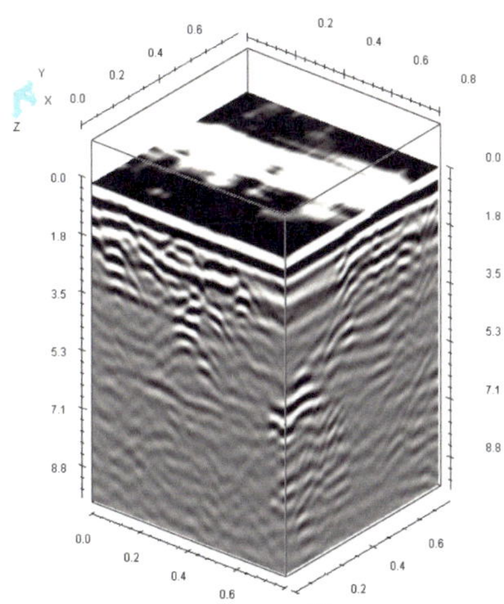

Fig. 23. 3D presentation of underneath beam

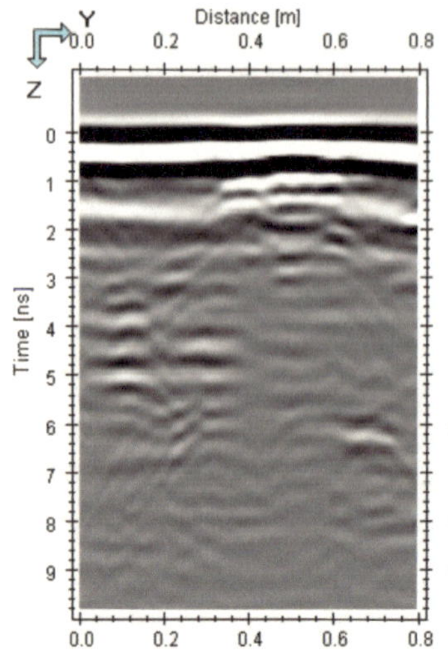

Fig. 24. Width of the wooden beam

To summarize, the deterioration of delamination within the wall paintings is located mainly at the top of the wooden beam and under the plaster layers, and it is 3.2 m in length, 0.28 m in width, and 5 cm in depth from the wall painting surface.

3.3 Detection in Lashao Temple

The detection area, 0.8 m long and 1.2 m high, in Lashao Temple of Wushan Shuilian Caves is relatively flat, on the north of the lion relief of the Buddha figure (Fig. 25). Altogether, 17 horizontal and 13 vertical profiles (Fig. 26) have been taken, and the distance between adjacent profiles is 0.1 m.

Fig. 25. Overall view of the field test in Lashao Temple

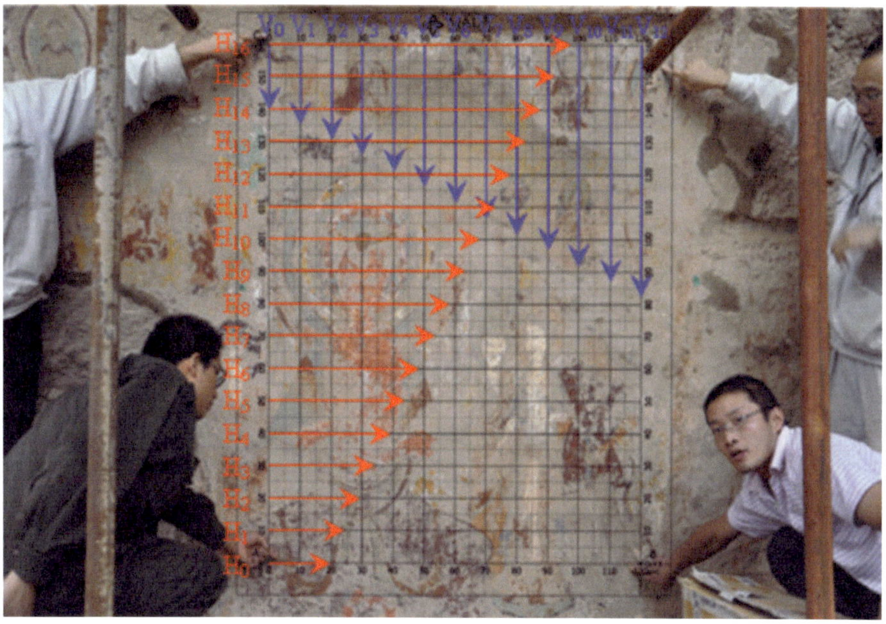

Fig. 26. In-situ detection of delamination in wall paintings in Lashao Temple

As a rule of thumb, the relative dielectric constant of mural plaster is about 1.5, so that electromagnetic wave travels in plaster layer at the speed of about 2.45×10^8 m/s (245 m/μs or 245 mm/ns). In the dialog box of parameter setting, the window of two-way travel time is input as 3 ns, which is equal to set the effective detection depth as 36.75 cm, the sampling frequency is chosen as 142 GHz, the interval of impulse triggering time is set as 0.1 s, and the function of automatic stacking is turned on. During the operation of signal processing, the following four filters are loaded: DC Removal, Substract Mean Trace, Band Pass, and Running Average.

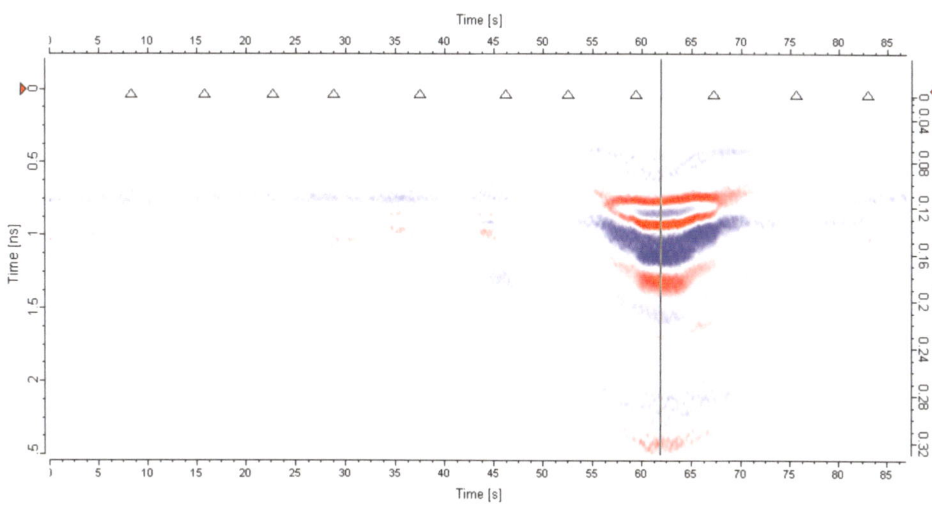

Fig. 27. Presentation of GPR profile H_3

It is shown in the interpretation results (Fig. 27, Fig. 28, Fig. 29) that the delamination in wall paintings in the detection region is mainly located at the lower left corner, which is in consistence of the area where the loss of mural plasters is serious. In addition, delamination is also serious in the lower right corner and the upper part of detection area. Taking into account that the vertical resolution of RAMAC ground penetrating radar is about 5 mm, delamination in detection area should be more serious (Fig. 30).

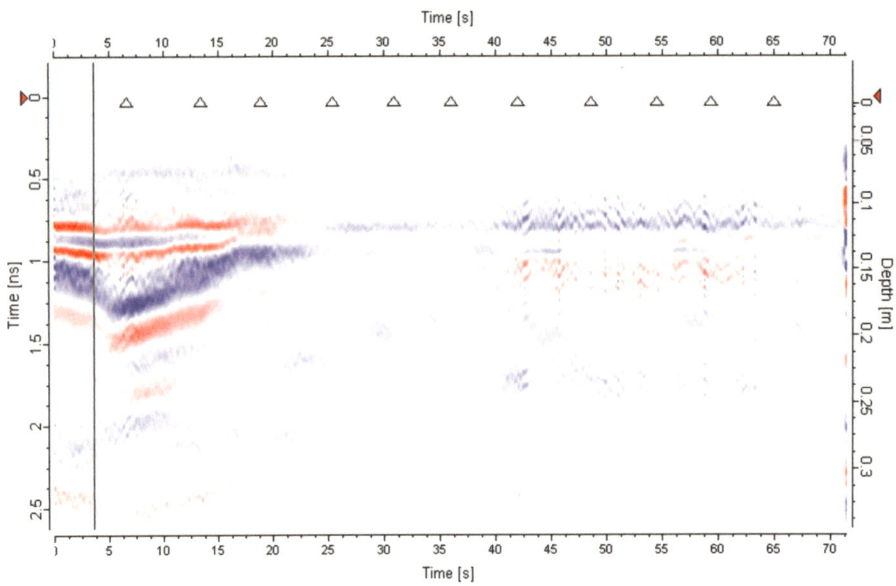

Fig. 28. Presentation of GPR profile H_5

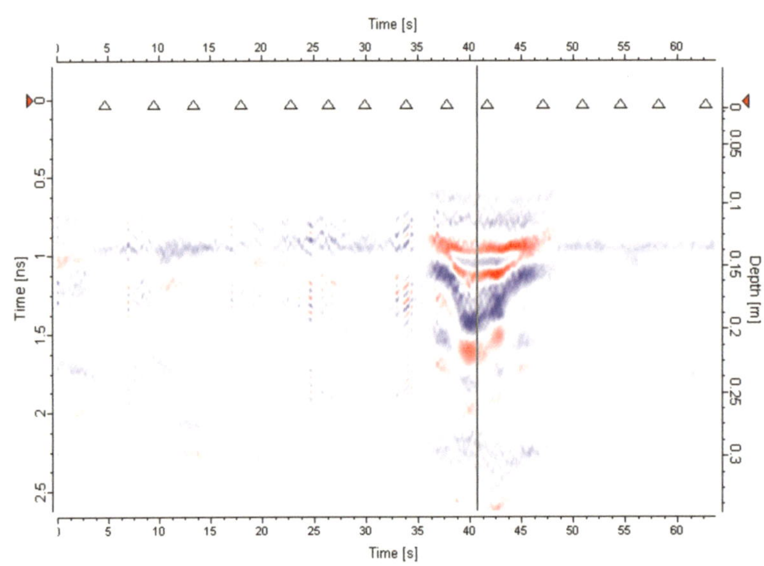

Fig. 29. Presentation of GPR profile V_1

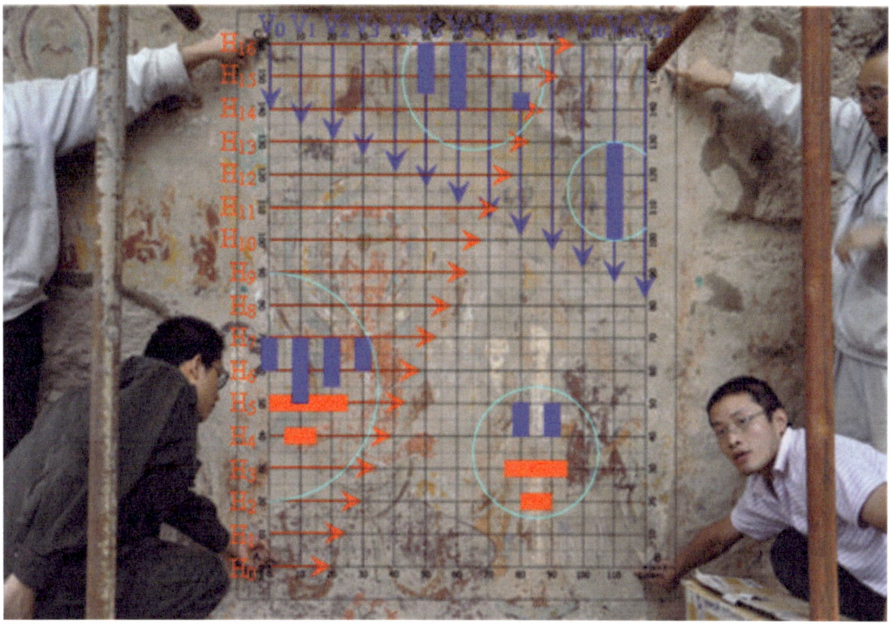

Fig. 30. Comprehensive interpretation of delamination in wall paintings, marked by arc

4. Conclusion

Focusing on the propagation of high frequency pulse electromagnetic waves in layered lossy and dispersive medium and after the physical forward modeling experiment, this chapter has successfully located delamination in polished wall paintings by wall coupling antennas using RAMAC ground penetrating radar. It is shown that the ultra-wide band ground penetrating radar is capable of detecting delamination in vertical resolution of about 5 mm when it is equipped with a transmitting antenna of 1.6 GHz central frequency.

5. Acknowledgment

This project is jointly sponsored by State Administration of Cultural Heritage, People's Republic of China in Cultural Heritage Conservation Science and Technology Foundation (No.200101).

The author wishes to thank Dr. Tao YANG of Lanzhou University and Xuebing BAI of Beijing Xing Heng Yun Science & Trade Co., Ltd. , China for their collaboration and support with the case work. The author is also indebted to Professor Yi SU of National University of Defense Technology, China and Professor Zheng'ou ZHOU of University of Electronic Science and Technology of China for their kind and helpful comments in the preparation of this chapter.

The author is grateful to Professor Zuixiong LI & Dr. Liyi ZHAO of Dunhuang Academy and all the staff of Conservation Institute and Technical & Service Center for Protection of Cultural Heritage helped to carry out the lab test and the field test, they are greatly acknowledged for their invaluable logistic support.

6. References

[1] Z. Li, W. Wang, X. Wang, J. Chen, and G. Qiang Ba, *Report on Wall Painting Conservation and Restoration Project of Potala Palace, Tibet.* Beijing, China: Cultural Relics Press, 2008. ISBN 9787501024704.

[2] W. Wang, Z. Ma, Z. Li, T. Yang, and Y. Fu, Consolidating of Detached Murals through Grouting Techniques, *Sciences of Conservation and Archaeology,* vol. 18, no. 1, pp. 52-59, Mar. 2006. ISSN 1005-1538.

[3] W. Wang, L. Zhao, T. Yang, Z. Ma, Z. Li, and Z. Fan, Preliminary Detection of Grouting Effect on Delaminated Wall Paintings in Tibet Architecture, *Chinese Journal of Rock Mechanics and Engineering,* vol. 28, supp. 2, pp. 3776-3781, Sep. 2009. ISSN 1000-6915.

[4] L. Kong, and Z. Zhou, A Effective Method of Improved Resolution for Imaging of Subsurface Ground Penetrating Radar, *Signal Processing*, vol. 18, no. 6, pp. 505-508, Jun. 2002. ISSN 1003-0530.

[5] Y. Su, C. Huang, and W. Lei, *Theory and Application of Ground Penetrating Radar.* Beijing, China: Science Press, 2006. ISBN 7030172833.

[6] Z. Li, T. Yang, and W. Wang, Forward Replica Modeling for Detection of Delamination in Wall Paintings with Ground Penetrating Radar, *Journal of Engineering Geology,* vol. 17, no. 5, pp. 675-681, Oct. 2009. ISSN 1004-9665.

[7] Z. Li, T. Yang, W. Wang, and W. Chen, Detection of Delamination in Tibetan Wall Paintings by Using Ground Penetrating Radar, *Journal of University of Electronic Science and Technology of China,* vol. 39, no. 6, pp. 865-870, Dec. 2010. ISSN 1008-8105.

8

Interaction of Electromagnetic Radiation with Substance

Andrey N. Volobuev
Samara State University
Russia

1. Introduction

1.1 Distribution of Electromagnetic Field Momentum in dielectrics in stipulation of self-induced transparency

Interaction of quantums of electromagnetic radiation with substance can be investigated both from a wave position, and from a quantum position. From a wave position under action of an electromagnetic wave there are compelled fluctuations of an electronic orbit and nucleus of atoms. The energy of electromagnetic radiation going on oscillation of nucleus passes in heat. Energy of fluctuations of an electronic orbit causes repeated electromagnetic radiation with energy, smaller, than initial radiation.

From a quantum position character of interaction is more various. Interaction without absorption of quantums is possible: resonant absorption, coherent dispersion. The part of quantums is completely absorbed. Quantums can be absorbed without occurrence secondary electrons. Thus all energy of quantums is transferred fonons - to mechanical waves in a crystal lattice, and the impulse is transferred all crystal lattice of substance. At absorption of quantums can arise secondary electrons, for example, at an internal photoeffect. Absorption of quantums with radiation of secondary quantums of smaller energy and frequency is possible, for example, at effect of Compton or at combinational dispersion.

All these processes define as formation of impulses of electromagnetic radiation in substance, and absorption of radiation by substance.

1.2 Coordination of the electromagnetic impulse with the substance

Firstly, consider the one-dimensional task the electric part of electromagnetic field momentum with the dielectric substance, which posses a certain numerical concentration n of centrosymmetrical atoms - oscillators. For the certainty of the analysis we suggest the atom to be one-electronic. It is also agreed, that no micro current or free charge are present in the medium. The peculiarities of interaction between magnetic aspect of momentum and the atoms will be considered later.

We accept that there takes place the interaction of quantum of electromagnetic radiation with nuclear electrons, thus quantum are absorbed by the electrons. By gaining the energy of quantum the electrons shift to the advanced power levels. Further, by means of resonate shift of electrons back, appears the quantum radiation forward. The considered medium lacks non-radiating shift of electrons, i.d. the power of quantum is not transfered to the atom.

Thus, the absorption of electromagnetic radiation in the case of its power dissipation in the substance, owing to SIT, is disregarded. There appears the atomic sypraradiation of quantum. Thus, the forefront of momentum passes the power on to the atomic electrons of the medium, forming its back front.
The probabilities of quantum's absorption and radiation by the electrons in the unity of time, with a large quantity of quantum in the impulse, according to Einstein, can be referred to as the approximately identical [6]. For the separate interaction of the with the electron this very probability is the same and is proportional to the cube of the fine-structure constant ~ $(1/137)^3$ [7]. Consider a random quantity - the number of interactions of quantum with atomic electrons in the momentum. In accordance with the Poisson law of distribution, the probability of that will not be swallowed up any quantum atomic's electrons (will not take place any interaction), at rather low probability of separate interaction, is equal an exponent from the mathematical expectation of a random variable – an average quantity of interactions λ of quantums and electrons in impulse, taken with the minus $p = \exp(-\lambda)$. Therefore, as it will be explained further, it is possible that the intensity of non-absorbed power of impulse by the atomic electrons of the medium in it forefront is determined by the exponential Bouguer law [3] (in German tradition - Beer law)

$$I = I_0 \exp(-\alpha l), \tag{1.1}$$

where α - index of electromagnetic wave and substance interaction, l – length of interaction layer, I_0 - intensity of incident wave. Thus, the intensity of atomic electron's power recoil into impulse on its back front could be described with the help of the Bouguer law with the negative index of absorption [8].
The index of interaction is $\alpha = \sigma n$, where α – effective section of atom-oscillator interaction with the wave. Hence,

$$\alpha l = \sigma n l = n V_{eff} = n V \frac{V_{eff}}{V} = M \frac{V_{eff}}{V} = MN, \tag{1.2}$$

where V_{eff} - the effective volume of interaction. In defying (1.2) the right part of the formula is multiplied and divided by the geometric volume V, in which there is M of particles interacting with the radiation. The ratio $\frac{V_{eff}}{V} = N$. The ratio of effective volume of interaction to the geometric volume characterizes the medium possibility of electromagnetic radiation's interaction with the atom. Hence, by exponential function in the Bouguer law (1.1) the mathematical expectation of random variable is supposed, which subdues to the Poisson law distribution - average variable of atoms interacting with the electromagnetic radiation in the area of impulse influence $\lambda = NM$.

Taking into account that the wave intensity is $I \sim \begin{pmatrix} E^2 \\ H^2 \end{pmatrix}$ we shall have

$$\begin{pmatrix} |E| \\ |H| \end{pmatrix} = \begin{pmatrix} |E_0| \\ |H_0| \end{pmatrix} \exp\left(-\frac{\alpha}{2} l\right), \tag{1.3}$$

where $|E_0|$, $|H_0|$ – the amplitudes of electric and magnetic fields' strength of the impulse on longitudinal coordinate $X = 0$.

In the formula (1.3) and further the upper variables in parentheses are referred to electric field, and lower – to the magnetic field of impulse.

By the ratio (1.2) it is possible to find

$$N = -\frac{2}{M}\ln\left|\frac{E}{E_0}\right| = -\frac{2}{M}\ln\left|\frac{H}{H_0}\right|. \qquad (1.4)$$

The formula (1.4) demands some further consideration. If $E<E_0$, that reflects the process of wave absorption by atomic electrons $N>0$ and classical consideration of electromagnetic wave interaction with the atom is quite admissible. The case when $E>E_0$ reflects the process of wave over-radiation. Thus, $N<0$ and variable N can not be considered as the probability of electromagnetic wave interaction with the atom. In this case we speak about the quantum-mechanical character of the process of interaction between the quantum and the bi-level power system of the atom, provided that the power transition's radiation is reversed. Variable N in this case possess the notion of united average of filling by atom ($-1<N<1$). Due to the use of the average of filling to raise the atom and bend of its magnetic moment in the magnetic field of the impulse, the existence of bi-level quantum system by magnetic quantum numbers. Thus, the variable N provides with the measure of inversion of the system of atom-radiators by the raised atoms [2] as well as the measure of inversion of the magnetic moment of the atom's system by magnetic quantum numbers. If $N=-1$ all the atoms occur in the basic condition [3].

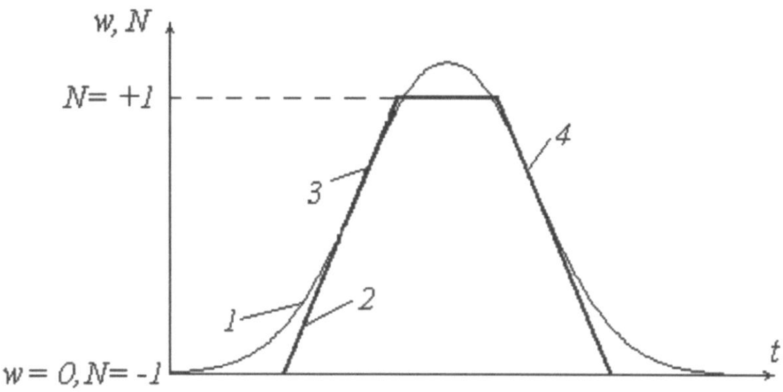

Fig. 1. Dependence of volumetric density of energy of electromagnetic radiation impulse w (curve 1) and average on atoms of number of filling N (curve 2) from time; 3 and 4 - points of an excess of function $w(t)$

We consider the dependence of the average of filling on the time $N(t)$. If to accept the proportion of polarization of separate bi-level atom to the intensity of electric field in the impulse, then, in accordance with the Maxwell-Bloch equations, the average by atoms of considered volume, the filling number is proportional to the volumetric density of electromagnetic wave power $N \sim w$ [3]. However such a monotonous dependence between these variables can not remain on the whole extent of the impulse. Firstly, by the high volumetric density of impulse power *w*, typical of SIT, when the central part of impulse power is higher than any variable *w*, there exists energetic saturation of the medium. The average filling number thus *N=1*, all the atoms are raised, fig. 1 (curve 1 - the dependence *w* of time, thicker curve 2 – the considered dependence *N* of time). The violation of proportion *N~w* in the central part of impulse is the basic drawback of frequently used system of Maxwell-Bloch equations for the SIT description.

Secondly, the period of variable *N* relaxation is not less than 1 *ns* [2] that is why the dependence *N(t)* can not repeat high-frequently oscillations on both fronts of the impulse. The dependence *N~w* could characterize the proportion of average filling number and envelope *w* (curve 1) in the impulse. However, in two points of the fold (3 and 4 fig. 1) on the sites of increase and decrease if the envelope *w* the variable $\partial^2 w / \partial t^2 = 0$ hence, also $\partial^2 N / \partial t^2 = 0$. Besides, the dependence *N(t)* has the symmetrical character as at the SIT impulse becomes the conservative system (there is no reverse dispersion and dissipation of power) [2]. Therefore, it could be thoroughly concerned that on the whole extent of impulse, except the points of curve's *N(t)* fold, the condition remains

$$\frac{\partial^2 N}{\partial t^2} = 0 , \tag{1.5}$$

while the dependence *N(t)* has the character as shown on the fig. 1, curve 2. it could be also highlighted the high generality of formula (1.5), which is possible for any piecewise linear function *N(t)*. Thus, the points of function break are excluded, as the derivates undergo the break.

1.3 Non-linear Schrödinger equation

One-dimensional wave equation for electric and magnetic aspects of electromagnetic field for the considered problem is [2]

$$\frac{\partial^2}{\partial X^2}\begin{pmatrix} E \\ H \end{pmatrix} - \frac{1}{c^2}\begin{pmatrix} \mu \\ \varepsilon \end{pmatrix}\frac{\partial^2}{\partial t^2}\begin{pmatrix} E \\ H \end{pmatrix} = \frac{1}{c^2}\begin{pmatrix} \mu / \varepsilon_0 \\ \varepsilon \end{pmatrix}\frac{\partial^2}{\partial t^2}\begin{pmatrix} P \\ J \end{pmatrix}, \tag{1.6}$$

where $E \equiv E_Y$ or $E \equiv E_Z$, $H \equiv H_Y$ or $H \equiv H_Z$, X and t - accordingly the coordinate alongside of which the impulse and the time are distributed, P – polarization of substance, J - its magnetization, ε_0 and μ_0 – electrical and magnetic constant, ε – relative static permittivity of substance, μ - relative magnetic permittivity, $c = 1/\sqrt{\varepsilon_0 \mu_0}$ - speed of light in vacuum.

We introduce the transformation of electric field intensity be formula

$$\begin{pmatrix} E(X,t) \\ H(X,t) \end{pmatrix} = \Phi(X,t)\exp(-i\omega_0 t) . \tag{1.7}$$

The function $\Phi(X, t)$ is less rapidly changing one in time then *E(X,t)* or *H(X,t)*, ω_0 - aspect of cyclic frequency of high-frequent oscillations of the field.

By substituting (1.7) and (1.6) we get

$$\left[\frac{\partial^2 \Phi}{\partial X^2} - \frac{1}{c^2}\begin{pmatrix} \mu \\ \varepsilon \end{pmatrix}\left(\frac{\partial^2 \Phi}{\partial t^2} - 2i\omega_0 \frac{\partial \Phi}{\partial t} - \omega_0^2 \Phi\right)\right]\exp(-i\omega_0 t) = \frac{1}{c^2}\begin{pmatrix} \mu / \varepsilon_0 \\ \varepsilon \end{pmatrix}\frac{\partial^2}{\partial t^2}\begin{pmatrix} P \\ J \end{pmatrix} . \tag{1.8}$$

We estimate the relative variable of first and second items in the parenthesis of the left side (1.8). for this purpose we would introduce the scales of variables time t and Φ

$$T = t / t^*, \quad \Phi_0 = \Phi / \Phi^*,$$

where the asterisk designates dimensionless parameters. For the time scale the duration (period) of impulse T should be logically chosen. The scale Φ_0 is chosen from a condition that dimensionless second derivative $\frac{\partial^2 \Phi^*}{\partial t^{*2}}$ and the dimensionless function Φ^* are in the same order. Hence, the the first item in round brackets (1.8) is $\frac{\Phi_0}{T^2}\frac{\partial^2 \Phi^*}{\partial t^{*2}}$, and the last one $\omega_0^2 \Phi_0 \Phi^*$. Instead of impulse T period we introduce cyclic frequency of impulse $\omega = \frac{2\pi}{T}$. By comparing these items, it is realized, that $\frac{\Phi_0 \omega^2}{4\pi^2}\frac{\partial^2 \Phi^*}{\partial t^{*2}} << \omega_0^2 \Phi_0 \Phi^*$ as the cyclic frequency of impulse is far less than infrequences of field's oscillations, especially when $\omega^2 << \omega_0^2$. Similarly, it can be presented that the second item in the round brackets (1.8) is far more that the first one.
Hence, by disregarding the small item in (1.8), we observe

$$\left[\frac{\partial^2 \Phi}{\partial X^2} - \frac{1}{c^2}\left(\frac{\mu}{\varepsilon}\right)\left(2i\omega_0 \frac{\partial \Phi}{\partial t} + \omega_0^2 \Phi\right)\right]\exp(-i\omega_0 t) = \frac{1}{c^2}\left(\frac{\mu / \varepsilon_0}{\varepsilon}\right)\frac{\partial^2}{\partial t^2}\begin{pmatrix} P \\ J \end{pmatrix}. \tag{1.9}$$

By accepting vector of polarization P or magnetizing J to be directly proportional, accordingly, to the electric and magnetic fields strength, we could derive the wave equation from (1.6), which is possible to any form of the wave. However, there exists a physical mechanism, which restrict the wave form. This mechanism is connected with the way of over-radiating of electromagnetic impulse with the atomic electrons. This process is precisely considered further.
We consider the strength of electric and magnetic fields of impulse as

$$\begin{pmatrix} E(X,t) \\ H(X,t) \end{pmatrix} = \begin{pmatrix} |E(X,t)| \\ |H(X,t)| \end{pmatrix} \exp[i(rX - \delta t)], \tag{1.10}$$

where r and δ - are constants, $|E(X,t)|$ and $|H(X,t)|$ are the modules of functions $E(X,t)$ and $H(X,t)$.
Formulas (1.4) and (1.5) reflect the offered physical model of electric and magnetic field of impulse interaction with atoms in SIT.
Hence, taking into account (1.4) and (1.5) there is

$$\frac{\partial^2 \ln\left|\frac{E}{E_0}\right|}{\partial t^2} = \frac{\partial^2 \ln\left|\frac{H}{H_0}\right|}{\partial t^2} = 0. \tag{1.11}$$

By transforming (1.11) we have

$$\frac{\partial^2 |E|}{\partial t^2} = \left(\frac{\partial \ln|E|}{\partial t}\right)^2 |E|. \tag{1.12}$$

The similar ratio can be also referred to the function $|H|$. These ratios should not be regarded as the equations to define the module of electric and magnetic aspect of impulse. It is the approximate expression of the second derivative $\frac{\partial^2 |E|}{\partial t^2}$ or $\frac{\partial^2 |H|}{\partial t^2}$ for the considered physical model and reflects several non-linear effects of interaction between electromagnetic radiation and substance. The approximate ratio (1.12) defines the connection of medium polarization P with the strength of impulse electric field (similarly to the magnetization J with the magnetic field strength), that would be considered further. The electromagnetic field impulse strengths should be estimated from the equation (1.6) taking into account the ratio (1.12).

In accordance with (1.10), $\begin{pmatrix} |E(X,t)| \\ |H(X,t)| \end{pmatrix} = \begin{pmatrix} E(X,t) \\ H(X,t) \end{pmatrix} \exp[-i(rX - \delta t)]$, hence, from (1.12) we estimate equation for the electromagnetic field impulse

$$\frac{\partial^2 E}{\partial t^2} = -2i\delta \frac{\partial E}{\partial t} + \left[\delta^2 + \left(\frac{\partial \ln \left| \frac{E}{E_0} \right|}{\partial t} \right)^2 \right] E . \tag{1.13}$$

The same ratio exists for the magnetic field also. Passing over to (1.13) to the function $\Phi(X,t)$ by formula (1.7) and by concerning $P = \varepsilon_0 \chi E$, where χ – relative dielectric permittivity of substance, we have

$$\frac{\partial^2 P}{\partial t^2} = \left[-2i\delta\varepsilon_0\chi \frac{\partial \Phi}{\partial t} - 2\delta\varepsilon_0\chi\omega_0\Phi + \varepsilon_0\chi \left[\delta^2 + \left(\frac{\partial \ln \left| \frac{\Phi}{\Phi_0} \right|}{\partial t} \right)^2 \right] \Phi \right] \exp(-i\omega_0 t) , \tag{1.14}$$

For the variable $\frac{\partial^2 J}{\partial t^2}$ by using $J = \chi H$, where χ – relative magnetic permittivity of substance, we get the ratio, similar to (1.14), except that the right part lacks ε_0.

The variables $\begin{pmatrix} E_0 \\ H_0 \end{pmatrix} = \Phi_0 \exp(-i\omega_0 t)$. By comparing (1.7) and (1.10) we state $\begin{pmatrix} |E| \\ |H| \end{pmatrix} = |\Phi|$, $\begin{pmatrix} |E_0| \\ |H_0| \end{pmatrix} = |\Phi_0| = const$.

By substituting (1.14) into (1.9)

$$2i(\omega_0 + \chi\delta)\frac{\partial \Phi}{\partial t} + c^2 \begin{pmatrix} 1/\mu \\ 1/\varepsilon \end{pmatrix} \frac{\partial^2 \Phi}{\partial X^2} + \left(\omega_0^2 + 2\chi\delta\omega_0 - \chi\delta^2\right)\Phi = \chi \left(\frac{\partial \ln \left| \frac{\Phi}{\Phi_0} \right|}{\partial t} \right)^2 \Phi . \tag{1.15}$$

In the equation (1.15) the variable χ is meaningful to dielectric permittivity for electric and magnetic permittivity for the magnetic aspects of electromagnetic field.
The non-linear Schrödinger equation with complicated type of linearity is received. We introduce the signs: $\alpha = \omega_0 + \chi\delta$, $\gamma = \omega_0^2 + 2\chi\delta\omega_0 - \chi\delta^2 = \alpha^2 - \begin{pmatrix}\varepsilon \\ \mu\end{pmatrix}\chi\delta^2$, where $\begin{pmatrix}\varepsilon \\ \mu\end{pmatrix} = 1 + \chi$ - relative permittivities of the substance. Hence, the equation (1.15) will be

$$2i\alpha\frac{\partial\Phi}{\partial t} + c^2\begin{pmatrix}1/\mu \\ 1/\varepsilon\end{pmatrix}\frac{\partial^2\Phi}{\partial X^2} + \gamma\Phi = \chi\left(\frac{\partial \ln\left|\frac{\Phi}{\Phi_0}\right|}{\partial t}\right)^2\Phi. \tag{1.16}$$

We shall find the solution to the non-linear Schrödinger equation (1.16) as in [9]

$$\Phi = \Phi_0 f(kX - \omega t)\exp\left[i\left(rX - \delta^* t\right)\right], \tag{1.17}$$

where the type of the function $f(kX - \omega t)$ is still unknown. The variables k, ω and δ^* - constants. By marking $\zeta = kX - \omega t$, and substituting (1.17) in (1.16) and concerning $|\Phi| = \Phi_0 f(\zeta)$ we get

$$c^2\begin{pmatrix}1/\mu \\ 1/\varepsilon\end{pmatrix}k^2\frac{d^2 f}{d\zeta^2} + 2i\frac{df}{d\zeta}\left(krc^2\begin{pmatrix}1/\mu \\ 1/\varepsilon\end{pmatrix} - \alpha\omega\right) + f\left(\gamma + 2\alpha\delta^* - r^2c^2\begin{pmatrix}1/\mu \\ 1/\varepsilon\end{pmatrix}\right) = \chi\omega^2\left(\frac{d\ln f}{d\zeta}\right)^2 f \tag{1.18}$$

If to permit that $krc^2 = \alpha\begin{pmatrix}\mu \\ \varepsilon\end{pmatrix}\omega$ as there should not be any imaginary items in (1.18), this equation is transformed to

$$c^2\begin{pmatrix}1/\mu \\ 1/\varepsilon\end{pmatrix}k^2\frac{d^2 f}{d\zeta^2} + f\left(\gamma + 2\alpha\delta^* - r^2c^2\begin{pmatrix}1/\mu \\ 1/\varepsilon\end{pmatrix}\right) = \chi\omega^2\left(\frac{d\ln f}{d\zeta}\right)^2 f. \tag{1.19}$$

We consider the solution of the equation (1.19) by

$$f = C_1\exp\left[\frac{C_2\zeta^2}{4}\right], \tag{1.20}$$

where C_1 and C_2 - constants. By substituting (1.20) into (1.19) we get that the constant C_1 could be the arbitrary variable, $\chi\begin{pmatrix}\mu \\ \varepsilon\end{pmatrix}\omega^2 = k^2c^2$.

The constant C_2 could not depend upon the parameters of equation. It is accepted that C_2=-1. Then the frequency and the wave number in (1.17), accordingly, are

$$\delta^* = \frac{c^2\begin{pmatrix}1/\mu \\ 1/\varepsilon\end{pmatrix}\left(r^2 + k^2/2\right) - \gamma}{2\alpha}; \qquad r = \frac{\alpha\begin{pmatrix}\mu \\ \varepsilon\end{pmatrix}\omega}{kc^2}. \tag{1.21}$$

The formulas (1.21) associate the frequency and the wave number of oscillations of function *Φ(X,t)* with the parameters of substance and electromagnetic field impulse.
The most simple ratios between the parameters are gained, when $\delta = \omega_0$. In this case $\alpha = \begin{pmatrix}\varepsilon\\ \mu\end{pmatrix}\delta$, $\gamma = \begin{pmatrix}\varepsilon\\ \mu\end{pmatrix}\delta^2$. From the equations in (1.21), and concerning $\chi\begin{pmatrix}\mu\\ \varepsilon\end{pmatrix}\omega^2 = k^2c^2$ there is

$$r = \frac{\alpha\begin{pmatrix}\mu\\ \varepsilon\end{pmatrix}\omega}{kc^2} = \frac{\alpha k}{\chi\omega} = \frac{\begin{pmatrix}\varepsilon\\ \mu\end{pmatrix}\delta k}{\chi\omega}, \quad \delta^* = \frac{\alpha}{2\chi} + \frac{\chi\omega^2}{4\alpha} - \frac{\gamma}{2\alpha} = \frac{2\begin{pmatrix}\varepsilon\\ \mu\end{pmatrix}\delta^2 + \chi^2\omega^2}{4\chi\begin{pmatrix}\varepsilon\\ \mu\end{pmatrix}\delta}. \tag{1.22}$$

By concerning that $2\begin{pmatrix}\varepsilon\\ \mu\end{pmatrix}\delta^2 >> \chi^2\omega^2$, we have $\delta^* \approx \frac{\delta}{2\chi}$. This inequality is true, as for the rarefied gas *(n <* 10^{18} *atoms/cm*^{3}*)* $\begin{pmatrix}\varepsilon\\ \mu\end{pmatrix} >> \chi$ and the frequency of wave filling of impulse δ is far more than frequency of impulse envelope ω.

Taking into account (1.10), (1.20) and the $\begin{pmatrix}|E|\\ |H|\end{pmatrix} = |\Phi|$, we can find the laws of electromagnetic field strengths shifting by

$$\begin{pmatrix}E\\ H\end{pmatrix} = \begin{pmatrix}E_0\\ H_0\end{pmatrix}\exp\left[-\frac{(kX-\omega t)^2}{4}\right]\exp\left[i(rX-\delta t)\right]. \tag{1.23}$$

It should be stressed, that though, the ratios for the electric aspect of impulse in [1] and (1.23) are similar to each other and feature the same phases of oscillations, that is possible on some distance from the over-radiating atom, the non-linear Schrödinger equations are differ in type of non-linearity. The reason of this lies in the fact that in [1] the impulse was considered with regard to low intensity, the one that does not lead to the energetic saturation of medium, in which it is disseminated.

For the estimation, like in [1] we have $k = \frac{\omega}{c} = 2{,}1\cdot 10^4 m^{-1}$, $r = \frac{\delta}{c} = 2{,}1\cdot 10^5 m^{-1}$, $\omega = 6{,}28\cdot 10^{12}$ s^{-1}, $\delta = 6{,}28\cdot 10^{13}$ s^{-1}.

For instance, the result of strength estimation of the electric filed impulse by the coordinate *X*, calculated with the MathCAD system by formula (1.23), is shown in fig.2.

Taking in to account the reciprocal orthogonality of planes of vectors' envelopes of electric and magnetic fields impulse, we could gain the type of electromagnetic soliton, fig. 3.

Figure 4 shows the envelopes of electric field impulse in the SIT, based on formula (1.23), curve 1, and by formula (1.24), being the consequence of Maxwell-Bloch theory, curve 2. the impulse envelope of electric field strength in this theory is expressed as the first derivative of the *Sin*-Gordon equation solving and is

$$E = \frac{E_0}{\mathrm{ch}(kX - \omega t)}. \tag{1.24}$$

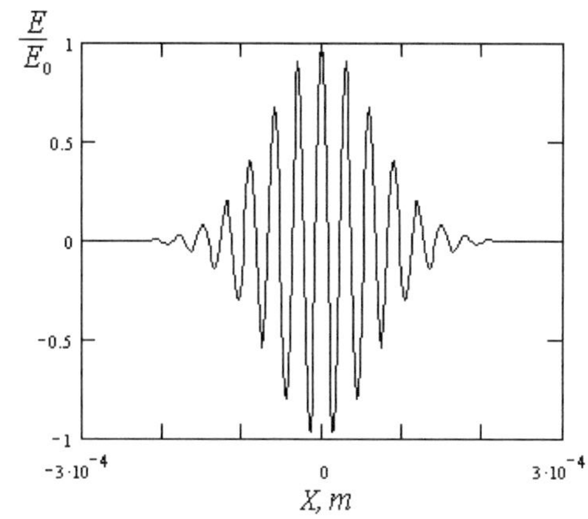

Fig. 2. Calculation of the electric component of electromagnetic radiation impulse in dielectric

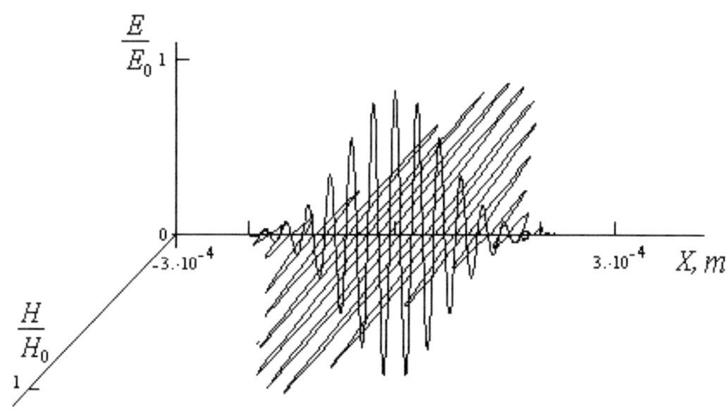

Fig. 3. Intensity of electric and magnetic fields electromagnetic solitone in dielectric in conditions of the self-induced transparency

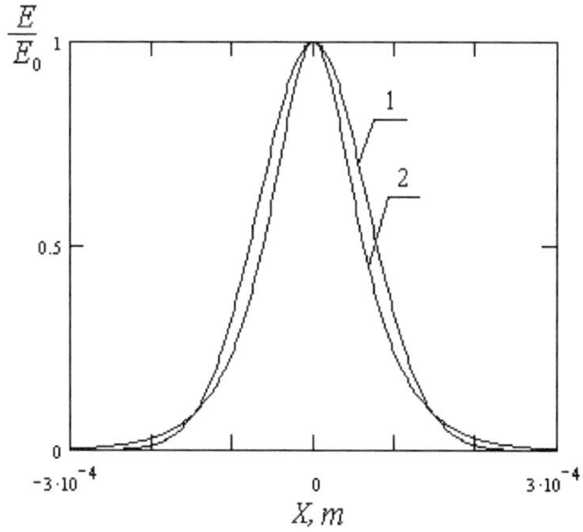

Fig. 4. Comparison bending around of the electromagnetic field impulse, received on the basis of the offered theory, a curve 1, and the equations the Maxwell - Bloch, curve 2

Evidently, the first derivative of *Sin*-Gordon equation solving is similar to the soliton envelope in the non-linear Schrödinger equation with cube non-linearity solving (27). Curves 1and 2 in fig. 4 are designed for the same parameters as the function in fig. 2. We can infer from fig. 4 that impulse, referred to formula (1.23), curve 1, is broader in its central part, but asymptotically shorter than impulse, inferred by the Maxwell-Bloch theory, curve 2. Evidently, its is bound with the energetic permittivity of medium in the central part of impulse.

2. Angular distribution of photoelectrons during irradiation of metal surface by electromagnetic waves

There is the problem of achieving of the maximum photoelectric flow during irradiation of the metal by flow of electromagnetic waves while designing of photoelectrons. The depth of radiation penetration into metal during irradiation of its surface is defined by the Bouguer low [10]:

$$I = I_0 \exp\left(-\frac{4\pi}{\lambda} n\chi z\right),$$

where I_0 – is the intensity of the incident wave, I – is the intensity on z-coordinate, directioned depthward the metal, λ – is the wavelength of radiation, $n\chi$ – is the product of refractive index by extinction coefficient.
Let's estimate the thickness of the metal at which intensity of light decreases in e = 2,718 times:

$$z = \frac{\lambda}{4\pi n\chi}$$

Average wavelength of a visible light for gold λ=550 nm, $n\chi = 2{,}83$, therefore z = 15,5 nm. Considering [11] that lattice constant for gold a = 0,408 nm, it is possible to deduce that electromagnetic radiation penetrates into the metal on 40 atomic layers.
Therefore radiation interaction occurs basically of the top layers of atoms and angular distribution of electron escape from separate atoms, i.e. during the inner photoemissive effect, it will appreciably have an impact on distribution of electron escape from the metal surface.
As a result it is interesting to consider angular distribution of photoelectrons during the inner photoemissive effect.

2.1 Nonrelativistic case

Although Einstein has explained the photoeffect nature in the early 20th century, various aspects of this phenomenon draw attention, till nowadays for example, the role of tunnel effect is investigated during the photoeffect [12].
In the description of angular distribution of the photoelectrons which are beaten out by photons from atoms, there are also considerable disagreements. For example it is possible to deduce that the departure of photoelectrons forward of movement of the photon and back in approach of the main order during the unitary photoeffect is absent, using the computational method of Feynman diagrams [13]. I is marked that photoelectrons don't take off in the direction of distribution of quantum [14]. This conclusion is made on the basis of positions which in the simplified variant are represented by the following.

The momentum of the taken off electron is defined basically by action produced by the electric vector of quantum of light on electron. If electron takes off in the direction of an electric vector of quantum it gets the momentum. On a plane set at an angle to a plane of polarization of quantum of light, (fig. 5) electron momentum value will be $p_{1m} = p_e \cos\phi$.

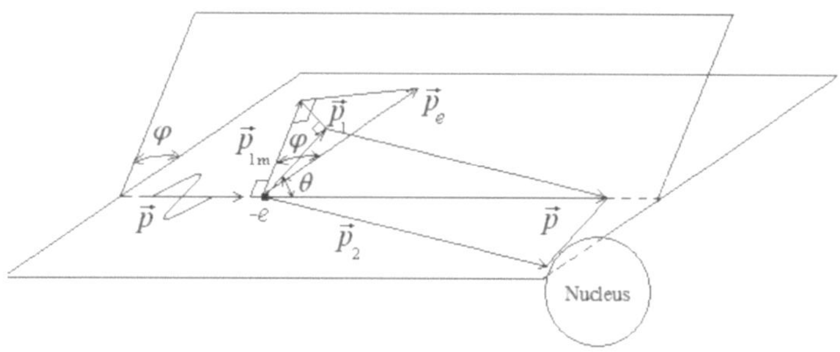

Fig. 5. Direction of vectors of particle momentum at the inner photoemissive effect

Besides, if the electron momentum is set at an angle θ to the direction of quantum of light its value will be:

$$p_1 = p_e \cos\phi \sin\theta \tag{2.1}$$

Therefore, photoelectron energy is equal to:

$$E_1 = \frac{p_1^2}{2m_1} = \frac{p_e^2 \cos^2\phi \sin^2\theta}{2m_1}, \tag{2.2}$$

where m_1 – is the electronic mass.
If $\theta = 0$ then photoelectron energy $E_1 = 0$. Photoelectrons take off readies its maximum in the direction of a light vector or a polarization vector, i.e. an electric field vector of quantum of light. The same dependence is offered in the work [7]. The formula (2.2) has the simplified nature in comparison with [7, 14], but convey correctly the basic dependence of distribution energy of a photoelectrons escape from the corners φ and θ.

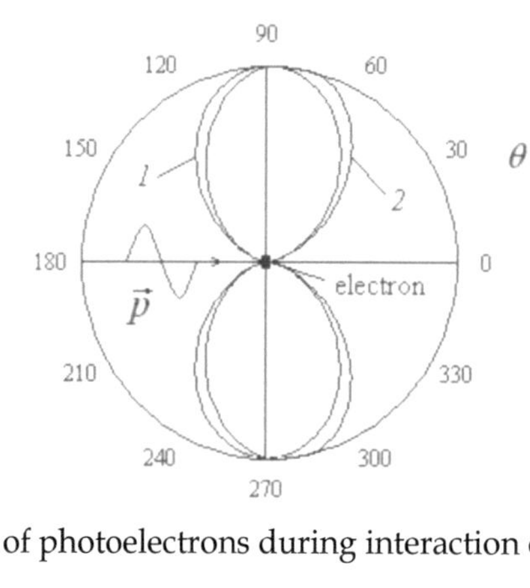

Fig. 6. Angular distribution of photoelectrons during interaction of orbital electron with the electromagnetic wave

The lack of dependence (2.2) is that at its conclusion the law of conservation of momentum, wasn't used and therefore there is no electron movement to the direction $\theta = 0$. Usage of the momentum conservation equation in [7, 14] can't be considered satisfactory since in the analysis made by the authors it has an auxiliary character. At the heart of the analysis [7, 14] is the passage of electron from a discrete energy spectrum to a condition of a continuous spectrum under the influence of harmonious indignation, i.e. the matrix element of the perturbation operator is harmonious function of time. In other words, the emphasis is on the wave nature of the quantum cooperating with electron. Angular distribution of electron energy in the relative units, made according the formula (2.2) is shown on fig. 6, a curve 1. Let's illustrate the correction to the formula (2.2) connected with presence of photon momentum, following [15].

Fig.7 demonstrates change of photoelectron momentum in the presence of a photon momentum. The conclusion made on the basis of the is $\theta = \theta' + \delta$. Let's find $\sin\theta = \sin\theta' \cos\delta + \sin\delta \cos\theta'$. Considering that δ is too small we find $\sin\theta = \sin\theta' \left(1 + \frac{\sin\delta \cos\theta'}{\sin\theta'}\right) = \sin\theta' \left(1 + \frac{p}{p_e}\cos\theta'\right)$. The law of sines for a triangle on fig. 7 is used.

Further consideration $\beta' = \frac{p}{p_e} = \frac{h\nu}{mVc} = \frac{1}{2}\beta + \frac{W}{mVc}$, where $\beta = \frac{V}{c}$ – is the relation of photoelectron speed to a speed of light in vacuum, W – is the work function of electrons from atom, we have $\sin\theta = \sin\theta'\left(1 + \beta' \cos\theta'\right)$. Taking for granted that β' is small we will transform (2.2) into $E_1 = \frac{p_e^2 \cos^2\phi \sin^2\theta'}{2m_1}\left(1 + 2\beta' \cos\theta'\right)$. Angular distribution of electron energy for $\beta' = 0{,}15$, made according to the (2.2) taking into account the correction is shown on fig. 7, a curve 2.

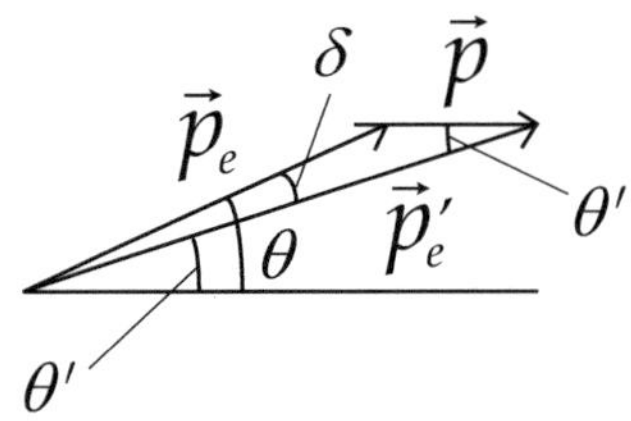

Fig. 7. The account of the momentum of quantum $\vec{p}$ using wave approach of electron interaction with the electromagnetic wave

Thus scattering indicatrix of photoelectrons has received some slope forward, but to the direction of quantum momentum, i.e. at $\theta = 0$ electrons don't take off as before.

The formula (2.2) is accounted as a basis of the wave nature of light. For the proof of this position we will consider interaction of an electromagnetic wave with orbital electron. The description of orbital movement electron is done on the basis of Bohr semiclassical theory since interacting process of electron with an electromagnetic wave is investigated from the positions of classical physics, fig. 8.

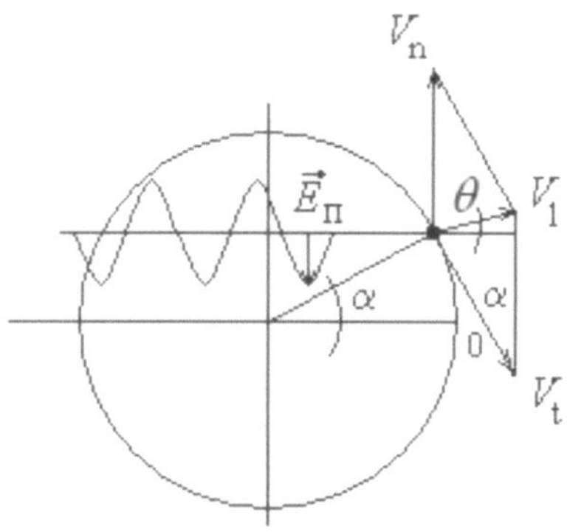

Fig. 8. Attitude of components velocity of orbital electron during its interaction with the electromagnetic wave

By the sine law from a triangle of speeds we find:

$$\sin\alpha = \frac{V_1}{V_t}\cos\theta \,, \tag{2.3}$$

where V_t - is the speed of electron movement round the nucleus, V_1 – is the total speed of electron considering the influence on it of an electromagnetic wave.
By the law of cosines we have:

$$V_1^2 = V_n^2 + V_t^2 - 2V_nV_t\cos\alpha = V_n^2 + V_t^2 - 2V_nV_t\sqrt{1-\left(\frac{V_1}{V_t}\cos\theta\right)^2} \,, \tag{2.4}$$

where V_n - is the component of the general speed of electron movement after its detachment from a nucleus which arises under the influence of dielectric field intensity $\vec{E}$ in the electromagnetic wave.
Solving (2.4) rather V_1, we find:

$$\left(V_1^2\right) = \left(V_n^2 + V_t^2 - 2V_n^2\cos^2\theta\right) + \sqrt{\left(V_n^2 + V_t^2 - 2V_n^2\cos^2\theta\right)^2 - \left(V_n^2 - V_t^2\right)^2} \,. \tag{2.5}$$

The condition of detachment electron from atom at any position of electron $V_n \geq V_t$.
In case of equality of speeds $V_n = V_t$ we have:

$$V_1 = 2V_n\sin\theta \,. \tag{2.6}$$

Distribution of speeds (2.6) corresponds to (2.2) and fig. 6, a curve 1. Thus, the parity (2.6) arises if to consider only the wave nature of the electromagnetic wave cooperating with orbital electron.
In [6] distribution of an angle of the electron escape is investigated only for a relativistic case. It is thus received that electrons are emanated mainly to a direction of photon distribution. However the done conclusion is also actually based on the formula (2.1). Therefore the drawback of the conclusion [6] is in absence in definitive formulas of angular

distribution of electrons of nuclear mass m2. And after all the nuclear mass defines a share of the photon momentum which can incur a nuclear.

Let's consider the phenomenon of the inner photoemissive effect from positions of corpuscular representation of quantum of light, fig. 5. The quantum of light by momentum $\vec{p}$ and energy E beats out electron from atom, making A a getting out. Thus both laws of conservation of energy should be observed:

$$E = A + E_1 + E_2 , \tag{2.7}$$

Where E_1 – is the kinetic energy of taken off electron, E_2 – is the kinetic energy of nucleus as well as the law of conservation of momentum:

$$\vec{p} = \vec{p}_1 + \vec{p}_2 , \tag{2.8}$$

Where $\vec{p}_1$ – is the momentum of taken off electron, $\vec{p}_2$ – is the momentum transferred to a nucleus.

The formula (2.7) differs from Einstein's standard formula $E = A + E_1$. The point is that Einstein's formula means the absence of angular distribution of photoelectrons speed. Really, if energy of photon E is set and work function A for the given chemical element is determined certain speed of the electron escape from atom is thereby set. It means that speeds of electrons, taking off to every possible directions are identical, and the problem of finding out their angular distribution is becoming incorrect.

The value of the momentum transferred to a nucleus can be found using the formula, following (2.8):

$$p_2^2 = p^2 + p_1^2 - 2pp_1\cos\theta . \tag{2.9}$$

The system of equations (2.7) and (2.9) to obtain a combined solution and the equation (2.9) are convenient to express through energy. Taking into account $E = pc$, where c – is the speed of light in vacuum, $p_1^2 = 2m_1E_1$ and $p_2^2 = 2m_2E_2$, we find:

$$2m_2E_2 = \left(\frac{E}{c}\right)^2 + 2m_1E_1 - 2\frac{E}{c}\sqrt{2m_1E_1}\cos\theta , \tag{2.10}$$

where m_1 – is the electronic mass, m_2 – is the nuclear mass.

Substituting in (2.10) kinetic energy of nuclear E_2 by (2.7), we have:

$$E - A - E_1 = \frac{1}{2m_2}\left(\frac{E}{c}\right)^2 + \frac{m_1}{m_2}E_1 - \frac{E}{m_2c}\sqrt{2m_1E_1}\cos\theta . \tag{2.11}$$

Let us introduce the following notation $G = \sqrt{E_1}$, $\alpha = \frac{E}{m_2c}\sqrt{2m_1}$, $\sigma = 1 + \frac{m_1}{m_2}$, $\gamma = \frac{1}{2m_2}\left(\frac{E}{c}\right)^2 + A - E$. Then the equation (2.11) will be transformed into:

$$\beta G^2 - \alpha G\cos\theta + \gamma = 0 . \tag{2.12}$$

Solving quadratic equation (2.12) provided $\sigma \approx 1$ (electronic mass is much less that nuclear mass), we find:

$$G_{1,2} = \frac{\alpha}{2}\cos\theta \pm \sqrt{\frac{\alpha^2}{4}\cos^2\theta - \gamma}\,. \tag{2.13}$$

Substituting in (2.13) accepted notation we have:

$$G_{1,2} = \sqrt{\frac{m_1}{2}}\frac{E}{m_2 c}\cos\theta \pm \sqrt{\frac{E^2}{2m_2c^2}\left(\frac{m_1}{m_2}\cos^2\theta - 1\right) + E - A}\,. \tag{2.14}$$

Considering that $G = \sqrt{E_1} = V_1\sqrt{\frac{m_1}{2}}$, where V_1 - speed of photoelectrons provided $\frac{m_1}{m_2}\cos^2\theta << 1$, we find:

$$V_1 = \frac{E}{m_2 c}\cos\theta \pm \sqrt{\left(\frac{m_1}{m_2}\cos^2\theta - 1\right)\frac{m_2}{m_1}\left(\frac{E}{m_2 c}\right)^2 + \frac{2(E-A)}{m_1}}\,. \tag{2.15}$$

Provided that nuclear mass is aiming to infinity $m_2 \to \infty$ the formula (2.15) is transformed into Einstein's standard law for the photoeffect. Besides, this, as if it has been specified earlier, angular distribution of speed of photoelectrons disappears.
The condition $m_2 \to \infty$ is fair in outer photoemissive effect when the photon momentum is transferred to the whole metal through single atoms. Therefore for an outer photoemissive effect, i.e. for interaction of the solid and the photon, Einstein's formula $E = A + E_1$ is applicable absolutely.
For the inner photoemissive effect in the formula (2.15) it is necessary to use effective nuclear mass $m_{2eff} > m_2$, considering attractive powers between atoms in substance.
Transforming the formula (2.15), we get:

$$V_1 = \frac{E}{m_2 c}\left(\cos\theta \pm \sqrt{\cos^2\theta + \frac{m_2}{m_1}\left(2(E-A)\frac{m_2c^2}{E^2} - 1\right)}\right). \tag{2.16}$$

Let us nominate $\Delta = E - A$. Distribution of photoelectrons will arise at $\Delta \geq \frac{E^2}{2m_2c^2}$. In the right part of the received inequality there is a very small value, therefore distribution of photoelectrons will arise practically at $E > A$.

Let us nominate $\Delta = \eta\frac{E^2}{2m_2c^2}$, where $\eta \geq 1$ characterizes the value of exceedance of photon energy over work function in relative units. Thus the formula (2.16) takes the form:

$$V_1 = \frac{E}{m_2 c}\left(\cos\theta + \sqrt{\cos^2\theta + \frac{m_2}{m_1}(\eta - 1)}\right). \tag{2.17}$$

The analysis of the formula (2.17) shows that the root must to taking a plus since otherwise electron scattering basically goes aside, contrary to the direction of a falling photon. Angular distribution of the electron escape during the inner photoemissive effect in the relative units $\frac{V_1}{E/m_2c}$ is shown on fig. 9, made according to the formula (2.17) with several values η for copper.

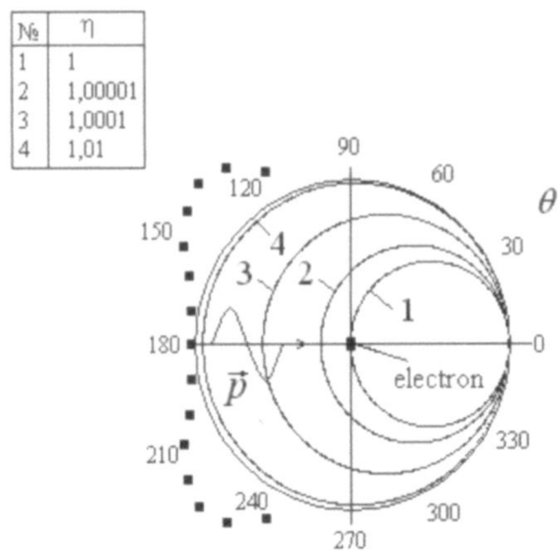

Fig. 9. Angular distribution of photoelectrons during the inner photoemissive effect depending on parameter η during interaction of orbital electrons with light quantum. The results of experiments [16] are shown by black small squares

The figure makes it evident that speeds of photoelectrons become almost identical in all directions already at $\eta \geq 1{,}01$. Then Einstein's formula $E = A + E_1$ becomes fair and for the inner photoemissive effect. Considering that, for example, for copper the relation $\frac{E^2}{2m_2c^2} \Big/ E \approx 4{,}2 \cdot 10^{-11}$ equivalent as far as the order of value is concerned $\frac{E^2}{2m_2c^2} \Big/ \Delta = \frac{1}{\eta}$ in the field of red photoelectric threshold (λ_r = 250 nm), it is possible to draw the conclusion that the evident difference of distribution of photoelectrons speeds from spherical, i.e. actually formula is violated $E = A + E_1$, can be observed only in very short wave part of spectrum γ-radiations.

The observed data of angular distribution of the photoelectrons which have been beaten out from a monolayer of atoms of copper by covering the nickel surface are shown in fig. 9 by black small squares [16]. The wavelength of quanta allowed observing the photoeffect with 2*p*-atom shell of copper, but the photoeffect on nickel thus was absent. Experimental distribution of photoelectrons contradicts calculated distribution in fig. 6. Moreover, in distinction in fig. 6, small maxima of indicatrix of the distributions directed to an opposite direction of flight of light quanta at an angle of approximately 45° to the direction of light flux are observed. In [16] these maxima are explained by focusing properties of all population of atoms of the surface. The amplitude of maxima ascends with the increase of quantity of the monolayers of copper atoms on nickel.

Thus, angular distribution of photoelectrons will be absolutely various depending on whether what properties, wave or corpuscular are reveal by the light quantum in interaction with orbital electron. Only experiment can give the answer to the question what distribution it is true, fig. 6 or fig. 9. However existence of electron flux from an illuminated surface at normal light incidence [16], in the direction opposite to intensity of light, shows at the prevalence of corpuscular properties of light in its interaction with atoms.

2.2 Relativistic case

Dealing with relativistic case of the inner photoemissive effect, the law of conservation of energy needs to be written down as:

$$E = A + E_k + E_2 , \tag{2.18}$$

Where E_k - is the kinetic energy of photoelectron.

The law of conservation of momentum remains in the form (2.9). Using relativistic relation between the energy and the momentum for electron:

$$E_1^2 = p_1^2 c^2 + m_1^2 c^4 , \tag{2.19}$$

where E_1 - is the total energy of electron, m_1 - is the electron rest mass, we will express the momentum of electron from (2.19)and we will substitute in (2.9). For convenience of the further transformations we will write down (2.19) into:

$$p_1^2 = \frac{E_1^2 - m_1^2 c^4}{c^2} = \frac{\left(E_1 + m_1 c^2\right) E_k}{c^2} . \tag{2.20}$$

Formulating (2.20) the relation has been used:

$$E_k = E_1 - m_1 c^2 . \tag{2.21}$$

The equation (2.9) will be transformed into:

$$2m_2 c^2 E_2 = E^2 + \left(E_1 + m_1 c^2\right) E_k - 2E\sqrt{\left(E_1 + m_1 c^2\right) E_k} \cos\theta . \tag{2.22}$$

Because of that the nucleus that has a big mass and a relatively low speed after interaction with the photon, expression for relation of the momentum of the nucleus with its kinetic energy E_2 is used in the nonrelativistic form.

Substituting value E_2 in (2.22) from the equation (2.18), we get:

$$2m_2 c^2 \left(E - A - E_k\right) = E^2 + \left(E_1 + m_1 c^2\right) E_k - 2E\sqrt{\left(E_1 + m_1 c^2\right) E_k} \cos\theta . \tag{2.23}$$

Let us nominate:

$$\alpha = \frac{E\sqrt{E_1 + m_1 c^2}}{m_2 c^2}; \qquad G = \sqrt{E_k} . \tag{2.24}$$

As a result (2.23) will be transformed into:

$$\delta G^2 - \alpha G \cos\theta + \gamma = 0 . \tag{2.25}$$

The notation $\gamma = \frac{1}{2m_2}\left(\frac{E}{c}\right)^2 + A - E = \frac{1}{2m_2}\left(\frac{E}{c}\right)^2 - \Delta = \frac{1}{2m_2}\left(\frac{E}{c}\right)^2 (1-\eta)$ corresponds to item 1 section.

The value

$$\delta = 1 + \frac{m_2 c^2}{2}\left(\frac{\alpha}{E}\right)^2 = 1 + \frac{m_1}{2m_2} + \frac{E_1}{2m_2 c^2} = 1 + \frac{m_1}{2m_2} + \frac{E_k - m_1 c^2}{2m_2 c^2} = 1 + \frac{E_k}{2m_2 c^2} \approx 1 \quad .$$

It is thus accounted for that $E_k << 2m_2 c^2$.

Solving the equation (2.25), we get:

$$G_{1,2} = \frac{\alpha}{2}\cos\theta \pm \sqrt{\frac{\alpha^2}{4}\cos^2\theta - \gamma} \, . \qquad (2.26)$$

Substituting notations, we find:

$$E_k = \left(\frac{\alpha}{2}\right)^2 \left(\cos\theta + \sqrt{\cos^2\theta + \frac{2}{\alpha^2}\left(\frac{E^2}{m_2 c^2}\right)(\eta - 1)} \right)^2 . \qquad (2.27)$$

In contrast to the nonrelativistic case, the formula (2.17), formula (2.27) possesses in its right part value $\alpha = \frac{E\sqrt{E_1 + m_1 c^2}}{m_2 c^2}$ which depends on the total energy of electron E_1 the structure of which includes also kinetic energy E_k. But dependence of value α on E_k not strong as the total energy structure includes rather big rest energy of electron $m_1 c^2$.

Considering that $E_1 = \frac{m_1 c^2}{\sqrt{1-\beta^2}}$, where $\beta = \frac{V_1}{c}$ is the relative speed of the photoelectron, we find:

$$\frac{\alpha^2}{2} = \left(\frac{E}{m_2 c}\right)^2 \frac{m_1}{2}\left(\frac{1}{\sqrt{1-\beta^2}} + 1\right) . \qquad (2.28)$$

Substituting the equation (2.28) in the equation (2.27) and considering that $E_k = m_1 c^2\left(\frac{1}{\sqrt{1-\beta^2}} - 1\right)$, we get:

$$2c\sqrt{(1-\mu)} = \frac{E}{m_2 c}\left(\cos\theta + \sqrt{\cos^2\theta + \mu\frac{m_2}{m_1}(\eta - 1)} \right), \qquad (2.29)$$

where $\mu = \frac{2}{\frac{1}{\sqrt{1-\beta^2}} + 1}$.

Considering that $2c\sqrt{(1-\mu)} = \sqrt{\frac{2E_k}{m_1} \Big/ \left(1+\frac{E_k}{2m_1c^2}\right)} \approx V_1$, at $1 >> \frac{E_k}{2m_1c^2}$, we find:

$$V_1 \approx \frac{E}{m_2 c}\left(\cos\theta + \sqrt{\cos^2\theta + \mu\frac{m_2}{m_1}(\eta - 1)}\right) \tag{2.30}$$

The formula (2.30) allows to consider relativistic effects at the photoeffect, in case of rather big speeds of photoelectrons. Thus, in contrast to (2.17), relativistic coefficient μ is introduced under the root. The calculation of dependence $\mu(\beta)$ shows on fig. 10, relativistic effects while calculating distribution of photoelectrons escape, can be neglected and (2.17) can be used while the photoelectron speeds read approximately half the value of the light speed in the vacuum.

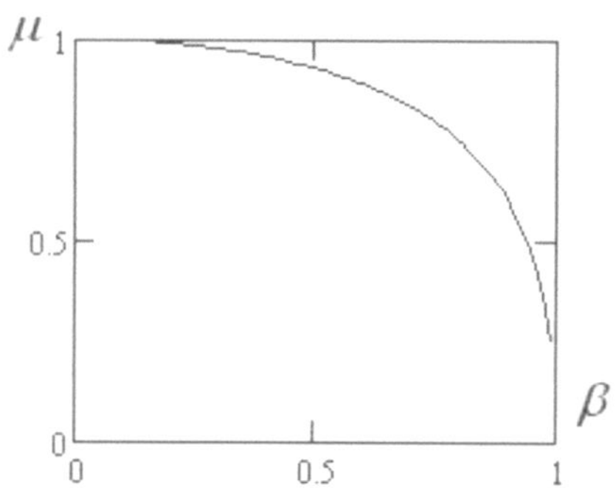

Fig. 10. Dependence of relativistic coefficient β on relative speed of photoelectrons $\beta = \frac{V}{c}$

3. Conclusion

The laws of formation of the impulse of electromagnetic radiation in dielectric environment for conditions self-induced transparency are considered. The insufficiency of the description of such impulse with the help of the equations Maxwell - Bloch are shown. The impulse of electromagnetic radiation in conditions of a self-induced transparency submits to nonlinear equation of Schrödinger with logarithmic nonlinearity. The way of connection of an average number filling and energy of the impulse taking into account energy saturation of environment are offered. The calculation of a electrical component of the impulse is submitted.

Angular distribution of photoelectrons is investigated during the inner photoemissive effect for two variants: quantum of light basically reveals wave and basically corpuscular properties interacting with orbital electron. Distinction in angular distribution of photoelectrons for these variants is demonstrated. If electromagnetic radiation shows basically quantum properties during a photoeffect there is an emission of photoelectrons on a direction of movement of quantums. It corresponds Einstein's to formula. In Einstein's formula there is no corner of a start of photoelectrons. Angular distribution in the second variant is investigated for the nonrelativistic and relativistic cases.

4. References

[1] Volobuev A.N., Neganov V.A. The Electromagnetic Envelope Soliton Propagating in Dielectric. *Technical Physics Letters*. S-Petersburg. (2002). Vol. 28. No. 2, pp. 15-20.
[2] R.K. Dodd, J.C. Eilbeck, J.D. Gibbon, H.C. Morris. Solitons and Nonlinear Wave Equations. Harcourt Brace Jovanovich, Publishers. London, New York, Toronto. (1982), pp. 545, 588, 601.
[3] M.J. Ablowitz, H. Segur. Solitons and the Inverse Scattering Transform. *SIAM*. Philadelphia. (1981), pp. 374-378.
[4] Volobuev A.N. Modeling of Physical Processes, to Describe by Nonlinear Schrodinger Equation. *Mathematical modelling*. Moscow. (2005). Vol. 17. No. 2, pp. 103 - 108.
[5] G.L. Lamb, Jr., D.W. McLaughlin. Aspects of Solitons Physics. In collection under edition of [6] R. K. Bulllough, P.J. Caudrey. *Solitones*. Springer-Verlag, Berlin-New York, (1980), pp. 59 - 102.
[7] Berestetskij V.B., Lifshits E.M., Pitaevskij L.P. Quantum electrodynamics. Moscow, *Science*, (1989), p. 192, 249.
[8] Davidov A.S. Quantum mechanics. Moscow. *Physmatlit*. (1963), p. 335, 366.
Birnbaum J.P. Optical quantum generators. Moscow. Soviet radio, (1967), pp. 49,50.
[9] G.B. Whitham F. R. S. Linear and Nonlinear Waves. A Wiley-Interscience Publncation. John Wiley & Sons, London, New York, Toronto. (1974), p. 575.
[10] Ditchbern R. Physical optics. M: Science. (1965). P. 413, 419, 497.
[11] Ashcroft N., Mermin N. Solid-state physics. Moscow, World. (1979). V.1. P. 82.
[12] Nolle E.L. Tunneling mechanism of the photoeffect in metal nanoparticles activated by caesium and oxygen. Moscow, UFN. (2007). V.177. № 10. P. 1133–1137.
[13] Mihajlov A.I., Mihajlov I.A. A double nuclear photoeffect in relativistic region. Angular and energy distributions of photoelectrons. Moscow, JETP. (1998). V. 114. №. 5. P. 1537 – 1554.
[14] Levich V. G. *The Course of theoretical physics*. V.2. Moscow, PhysMathGiz. (1962). P. 658.
[15] Blohin M. A Physics of Roentgen rays. Moscow, State Publishing House. (1957). P. 261.
[16] Steigerwald D.A., Egelhoff W.F., jr. *Phys. Rev. Lett.* (1988). Vol. 60. P. 2558.

9

Ultrafast Electromagnetic Waves Emitted from Semiconductor

YiMing Zhu and SongLin Zhuang
Engineering Research Center of Optical Instrument and System, Ministry of Education, University of Shanghai for Science and Technology
China

1. Introduction

Semiconductor devices have become indispensable for generating electromagnetic radiation in every day applications. Visible and infrared diode lasers are at the core of information technology, and at the other end of the spectrum, microwave and radio frequency emitters enable wireless communications. But the ultrafast electromagnetic waves, whose frequency locates in terahertz (THz) region (0.3 - 30 THz; 1 THz = 10^{12} Hz), has remained largely underdeveloped, despite the identification of various possible applications. One of the major applications of THz spectroscopy systems is in material characterization, particularly of lightweight molecules and semiconductors [1] [2]. Furthermore, THz imaging systems may find important niche applications in security screening and manufacturing quality control [3] - [5]. An important goal is the development of three dimensional (3-D) tomographic T-ray imaging systems. THz systems also have broad applicability in a biomedical context, such as the T-ray biosensor [6]. A simple biosensor has been demonstrated for detecting the glycoprotein avidin after binding with vitamin H (biotin) [7].

However, progresses in these areas have been hampered by the lack of efficient ultrafast electromagnetic wave / THz wave sources. As shown in Fig. 1, transistors and other electronic devices based on electron transport are limited to about ~ 300 GHz (~ 50 GHz being the rough practical limit; devices much above that are extremely inefficient) [8]. On the other hand, the wavelength of semiconductor lasers can be extended down to only ~ 10 μm (about ~ 30 THz) [9]. Between two technologies, lie the so called terahertz gap, where no semiconductor technology can efficiently convert electrical power into electromagnetic radiation. The lack of a high power, low cost, portable room temperature THz source is the most significant limitation of modern THz systems. A number of different mechanisms have been exploited to generate THz radiation, such as photocarrier acceleration in photoconducting antennas, second order nonlinear effects in electro-optical (EO) crystals and quantum cascade laser. Currently, conversion efficiencies in all of these sources are very low, and consequently, average THz beam powers tend to be in the nanowatt to microwatt range, whereas the average power of the femtosecond optical source is in the region of ~ 1 W. There is still a long way to go before commercial devices based on this principle can be mass-produced.

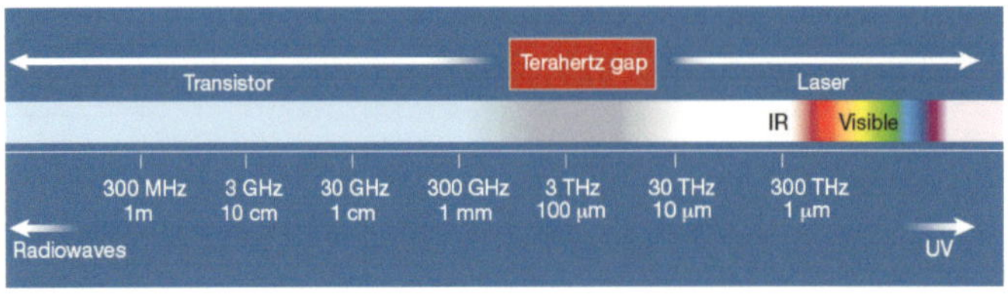

Fig. 1. The THz gap. The gap, lying roughly between 300 GHz and 30 THz, exists because the frequencies generated by transistors and lasers, typical semiconductor devices, do not overlap. No current semiconductor technology can efficiently convert electrical power into electromagnetism in this range

Photoconduction is one of the most common approaches for generating broadband pulsed ultrafast electromagnetic beams/THz beams. The photoconductive approach uses high speed photoconductors as transient current sources for radiating antennas [10]. Typical photoconductors include high resistivity GaAs, InP and radiation damaged silicon wafers. Metallic electrodes are used to bias the photoconductive gap and form an antenna. The physical mechanism for THz beam generation in photoconductive antennas begins with an ultrafast laser pulse (with a photon energy larger than the bandgap of the material $h\upsilon \geq E_g$), which creates electron and hole pairs in the photoconductor. The free carriers are then accelerated by the static bias field and form a transient photocurrent, and the fast, time varying current radiates electromagnetic waves. The THz radiation E_{THz}, emitted from the samples should be proportional to the change of the photocurrent J:

$$E_{\mathrm{THz}} \propto \frac{\partial J}{\partial t} \tag{1}$$

Figure 2 shows the principle of generating THz radiation from antenna by photoconduction method. Several material parameters affect the intensity and the bandwidth of the resultant THz radiation. For efficient THz radiation, it is desirable to have rapid photocurrent rise and decay time. Thus, semiconductors with small effective electron masses such as InAs and InP are attractive. The maximum drift velocity is also an important material parameter, but it is generally limited by the intraband scattering rate or by intervalley scattering in direct semiconductors such as GaAs [11] [12]. Because the radiating energy mainly comes from stored surface energy in the form of the static bias field, the THz radiation energy scales up with the bias and optical fluency. The breakdown field of the material is another important parameter because this determines the maximum bias that may be applied. Photoconductive emitters are capable of relatively large average THz powers in excess of ~ 40μW [13] and bandwidths as high as ~ 4 THz [14].

As described above, impulsive currents in semiconductors excited by femtosecond optical pulses radiate ultrafast electromagnetic waves in the THz frequency range, serving, for example, as an ultrafast electromagnetic wave source for THz spectroscopy. As described in Eq. (1), the ultrafast electromagnetic wave in the far field is proportional to the time derivative of the current, $\partial J/\partial t$. In a simplified view, since J is proportional to carrier velocity which can be written as $J = nev$, where, e, n, and v are the elementary electric charge, total number of photoexcited carriers and there velocity, respectively, E_{THz} is proportional to carrier acceleration by assuming that the carriers are excited by sufficiently short laser pulses and their lifetime is long enough (n is regarded as being constant). Hence, in principle, the observation of such THz electromagnetic waves provides us with a way of looking into the temporal variation of v, namely the acceleration / deceleration dynamics of photoexcited carriers [11] [12] [15] [16]. Figure 3 shows the illustration of ultrafast electromagnetic waves, THz waves, emitted from intrinsic bulk GaAs photoexcited by femtosecond laser pulses under bias electric fields. By understanding the nonequilibrium transport of carriers in bulk GaAs, the gain region in the THz range, can be clarified, which is the necessary condition for the microwave oscillation [17] [18].

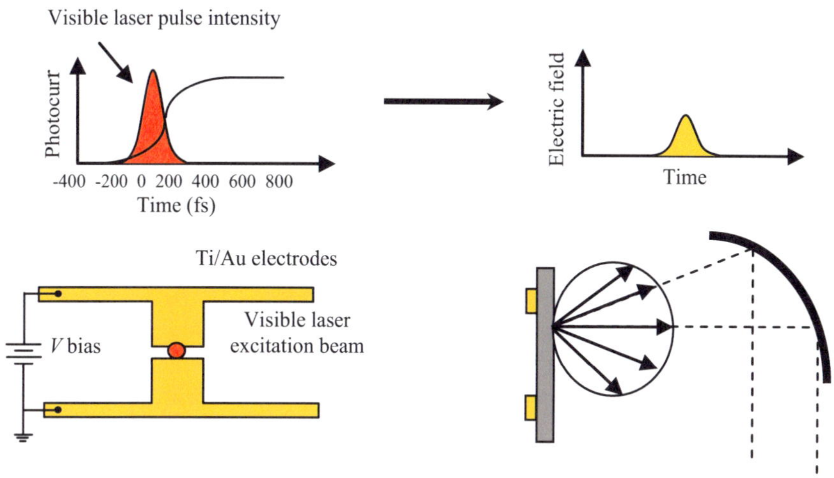

Fig. 2. The principle of generating THz radiation from antenna by photoconduction method

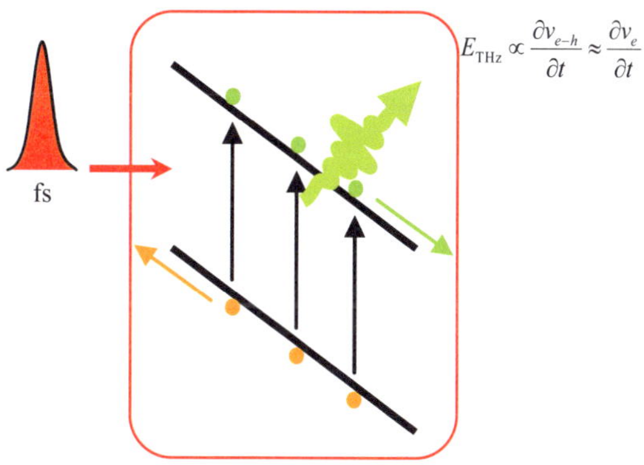

Fig. 3. The illustration of generating THz waveforms from intrinsic bulk GaAs photoexcited by femtosecond laser pulses under bias electric fields

The novel experimental technique of the time domain THz spectroscopy for observing ultrafast carrier motion in the femtosecond time regime together with electromagnetic wave radiation that is proportional to $\mathrm{d}v/\mathrm{d}t$, is introduced in section 2. Invaluable information, femtosecond acceleration of carriers in bulk GaAs under very high electric fields investigated by time domain THz spectroscopy, contributing to better understanding of device physics in existing high speed electron devices, is discussed in section 3. Finally, the present insights on the gain in GaAs due to electrons intervalley transfer under high electric fields, which is of practical importance for its exploitation in ultrafast electromagnetic wave oscillators, are discussed in section 4.

2. Time domain terahertz spectroscopy by electro-optic sampling method

Time domain terahertz (THz) spectroscopy affords a powerful technique for the research of semiconductors of current industrial interest [2]; the time domain studies of the emitted ultrafast wave / THz wave from various semiconductor structures provide a unique way of looking directly into the temporal as well as spatial evolution of the excited carriers on the sub-picosecond time scale. In this section, time domain THz electro-optic (EO) sampling via the Pockels effect is discussed in detail, including the EO effect, the response spectrum of the EO crystal, and the experimental setup for EO sampling.

Recently, the technique called time domain THz spectroscopy has been attracting wide interest. Time domain THz spectroscopy uses short pulses of broadband THz radiation, which are typically generated using ultrafast laser pulses. This technique has been developed from a work in 1980's at AT&T Bell Labs and IBM T. J. Watson Research Center [19] [20]. Historically, freely propagating THz pulses were measured by means of either photoconductive antennas [21] [22] or far infrared interferometic techniques using incoherent detectors such as bolometers [23] [24]. Although the photoconductive antennas have excellent sensitivity, their frequency response is limited by resonant behavior of the Hertzian dipole structure. For the interferometric techniques, the sensitivity is far worse than that of the photoconductive antennas, because the measurement is coherent and its sensitivity is ultimately limited by the thermal background. In addition, bolometers usually require liquid helium cooling.

On the other hand, ultrafast EO sampling has been widely used in the measurement of local transient electric fields in materials [25] [26]. There existed a need to extend the local electric fields measurement to free space. In 1995, three groups reported their first results using free space EO sampling independently, nearly at the same time [27]-[29]. Although the preliminary results were very poor, rapid progress has been made in the intervening years. It turns out that free space EO sampling is a powerful tool for THz pulse measurement, providing many advantages, such as high sensitivity, ultrabroad frequency response, ease of use, and parallel measurement capability [30]-[35].

Ultrafast carrier motion in the femtosecond time regime accompanies electromagnetic wave radiation, the electric field component of which is proportional to dv/dt. Therefore, the investigation of such electromagnetic wave (or THz radiation) allows us a very unique opportunity of looking directly into the acceleration/deceleration dynamics of carriers in semiconductors. In this chapter, time domain THz spectroscopy based on EO sampling via the Pockels effect is discussed. Furthermore, the response spectrum of the EO crystals are also discussed and 100 μm-thick ZnTe, whose response spectrum is flat up to the frequency around 4 THz, is chosen as the EO crystal for this experiment.

2.1 Electro-optic effect

EO sampling is a phase sensitive detection technique of electromagnetic radiation which measures a birefringence in an EO crystal induced by the incident electromagnetic radiation. This birefringence in the crystal is probed as a phase shift of an optical probe beam. In the time resolved detection of THz pulse, the transient birefringence is probed with the optical pulses at different time delays.

The principle of EO sampling can be explained as follows. Suppose that the probe beam is propagating in the z-direction and x, y are the crystal axes of the EO crystal as shown in Fig. 4. When an electric field is applied to the EO crystal, the electrically induced birefringence axes x' and y' are at an angle of 45° with respect to x and y. If the input beam is polarized along x, then the output light can be obtained from the following expression:

$$\begin{pmatrix} E_x \\ E_y \end{pmatrix} = \begin{pmatrix} \cos(\frac{\pi}{4}) & -\sin(\frac{\pi}{4}) \\ \sin(\frac{\pi}{4}) & \cos(\frac{\pi}{4}) \end{pmatrix} \begin{pmatrix} \exp(i\Delta) & 0 \\ 0 & 1 \end{pmatrix} \begin{pmatrix} \cos(\frac{\pi}{4}) & \sin(\frac{\pi}{4}) \\ -\sin(\frac{\pi}{4}) & \cos(\frac{\pi}{4}) \end{pmatrix} \tag{2}$$

where $\Delta = \Gamma_0 + \Gamma$ is the phase difference between the x' and y' polarizations, including both the dynamic (Γ, THz induced) and the static (Γ_0, from the intrinsic or residual birefringence of the EO crystal and the compensator) phase difference. Following Eq. (2), the light intensity in the x and y polarizations is

$$\begin{cases} I_x = |E_x|^2 = I_0 \cos^2 \dfrac{\Gamma_0 + \Gamma}{2} \\ I_y = |E_y|^2 = I_0 \sin^2 \dfrac{\Gamma_0 + \Gamma}{2} \end{cases} \tag{3}$$

where $I_0 = E_0^2$ is the input intensity. It can be seen that I_x and I_y are complementary, i.e. $I_x + I_y = I_0$. This is the result of energy conservation and follows from the fact that absorption in the crystal has been ignored. To extract the light in the x and y polarizations separately, a Wollaston prism is usually used.

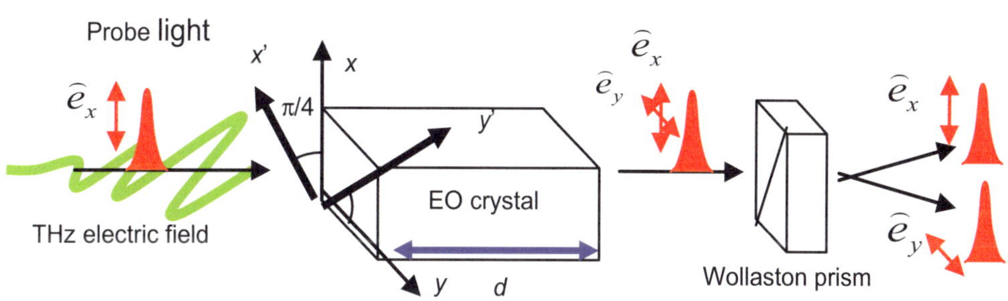

Fig. 4. The principle for EO sampling. x' and y' are the birefringence axes of the EO crystal, d is the thickness of the EO crystal

The static phase term Γ_0 (or called the optical bias) is often set equal to $\pi/2$ for balanced detection. For an EO crystal without intrinsic birefringence, such as ZnTe, a quarter wave plate is often used to provide this optical bias. Because $|\Gamma| << 1$ in most cases in EO sampling, we have

$$\begin{cases} I_x = \dfrac{I_0}{2}(1 - \Gamma) \\ I_y = \dfrac{I_0}{2}(1 + \Gamma) \end{cases} . \tag{4}$$

In the two beams, the signals have the same magnitudes but opposite signs. For balanced detection, the difference between I_y and I_x is measured, giving the signal

$$I_s = I_y - I_x = I_0\Gamma \quad (5)$$

This signal is proportional to the THz induced phase change, Γ, and, Γ, in turn, is proportional to the electric field of the THz pulse, E_{THz}. For a (110)-oriented ZnTe crystal, the following relation holds:

$$\Gamma = \frac{\pi n_g(\lambda_0)^3 \gamma_{41} d}{\lambda_0} E_{THz} , \quad (6)$$

where d is the thickness of the crystal, $n_g(\lambda_0)$ is the refractive index of the crystal at the wavelength of the optical probe beam λ_0, and γ_{41} is the EO coefficient. By detecting the phase retardation Γ induced in the EO crystal by balanced detection as shown in Eq. (5), E_{THz} can be accurately detected (Eq. (6)).

2.2 The response spectrum of the electro-optic crystal

For a transient THz pulse, phase matching should be considered. When the probe pulse has a group velocity different from that of a THz pulse (so-called GVM), the probe does not always sample the same position on the THz pulse. Instead, it scans across the THz pulse as the two propagating through the crystal, leading to broadening of the measured THz waveform. Furthermore, the bandwidth of EO crystal is limited by reststrahlen band of the crystal. To estimate frequency response and upper limit of the EO sensor, we performed a calculation of the theoretical response function of EO crystals, which shows that the importance of properly accounting for the dispersion of the EO coefficient and the GVM.
The complete frequency response function R(ω) of EO sensors is obtained with the following model: The influences of linear optics such as the GVM between the group velocity of the optical probe pulse and that of the THz pulse, reflection losses at the detector surface, and absorption inside the crystal have been described recently [36] [37].
After a probe pulse and THz wave co-propagate through a sensor of thickness d, the accumulated GVM time is

$$\delta(\omega) = \frac{n_g(\lambda_0) - n(\omega)}{c} d , \quad (7)$$

where c is the speed of the light and $n(\omega)$ is the complex refractive index of the THz radiation of frequency ω.
Then, the amplitude and its phase of the EO modulation of the optical probe pulse induced by the THz wave is proportional to d as well as the time average of the electric field across the GVM time, $\delta(\omega)$. It has the form as [37],

$$G(\omega) = \frac{t(\omega)}{\delta(\omega)} \int_0^{\delta(\omega)} \exp[i2\pi\omega t]dt = t(\omega) \frac{\exp[i2\pi\omega\delta(\omega)] - 1}{i2\pi\omega\delta(\omega)} , \quad (8)$$

where $t(\omega) = 2/[n(\omega)+1]$ is the Fresnel transmission coefficient.

The group refractive index $n_g(\lambda_0)$ at the wavelength of the optical probe pulse is obtained from dispersion data in the near infrared range. The complex refractive index $n(\omega)$ of the THz radiation can be expressed as:

$$n(\omega)=\sqrt{\left[1+\left\{\frac{(\hbar\omega_{\mathrm{LO}})^2-(\hbar\omega_{\mathrm{TO}})^2}{(\hbar\omega_{\mathrm{TO}})^2-(\hbar\omega)^2-i\hbar\gamma\omega}\right\}\right]\times\varepsilon_\infty} \tag{9}$$

The transverse optical (TO) phonon and longitudinal optical (LO) phonon energies, and the lattice damping are denoted by $\hbar\omega_{\mathrm{TO}}$, $\hbar\omega_{\mathrm{LO}}$, and γ, respectively. ε_∞ is the high frequency dielectric constant.

$G(\omega)$ describes the detector response if the EO coefficient γ_{41} of the sensor material is frequency independent. However, γ_{41} exhibits a strong dispersion and resonant enhancement due to the lattice resonance. This effect accounts for very pronounced features in the response functions. An analytic expression for $\gamma_{41}(\omega)$ has been derived in [38], Eq. (5):

$$\gamma_{41}(\omega)=r_e\times\left[1+C\left(1-\frac{(\hbar\omega)^2+i\hbar\omega\gamma}{(\hbar\omega_{\mathrm{TO}})^2}\right)^{-1}\right] \tag{10}$$

The Faust-Henry coefficient C represents the ratio between the ionic and the electronic part of the EO effect at $\omega = 0$. The purely electronic nonlinearity r_e is assumed to be constant at the mid and far infrared frequencies ω.

The full complex response function $R(\omega)$ of EO sensor is, then, given by

$$R(\omega)=G(\omega)\times\gamma_{41}(\omega)\,. \tag{11}$$

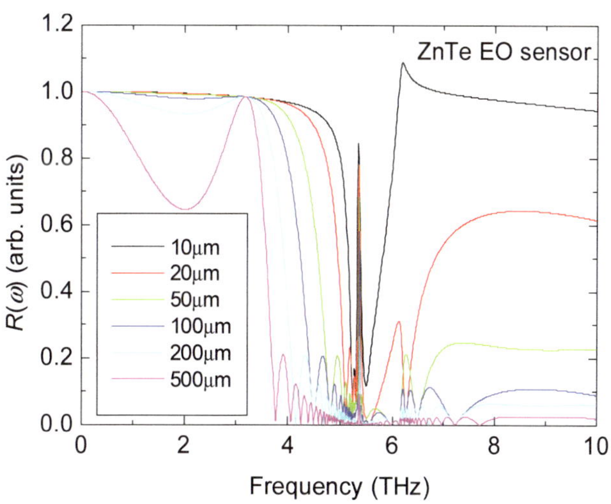

Fig. 5. The calculated amplitude response $R(\omega)$ of the ZnTe EO sensor with different thickness, normalized to unity at low frequencies

Figure 5 displays the calculated spectra of $|R(\omega)|$ of ZnTe sensors with various thickness, normalized to unity at low frequencies. As shown in Fig. 5, a structured dip in the spectra appears in the reststrahl region around 5.3 THz. This feature reflects the high absorption of the material and the large velocity mismatch near the lattice resonance. At the TO phonon frequency (6.2 THz in ZnTe), the nonlinear coefficient $\gamma_{41}(\omega)$ is resonantly enhanced. We can clearly see, with decreasing the thickness of ZnTe crystal, the upper limit of the frequency increases. However, the sensitivity of the EO sensor is proportional to the thickness of the crystal, as shown in Eq. (6) in section 2.1. Therefore, a trade-off exists between broadband response and high sensitivity.

2.3 Time domain terahertz spectroscopy system

Figure 6 shows the setup for free space THz EO sampling measurements. The ulrafast laser pulse is split by a beam splitter into two beams: a pump beam (strong) and a probe beam (weak). The pump beam illuminates the sample and generates the THz radiation. The generated THz radiation is a short electromagnetic pulse with a duration on the order of one picosecond, so the frequency is in the THz range. The THz beam is focused by a pair of parabolic mirrors onto an EO crystal. The beam transiently modifies the index ellipsoid of the EO crystal via the Pockels effect, as discussed in detail in section 2.1. The linearly polarized probe beam co-propagates inside the crystal with THz beam and its phase is modulated by the refractive index change induced by the THz electric field, E_{THz}. This phase change is converted to an intensity change by a polarization analyzer (Wollaston prism). A pair of balanced detectors is used to suppress the common laser noise. This also doubles the measured signal. A mechanical delay line is used to change the time delay between the THz pulse and the probe pulse. The waveform of E_{THz} can be obtained by scanning this time delay and performing a repetitive sampling measurement. The principle of THz signal sampled by the femtosecond pulse is shown in Fig. 7. By this sampling method, we do not need the instruments for high speed measurement because the THz signals are converted to the electrical signals after the sampling by the femtosecond probe pulse. To increase the sensitivity, the pump beam is modulated by a mechanical chopper and E_{THz} induced modulation of the probe beam extracted by a lock-in amplifier.

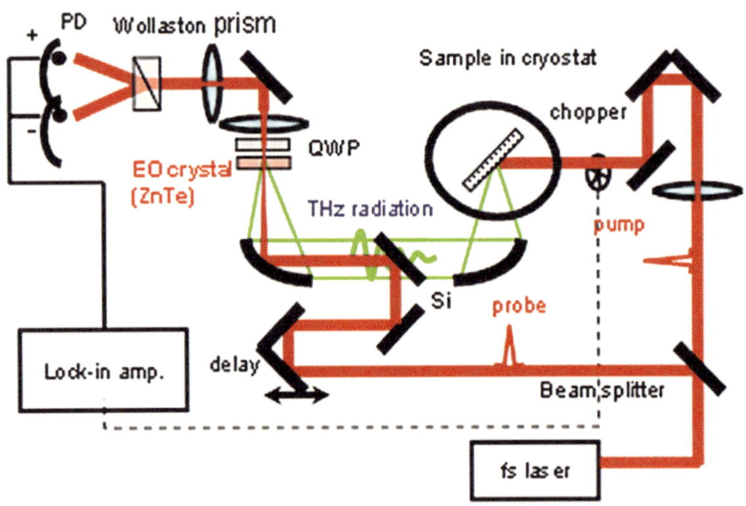

Fig. 6. Experimental setup for free space EO sampling

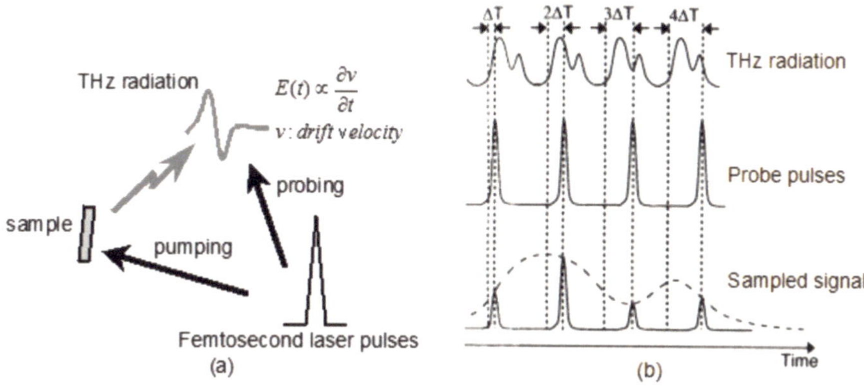

Fig. 7. The principle of THz signals sampled by the femtosecond probe beam

3. Femtosecond acceleration of carriers in bulk GaAs

Ultrafast nonequilibrium transport of carriers in semiconductors biased at high electric fields is of fundamental interest in semiconductor physics. From a fundamental point of view, the detailed understanding of the femtosecond dynamics of carriers in an electric field is a key issue in many body physics. As an example, one may ask whether the well-developed semiclassical picture of electronic conduction breaks down on ultrafast time scales and at very high electric fields [39]. Furthermore, clarifying carrier dynamics under extremely nonequilibrium conditions is also strongly motivated by the need to obtain information relevant for the design of ultrahigh speed devices. Indeed, transit times even less than 1 ps have been reported for compound semiconductor field effect transistors [40]. In such ultrashort channel transistors, carriers experience very few scattering events in the channel and drift in a very nonstationary manner as schematically illustrated in the Fig. 8. Consequently, the performance of such ultrafast transistors is not mainly determined by the steady-state properties, such as saturation velocities and nobilities, but is governed by the nonstationary carrier transport subjected to high electric fields. It is, therefore, essential to understand nonequilibrium transport of carriers subjected to high electric fields in such devices. However, it has been difficult to characterize such very fast phenomena by using conventional electronics, such as sampling oscilloscopes, because of their limited bandwidth. Consequently, Monte Carlo calculations have been the only tool for discussing transient carrier transport.

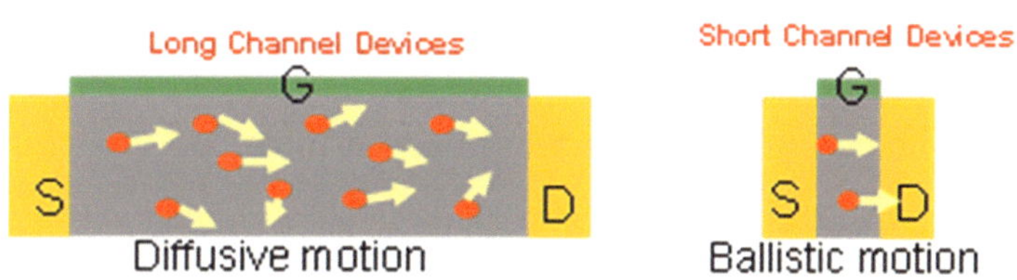

Fig. 8. Carrier transport in (a) long diffusive channels (b) very short ballistic channels. (a) In conventional long-channel transistors, their performance is determined by steady-state properties, such as saturation velocities and nobilities. However, (b) in very short channel transistors, carrier transport is quasi-ballistic and their switching speed is governed by how fast carriers are accelerated

So far, only a few experimental studies have been done on the femtosecond carrier acceleration under nonequilibrium in compound semiconductor [11] [12] [15]. Leitenstorfer, et al. presented the first experimental results on the ultrafast electromagnetic wave / terahertz wave emission from electrons and ions in GaAs and InP accelerated by very high electric fields and showed firm experimental evidence of velocity overshoot of electrons in the femtosecond time range [11] [12]. They separated the contributions of electrons and lattices, and determine the transient acceleration and velocity of carriers with a time resolution in the order of 10 fs. The maximum velocity overshoots and traveling distances during the nonequilibrium regime have been determined for the first time.

In this section, we have investigated nonequilibrium acceleration of carriers in bulk GaAs subjected to very high electric fields by time domain THz spectroscopy [16]. It is found that THz emission waveforms have a bipolar feature; i.e., an initial positive peak and a subsequent negative dip. This feature arises from the velocity overshoot. The initial positive peak has been interpreted as electron acceleration in the bottom of Γ the valley in GaAs, where electrons have a light effective mass, while the subsequent negative dip has been attributed to intervalley transfer from the Γ valley to the X and L valleys.

3.1 GaAs *m-i-n* diode & short channel GaAs *m-i-n* diode

Undoped bulk GaAs sample grown by molecular beam epitaxy was used. Sample #1 had an *m-i-n* geometry with a 1 μm-thick intrinsic GaAs layer. An ohmic contact was fabricated by depositing and annealing AuGeNi alloy on the back side of the sample. A semitransparent NiCr Schottky film was deposited on the surface to apply a DC electric field, F, to the intrinsic GaAs layer. The sample was mounted on a copper plate, which was put in the cryostat, for the time domain THz measurements. The illuminated surface area (window area) was approximately 1.5×1 mm^2. The special design of the window size was adopted to avoid the field screening effect.

3.2 Femtosecond acceleration of carriers in GaAs *m-i-n* diode & short channel GaAs *m-i-n* diode

Femtosecond laser pulses from a mode locked Ti: sapphire laser operated at a repetition rate of 76 MHz was used for time domain THz spectroscopy. The full width at half maximum (FWHM) of spectral bandwidth of the femtosecond laser pulses was approximately 20 meV. The central photon energies of the light pulses were set to 1.422 eV at 300 K and 1.515 eV at 10 K, in such a way that electrons were created near the bottom of the conduction band, as well as holes near the top of the valence band. The free space EO sampling technique was used to record temporal waveforms of THz electric fields emitted from the samples [41] [42], as described in chapter 2. The EO sensor used in this experiment was a 100 μm-thick (110)-oriented ZnTe crystal. The spectral bandwidth of this sensor was approximately 4 THz [11] [37], as shown in Fig. 5 in section 2.2. The corresponding resolution is ~ 250 fs.

Figures 9 show temporal waveforms of the THz electric fields emitted from samples #1 (an *m-i-n* diode with a 1 μm-thick intrinsic GaAs layer) [Figs. 9 (a), 9 (b)] for various F at 300 K and 10 K, respectively. In [43], the position of $t = 0$ was empirically determined by choosing a position which does not cause artificial jumps in the phase of Fourier spectra of time domain THz data. To determine the position of $t = 0$ more accurately, we adopted a newly developed method, i.e., the maximum entropy method (MEM) [44]-[46]. The position of $t = 0$ set in Figs. 9 has been determined by MEM within an error of ±30 fs, which is limited by the time interval of the recorded points.

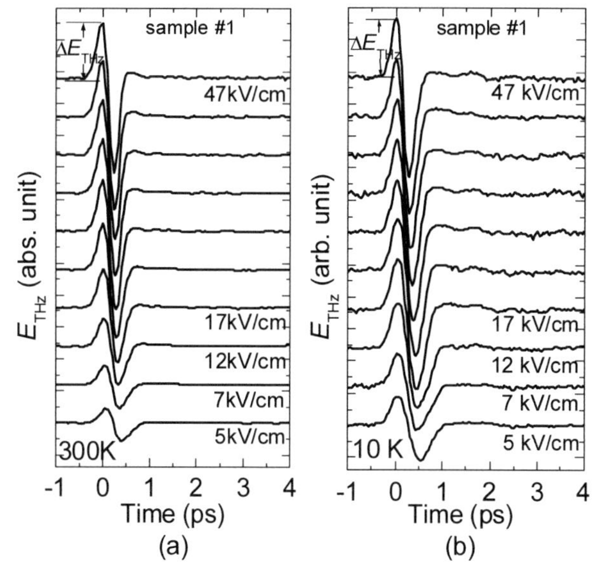

Fig. 9. Bias field dependence of the temporal waveforms of THz electric field (E_{THz}) emitted from (a) an *m-i-n* diode with 1 μm-thick intrinsic GaAs layer at 300 K; (b) an *m-i-n* diode with 1 μm-thick intrinsic GaAs layer at 10 K;

From the Maxwell equations,

$$\begin{cases} \nabla \times E = -\mu_0 \dfrac{\partial H}{\partial t} \\ \nabla \times H = J + \dfrac{\partial D}{\partial t} \end{cases} \tag{12}$$

THz electric field, E_{THz}, induced by transient current, J, due to photoexcited carriers is given by

$$E_{\mathrm{THz}} \propto \frac{\partial v}{\partial t} \tag{13}$$

E_{THz} is the emitted ultrafast electromagnetic wave, which is proportional to carrier acceleration. The interband transitions induced by the ultrashort excitation pulses create electrons close to the bottom of the Γ valley in the conduction band as well as holes near the top of the valence band. In an ideal case without scattering, these electrons are ballistically accelerated by the applied electric field in the Γ valley. When electrons in the central valley have gained energy from the applied electric field comparable to the energetic distance between the bottoms of the Γ and the satellite valleys, the electrons are scattered to the satellite valleys by longitudinal optical (LO) phonon. Because the effective mass of electrons in the satellite valley is much heavier than that in the central valley, a sudden decrease of drift velocity is expected. Figure 10 (b) shows the ideal acceleration of electrons in bulk GaAs when the electrons are in the band as shown in Fig. 10 (a). The initial positive part shown in Fig. 10 (b) corresponds to ballistic acceleration of electrons in the Γ valley and the subsequent sudden dip expresses deceleration due to intervalley transfer from the Γ valley to the satellite valleys.

As shown in Fig. 9, the leading edge of the E_{THz} observed in the experiment is due both to the duration of the femtosecond pulses and to the limitation of the spectral bandwidth of the EO crystal detector. From the experimental data, we can clearly see the THz emission waveforms have a bipolar feature; i.e., an initial positive peak and a subsequent negative dip. This feature arises from the velocity overshoot [12]. The initial positive peak has been interpreted as electron acceleration in the Γ valley, where electrons have a light effective mass, while the subsequent negative dip has been attributed to the electron deceleration due to intervalley transfer from the Γ valley to the X and L valleys. From Figs. 9 (a) and (b), we can find at 10 K, the duration time of the waveforms is longer than that at 300 K at any bias electric fields, F, for both samples. The longer duration time of the THz waveforms at 10 K is due to the lower scattering rate of the LO phonon at 10 K, which will be discussed in detail in section 4.2. It is also clearly shown in the Figs. 9 (a) and (b), E_{THz} increases more abruptly with increasing electric field and its bipolar feature becomes pronounced.

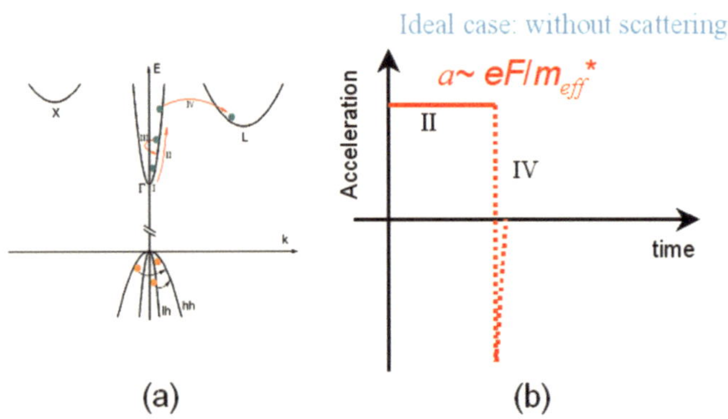

Fig. 10. (a) The band diagram of the GaAs; I: Electrons generated at the bottom of the Γ valey; II: Ballistic acceleration; III: Intravalley scattering; IV: Intervalley scattering; (b) ideal acceleration of electrons in bulk GaAs

Taking advantage of the novel experimental method, invaluable information on nonequilibrium carrier transport in the femtosecond time range, which has previously been discussed only by numerical simulations, has been obtained. The present insights on the nonstationary carrier transport contribute to better understanding of device physics in existing high speed electron devices and, furthermore, to new design of novel ultrafast electromagnetic wave oscillators.

4. Power dissipation spectrum under step electric field in Bulk GaAs in terahertz region

It is well known that negative differential conductivities (NDCs) due to intervalley transfer appear in many of the compound semiconductors, such as GaAs, under high electric fields. NDCs are of practical importance, notably for its exploitation in microwave and ultrafast electromagnetic wave / THz wave oscillators [47]. However, since such NDC gain has a finite bandwidth, it gives intrinsic upper frequency limit to ultrafast electromagnetic wave oscillators [48]. For this reason, a number of works have been done to investigate the mechanism, which limits the bandwidth of the gain, and found that it is mainly controlled by the energy relaxation time [49]-[51]. Figure 11 shows the real and imaginary parts of the differential mobility spectra, Re[$\mu(\omega)$] and Im[$\mu(\omega)$], in bulk GaAs for various electric fields at 300 K calculated by using Monte Carlo method [52]. Monte Carlo results predicted that the real part of the negative differential mobility (i.e., gain) can persist up to a few hundred GHz, as seen in Fig. 11. Together with theoretical works, efforts have been devoted to push the high frequency limit into the THz frequency range. Gunn and IMPATT diodes have already been operated at frequencies of 100-200 GHz [53]-[55]. However, in those experiments, high frequency operation was not possible due to electronic circuits; i.e., the cutoff frequency, ν_c, is mainly governed by stray capacitance, inductance, and more fundamentally by the length of the samples, not by the material properties themselves. Therefore, it is extremely important to characterize the gain bandwidth of materials in the THz range.

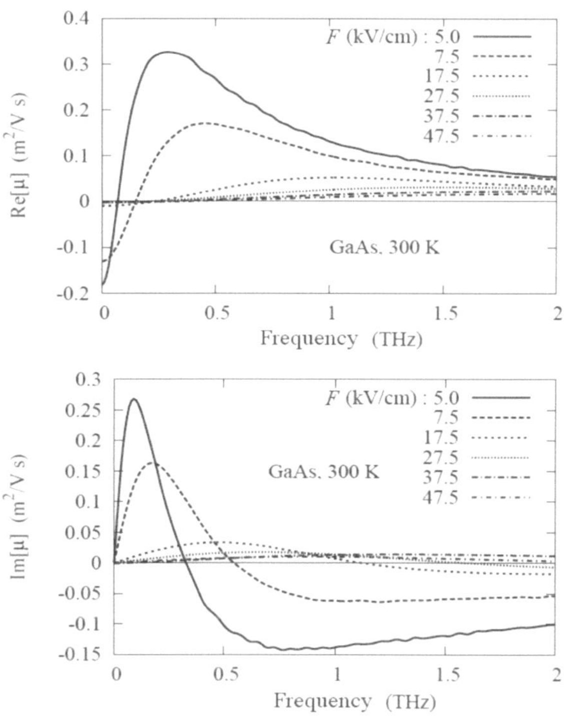

Fig. 11. The real and imaginary parts of the differential mobility spectra, Re[$\mu(\omega)$] and Im[$\mu(\omega)$], in bulk GaAs in THz range for various electric field at 300 K calculated by using Monte Carlo method [52]

The time domain THz spectroscopy has provided us with a unique opportunity to observe the motion of electron wave packet in the sub-picosecond time range and inherently measuring the response of the electron system to the applied bias electric field [11][12][15]. The Fourier spectra of the THz emission give us the power dissipation spectra under step electric field in THz range, from which we can find the gain region of material in THz range. In this section, we present the measured THz radiation from the intrinsic bulk GaAs under strong biased electric field. The power dissipation spectra under step electric field in THz range have been obtained by using the Fourier transformation of the time domain THz traces [17] [18]. From the power dissipation spectra, the cutoff frequency for negative power dissipation of the bulk GaAs has also been found. It is found that the cutoff frequency for the gain gradually increases with increasing electric fields up to 50 kV/cm and saturates at around 1 THz at 300 K / 750 GHz at 10 K. We also investigated the temperature dependence of cutoff frequency for negative power dissipation, from which we find that this cutoff frequency is governed by the energy relaxation process of electrons from the L valley to the Γ valley via successive optical phonon emission.

4.1 Power dissipation spectrum under step electric field in GaAs

Firstly, we recall the fact that the time domain THz emission experiments inherently measure the step response of the electron system to the applied electric field, as described in more detail in (Shimad et al., 2003). In the THz emission experiment, we first set a DC electric field, *F*, in the samples and, then, shoot a femtosecond laser pulse to the sample at *t* = 0 to create a step-function-like carrier density, nΘ(t), as shown in Fig. 12 (a)-(c). Subsequently, THz radiation is emitted from the accelerated photoexcited electrons [Fig. 12 (d)]. Now, by using our imagination, we view the experiment in a different way. Let's perform the following thought experiment, as shown in Figs. 12 (e)-(h); we first set electrons in the conduction band under a flat band condition [Fig. 12 (f)] and, at *t* = 0, suddenly apply a step-function-like bias electric field, FΘ(t) [Fig. 12 (g)]. We notice that electrons in the thought experiment emit the same THz radiation as in the actual experiment [Fig. 12 (h)]. This fact means that what we do in our femtosecond laser pulse measurement is equivalent to application of a step-function-like electric field to electrons. The step like electric field can be described as,

$$F(t) = F_0\Theta(t) = \begin{cases} 0 & (t<0) \\ F_0 & (t>0) \end{cases} \tag{14}$$

By noting this important implication, the power dissipation spectra under step-function-like electric fields in THz range can be obtained from the thought experimental scheme, as shown in the following;

In time domain, the power density is defined as $P(t) = \frac{\partial \varepsilon}{\partial t} = J(t)F(t)$, i.e., power density equals the product of the current density, *J(t)*, and electric field, *F(t)*. If the direction of *J(t)* is the same as the direction of *F(t)*, this gives a Joule heating. However, if P(t) is negative, this is a gain. Similarly, if the power dissipation spectrum, $S(\omega) = \frac{\partial \varepsilon}{\partial \omega}$, is negative at frequency ω, it means the system gains the energy at this frequency; i.e., gain in the frequency domain. The power dissipation spectrum, $S(\omega)$, is closely related to the differential conductivity spectrum, as

$$S(\omega) = \sigma(\omega)\left|F(\omega)\right|^2, \tag{15}$$

in the linear response regime. However, this formulation which use small signal conductivity, $\sigma(\omega)$, is not strictly correct in nonlinear response situation. Therefore, we will avoid the formulation using $\sigma(\omega)$ and derive the power dissipation spectrum from the time domain THz data.

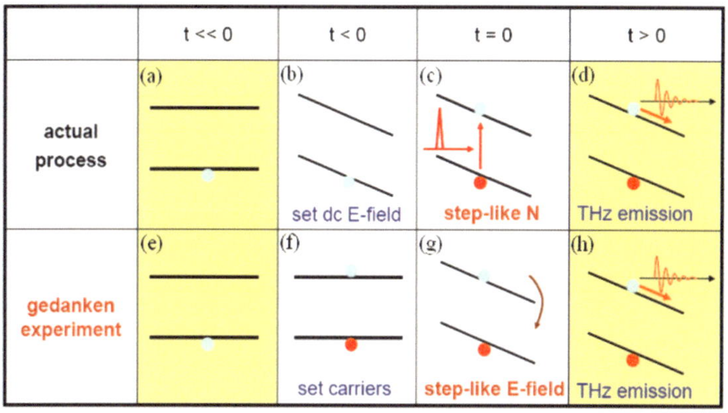

Fig. 12. The comparison between the real experiment [(a)-(d)] and thought experiment [(e)-(h)]. Electrons in the thought experiment emit the same THz radiation as in the actual experiment, which means the time domain THz emission experiments inherently measure the step response of the electron system to the applied electric field

Mathematically, the power dissipation, ε, can be expressed as,

$$\varepsilon = \int_{-\infty}^{+\infty} P(t)dt = V\int_{-\infty}^{+\infty} J(t)F(t)dt\,, \tag{16}$$

where V is the volume of the sample, $J(t)$ and $F(t)$ are the time dependent current density and applied electric field, respectively. On the other hand, the energy can also be expressed in the frequency domain as,

$$\varepsilon = V\int_{-\infty}^{+\infty} J(t)F(t)dt = \frac{V}{2\pi}\int_{-\infty}^{+\infty} J(\omega)\bar{F}(\omega)d\omega\,, \tag{17}$$

where $J(\omega)$ and $F(\omega)$ are the Fourier spectra of $J(t)$ and $F(t)$, respectively. Then, the power dissipation spectrum is obtained as,

$$S(\omega) = \frac{\partial\varepsilon}{\partial\omega} = \frac{V}{2\pi}J(\omega)\bar{F}(\omega) \tag{18}$$

From simple mathematics, $J(\omega)$ can be obtained from the Fourier transformation of the time domain THz traces as,

$$\begin{aligned} J(\omega) &= \int_{-\infty}^{+\infty} J(t)\exp(-i\omega t)dt \propto \int_{-\infty}^{+\infty}[\int_{-\infty}^{t} E_{\mathrm{THz}}(t)dt]\exp(-i\omega t)dt \\ &= \frac{1}{i\omega}\int_{-\infty}^{+\infty} E_{\mathrm{THz}}(t)\exp(-i\omega t)dt + \pi E_{\mathrm{THz}}(0)\delta(\omega) = \frac{1}{i\omega}E_{\mathrm{THz}}(\omega) + \pi E_{\mathrm{THz}}(0)\delta(\omega) \end{aligned} \tag{19}$$

As mentioned in section 4.3, the creation of step-function-like carrier density by femtosecond laser pulses in the actual experiment can be replaced with the application of step-function-like electric field in the thought experiment scheme, then, the $F(\omega)$ can be expressed as,

$$F(\omega) = \int_{-\infty}^{+\infty} F(t)\exp(-i\omega t)dt = F_0[\frac{1}{i\omega} + \pi\delta(\omega)]\,, \tag{20}$$

where F_0 is the magnitude of the internal electric field applied on the sample.

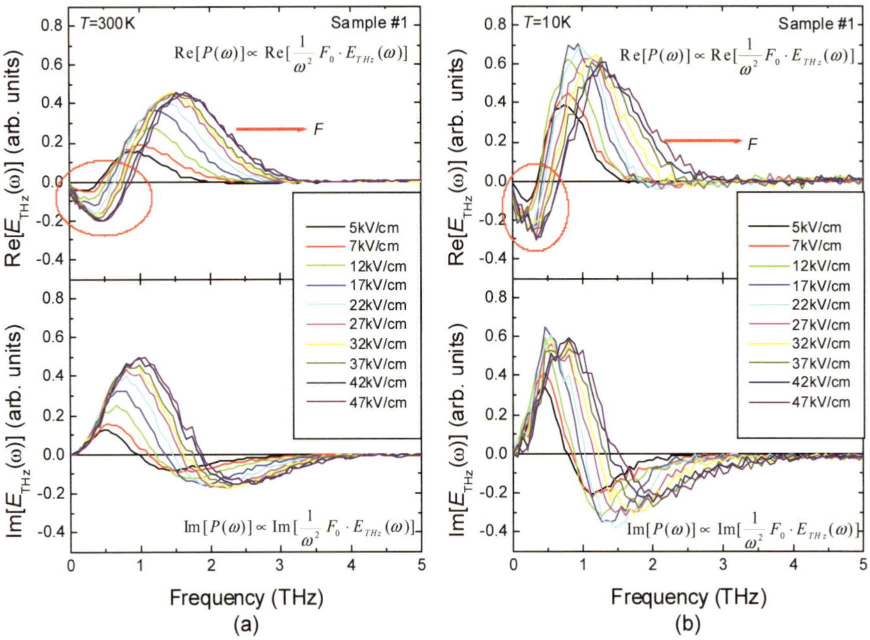

Fig. 13. The Re[$E_{THz}(\omega)$] and Im[$E_{THz}(\omega)$], which are proportional to the real and the imaginary parts of the power dissipation spectra in bulk GaAs for various F at 300K and 10K come from the Fourier transformation of THz traces emitted from sample #1, an *m-i-n* diode sample with a 1 μm-thick intrinsic GaAs layer

Simply substituting Eq. (19) and (20) into Eq. (18), the power dissipation spectrum under step-function-like electric field can be written as,

$$S(\omega) = \frac{F_0}{\omega^2} E_{\mathrm{THz}}(\omega) \quad \text{for } \omega \neq 0 \tag{21}$$

Then, we can obtain an important message from Eq. (21) that the real and imaginary part of the Fourier spectra of $E_{THz}(t)$ (i.e., Re[$E_{THz}(\omega)$] and Im[$E_{THz}(\omega)$]) are proportional to Re[S(ω)] and Im[S(ω)], respectively.

From the discussion above, we know from the measured temporal waveforms of THz electric fields emitted from GaAs samples shown in Figs. 9, the power dissipation spectra in intrinsic bulk GaAs under step-function-like electric fields can be determined by using Fourier transformation of $E_{THz}(t)$. Figures 13 (a) and (b) show Re[$E_{THz}(\omega)$] and Im[$E_{THz}(\omega)$] for sample #1, a metal-intrinsic-n type semiconductor (*m-i-n*) diode sample with a 1 μm-thick intrinsic GaAs layer under various electric fields at 300 K and 10 K, respectively. In the Fourier transformation process, the position of t = 0 is very crucial. Here, we determined t = 0 by using the maximum entropy method [44]-[46]. t = 0 can be determined within an error of ±30 fs, which is limited by the time interval of the recorded points. The phase misplacement can be written as $\Delta\theta = \omega\Delta t$. Therefore, the phase error is estimated within an error of < $\pm\pi/7$, assuming that the frequency of the signal, ω, is mainly around ~ 2.5 THz.

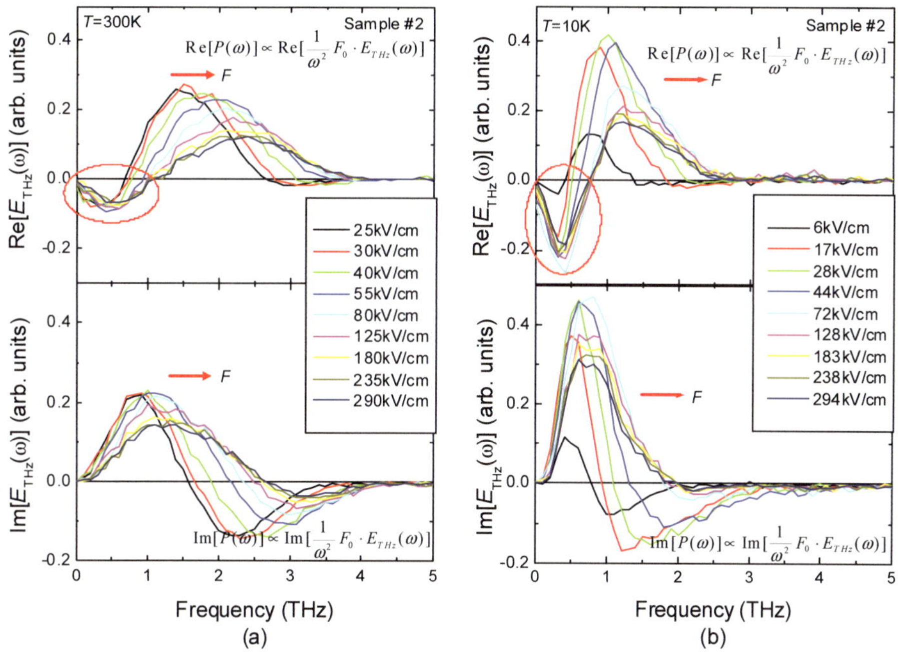

Fig. 14. The Re[$E_{THz}(\omega)$] and Im[$E_{THz}(\omega)$], which are proportional to the real and the imaginary parts of the power dissipation spectra in bulk GaAs for various F at 300K and 10K come from the Fourier transformation of THz traces emitted from sample #2, an *m-i-n* diode sample with a 180 nm-thick intrinsic GaAs layer

From Re[$E_{THz}(\omega)$] plotted in Fig. 13, we can see the negative region of Re[$S(\omega)$] persists up to several hundred GHz (~ 800 GHz at 47 kV/cm at 300 K), indicating that GaAs has an intrinsic gain region up to the THz range. This frequency is much higher than the gain region predicted by Monte Carlo simulation shown in Fig. 11. The negative power dissipation turns to positive as the frequency increases. Comparing with Re[$E_{THz}(\omega)$] at 10 K and 300 K shown in Figs. 13 (a) and (b), we find magnitude of $S(\omega)$ at 10 K is larger and the width is narrower than those at 300 K at same applied electric fields.
In the case of collisions with phonons in the Γ valley, each collision redirects the momentum and restrains the drift velocity. Furthermore, intervalley scattering from the Γ valley to the L valley requires phonon with a large wave vector, and group theoretical selection rules show that the longitudinal optical (LO) phonon is the only one allowed because the L valley minimum lies on the edge of the Brillioun zone. Because the probability of LO phonon emission is proportional to $\langle N_{LO}+1\rangle$ (N_{LO} is the thermal equilibrium phonon population per unit volume; $N_{LO} = \dfrac{1}{\exp(\hbar\omega_{LO} / k_B T) - 1}$), which is temperature dependent, we expect the scattering time is shorter at higher temperatures. The calculation result by Ruch and Fawcett's also showed different magnitude of drift velocity at low and high temperatures [56].
Furthermore, to investigate the power dissipation spectra under extremely large electric fields, we measured the time domain THz traces emitted from sample #2, an *m-i-n* diode with a 180 nm-thick intrinsic GaAs layer, at extremely high electric fields at 300 K and 10 K. Figures 14 (a) and (b) shows the real and imaginary parts of the Fourier spectra of the time

domain THz traces emitted from samples #2. We can see even at very high electric fields, for example, F = 300 kV/cm, the negative Re[$E_{THz}(\omega)$] still exists at low frequency region. At high frequency, the Re[$E_{THz}(\omega)$] changes its sign from negative to positive.

4.2 The cutoff frequency for negative power dissipation in GaAs

4.2.1 Electric field dependence of the cutoff frequency

From the real part of the power dissipation spectra, we can find the cutoff frequencies for the gain region, ν_c, at various F. The solid circles and the solid triangles in Fig. 15 show ν_c obtained from sample #1 as a function of F at 300 K and 10 K. The open circles and the open triangles correspond to ν_c obtained from sample #2 at 300 K and 10 K, respectively. The cutoff frequency increases with increasing F for F < 50 kV/cm. However, above 50 kV/cm, ν_c starts to saturate. The sharp kink of ν_c at 50 kV/cm shown in Fig. 15 strongly suggests that this is a very critical electric field. This point will be discussed in section 4.3.

4.2.2 Temperature dependence of the cutoff frequency

To clarify the mechanism of the cutoff frequency, ν_c, we also investigated the temperature dependence of ν_c. Figure 16 shows the temperature dependence of ν_c obtained for sample #1 at 7 kV/cm and 27 kV/cm by solid triangles and circles, respectively. We can clearly see that the temperature dependence of ν_c agrees with that of the emission rate of LO phonons, $<N_{LO}+1>$ (plotted by a dashed line in Fig. 16). The agreement between the experimental results and the LO phonon emission rate strongly suggests that ν_c is governed by the energy relaxation process of electrons from the L valley to the Γ valley via optical phonon emission. At lower temperature, LO phonon emission rate decreases and a longer energy relaxation time is expected, which results in a lower ν_c at 10 K than at 300 K, as shown in Fig. 15.

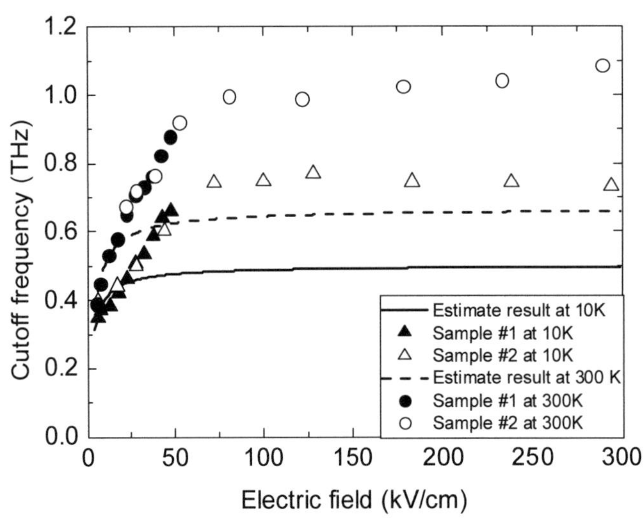

Fig. 15. The cutoff frequencies of the negative region of Re[$E_{THz}(\omega)$] obtained for sample #1, as a function of the electric field F at 300 K and 10 K are shown by solid circles and triangles, respectively. The open circles and triangles correspond to the cutoff frequencies obtained from sample #2, at 300 K and 10 K, respectively. The estimated cutoff frequencies are also plotted by a dashed line and a solid line at 300K and 10K, respectively

4.3 Mechanism for the cutoff frequency

When a small step-function electric field is added on a large DC electric field, the electron velocity is predominantly redirected, which causes a general increase of the kinetic energy, or a heating, of the electron system until the mean collision rate with the lattice has risen sufficiently to balance the increasing supply of energy to the electrons from the electric field. The characteristic time for this thermal readjustment is known as the energy relaxation time, which is the limiting process for the upper frequency limit of transferred electron devices [49]. However, the mechanism for the cutoff frequency for gain, ν_c, under extremely nonequilibrium condition has been still argued for more than 30 years.

4.3.1 Energy relaxation and momentum relaxation times govern ν_c.

Monte Carlo simulation has been used as a tool to estimate ν_c [52] [57], as shown in Fig. 11. Gruzinskis [58] and Reggiani, et al. [52] calculated the energy relaxation time τ_ε by Monte Carlo simulation and found that τ_ε is ~ 1.5 ps, as shown in Fig. 17. Also, they discussed that ν_c is determined by both energy relaxation time τ_ε and momentum relaxation time τ_v (τ_v is field dependent and $\tau_v << \tau_\varepsilon$, as shown in Fig. 17) [58]. The calculated ν_c is below 500 GHz, as shown in Fig. 11, which is much lower than ν_c obtained by our experiment.

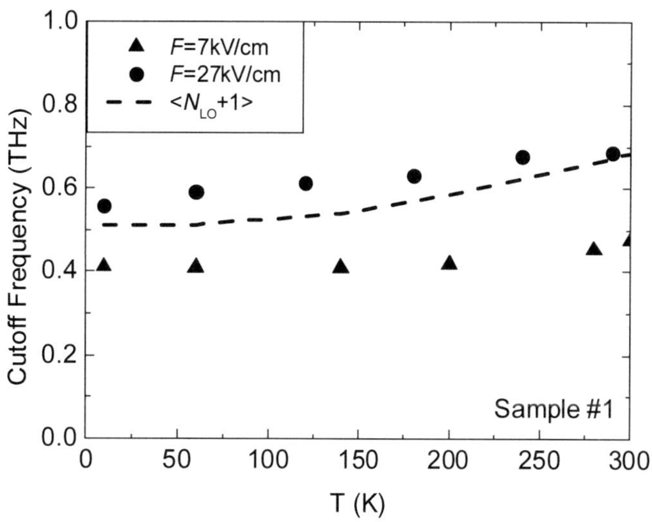

Fig. 16. The cutoff frequencies of the negative region of Re[$E_{THz}(\omega)$] obtained for sample #1, as a function of the temperature, at 7 kV/cm and 27 kV/cm are shown by solid triangles and circles, respectively. The reciprocal of the energy relaxation time due to the LO scattering in Γ valley as a function of temperature *T* is also plotted by a dashed line

4.3.2 Relaxation process of electrons in the L valley governs ν_c.

Another model is that the energy relaxation time is governed by relaxation process of electrons in the L valley [50]. The electrons start to be accelerated at the bottom of the Γ valley, scattered into the L valley, relax and dwell there, and, then, scattered back to the Γ valley. In this model, ν_c is mainly governed by relaxation time of electrons in the L valley and estimated to be ~ 200 GHz, which is also much lower than our experimental result.

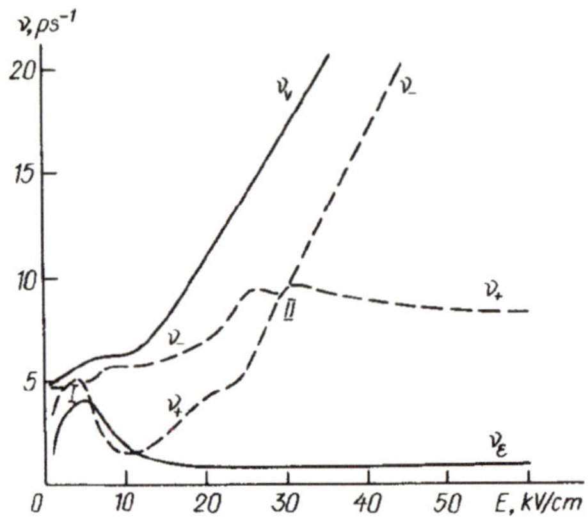

Fig. 17. The calculated energy and velocity relaxation rates (solid line) in n-InP [58]

4.3.3 Relaxation process of electrons in the Γ valley governs ν_c.

The temperature dependence of ν_c, strongly suggests that ν_c is mainly governed by the energy relaxation process of electrons from the L valley to the Γ valley via successive optical phonon emission. Here, we roughly estimate ν_c.

In the ε-k diagram, the electrons experience a cyclic motion in the k space; i.e., be accelerated from the bottom of the Γ valley, go to the L valley by intervalley scattering, be scattered back, and then relax to the bottom of the Γ valley again, as shown in Fig. 18 (a).

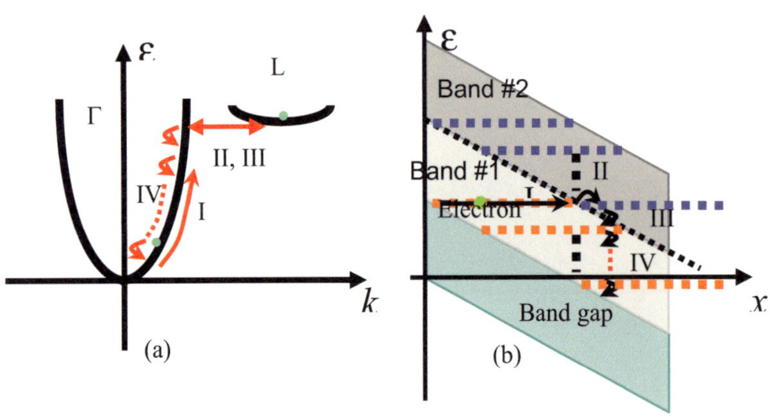

Fig. 18. (a) Electrons experience a cyclic motion in the k space; I: acceleration from the bottom of Γ valley; II: intervalley scattering to the L valley; III: intervalley scattering back to the Γ valley; IV: relaxation to the bottom of the Γ valley; (b) the real space picture; I: acceleration in the band #1 (associated with the Γ valley); II: reach the edge of the band #2 (the L valley related states) and experience intervalley scattering; III: scattering back; IV: relaxation to the bottom of the conduction band

For this order estimation, we assume electrons are ballistically accelerated by the electric field F in the Γ valley. In addition, we also neglect the acoustic scattering, e-e scattering. Assuming the simple Newton's law

$$\frac{\hbar dk}{dt} = -eF \, , \tag{22}$$

where e is the elementary charge, $\hbar = h/2\pi$ (h is the Planck constant), k is the wave vector of electrons, the time for the acceleration process can be estimated by assuming the energy separation between the Γ minimum and the bottom of L valley $\varepsilon_{\Gamma\text{-L}}$ = 0.29 eV. The subsequent intervalley scattering time is about 50 fs [59].

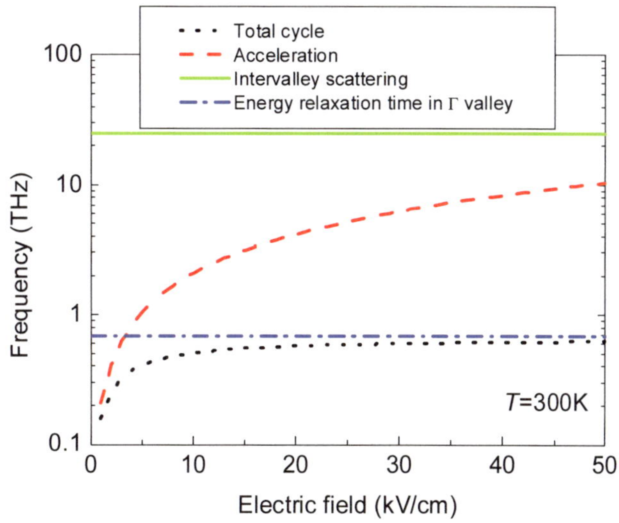

Fig. 19. The estimated cutoff frequencies of the negative region of Re[$E_{THz}(\omega)$]as a function of the electric field F at 300 K are shown by black dashed line. The red dashed line, green solid line and blue dash-dotted line correspond to ballistic acceleration process, the intervalley transfer process and the energy relaxation process in Γ valley, at 300 K, respectively

The energy relaxation process in the Γ valley is dominated by the LO phonon scattering. Since the LO phonon energy is 36.5 meV, 8 LO phonons must be successively emitted in the process for electrons to relax from the bottom of the L valley to the bottom of Γ valley ($\varepsilon_{\Gamma\text{-L}}$ = 0.29 eV). The relaxation time for one polar scattering process by LO phonon emission in bulk is expressed [49],

$$\frac{1}{\tau_{(e)}(\hat{k})} = \frac{2\pi}{\hbar}\frac{1}{(2\pi)^3}\frac{e^2\hbar\omega_{LO}}{2\varepsilon_0}\langle N_{LO}+1\rangle\left(\frac{1}{\varepsilon_\infty}-\frac{1}{\varepsilon_r}\right)\frac{m_{eff}{}^*}{\hbar^2\sqrt{\left|\hat{k}\right|^2-\frac{2m_{eff}{}^*\hbar\omega_{LO}}{\hbar^2}}}\times$$
$$\int_0^{+\infty}\int_0^{\pi}\int_0^{2\pi}\left|k'\right|^2\sin\theta\frac{1}{\left|\hat{k}'-\hat{k}\right|^2}\delta\left(\left|\hat{k}\right|'-\sqrt{\left|\hat{k}\right|^2-\frac{2m_{eff}{}^*\hbar\omega_{LO}}{\hbar^2}}\right)d\left|\hat{k}\right|'d\theta d\phi \, , \tag{23}$$

where ε_0 is the dielectric permittivity of the vacuum, ε_s is the static dielectric constant, ε_∞ is the high frequency dielectric constant, N_{LO} is the thermal equilibrium phonon population in per unit volume, $\hbar = h/2\pi$ (h is the Planck constant), ω_{LO} is the frequency of LO phonon, $m_{eff}{}^*$ is the effective mass of electron, and k, k' are the wave vector of electrons before and after scattering, respectively.

By using Eq. (23), the energy relaxation time in the Γ valley is estimated to be 1.46 ps at 300 K and 1.95 ps at 10 K, by summing up emission times of 8 LO phonons. The temperature dependence of the energy relaxation time is due to the temperature dependence of the phonon emission process, $\langle N_{LO}+1\rangle$. The estimated cutoff frequency, ν_c, at 300 K, are plotted in Fig. 19 [ballistic acceleration (dashed line), intervalley transfer (solid line), energy relaxation (dash-dotted line), and the total (dotted line)]. From this estimation, it is concluded that the energy relaxation process in the Γ valley takes the longest time and governs the upper frequency limitation for the gain. Furthermore, the estimated results are also plotted by a dashed line and a straight line for 300 K and 10 K, respectively, in Fig. 15 for the comparison with the experimental results.

As shown in Fig. 15, the overall trend of this estimation of ν_c is the same as our experimental results. However, at very high electric fields, the magnitude is off by 40%. The discrepancy between experiment and calculation may come from the field dependent energy relaxation time of electrons in the Γ valley as calculated by Fischetti [59]. Fischetti's calculation results show that at very high electric fields, the energy relaxation process is two times faster than that at low electric fields [59]. Then, the dash-dotted line of Fig. 19 shifts to higher frequency range, which results in higher ν_c.

Under very high electric fields, Wannier Stark (WS) ladder may be formed, although it has never been observed experimentally. The cyclic motion of electrons in the k space can be expressed in the real space as Fig. 18 (b). Electrons are created in the WS ladder states immediately after short pulsed photoexcitation under DC electric field F, and accelerated in the conduction band (band #1 in Fig. 18 (b)), which corresponds to the ballistic acceleration in the Γ valley. When they reach the edge of the band #1, they are scattered into the states in the band #2 (L valley related states, as shown in Fig. 18 (b)). After electrons dwell in the upper states, they are scattered downward (intervalley transfer process), and then relax to the bottom of the conduction band (energy relaxation in the Γ valley process).

Based on our experimental data, we conclude that the cutoff frequency is governed by the energy relaxation process of electrons in the Γ valley via successive optical phonon emission. In our estimation, we neglect scattering while electrons are accelerated by the electric field in the band #1. This scattering in the acceleration process makes the sharp kink in the cutoff frequency, ν_c, around 50 kV/cm, as shown in Fig. 15. In the low electric field range (F < 50 kV/cm), the number of events of LO phonon scattering decreases with increasing electric field F. Each scattering takes ~ 0.13 ps. However, in the high electric field range (F > 50 kV/cm), electrons at the bottom of the conduction band can go into the band #2 without experiencing any scattering. The mean free path of electrons can be estimated by,

$$\lambda = \frac{1}{2}\frac{eF}{m_{eff}^{*}}\tau_v^{\ 2}, \tag{24}$$

where the momentum relaxation time of electrons in band #1 (Γ valley) is about $\tau_v \sim 0.13$ ps, mainly limited by LO phonon scattering. The mean free path is calculated to be ~ 60 nm at F = 50 kV/cm. On the other hand, the distance, l, for electrons starting from the bottom of conduction band to the bottom of band #2 is estimated from,

$$eFl = \varepsilon_{\Gamma-\mathrm{L}} = 0.29 \text{ eV} . \tag{25}$$

At 50 kV/cm, l is calculated to be ~ 60 nm, almost the same as the mean free path of electrons. Therefore, we can conclude that when $F < 50$ kV/cm, ν_c increases with increasing F because LO phonon scattering strongly affects the acceleration process, and that for F > 50 kV/cm, ν_c is almost F independent because no scattering takes place in the acceleration process. The illustration of this situation is shown in Figs. 20.

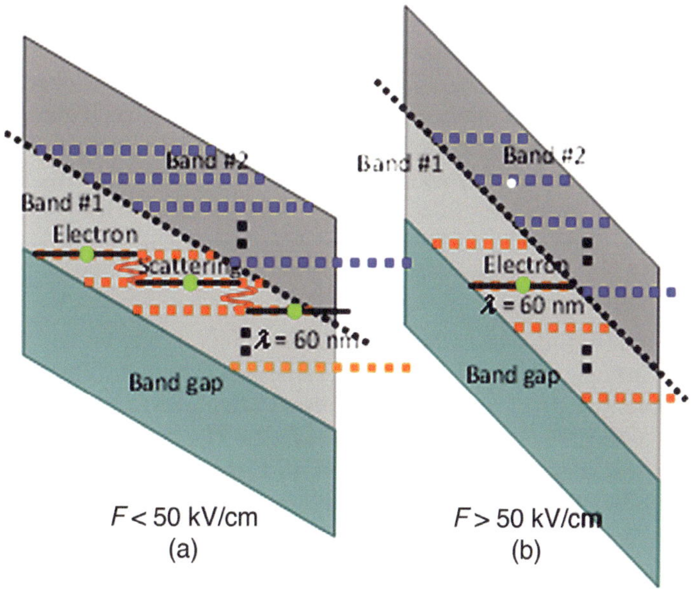

Fig. 20. (a) For $F < 50$ kV/cm, LO phonon scattering strongly affects the acceleration process; (b) $F > 50$ kV/cm, no scattering takes place in the acceleration process

The other possible explanation is that the strong band mixing gives the sharp kink in ν_c around 50 kV/cm, as shown in Fig. 15. For $F > 50$ kV/cm, the tunneling probability for electrons in the band #1 to the band #2 is significantly enhanced (Zener tunneling). Such Zener tunneling process may limit the increase in ν_c.

In summary, the power dissipation spectra under step electric field in bulk GaAs in THz range have been investigated by Fourier transformation of the ultrafast electromagnetic waveforms/THz waveforms emitted from intrinsic bulk GaAs photoexcited by femtosecond laser pulses under strong bias electric fields. The cutoff frequency for the gain region, that is of practical importance, notably for its exploitation in ultrafast electromagnetic wave generators, has been also found. The cutoff frequency gradually increases with increasing electric field within 50 kV/cm, saturates and reaches 1 THz (300 K) above 50 kV/cm. Furthermore, illustrated by the electric and temperature dependence of the cutoff frequency, we also find out that this cutoff frequency is mainly dominated by the energy relaxation process of electrons from the L to the Γ valley via successive optical phonon emission.

5. Conclusion

Ultrafast electromagnetic waves emitted from semiconductors under high electric fields, which are closely related with ultrafast nonequilibrium transport of carriers in semiconductor, are of fundamental interest in semiconductor physics. Furthermore, clarifying carrier dynamics under extreme nonequilibrium conditions is also strongly motivated by the need to obtain information relevant for the design of ultrahigh speed devices and ultrafast electromagnetic wave emitters. It is, therefore, essential to understand nonequilibrium transport of carriers subjected to high electric fields in such devices. The time domain terahertz (THz) spectroscopy gives us a unique opportunity of observing motions of electron wave packets in the sub-ps time range and measuring the response of electron systems to applied bias electric fields. By understanding the nonequilibrium transport of carriers in bulk GaAs, the intrinsic property of the negative differential conductivity (NDC) in the THz region, which means the gain, can be clarified.

In section 2, we describe the experimental technique of the time domain THz spectroscopy used in this work. Ultrafast carrier motion in the femtosecond time regime accompanies electromagnetic wave radiation that is proportional to dv/dt. Consequently, the investigation of such electromagnetic wave (or THz radiation) allows us a very unique opportunity of looking directly into the acceleration/deceleration dynamics of carriers in semiconductors.

In section 3, the femtosecond acceleration of carriers in bulk GaAs under very high electric fields, F, has been investigated by time domain THz spectroscopy. The observed ultrafast electromagnetic emission waveforms / THz emission waveforms have a bipolar feature; i.e., an initial positive peak and a subsequent negative dip. The initial positive peak has been interpreted as electron acceleration in the bottom of Γ the valley in GaAs, where electrons have a light effective mass, while the subsequent negative dip has been attributed to intervalley transfer from the Γ valley to the X and L valleys.

In section 4, we have investigated the gain due to intervalley transfer under high electric fields, which is of practical importance for its exploitation in ultrafast electromagnetic wave oscillators. The power dissipation spectra under step-function-like electric fields in THz range have been obtained by using the Fourier transformation of the time domain THz traces. From the power dissipation spectra, the cutoff frequencies for negative power dissipation (i.e., gain) in bulk GaAs have been determined. The cutoff frequency gradually increases with increasing electric field, F, for $F < 50$ kV/cm, saturates and reaches 1 THz (300 K) for $F > 50$ kV/cm. Furthermore, from the temperature dependence of the cutoff frequency, we found that it is governed by the energy relaxation process of electrons from L to Γ valley via successive optical phonon emission.

Finally, section 5 summarizes this chapter. As demonstrated in this chapter, taking advantage of the novel experimental method, invaluable information on nonequilibrium carrier transport in the femtosecond time range, which has previously been discussed only by numerical simulations, has been obtained. The present insights on the nonstationary carrier transport contribute to better understanding of device physics in existing high speed electron devices. Furthermore, the gain in GaAs due to electrons intervalley transfer under high electric fields obtained from the Fourier transformation of the time domain THz traces is also discussed, which is of practical importance for its exploitation in ultrafast electromagnetic wave oscillators.

6. Acknowledgment

The authors thank Mr. JunAn Zhu, a student from the University of Shanghai for Science and Technology, for editing the manuscript of this chapter.

7. References

[1] M. van Exter, Ch. Fattinger, and D. Grischkowsky, Terahertz time-domain spectroscopy of water vapour, *Opt. Lett.* Vol. 14, No. 20, pp. 1128-1130 (1989).

[2] D. Grischkowsky, S. Keiding, M. van Exter, and Ch. Fattinger, Far-infrared time-domain spectroscopy with terahertz beams of dielectrics and semiconductors, *J. Opt. Soc. Am. B*, Vol. 7, No. 10, pp. 2006-2015 (1990).

[3] R. A. Cheville, and D. Grischkowsky, Far-infrared terahertz time-domain spectroscopy of flames, *Opt. Lett.*, Vol. 20, No. 15, pp. 1646-1648 (1995).

[4] B. Ferguson, S. Wang, D. Gray, D. Abbott, and X.-C. Zhang, T-ray computed tomography, *Opt. Lett.*, Vol. 27, No. 15, pp. 1312-1314 (2002).

[5] B. B. Hu, and M. C. Nuss, Imaging with terahertz waves, *Opt. Lett.* Vol. 20, No. 16, pp. 1716-1718 (1995).

[6] S.P. Mickan, A. Menikh, H. Liu, C.A. Mannella, R. MacColl, D. Abbott, J. Munch, and X.-C. Zhang, Label-free bioaffinity detection using terahertz technology, *Physics in Medicine and Biology (IOP)*, Vol. 47, No. 21, pp. 3789-3795 (2002).

[7] S.P. Mickan, D. Abbott, J. Munch, and X-C Zhang, Noise reduction in terahertz thin film measurements using a double modulated differential technique, *Fluctuation and Noise Letters (Elsevier)*, Vol. 2, No. 1, pp. R13-R28 (2002).

[8] M. J. W. Rodwell, M. Urteaga, T. Mathew, D. Scott, D. Mensa, Q. Lee, J. Guthrie, Y. Betser, S. C. Martin, R. P. Smith, S. Jaganathan, S. Krishnan, S. I. Long, R. Pullela, B. Agarwal, U. Bhattacharya, L. Samoska, and M. Dahlstrom, Submicron scaling of HBTs, *IEEE Trans. Electron Devices,* Vol. 48, No. 11, pp. 2606-2624 (2001).

[9] M. Beck, D. Hofstetter, T. Aellen, J. Faist, U. Oesterle, M. Ilegems, E. Gini, and H. Melchior, Continuous Wave Operation of a Mid-Infrared Semiconductor Laser at Room Temperature, *Science,* Vol. 295, No. 5553, pp. 301-305 (2002).

[10] G. A. Mourou, C. V. Stancampiano, A. Antonetti, and A. Orszag, Picosecond microwave pulses generated with a subpicosecond laser-driven semiconductor switch, *Appl. Phys. Lett.* Vol. 39, No. 4, pp. 295-296 (1981).

[11] A. Leitenstorfer, S. Hunsche, J. Shah, M. C. Nuss, and W. H. Knox, Femtosecond Charge Transport in Polar Semiconductors, *Phys. Rev. Lett.,* Vol. 82, No. 25, pp. 5140–5142 (1999).

[12] A. Leitenstorfer, S. Hunsche, J. Shah, M. C. Nuss, and W. H. Knox, Femtosecond high-field transport in compound semiconductors, *Phys. Rev. B,* Vol. 61, No. 24, pp. 16642–16652 (2000).

[13] G. Zhao, R. N. Schouten, N. van der Valk, W. T. Wenchebach, and P. C. M. Planken, Design and performance of a THz emission and detection setup based on a semi-insulating GaAs emitter, *Rev. Sci. Instrum.,* Vol. 73, No. 4, pp. 1715-1719 (2002).

[14] N. Katzenellenbogen, and D. Grischkowsky, Efficient generation of 380 fs pulses of THz radiation by ultrafast laser pulse excitation of a biased metal-semiconductor interface, *Appl. Phys. Lett.*, Vol. 58, No. 3, pp. 222-224 (1991).

[15] M. Abe, S. Madhavi, Y. Shimada, Y. Otsuka, and K. Hirakawa, Transient carrier velocities in bulk GaAs: Quantitative comparison between terahertz data and ensemble Monte Carlo calculations, *Appl. Phys. Lett.*, Vol. 81, No. 4, pp. 679-681 (2002).

[16] Y. M. Zhu, T. Unuma, K. Shibata, and K. Hirakawa, Femtosecond acceleration of electrons under high electric fields in bulk GaAs investigated by time-domain terahertz spectroscopy, *Appl. Phys. Lett.*, Vol. 93, No. 4, pp. 042116-042118 (2008).
[17] Y. M. Zhu, T. Unuma, K. Shibata, and K. Hirakawa, Power dissipation spectra and terahertz intervalley transfer gain in bulk GaAs under high electric fields, *Appl. Phys. Lett.*, Vol. 93, No. 23, pp. 232102-232104 (2008).
[18] Y. M. Zhu, X. X. Jia, L. Chen, D. W. Zhang, Y. S. Huang, B. Y. He, S. L. Zhuang, and K. Hirakawa, Terahertz power dissipation spectra of electrons in bulk GaAs under high electric fields at low temperature, *Acta Phys. Sinica,* Vol. 58, No. 4, pp. 2692-2695 (2009).
[19] D. H. Auston, K. P. Cheung, J. A. Valdmanis, and D. A. Kleinman, Cherenkov Radiation from Femtosecond Optical Pulses in Electro-Optic Media, *Phys. Rev. Lett.*, Vol. 53, No. 16, pp. 1555–1558 (1984).
[20] Ch. Fattinger, and D. Grischkowsky, Point source terahertz optics, *Appl. Phys. Lett.*, Vol. 53, No. 16, pp. 1480–1482 (1988).
[21] D. H. Auston, K. P. Cheung, and P. R. Smith, Picosecond photoconducting Hertzian dipoles, *Appl. Phys. Lett.*, Vol. 45, No. 3, pp. 284-286 (1984).
[22] A. P. DeFonzo, M. Jarwala, and C. R. Lutz, Transient response of planar integrated optoelectronic antennas, *Appl. Phys. Lett.*, Vol. 50, No. 17, pp. 1155–1157 (1987).
[23] C. Johnson, F. J. Low, and A. W. Davidson, Germanium and germanium-diamond bolometers operated at 4.2 K, 2.0 K, 1.2 K, 0.3 K, and 0.1 K, *Opt. Eng.*, Vol. 19, pp. 255-258 (1980).
[24] R. C. Jones, The ultimate sensitivity of radiation detectors, *J. Opt. Soc. Am.* Vol. *37*, No. 11, pp. *879*–890 (1947).
[25] D. R. Dykaar, T. Y. Hsiang, and G. A. Mourou, An application of picosecond electro-optic sampling to superconducting electronics, *IEEE Trans. Magn.,* Vol. 21, No. 2, pp. 230-233 (1985).
[26] J. A. Valdmanis, and G. A. Mourou, Sub-Picosecond Electro-Optic Sampling: Principles and Applications, *IEEE J. Quant. Electron.*, Vol. 22, No. 1, pp. 69-78 (1986).
[27] P. U. Jepsen, C. Winnewisser, M. Schall, V. Schya, S. R. Keiding, and H. Helm, Detection of THz pulses by phase retardation in lithium tantalite, *Phys. Rev. E* Vol. 53, No. 4, pp. 3052-3054 (1996).
[28] A. Nahata, D. H. Auston, T. F. Heinz, and C. Wu, Coherent detection of freely propagating terahertz radiation by electro-optic sampling, *Appl. Phys. Lett.*, Vol. 68, No. 2, pp. 150-152 (1996).
[29] Q. Wu, and X.-C. Zhang, Free-space electro-optic sampling of terahertz beams, *Appl. Phys. Lett.* Vol. 67, No. 24, pp. 3523-3525 (1995).
[30] G. Gallot, and D. Grischkowsky, Electro-optic detection of terahertz radiation, *J. Opt. Soc. Am. B,* Vol. 16, No. 8, pp. 1204-1212 (1999).
[31] A. Leitenstrofer, S. Hunsche, J. Shah, M. C. Nuss, and W. H. Know, Detectors and sources for ultrabroadband electro-optic sampling: Experiment and theory, *Appl. Phys. Lett.,* Vol. 74, No. 11, pp. 1516-1518 (1999).
[32] Z. G. Lu, P. Campbell, and X.-C. Zhang, Free-space electro-optic sampling with a high-repetition-rate regenerative amplified laser, *Appl. Phys. Lett.*, Vol. 71, No. 5, pp. 593-595 (1997).

[33] C. Winnewisser, P. U. Jepsen, M. Schall, V. Schya, and H. Heim, Electro-optic detection of THz radiation in $LiTaO_3$, $LiNbO_3$ and ZnTe, *Appl. Phys. Lett.*, Vol. 70, No. 23, pp. 3069-3071 (1997).
[34] Q. Wu, and X.-C. Zhang, Electrooptic sampling of freely propagating terahertz fields, *Opt. Quant. Electron.*, Vol. 28, No. 7, pp. 945-951 (1996).
[35] Q. Wu, M. Litz, and X.-C. Zhang, Broadband detection capability of ZnTe electro-optic field detectors, *Appl. Phys. Lett.*, Vol. 68, No. 21, pp. 2924-2926 (1996).
[36] H. J. Bakker, G. C. Cho, H., Kurz Q. Wu, and X.-C. Zhang, Distortion of terahertz pulses in electro-optic sampling, *J. Opt. Soc. Am. B*, Vol. 15, No. 6, pp. 1795-1801 (1998).
[37] Q. Wu, and X.-C. Zhang, 7 terahertz broadband GaP electro-optic sensor, *Appl. Phys. Lett.*, Vol. 70, No. 14, pp. 1784-1786 (1997).
[38] W. L. Faust, and C. H. Henry, Mixing of Visible and Near-Resonance Infrared Light in GaP, *Phys. Rev. Lett.*, Vol. 17, No. 25, pp. 1265–1268 (1966).
[39] K. K. Thornber, *Path Integrals and Their Applications in Quantum, Statistical and Solid-State physics*, Springer, ISBN-10: 0306400170, Plenum, New York (October 1, 1978).
[40] Y. Yamashita, A. Endoh, K. Shinohara, K. Hikosaka, T. Matsui, S. Hiyamizu, and T. Mimura, Pseudomorphic $In_{0.52}Al_{0.48}As/In_{0.7}Ga_{0.3}As$ HEMTs with an ultrahigh fT of 562 GHz, *IEEE Electron Dev. Lett.*, Vol. 23, No. 10, pp. 573-575 (2002).
[41] P.C.M. Planken, H.K. Nienhuys, H.J. Bakker, and T. Wenckebach, Measurement and calculation of the orientation dependence of terahertz pulse detection in ZnTe, *J. Opt. Soc. Am. B*, Vol. 18, No. 3, pp. 313-317 (2001).
[42] Q. Wu, and X.-C. Zhang, Free-space electro-optics sampling of mid-infrared pulses, *Appl. Phys. Lett.*, Vol. 71, No. 10, pp. 1285-1286 (1997).
[43] N. Sekine, and K. Hirakawa, Dispersive Terahertz Gain of a Nonclassical Oscillator: Bloch Oscillation in Semiconductor Superlattices, *Phys. Rev. Lett.*, Vol. 94, No. 5, pp. 057408-067411 (2005).
[44] E. Gornov, E. M. Vartiainen, and K.-E. Peiponen, Comparison of subtractive Kramers-Kronig analysis and maximum entropy model in resolving phase from finite spectral range reflectance data, *Appl. Opt.*, Vol. 45, No. 25, pp. 6519-6524 (2006).
[45] E. M. Vartianen, K.-E., Peiponen and T. Asakura, Phase Retrieval in Optical Spectroscopy: Resolving Optical Constants from Power Spectra, *Appl. Spectrosc.*, Vol. 50, No. 10, pp. 1283-1289 (1996).
[46] E. M. Vartiainen, Y. Ino, R. Shimano, M. Kuwata-Gonokami, Y. P. Svirko, and K.-E. Peiponen, Nemerical phase correction method for terahertz time-domain reflection spectroscopy, *J. Appl. Phys.*, Vol. 96, pp. 4171-4175 (2004).
[47] J. B. Gunn, and C. A. Hogarth, A Novel Microwave Attenuator Using Germanium, *J.Appl.Physics*, Vol. 26, pp. 353-354 (1955).
[48] P. Das, and P. W. Staecker, *Proc. of Sixth Int. Con. on Micr. Gen. & Ampl.*, Cambridge (1966).
[49] P. J. Bulman, G. S. Hobson, and B. C. Taylor, *Transferred Electron Devices*, Academic Press, ISBN-10: 0121408507, London and New York (1972).
[50] P. Das, and R. Bharat, Hot electron relaxation times in two-valley semiconductors and their effect on bulk-microwave oscillators, *Appl. Phys. Lett.*, Vol. 11, No. 12, pp. 386-388 (1967).

[51] A. F. Gibson, T. W. Granville, and E. G. S. Paige, A study of energy-loss processes in germanium at high electric fields using microwave techniques, *J. Phys. Chem. Solid* Vol. 19, No. 3-4, pp. 198-217 (1961).
[52] L. Reggiani, INFM - National Nano-technology Laboratory, Dipartimento di Ingegneria dell's Innovazione, Universita di Lecce. (Calculated by Professor Reggiani L.)
[53] C. Benz, and J. Freyer, 200 GHz pulsed GaAs-IMPATT diodes, *Electron. Lett.*, Vol. 34, No. 24, pp. 2351-2353 (1998).
[54] R. Kamoua, H. Eisele, and G. I. Haddad, D-band (110-170 GHz) InP Gunn devices D-Band (110-170GHz) InP Gunn Devices, *Solid-State Electron.*, Vol. 36, No. 11, pp. 1547-1555 (1993).
[55] M. F. Zybura, S. H. Jones, B. W. Lim, J. D. Crowley, and J. E. Carlstrom, 125–145 GHz stable depletion layer transferred electron oscillators, *Solid-State Electron.*, Vol. 39, No. 4, pp. 547-553 (1996).
[56] J. G. Ruch, and W. Fawcett, Temperature dependence of transport properties of Gallium Arsenide determined by a Monte-carlo method, *J. Appl. Phys.*, Vol. 41, No. 9, pp. 3843-3865 (1970).
[57] H. D. Rees, Hot electron effects at microwave frequencies in GaAs, *Soli. Stat. Commun.*, Vol. 7, No. 2, pp. 267-270 (1969).
[58] V. Gruzinskis, E. Starikov, and P. Shiktorov, Conservation equations for hot carriers—II. Dynamic features, *Solid-State Electron.* Vol. 36, No. 7, pp. 1067-1075 (1993).
[59] M. V. Fischetti, Monte-carlo simulation of transport in technologically significant semiconductors of the diamond and zincblende structures. 1. Homogeneous transport, *IEEE Trans. Electron Devices,* Vol. 38, No. 3, pp. 634-649 (1991).

10

Electromagnetic Wave Propagation in Ionospheric Plasma

Ali Yeşil[1] and İbrahim Ünal[2]
[1]*Fırat University, Elazığ*
[2]*İnönü University, Malatya*
Turkey

1. Introduction

Ionosphere physics is related to plasma physics because the ionosphere is, of course, a weak natural plasma: an electrically neutral assembly of ions and electrons [1]. The ionosphere plays a unique role in the Earth's environment because of strong coupling process to regions below and above [2]. The ionosphere is an example of naturally occurring plasma formed by solar photo-ionization and soft x-ray radiation. The most important feature of the ionosphere is to reflect the radio waves up to 30 MHz. Especially, the propagation of these radio waves on the HF band makes the necessary to know the features and the characteristics of the ionospheric plasma media. Because, when the radio waves reflect in this media, they are reflected and refracted depending on their frequency, the frequency of the electrons in the plasma and the refractive index of the media and thus, they are absorbed and reflected by the media.

The understanding of the existence of a conductive layer in the upper atmosphere has been emerged in a century ago. The idea of a conductive layer affected by the variations of the magnetic field in the atmosphere has been put forward by the Gauss in 1839 and Kelvin in 1860. Newfoundland radio signal from Cornwall to be issued by Marconi in 1901, at the first experimental evidence of the existence of ionospheric, respectively. In 1902, Kennely and Heaviside indicated that the waves are reflected from a conductive layer on the upper parts of the atmosphere. In 1902, Marconi stated that changing the conditions of night and day spread. In 1918, high-frequency band has been used by aircraft and ships. HF band in the 1920s has increased the importance of expansion. Then, put forward the theory of reflective conductive region, has been shown by experiments made by Appleton and Barnet. The data from the 1930s started to get clearer about the ionosphere and The Radio Research Station, Cavendish Laboratory, the National Braun of Standards, the various agencies such as the Carnegie Institution began to deal with the issue. In the second half of the 20th century, the work of the HF electromagnetic wave has been studied by divided into three as the full-wave theory, geometrical optics and conductivity. Despite initiation of widespread use of satellite-Earth communication systems, the use of HF radio spectrum for civilian and military purposes is increasing. Collapse of communication systems, especially in case of emergency situations, communication is vital in this band.

In the ionosphere, a balance between photo-ionization and various loss mechanisms gives rise to an equilibrium density of free electrons and ions with a horizontal stratified structure. The density of these electrons is a function of the height above the earth's surface and is dramatically affected by the effects of sunrise and sunset, especially at the lower altitudes. Also, the many parameters in the ionosphere are the function of the electron density. The ionosphere is conventionally divided into the D, E, and F-regions. The D-region lies between 60 and 95 km, the E-region between 95 and 150 km, and the F-region lies above 150 km. During daylight, it is possible to distinguish two separate layers within the F-region, the F1 (lower) and the F2 (upper) layers. During nighttime, these two layers combine into one single layer. The combined effect of gravitationally decreasing densities of neutral atoms and molecules and increasing intensity of ionizing solar ultra-violet radiation with increasing altitudes, gives a maximum plasma density during daytime in the F-region at a few hundred kilometers altitude. During daytime, the ratio of charged particles to neutral particles concentration can vary from10^{-8} at 100 km to 10^{-4} at 300 km and 10^{-1} at 1000 km altitude. The main property of F-region consists of the free electrons.

As known, the permittivity and permeability parameters are related to electric and magnetic susceptibilities of material, on account of medium and moreover, the speed of electromagnetic wave and the characteristic impedance depend on any medium and the refractive index of medium gives detail information about any medium. Because of all of these reasons, ε is a measure of refractive index, reflection, volume and wave polarization of electromagnetic, impedance of medium.

Propagation of electromagnetic waves in the atmosphere is influenced by the spatial distribution of the refractive index of the ionosphere [3]. The theory of ionospheric conductivity was developed by many scientists and is now quite well understood, though refinements are still made from time to time. The ionosphere carries electric currents because winds and electric fields drive ions and electrons. The direction of the drift is at right angles to the geomagnetic field [4]. Furthermore, electrical conductivity is an important central concept in space science, because it determines how driving forces, such as electric fields and thermosphere winds, couple to plasma motions and the resulting electric currents. The tensor of electrical conductivity finds application in all the areas of ionospheric electrodynamics and at all the latitudes [5]. On the other hand, the most important parameter determining the behavior of any medium is the dielectric constant, which at any frequency determines the refractive index, the form of wave in medium, to be polarized, the state of wave energy and the propagation of wave.

In this chapter, the behavior of electromagnetic waves emitted from within the ionospheric plasma and the analytical solutions are necessary to understand the characteristics of the environment will be defined. Problems in plasma physics at the conductivity, dielectric constants and refractive index will be defined according to the media parameters. When these expressions, using Maxwell's equations expressed in the wave dispersion equation, wave propagation, depending on the parameters of the environment will be examined. These statements are expressed in terms of Maxwell's equations using the wave dispersion equation, wave propagation, depending on the parameters of the environment will be examined. By examining of the dispersion relation, the types of wave occurred in the media and relaxation mechanisms, polarizations and conflicts caused by ionospheric amplitude attenuation of these waves will be obtained analytically. Thus, resolving problems of ionospheric plasma in the emitted radio waves, the basic information that will be understood.

2. Conductivity for ionospheric plasma

One of the main parameters affecting the progression of the electromagnetic wave in a medium is the conductivity of the media. Conductivity of the media statement is obtained from motion equation of the charged particle that is taken account in the total force acting on charged particle in ionospheric plasma. Accordingly, the forces acting on charged particle is given as follows [6].

$$\begin{aligned}\text{Mass xacceleration} = &\text{ Electrical forces}\\ &+ \text{Magnetic forces}\\ &+ \text{Shooting (gravitational) forces}\\ &+ \text{Pressure changing forces}\\ &+ \text{Collision forces}\end{aligned} \tag{1}$$

The plasma approach, which the thermal motion due to temperatures of particles, are neglected, is called cold plasma. In this study, because of the cold plasma approximation, any force does not effect on charged particles due to the temperature and therefore the term of pressure changes is ignored [7]. Likewise, the gravitational force from the gravity is negligible due to so small according to the electric and magnetic forces. In addition, due to the electron mass (m_e) is very small according to the mass of the ion (m_i) ($m_e << m_i$) and thus the electron collision frequency is greater than the ion collision frequency, the movement of ions in the plasma is ignored near the electron movement. Accordingly, Equation (1) is written as the following equation for the electron:

$$m_e \frac{d\mathbf{V}_e}{dt} = -e(\mathbf{E} + \mathbf{V}_e \times \mathbf{B}) - m_e \nu_e \mathbf{V}_e \tag{2}$$

This expression is called the Langevin equation [8]. Where, the electron collision frequency ν_e is sum of the electron-ion collision frequency (ν_{ei}) and electron-neutral particle collision (ν_{en}) frequency.

2.1 DC conductivity

The first application of the Langevin equation is steady flow caused by of the electric field. For this, $\dot{\mathbf{V}}_e = 0$. Accordingly, the Langevin equation for electrons can be defined as follows:

$$0 = -e(\mathbf{E} + \mathbf{V}_e \times \mathbf{B}) - m_e \nu_e \mathbf{V}_e \tag{3}$$

Taking into consideration that the current density is $\mathbf{J_e} = -eN_e\mathbf{V_e}$, in order to obtained the current density, both sides of the expression is multiplied by the term of $-eN_e$. In this case,

$$e^2 N_e(\mathbf{E} + \mathbf{V}_e \times \mathbf{B}) = -m_e e N_e \nu_e \mathbf{V}_e \tag{4}$$

and the expression is transformed as follows:

$$e^2 N_e(\mathbf{E} + \mathbf{V}_e \times \mathbf{B}) = m_e \nu_e \mathbf{J} \tag{5}$$

For the current density, the following equation is obtained:

$$\mathbf{J} = \frac{e^2 N_e}{m_e \nu_e}\left(\mathbf{E} + \mathbf{V}_e \times \mathbf{B}\right) \tag{6}$$

Taking into consideration that $\mathbf{J} = \sigma \cdot \mathbf{E}$, the generalized Ohm Law can be defined by,

$$\mathbf{J} = \sigma_0\left(\mathbf{E} + \mathbf{V}_e \times \mathbf{B}\right) \tag{7}$$

Where, $\sigma_0 = \dfrac{e^2 N_e}{m_e \nu_e}$ is the dc conductivity of te plasma. If $\nu_e \to 0$, the term σ_0 is being $\sigma_0 \to \infty$. In this case, if the $\mathbf{E} + \mathbf{V}_e \times \mathbf{B} = 0$, the current is finite. For this reason, the case of $\sigma_0 = \infty$, the state $\mathbf{E} + \mathbf{V}_e \times \mathbf{B} = 0$ is called OHM law for the plasma. The geometry of the velocity and electric field and magnetic field is as shown in Figure 1, namely $\mathbf{V} = \hat{x}V_x + \hat{y}V_y + \hat{z}V_z$, $\mathbf{E} = \hat{x}E_x + \hat{y}E_y + \hat{z}E_z$ and $\mathbf{B} = \hat{z}B$, the $\mathbf{V}_e \times \mathbf{B}$ term in Equation (7) is being as follows:

$$\mathbf{V}_e \times \mathbf{B} = \begin{vmatrix} \hat{x} & \hat{y} & \hat{z} \\ V_x & V_y & V_z \\ 0 & 0 & B \end{vmatrix} = \hat{x}\left(V_y B\right) - \hat{y}\left(V_x B\right) \tag{8}$$

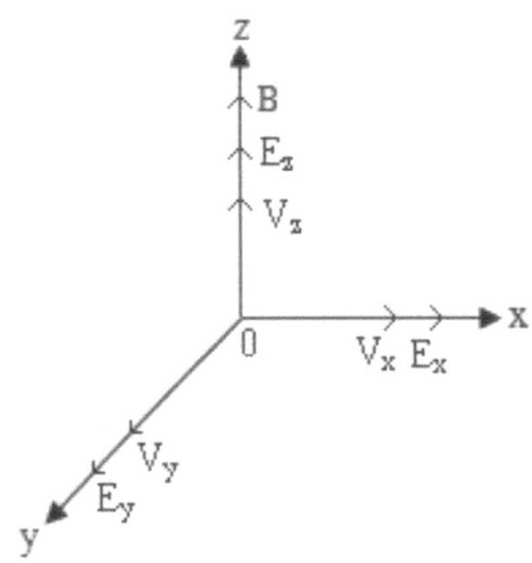

Fig. 1. The geometry of the velocity, electric field and magnetic field

Here, by the necessary mathematical operations are made; the following expressions are obtained for three components of current density:

$$J_x = \frac{e^2 N_e}{m_e \nu_e}\left(E_x + V_y B\right) = \frac{e^2 N_e}{m_e \nu_e} E_x + \frac{e^2 N_e V_y B}{m_e \nu_e} \tag{9}$$

$$J_y = \frac{e^2 N_e}{m_e \nu_e}\left(E_y - V_x B\right) = \frac{e^2 N_e}{m_e \nu_e} E_y - \frac{e^2 N_e V_x B}{m_e \nu_e} \tag{10}$$

$$J_z = \frac{e^2 N_e}{m_e \nu_e} E_z = \sigma_0 E_z \tag{11}$$

Taking into consideration that cyclotron frequency $\omega_{ce} = -\frac{eB}{m_e}$ for the electron, if the second terms on the right side of this expression is written in terms of current density **J** and the necessary procedures are made, x and y components of current density are obtained as follows:

$$J_x = \frac{e^2 N_e \nu_e}{m_e\left(\nu_e^2 + \omega_{ce}^2\right)} E_x + \frac{e^2 N_e \omega_{ce}}{m_e\left(\nu_e^2 + \omega_{ce}^2\right)} E_y \tag{12}$$

$$J_y = \frac{e^2 N_e \omega_{ce}}{m_e\left(\nu_e^2 + \omega_{ce}^2\right)} E_x + \frac{e^2 N_e \nu_e}{m_e\left(\nu_e^2 + \omega_{ce}^2\right)} E_y \tag{13}$$

The expressions J_x, J_y and J_z in the tensor form of $([\mathbf{J}]=[\sigma]\cdot[\mathbf{E}])$ can be written as follows:

$$\begin{bmatrix} J_x \\ J_y \\ J_z \end{bmatrix} = \begin{bmatrix} \frac{e^2 N_e \nu_e}{m_e\left(\nu_e^2 + \omega_{ce}^2\right)} & \frac{e^2 N_e \omega_{ce}}{m_e\left(\nu_e^2 + \omega_{ce}^2\right)} & 0 \\ -\frac{e^2 N_e \omega_{ce}}{m_e\left(\nu_e^2 + \omega_{ce}^2\right)} & \frac{e^2 N_e \nu_e}{m_e\left(\nu_e^2 + \omega_{ce}^2\right)} & 0 \\ 0 & 0 & \frac{e^2 N_e}{m_e \nu_e} \end{bmatrix} \cdot \begin{bmatrix} E_x \\ E_y \\ E_z \end{bmatrix} \tag{14}$$

Here, the conductivity tensor is called as dc conductivity tensor. As can be seen in this expression, $\sigma_0 = \frac{e^2 N_e}{m_e \nu_e}$ term provides the flow of the current in **B** direction. If the conductivity tensor is defined as follows:

$$\sigma = \begin{bmatrix} \sigma_1 & \sigma_2 & 0 \\ -\sigma_2 & \sigma_1 & 0 \\ 0 & 0 & \sigma_0 \end{bmatrix} \tag{15}$$

$\sigma_1 \left(= \frac{e^2 N_e \upsilon_e}{m_e\left(\nu_e^2 + \omega_{ce}^2\right)} \right)$ Conductivity is called as perpendicular conductivity ($\sigma_\perp$). When the **E** is perpendicular to the **B**, this conductivity provides to flow the current in the direction of **E**. The conductivity defined by $\sigma_2 \left(= \frac{e^2 N_e \omega_{ce}}{m_e\left(\nu_e^2 + \omega_{ce}^2\right)} \right)$ is called as Hall conductivity (σ_H) and it provides the current, which is perpendicular to the both of the **E** and **B** (Figure 2).

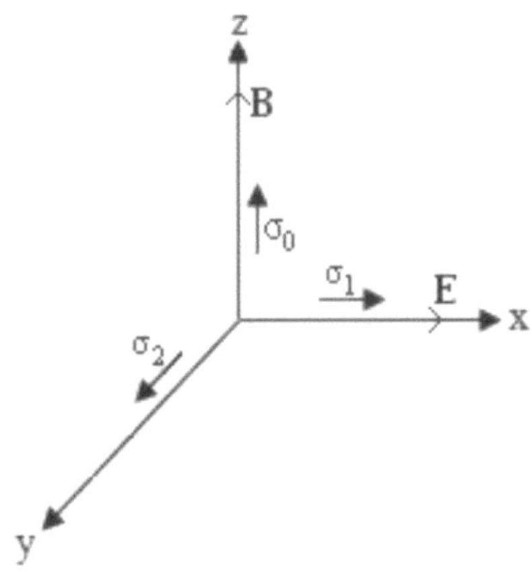

Fig. 2. The direction of the currents provided by the conductivities σ_0, σ_1 and σ_2 according to the geometry of **E** and **B**

If the $\sigma_1 = \dfrac{\sigma_0 \nu_e^2}{\nu_e^2 + \omega_{ce}^2} = \dfrac{\sigma_0}{1 + \dfrac{\omega_{ce}^2}{\nu_e^2}}$ and $\sigma_2 = \dfrac{\sigma_0 \nu_e \omega_{ce}}{\nu_e^2 + \omega_{ce}^2} = \dfrac{\sigma_0 \dfrac{\omega_{ce}}{\nu_e}}{1 + \dfrac{\omega_{ce}^2}{\nu_e^2}}$, the variation of the conductivities is being as shown in Figure 3. As shown in Figure 3, after the $\dfrac{\omega_{ce}}{\nu_e} \rangle 1$ σ_1 term decreases rapidly. Thus, the current perpendicular to the magnetic field is very small [8].

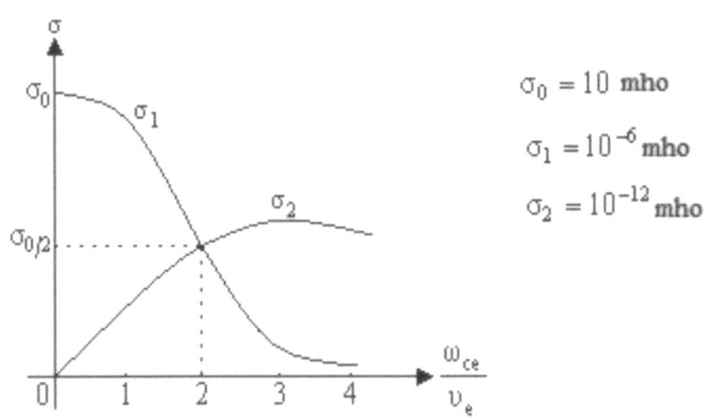

Fig. 3. The variation of σ_1 and σ_2 conductivities

2.2 AC conductivity

There is no steady-state in the ac conductivity. Taking into consideration that the fields vary as $e^{-i\omega t}$, the Equation (2) is transformed as follows:

$$\text{-}m_e i\omega \mathbf{V_e} = \text{-}e\left(\mathbf{E} + \mathbf{V_e} \times \mathbf{B}\right) - m_e \nu_e \mathbf{V_e} \tag{16}$$

According to the geometry given in Figure 1, this expression in terms of current density can be defined as follows:

$$\mathbf{J}(\nu_e - i\omega) = \frac{e^2N_e}{m_e}\mathbf{E} - \frac{e^2N_e}{m_e}(\mathbf{V}_e \times \mathbf{B}) \tag{17}$$

Thus, current density components are obtained as follows:

$$J_x = \frac{e^2N_e}{m_e(\nu_e - i\omega)}E_x + \frac{e^2N_e}{m_e(\nu_e - i\omega)}V_yB \tag{18}$$

$$J_y = \frac{e^2N_e}{m_e(\nu_e - i\omega)}E_y - \frac{e^2N_e}{m_e(\nu_e - i\omega)}V_xB \tag{19}$$

$$J_z = \frac{e^2N_e}{m_e(\nu_e - i\omega)}E_z \tag{20}$$

Considering the ω_{ce} cyclotron frequency, the components J_x and J_y can be obtained as follows:

$$J_x = \frac{e^2N_e(\nu_e - i\omega)}{m_e\left[\omega_{ce}^2 + (\nu_e - i\omega)^2\right]}E_x + \frac{e^2N_e\omega_{ce}}{m_e\left[\omega_{ce}^2 + (\nu_e - i\omega)^2\right]}E_y \tag{21}$$

$$J_y = -\frac{e^2N_e\omega_{ce}}{m_e\left[\omega_{ce}^2 + (\nu_e - i\omega)^2\right]}E_x + \frac{e^2N_e(\nu_e - i\omega)}{m_e\left[\omega_{ce}^2 + (\nu_e - i\omega)^2\right]}E_y \tag{22}$$

If the expressions (20)-(22) in the tensor form and the conductivity tensor σ' are defined as follows,

$$\begin{bmatrix} J_x \\ J_y \\ J_z \end{bmatrix} = \begin{bmatrix} \sigma_1' & \sigma_2' & 0 \\ -\sigma_2' & \sigma_1' & 0 \\ 0 & 0 & \sigma_0' \end{bmatrix} \cdot \begin{bmatrix} E_x \\ E_y \\ E_z \end{bmatrix} \tag{23}$$

$$\sigma' = \begin{bmatrix} \sigma_1' & \sigma_2' & 0 \\ -\sigma_2' & \sigma_1' & 0 \\ 0 & 0 & \sigma_0' \end{bmatrix} \tag{24}$$

The conductivities σ_0', σ_1' and σ_2' will be $\sigma_0' = \frac{e^2N_e}{m_e(\nu_e - i\omega)}$, $\sigma_1' = \frac{e^2N_e(\nu_e - i\omega)}{m_e\left[\omega_{ce}^2 + (\nu_e - i\omega)^2\right]}$, $\sigma_2' = \frac{e^2N_e\omega_{ce}}{m_e\left[\omega_{ce}^2 + (\nu_e - i\omega)^2\right]}$. Considering the oscillations frequency of the plasma for the electron as $\omega_{pe}^2\left(=\dfrac{e^2N_e}{m_e\varepsilon_0}\right)$, these equations can be written as $\sigma_0' = \dfrac{\omega_{pe}^2\varepsilon_0}{(\nu_e - i\omega)}$, $\sigma_1' = \dfrac{\omega_{pe}^2\varepsilon_0(\nu_e - i\omega)}{\left[\omega_{ce}^2 + (\nu_e - i\omega)^2\right]}$, $\sigma_2' = \dfrac{\omega_{pe}^2\varepsilon_0\omega_{ce}}{\left[\omega_{ce}^2 + (\nu_e - i\omega)^2\right]}$, respectively. Where, as in dc conductivity σ_0' is the conductivity of the magnetic field direction, σ_1' is the conductivity perpendicular to the magnetic field and σ_2' is the conductivity perpendicular to both of the electrical and the magnetic fields, respectively. If $\sigma_0' > \sigma_1' > \sigma_2'$ and $(\nu_e - i\omega)$, the ac conductivity tensor transforms to the dc conductivity tensor.

Example: ac conductivity tensor according to the geometry given in Figure 4 is obtained as in Equation (25):

$$\sigma' = \begin{bmatrix} \dfrac{\omega_{pe}^2\varepsilon_0(\nu_e - i\omega)}{\left[\omega_{ce}^2 + (\nu_e - i\omega)^2\right]} & 0 & -\dfrac{\omega_{pe}^2\varepsilon_0\omega_{ce}}{\left[\omega_{ce}^2 + (\nu_e - i\omega)^2\right]} \\ 0 & \dfrac{\omega_{pe}^2\varepsilon_0}{(\nu_e - i\omega)} & 0 \\ \dfrac{\omega_{pe}^2\varepsilon_0\omega_{ce}}{\left[\omega_{ce}^2 + (\nu_e - i\omega)^2\right]} & 0 & \dfrac{\omega_{pe}^2\varepsilon_0(\nu_e - i\omega)}{\left[\omega_{ce}^2 + (\nu_e - i\omega)^2\right]} \end{bmatrix} \tag{25}$$

If the magnetic field is only $\mathbf{B} = \hat{x}B$ direction, ac conductivity tensor becomes as follows:

$$\sigma' = \begin{bmatrix} \dfrac{\omega_{pe}^2\varepsilon_0}{(\nu_e - i\omega)} & 0 & 0 \\ 0 & \dfrac{\omega_{pe}^2\varepsilon_0(\nu_e - i\omega)}{\left[\omega_{ce}^2 + (\nu_e - i\omega)^2\right]} & \dfrac{\omega_{pe}^2\varepsilon_0\omega_{ce}}{\left[\omega_{ce}^2 + (\nu_e - i\omega)^2\right]} \\ 0 & -\dfrac{\omega_{pe}^2\varepsilon_0\omega_{ce}}{\left[\omega_{ce}^2 + (\nu_e - i\omega)^2\right]} & \dfrac{\omega_{pe}^2\varepsilon_0(\nu_e - i\omega)}{\left[\omega_{ce}^2 + (\nu_e - i\omega)^2\right]} \end{bmatrix} \tag{26}$$

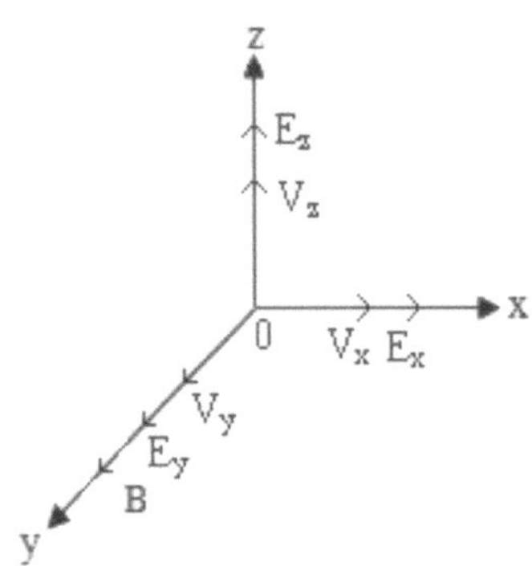

Fig. 4. The geometry of the velocity and electric field and magnetic field

Example: If the magnetic field has two-dimensional geometry as given in Figure 5 $\left(\mathbf{B} = \hat{y}B_y + \hat{z}B_z = \hat{y}\,BCos\theta + \hat{z}\,BSin\theta\right)$, the conductivity tensor are defined as in Equation (27):

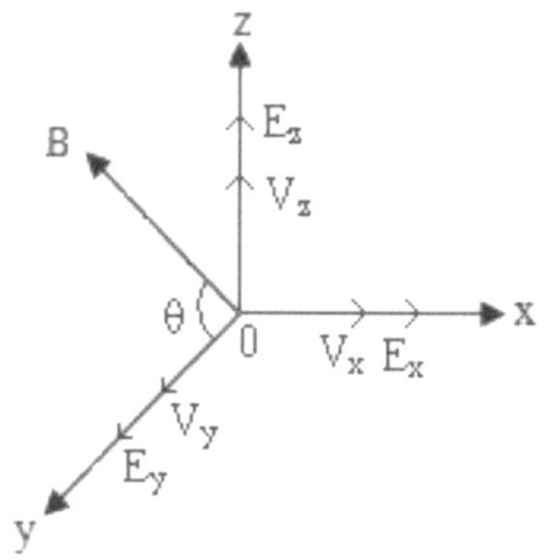

Fig. 5. The geometry of the velocity and electric field and magnetic field

$$\sigma' = \begin{bmatrix} \sigma_1' & \sigma_2'\,Sin\theta & -\sigma_2'\,Cos\theta \\ \sigma_2'\,Sin\theta & \sigma_0' + \left(\sigma_0' - \sigma_1'\right)\,Sin^2\theta & -\left(\sigma_0' - \sigma_1'\right)\,Cos\theta Sin\theta \\ \sigma_2'\,Cos\theta & \left(\sigma_0' - \sigma_1'\right)\,Cos\theta Sin\theta & \sigma_0' + \left(\sigma_0' - \sigma_1'\right)\,Cos^2\theta \end{bmatrix} \tag{27}$$

2.3 Earth's magnetic field and ionospheric conductivity

In this section, the Langevin equation defined by Equation (16) for ac conductivity will be discussed. However, the true magnetic field in the Earth's northern half-sphere given in Figure 6 will be dealt with. Accordingly, in the selected Cartesian coordinate system, the x-axis represents the geographic east, the y-axis represents the the geographic North and z-axis represents the up in the vertical direction, the magnetic field can be defined as follows [9]:

$$\mathbf{B} = \hat{x}\,BCosISinD + \hat{y}\,BCosICosD - \hat{z}\,BSinI \tag{28}$$

Where, I is the dip angle and D is the declination angle (between the magnetic north and the geographic north).
Thus, the term of $\mathbf{V}_e \times \mathbf{B}$ in Equation (16) can be obtained as in Equation (29).

$$\mathbf{V}_e \times \mathbf{B} = \begin{vmatrix} \hat{x} & \hat{y} & \hat{z} \\ V_x & V_y & V_z \\ BCosISinD & BCosICosD & -BSinI \end{vmatrix} = \hat{x}\left(-V_yBSinI - V_zBCosICosD\right)$$
$$-\hat{y}\left(-V_xBSinI - V_zBCosISinD\right) \tag{29}$$
$$+\hat{z}\left(V_xBCosICosD - V_yBCosISinD\right)$$

If this expression is written in Equation (16) and the necessary mathematical manipulations are made, three equations are obtained for x, y and z directions as follows:

$$-\frac{e}{m_e\left(\nu_e - i\omega\right)}E_x = V_x + \frac{\omega_{ce}\,SinI}{\left(\nu_e - i\omega\right)}V_y + \frac{\omega_{ce}\,CosICosD}{\left(\nu_e - i\omega\right)}V_z \tag{30}$$

$$-\frac{e}{m_e(\nu_e - i\omega)}E_y = V_y - \frac{\omega_{ce}\,SinI}{(\nu_e - i\omega)}V_x - \frac{\omega_{ce}\,CosISinD}{(\nu_e - i\omega)}V_z \tag{31}$$

$$-\frac{e}{m_e(\nu_e - i\omega)}E_z = V_z - \frac{\omega_{ce}\,CosICosD}{(\nu_e - i\omega)}V_x + \frac{\omega_{ce}\,CosISinD}{(\nu_e - i\omega)}V_y \tag{32}$$

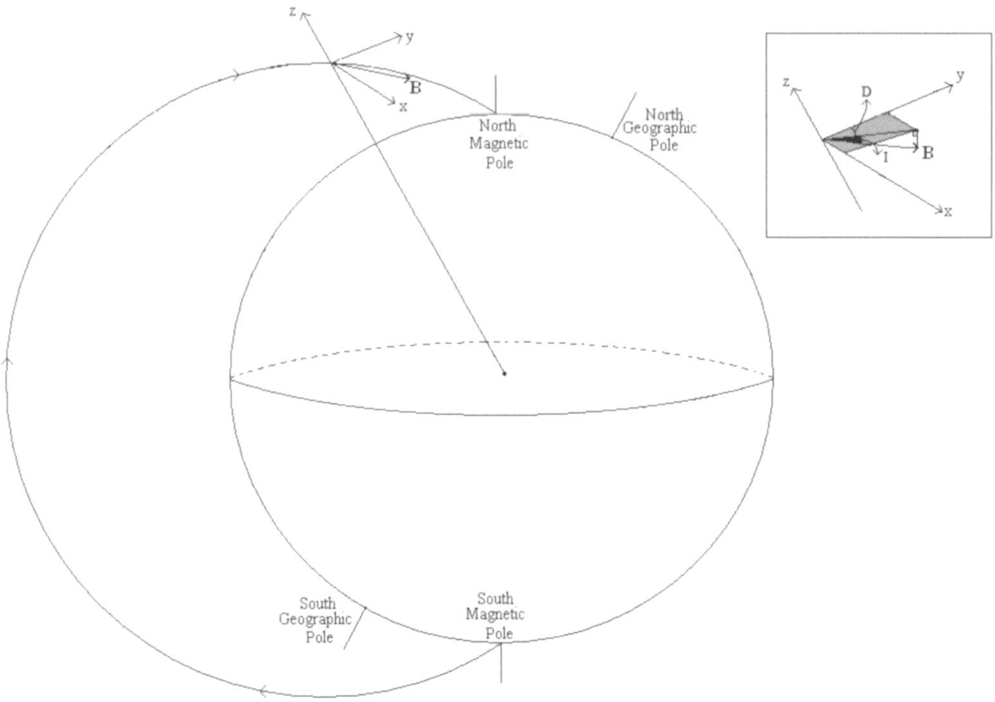

Fig. 6. Earth's magnetic field on north hemisphere

In order to obtain the current density, if the both sides of the expression is multiplied by $-eN_e$, the Equations (30)-(32) transform to Equations (33)-(35).

$$\frac{e^2N_e}{m_e(\nu_e - i\omega)}E_x = J_x + \frac{\omega_{ce}\,SinI}{(\nu_e - i\omega)}J_y + \frac{\omega_{ce}\,CosICosD}{(\nu_e - i\omega)}J_z \tag{33}$$

$$\frac{e^2N_e}{m_e(\nu_e - i\omega)}E_y = -\frac{\omega_{ce}\,SinI}{(\nu_e - i\omega)}J_x + J_y - \frac{\omega_{ce}\,CosISinD}{(\nu_e - i\omega)}J_z \tag{34}$$

$$\frac{e^2N_e}{m_e(\nu_e - i\omega)}E_z = -\frac{\omega_{ce}\,CosICosD}{(\nu_e - i\omega)}J_x + \frac{\omega_{ce}\,CosISinD}{(\nu_e - i\omega)}J_y + J_z \tag{35}$$

These equations are first order linear equation in three unknowns. It is impossible to obtain this expression in the solution of each other. Therefore, whether or not the solution, it is necessary to express by using, "Cramer's Method". This method gives the solution of linear equation system which has coefficients matrix as square matrix. According to this method, if the determinant of the coefficients matrix of the system is non-zero, the equation has a single solution. Let accept the $\frac{e^2 N_e}{m_e(\nu_e - i\omega)} = \sigma_0'$, $\frac{\omega_{ce}\,\mathrm{SinI}}{(\nu_e - i\omega)} = a$, $\frac{\omega_{ce}\,\mathrm{CosICosD}}{(\nu_e - i\omega)} = b$ and $\frac{\omega_{ce}\,\mathrm{CosISinD}}{(\nu_e - i\omega)} = c$ in Equations (33)-(35). Thus, these tree equations can be defined as follows:

$$\begin{bmatrix} \sigma_0' E_x \\ \sigma_0' E_y \\ \sigma_0' E_z \end{bmatrix} = \begin{bmatrix} 1 & a & b \\ -a & 1 & -c \\ -b & c & 1 \end{bmatrix} \cdot \begin{bmatrix} J_x \\ J_y \\ J_z \end{bmatrix} \tag{36}$$

Here, the coefficients matrix is named as A, by taking the determinant of this matrix, the Equation (37) can be obtained.

$$\det A = \begin{vmatrix} 1 & a & b \\ -a & 1 & -c \\ -b & c & 1 \end{vmatrix} = 1 + a^2 + b^2 + c^2 \neq 0 \tag{37}$$

This tells us that one solution of the equation. According to Cramer solution is as follows:

$$J_x = \frac{\det A_1}{\det A}, \quad J_y = \frac{\det A_2}{\det A}, \quad J_z = \frac{\det A_3}{\det A} \tag{38}$$

Here, the terms of A_1, A_2 and A_3 are defined as follows:

$$\det A_1 = \begin{vmatrix} \sigma_0' E_x & a & b \\ \sigma_0' E_y & 1 & -c \\ \sigma_0' E_z & c & 1 \end{vmatrix}, \quad \det A_2 = \begin{vmatrix} 1 & \sigma_0' E_x & b \\ -a & \sigma_0' E_y & -c \\ -b & \sigma_0' E_z & 1 \end{vmatrix}, \quad \det A_3 = \begin{vmatrix} 1 & a & \sigma_0' E_x \\ -a & 1 & \sigma_0' E_y \\ -b & c & \sigma_0' E_z \end{vmatrix} \tag{39}$$

In this statement, the x direction current density is resolved as in Equation (40).

$$J_x = \frac{\det A_1}{\det A} = \frac{\sigma_0'(1+c^2)}{1+a^2+b^2+c^2} E_x + \frac{\sigma_0'(bc-a)}{1+a^2+b^2+c^2} E_y + \frac{\sigma_0'(-ac-b)}{1+a^2+b^2+c^2} E_z \tag{40}$$

For the a, b and c, if the instead of the expressions given above are written, the J_x term is obtained as follows:

$$\begin{aligned} J_x = {} & \frac{\sigma_0'\left[(\nu_e - i\omega)^2 + \omega_{ce}^2 \mathrm{Cos}^2\mathrm{ISin}^2\mathrm{D}\right]}{\left[(\nu_e - i\omega)^2 + \omega_{ce}^2\right]} E_x + \frac{\sigma_0'\left[\omega_{ce}^2 \mathrm{Cos}^2\mathrm{ICosDSinD} - (\nu_e - i\omega)\,\omega_{ce}\mathrm{SinI}\right]}{\left[(\nu_e - i\omega)^2 + \omega_{ce}^2\right]} E_y \\ & + \frac{\sigma_0'\left[-\omega_{ce}^2 \mathrm{CosISinISinD} - (\nu_e - i\omega)\,\omega_{ce}\mathrm{CosISinD}\right]}{\left[(\nu_e - i\omega)^2 + \omega_{ce}^2\right]} E_z \end{aligned} \tag{41}$$

By the same way, y direction the current density is resolved as follows:

$$J_y = \frac{\det A_2}{\det A} = \frac{\sigma_0'(a+bc)}{1+a^2+b^2+c^2}E_x + \frac{\sigma_0'(1+b^2)}{1+a^2+b^2+c^2}E_y + \frac{\sigma_0'(c-ab)}{1+a^2+b^2+c^2}E_z \tag{42}$$

and the J_y term is obtained as follows:

$$\begin{aligned} J_y = {} & \frac{\sigma_0'\left[(\nu_e - i\omega)\omega_{ce}\mathrm{SinI} + \omega_{ce}^2 \mathrm{Cos}^2\mathrm{ICosDSinD}\right]}{\left[(\nu_e - i\omega)^2 + \omega_{ce}^2\right]}E_x + \frac{\sigma_0'\left[(\nu_e - i\omega)^2 + \omega_{ce}^2 \mathrm{Cos}^2\mathrm{ICos}^2\mathrm{D}\right]}{\left[(\nu_e - i\omega)^2 + \omega_{ce}^2\right]}E_y \\ & + \frac{\sigma_0'\left[(\nu_e - i\omega)\omega_{ce}\mathrm{CosISinD} - \omega_{ce}^2 \mathrm{CosISinICosD}\right]}{\left[(\nu_e - i\omega)^2 + \omega_{ce}^2\right]}E_z \end{aligned} \tag{43}$$

Likely, z direction the current density is resolved as follows:

$$J_z = \frac{\det A_3}{\det A} = \frac{\sigma_0'(b-ac)}{1+a^2+b^2+c^2}E_x + \frac{\sigma_0'(-c-ab)}{1+a^2+b^2+c^2}E_y + \frac{\sigma_0'(1+a^2)}{1+a^2+b^2+c^2}E_z \tag{44}$$

and the J_z term is obtained as follows:

$$\begin{aligned} J_z = {} & \frac{\sigma_0'\left[(\nu_e - i\omega)\omega_{ce}\mathrm{CosICosD} - \omega_{ce}^2 \mathrm{CosISinISinD}\right]}{\left[(\nu_e - i\omega)^2 + \omega_{ce}^2\right]}E_x \\ & + \frac{\sigma_0'\left[-(\nu_e - i\omega)\omega_{ce}\mathrm{CosISinD} - \omega_{ce}^2 \mathrm{CosISinICosD}\right]}{\left[(\nu_e - i\omega)^2 + \omega_{ce}^2\right]}E_y + \frac{\sigma_0'\left[(\nu_e - i\omega)^2 + \omega_{ce}^2 \mathrm{Sin}^2\mathrm{I}\right]}{\left[(\nu_e - i\omega)^2 + \omega_{ce}^2\right]}E_z \end{aligned} \tag{45}$$

After being current densities obtained in this manner, J_x, J_y and J_z terms can be edited again by considering the σ_1' an σ_2' conductivities. Also, J_x, J_y and J_z terms can be written in the form of tensor like the following:

$$\begin{bmatrix} J_x \\ J_y \\ J_z \end{bmatrix} = \begin{bmatrix} \sigma_{11} & \sigma_{12} & \sigma_{13} \\ \sigma_{21} & \sigma_{22} & \sigma_{23} \\ \sigma_{31} & \sigma_{32} & \sigma_{33} \end{bmatrix} \cdot \begin{bmatrix} E_x \\ E_y \\ E_z \end{bmatrix} \tag{46}$$

The conductivity tensor can defined as in Equation (47):

$$\sigma'' = \begin{bmatrix} \sigma_{11} & \sigma_{12} & \sigma_{13} \\ \sigma_{21} & \sigma_{22} & \sigma_{23} \\ \sigma_{31} & \sigma_{32} & \sigma_{33} \end{bmatrix} \tag{47}$$

and tensor components can be achieved in a simpler as follows [9, 10]:

$$\sigma_{11} = \sigma_1' + \left(\sigma_0' - \sigma_1'\right) \mathrm{Cos}^2 \mathrm{I}\, \mathrm{Sin}^2 \mathrm{D}$$

$$\sigma_{12} = -\sigma_2' \mathrm{SinI} + \left(\sigma_0' - \sigma_1'\right) \mathrm{Cos}^2 \mathrm{ICosDSinD}$$

$$\sigma_{13} = -\sigma_2' \mathrm{CosICosD} - \left(\sigma_0' - \sigma_1'\right) \mathrm{CosISinISinD}$$

$$\sigma_{21} = \sigma_2' \mathrm{SinI} + \left(\sigma_0' - \sigma_1'\right) \mathrm{Cos}^2 \mathrm{ICosDSinD}$$

$$\sigma_{22} = \sigma_1' + \left(\sigma_0' - \sigma_1'\right) \mathrm{Cos}^2 \mathrm{ICos}^2 \mathrm{D}$$

$$\sigma_{23} = \sigma_2' \mathrm{CosISinD} - \left(\sigma_0' - \sigma_1'\right) \mathrm{CosISinICosD}$$

$$\sigma_{31} = \sigma_2' \mathrm{CosICosD} - \left(\sigma_0' - \sigma_1'\right) \mathrm{CosISinISinD}$$

$$\sigma_{32} = -\sigma_2' \mathrm{CosISinD} - \left(\sigma_0' - \sigma_1'\right) \mathrm{CosISinICosD}$$

$$\sigma_{33} = \sigma_1' + \left(\sigma_0' - \sigma_1'\right) \mathrm{Sin}^2 \mathrm{I}$$

3. Dielectric constant for ionospheric plasma

Dielectric constant for ionospheric plasma could be founded by using Maxwell equations.

1) $$\nabla \cdot \mathbf{E} = 4\pi\rho = \frac{\rho}{\varepsilon_0} \tag{48}$$

2) $$\nabla \cdot \mathbf{B} = 0 \tag{49}$$

3) $$\nabla \times \mathbf{E} = -\frac{\partial \mathbf{B}}{\partial t} \tag{50}$$

4) $$\nabla \times \mathbf{B} = \frac{1}{c^2}\frac{\partial \mathbf{E}}{\partial t} + \mu_0 \mathbf{J} \tag{51}$$

Where $c = \frac{1}{\sqrt{\mu_0 \varepsilon_0}} = 3\times10^8$ is the light speed. According to fourth Maxwell equation,

$$\nabla \times \mathbf{B} = \mu_0 \varepsilon_0 \frac{\partial \mathbf{E}}{\partial t} + \mu_0 \mathbf{J} \tag{52}$$

and if the electric field is accepted the form change $e^{-i\omega t}$ then,

$$\nabla \times \mathbf{B} = \mu_0 \varepsilon_0 \frac{\partial}{\partial t}\left(\mathbf{E}_0 e^{-i\omega t}\right) + \mu_0 \mathbf{J} \tag{53}$$

From here

$$\nabla \times \mathbf{B} = -i\omega\mu_0\varepsilon_0\left[\mathbf{E} - \frac{\sigma\mathbf{E}}{i\omega\varepsilon_0}\right] \tag{54}$$

it is obtained as follow

$$\nabla \times \mathbf{B} = -i\omega\mu_0\varepsilon_0\left[1 - \frac{\sigma}{i\omega\varepsilon_0}\right]\mathbf{E} \tag{55}$$

or

$$\nabla \times \mathbf{B} = -i\omega\mu_0\mathbf{E}\left[\varepsilon_0\left(1 - \frac{\sigma}{i\omega\varepsilon_0}\right)\right] \tag{56}$$

According to the latest's equation, the dielectric constant of any medium

$$\varepsilon = \varepsilon_0\left(1 - \frac{\sigma}{i\omega\varepsilon_0}\right) \tag{57}$$

In which, because of (σ) the tensorial form $\underset{\sim}{1}$ is the unit tensor.

$$\underset{\sim}{1} = \begin{bmatrix} 1 & 0 & 0 \\ 0 & 1 & 0 \\ 0 & 0 & 1 \end{bmatrix} \tag{58}$$

Generally, the expression of ionospheric conductivity σ'' given in Equation (47) by using Equation (57), the dielectric structure of ionospheric plasma is shown by

$$\varepsilon = \varepsilon_0\left[\begin{bmatrix} 1 & 0 & 0 \\ 0 & 1 & 0 \\ 0 & 0 & 1 \end{bmatrix} - \frac{1}{i\omega\varepsilon_0}\begin{bmatrix} \sigma_{11} & \sigma_{12} & \sigma_{13} \\ \sigma_{21} & \sigma_{22} & \sigma_{23} \\ \sigma_{31} & \sigma_{32} & \sigma_{33} \end{bmatrix}\right] \tag{59}$$

4. The refractive index of the cold plasma

The refractive index (**n**) determines the behavior of electromagnetic wave in a medium and refractive index of the medium is founded by using Maxwell equations. If the curl $(\nabla\times)$ of the Equation (50) is taken,

$$\nabla \times \nabla \times \mathbf{E} = -\frac{\partial}{\partial t}(\nabla \times \mathbf{B}) \tag{60}$$

If $(\nabla \times \mathbf{B})$ term in this expression is re-written in (60),

$$\nabla \times \nabla \times \mathbf{E} = \mu_0\varepsilon_0\omega^2\left[\underset{\sim}{1} + \frac{i\sigma}{\varepsilon_0\omega}\right]\cdot\mathbf{E} \tag{61}$$

Since the electric field **E** varies as $e^{i(\mathbf{k}\cdot\mathbf{r}-\omega t)}$, it can be assumed that $\nabla = i\mathbf{k}$. In this case, the left side of the Equation (61) can be defined as follows:

$$\nabla\times\nabla\times\mathbf{E} = k^2\mathbf{E} - \mathbf{k}(\mathbf{k}\cdot\mathbf{E}) \tag{62}$$

If the Equations (61) and (62) are rearranged, then Equation (63) can be defined as follows:

$$k^2\mathbf{E} - \mathbf{k}(\mathbf{k}\cdot\mathbf{E}) = \frac{\omega^2}{c^2}\left[\underset{\sim}{1} + \frac{i\sigma}{\varepsilon_0\omega}\right]\cdot\mathbf{E} \tag{63}$$

If the wave vector k in this equation is written in terms of refractive index **n**, the Equation (63) can be defined in terms of refractive index as follows:

$$\mathbf{k} = \frac{\omega}{c}\mathbf{n} \tag{64}$$

$$n^2\mathbf{E} - \mathbf{n}(\mathbf{n}\cdot\mathbf{E}) = \left[\underset{\sim}{1} + \frac{i\sigma}{\varepsilon_0\omega}\right]\cdot\mathbf{E} \tag{65}$$

If the necessary procedures are used, the Equation (66) is obtained as follows:

$$\begin{bmatrix} n^2 & 0 & 0 \\ 0 & n^2 & 0 \\ 0 & 0 & 0 \end{bmatrix}\cdot\begin{bmatrix} E_x \\ E_y \\ E_z \end{bmatrix} = \begin{bmatrix} 1 & 0 & 0 \\ 0 & 1 & 0 \\ 0 & 0 & 1 \end{bmatrix} + \frac{i}{\varepsilon_0\omega}[\sigma]\cdot\begin{bmatrix} E_x \\ E_y \\ E_z \end{bmatrix} \tag{66}$$

By writing the conductivity tensor σ'' in the Equation (47) instead of the conductivity in the Equation (66), a relation for the cold plasma is obtained as follows:

$$\begin{bmatrix} n^2 - 1 - \frac{i\sigma_{11}}{\varepsilon_0\omega} & -\frac{i\sigma_{12}}{\varepsilon_0\omega} & -\frac{i\sigma_{13}}{\varepsilon_0\omega} \\ -\frac{i\sigma_{21}}{\varepsilon_0\omega} & n^2 - 1 - \frac{i\sigma_{22}}{\varepsilon_0\omega} & -\frac{i\sigma_{23}}{\varepsilon_0\omega} \\ -\frac{i\sigma_{31}}{\varepsilon_0\omega} & -\frac{i\sigma_{32}}{\varepsilon_0\omega} & -1 - \frac{i\sigma_{33}}{\varepsilon_0\omega} \end{bmatrix}\cdot\begin{bmatrix} E_x \\ E_y \\ E_z \end{bmatrix} = 0 \tag{67}$$

Here, since the electric fields do not equal to zero, the determinant of the matrix of the coefficients equals to zero. So:

$$\begin{vmatrix} n^2 - 1 - \frac{i\sigma_{11}}{\varepsilon_0\omega} & -\frac{i\sigma_{12}}{\varepsilon_0\omega} & -\frac{i\sigma_{13}}{\varepsilon_0\omega} \\ -\frac{i\sigma_{21}}{\varepsilon_0\omega} & n^2 - 1 - \frac{i\sigma_{22}}{\varepsilon_0\omega} & -\frac{i\sigma_{23}}{\varepsilon_0\omega} \\ -\frac{i\sigma_{31}}{\varepsilon_0\omega} & -\frac{i\sigma_{32}}{\varepsilon_0\omega} & -1 - \frac{i\sigma_{33}}{\varepsilon_0\omega} \end{vmatrix} = 0 \tag{68}$$

The matrix given by Equation (68) includes the information about the propagation of the wave. Since the matrix has a complex structure, it is impossible to solve the matrix in general. However, the numerical analysis can be done for the matrix. Therefore, solutions should be made to certain conditions, in terms of convenience.

4.1 k//B condition: plasma oscillation and polarized waves

When the progress vector of the wave (**k**) is parallel or anti-parallel to the earth's magnetic field or the any components of the earth's magnetic field, two cases can be observed for the ionospheric plasma depending on the refractive index of the medium. For example, if the wave propagates in the z direction as in the vertical ionosondas, the vertical component of the earth's magnetic field effects to the propagation of the wave. In this case, two waves occur from the solution of the determinant given by Equation (68). The first one is the vibration of the plasma defined as follows:

$$\omega_{pe}^2 = \omega^2\left(1 - n^2\right) \tag{69}$$

The second one is the polarized wave given by Equation (70) [9].

$$n_p^2 = 1 - \frac{X(1 \mp Y)}{(1 \mp Y)^2 + Z^2} + iZ\frac{X}{(1 \mp Y)^2 + Z^2} \tag{70}$$

The signs ± in the Equation (70) represents the right-polarized (-) and left-polarized (+) waves, respectively. In this equation, $X = \frac{\omega_{pe}^2}{\omega^2}$, $Y = \frac{\omega_{ce}}{\omega}$ and $Z = \frac{\nu_e}{\omega}$. The magnetic field in the expression Y is the cyclotron frequency caused by the z component of the earth's magnetic field $\left(Y = \frac{\omega_{ce}}{\omega}\operatorname{Sin}I\right)$. This equation is the complex expression. Since the refractive index of the medium determines the resonance ($n^2 \cong \infty$) and the cut-off ($n^2=0$) conditions of the wave, it is the most important parameter in the wave studies. If it is taken that Z=0, the equation becomes simple. The resonance and the reflection conditions of the polarized wave become different from the case of Z≠0.

4.2 k⊥B condition: ordinary and extra-ordinary waves

When the propagation vector of the wave (**k**) is perpendicular to the earth's magnetic field, two waves occur in the ionospheric plasma [9]. The first wave is the ordinary wave given as follows:

$$n_o^2 = 1 - \frac{X}{1 + Z^2} + iZ\frac{X}{1 + Z^2} \tag{71}$$

This wave does not depend on Earth's magnetic field. However it depends on the collisions. The collisions can change the resonance and the reflection frequencies of the wave. The second wave is the extra ordinary wave depending on the magnetic field. The refractive index of the extra ordinary wave can be defined as follows:

$$n_{ex}^2 = 1 - \frac{aX(1-X)+Z^2X(2-X)}{a^2+b^2} + iZ\frac{X(1-X)(2-X)-aX}{a^2+b^2} \tag{72}$$

Where $a = 1 - X - Y^2 - Z^2$ and $b = Z(2-X)$. This relation is valid for the y direction. So, the magnetic field in the Y is the cyclotron frequency caused by the y component of the earth's magnetic field. By the same way, the wave defined by this relation is also observed in the x direction. For the extra ordinary wave in the x direction, the cyclotron frequency, that is contained in Y, is the cyclotron frequency caused by the x component of the earth's magnetic field.

4.3 The Binom expansion $\left(\left(1+Z^2\right)^{-1} \approx 1 - Z^2\right)$ and the refractive indices

The solutions of the refractive indices mentioned above are complex and difficult. The solutions can be obtained by using the binom expansion. Accordingly, the refractive indices can be obtained by using the Binom expansion and the real part of refractive indices as follows:

1. For the right-polarized wave given by Equation (70):

$$\mu_p^2 \approx (1-X') + Z'^2 \frac{X'(4-3X')}{4(1-X')} \quad \text{for } X'\langle 1 \tag{73}$$

$$\mu_p^2 \approx Z'^2 \frac{X'2}{4(X'-1)} \quad \text{for } X'\rangle 1 \tag{74}$$

Where, $X' = \dfrac{X}{1-\text{YSinI}}$ and $Z' = \dfrac{Z}{1-\text{YSinI}}$.

2. For the extra ordinary wave given by Equation (71):

$$\mu_o^2 \approx (1-X) + Z^2 \frac{X(4-3X)}{4(1-X)} \quad \text{for } X\langle 1 \tag{75}$$

$$\mu_o^2 \approx Z^2 \frac{X^2}{4(X-1)} \quad \text{for } X\rangle 1 \tag{76}$$

3. For the extra ordinary wave given by Equation (72):

$$\mu_{ex}^2 \approx \frac{(1-X)^2 - Y^2\,\text{Cos}^2\text{I}\,\text{Cos}^2\text{D}}{1-X-Y^2\,\text{Cos}^2\text{I}\,\text{Cos}^2\text{D}} + Z^2 \frac{X^2\left[(1-X)^2 + Y^2\,\text{Cos}^2\text{I}\,\text{Cos}^2\text{D}\right]^2}{4\left[1-X-Y^2\,\text{Cos}^2\text{I}\,\text{Cos}^2\text{D}\right]^3\left[(1-X)^2 - Y^2\,\text{Cos}^2\text{I}\,\text{Cos}^2\text{D}\right]} \tag{77}$$

5. Relaxation mechanism of cold ionospheric plasma

5.1 Charge conservation

Maxwell added the displacement current to Ampere law in order to guarantee charge conversation. Indeed, talking the divergence of both sides of Ampere's law and using Gauss's law $\nabla \cdot \mathbf{D} = \rho$, we get:

$$\nabla \cdot \nabla \times \mathbf{H} = \nabla \cdot \mathbf{J} + \nabla \cdot \frac{\partial \mathbf{D}}{\partial t} = \nabla \cdot \mathbf{J} + \frac{\partial}{\partial t} \nabla \cdot \mathbf{D} = \nabla \cdot \mathbf{J} + \frac{\partial \rho}{\partial t} \tag{78}$$

Using the vector identity $\nabla \cdot \nabla \times \mathbf{H} = 0$, we obtain the differential from of the charge conversation law:

$$\nabla \cdot \mathbf{J} + \frac{\partial \rho}{\partial t} = 0 \text{ (charge conservation)} \tag{79}$$

Integrating both sides over a closed volume V surrounded by the surface S, and using the divergence theorem, we obtain the integrated

$$\oint_S \mathbf{J} \cdot d\mathbf{S} = -\frac{d}{dt} \int_V \rho \, dV \tag{80}$$

The left-hand represents the total amount of charge flowing outwards through the surface S per unit time. The right-hand side represents the amount by which the charge is decreasing inside the volume V per unit time. In other words, charge does not disappear into (or get created out of) nothingness-it decreases in a region of space only because it flows into other regions.

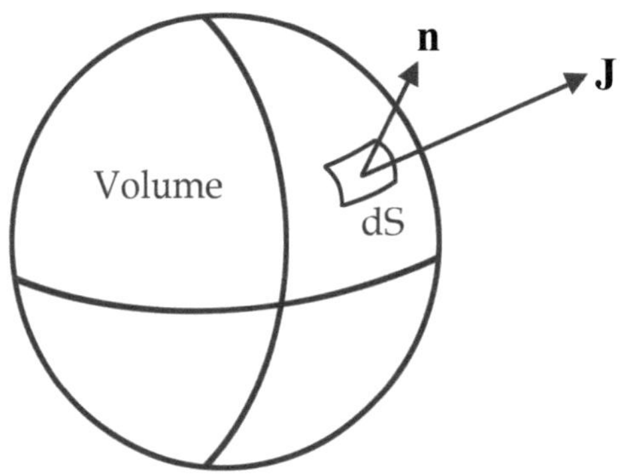

Fig. 7. Flux outwards through surface

Another consequence is that in good conductors, there cannot be any accumulated volume charge. Any such charge will quickly move to the conductor's surface and distribute itself such that to make the surface into an equipotential surface. Assuming that inside the conductor we have $\mathbf{D} = \varepsilon \cdot \mathbf{E}$ and $\mathbf{J} = \sigma \cdot \mathbf{E}$, we obtain

$$\nabla \cdot \mathbf{J} = \sigma \nabla \cdot \mathbf{E} = \frac{\sigma}{\varepsilon} \nabla \cdot \mathbf{D} = \frac{\sigma}{\varepsilon} \rho \tag{81}$$

Therefore,

$$\frac{d\rho}{dt}+\frac{\sigma}{\varepsilon}\rho=0 \tag{82}$$

with solution:

$$\rho(\mathbf{r},t)=\rho_0(\mathbf{r})e^{-\sigma t/\tau} \tag{83}$$

Where, $\rho_0(\mathbf{r})$ is the initial volume charge distribution. The solution shows that the volume charge disappears from inside and therefore it must accumulate on the surface of the conductors. For example, in copper,

$$\tau_{rel}=\frac{\varepsilon}{\sigma}=\frac{8.85\times10^{-12}}{5.7\times10^{7}}=1.6\times10^{-19}\,\text{sec} \tag{84}$$

By contrast, τ_{rel} is of the order of days in a good dielectric. For good conductors, the above argument is not quite correct because it is based on the steady-state version of Ohm's law, which must be modified to take into account the transient dynamics of the conduction charges.
It turn out that the relaxation time τ_{rel} is of the collision time, which is typically 10^{-14} sec.

5.2 Charge relaxation in conductors

We discuss the issue of charge relaxation in good conductors. Writing three-dimensionally and using, Ohm's law reads in the time domain:

$$J(r,t)=\omega_{pe}^2\int_{-\infty}^{t}e^{-\alpha(t-t')}\varepsilon_0E(r,t')dt' \tag{85}$$

Taking the divergence of both sides and using charge conversation, $\nabla\cdot\mathbf{J}+\dot{\rho}=0$, and Gauss's law, $\varepsilon_0\nabla\cdot\mathbf{E}=\rho$,we obtain the following integro-differential equation fort he charge density $\rho(\mathbf{r},t)$:

$$-\dot{\rho}(\mathbf{r},t)=\nabla\cdot\mathbf{J}(\mathbf{r},t)=\omega_{pe}^2\int_{-\infty}^{t}e^{-\alpha(t-t')}\varepsilon_0\nabla\cdot\mathbf{E}(\mathbf{r},t')dt'=\omega_{pe}^2\int_{-\infty}^{t}e^{-\alpha(t-t')}\rho(\mathbf{r},t')dt' \tag{86}$$

Differentiating both sides with respect to, we find that ρ satisfies the second-order differential equation:

$$\ddot{\rho}(\mathbf{r},t)+\alpha\dot{\rho}(\mathbf{r},t)+\omega_{pe}^2\rho(\mathbf{r},t)=0 \tag{87}$$

whose solution is easily verified to be a linear combination of:

$$e^{-\alpha t/2}\mathrm{Cos}(\omega_{rel}t),\ e^{-\alpha t/2}\mathrm{Sin}(\omega_{rel}t) \tag{88}$$

Where $\omega_{rel}=\sqrt{\omega_{pe}^2-\frac{\alpha^2}{4}}$. Thus the charge density is an exponentially decaying sinusoid with a relaxation time constant that is twice the collision time $\tau=1/\alpha$:

$$\tau_{rel} = \frac{2}{\alpha} = 2\tau \text{ (relaxation time constant)} \tag{89}$$

Typically, $\omega_{pe} \rangle\rangle \alpha$, so that ω_{rel} is practically equal to ω_{pe}. For example, using the numerical data of example, we find for copper $\tau_{rel} = 2\tau = 5\times10^{-14}$ sec. We calculate also: $f_{el} = \omega_{rel}/2\pi = 2.6\times10^{-15}$ Hz. In the limit $\alpha \to \infty$ or $\tau \to 0$, reduces to the naive relaxation.
In addition to charge relaxation, the total time depends on the time it takes fort he electric and magnetic fields to be extinguished from the inside of the conductor, as well as the time it takes fort he accumulated surface charge densities to setle, the motion of the surface charges being damped because of ohmic losses. Both of these times depend on the geometry and size of the conductor. The components of conductivity have given by Equation (47) and the permittivity of plasma has given by Equation (59).
As the conductivity of plasma has a tensor form, the permittivity of plasma is also a tensor. Moreover, the permittivity of plasma is frequency dependent and each component of the permittivity tensor has real and imaginer part. Consequently, the permittivity of plasma depends on the conductivity of plasma. The permittivity of plasma could be expressed as follow.

$$\varepsilon = \begin{bmatrix} \varepsilon_{11} & \varepsilon_{12} & \varepsilon_{13} \\ \varepsilon_{21} & \varepsilon_{22} & \varepsilon_{23} \\ \varepsilon_{31} & \varepsilon_{32} & \varepsilon_{33} \end{bmatrix} \tag{90}$$

The real and imaginary part of the permittivity is an indication of the ohmic power loss. Anisotropy is an inherent property of the atomic/molecular structure of the dielectric. It may also be caused by the application of external fields. For example, conductors and plasmas in the presence of a constant magnetic field -such as the ionosphere in the presence of the Earth's magnetic field- become anisotropic. The relaxation time in the ionospheric plasma has tensorial form as follow.

$$\tau = \begin{bmatrix} \tau_{11} & \tau_{12} & \tau_{13} \\ \tau_{21} & \tau_{22} & \tau_{23} \\ \tau_{31} & \tau_{32} & \tau_{33} \end{bmatrix} \tag{91}$$

The relaxation time in the ionospheric plasma is different at every direction and have real and imaginary parts [11].

6. Ionospheric absorption and attenuation of radio wave

When the radio waves propagate in the ionospheric plasma, they exhibit different behaviors related to their wave frequency, oscillation frequency of the electrons in the plasma medium and the refractive index of the medium. Depending on these behaviors, the wave is refracted, reflected or attenuated by absorption from medium. Radio-wave damping is due to movements in the ionosphere of electrons and ions are caused by collisions with other particles [12]. Due to the increase of collisions, absorption increases and field strength of the radio-wave decreases. As a result of this, amplitude of the radio-wave propagated in the ionosphere will decrease because of the absorption. Thus, the real part of the refractive

index effects to the phase velocity and the imaginary part of the refractive index is associated with spatial attenuation of the wave.
In accordance with the above information, in order to obtain the relation of the attenuation, amplitude of the wave, such as the electric field strength, is need to be expressed depending on the refractive index of the medium. Accordingly, the electric field strength of the wave can be defined as follows:

$$\mathbf{E} = \mathbf{E}_0\, e^{\mp i(\mathbf{k}\cdot\mathbf{r}-\omega t)} \tag{92}$$

Here, the wave vector **k** is written in terms of refractive index **n**, Equation (92) becomes as follows:

$$\mathbf{E} = \mathbf{E}_0\, e^{\mp i\left(\frac{\omega}{c}\mathbf{n}\cdot\mathbf{r}-\omega t\right)} \tag{93}$$

The refractive index in the ionospheric plasma is also defined as in equation (93).

$$n = \mu + i\chi \tag{94}$$

where, μ and χ represent the real and the imaginary part of the refractive index, respectively. The collision of the electron with the other particles effects to the real and the imaginary part of the refractive index. In the High Frequency (HF) waves, Z is very smaller than 1 $(Z\langle\langle 1)$. Therefore, Z can be defined as $\left(1+Z^2\right)^{-1} \approx 1-Z^2$. The real and the imaginary part of the refractive index and the phase velocity of the polarized ordinary and extra-ordinary wave can be determined by using this evolution.
Accordingly, if the refractive index given by Equation (94) is written in Equation (93), the electric field strength can be obtained as follows:

$$\mathbf{E} = \mathbf{E}_0\, e^{\mp i\left(\frac{\omega}{c}\mu r-\omega t\right)}\; e^{-\frac{\omega}{c}\chi r} \tag{95}$$

The part of the damping in the electric field strength in Equation (95) is the exponential term related to second χ in the right side of the equation. Thus, the attenuation of the wave is represented by χ. According to this, the refractive indexes of the polarized, ordinary and extra-ordinary waves given (70)-(72) equations for cold plasma could be showed as the real and imaginary parts as follow.

$$n^2 = F + iG \tag{96}$$

Accordingly, the electric field strength given in Equation (95) can be defined as follows:

$$\mathbf{E} = \mathbf{E}_0\, e^{\mp i\left(\frac{\omega}{c}\mu r-\omega t\right)}\; e^{-\frac{\omega}{c}\left[\frac{\pm\left(F^2+G^2\right)^{1/2}-F}{2}\right]^{1/2} r} \tag{97}$$

In the Equation (97), signs (+) and (-) in front of the first exponential expression represents the wave propagated to upward ($+\hat{z}$) and downward ($-\hat{z}$) after the reflection, respectively.

Here, by considering the electromagnetic wave that propagated to upward $+\hat{z}$, the term $e^{-\frac{\omega}{c}\chi r}$ can be defined as $e^{-\frac{\omega}{c}\chi z}$. Therefore, the damping term of the wave amplitude is defined with the height as follows:

$$\mathbf{E} = \mathbf{E}_0\, e^{\mp i\left(\frac{\omega}{c}\mu z - \omega t\right)}\, e^{-\frac{\omega}{c}\left[\frac{\pm\left(F^2+G^2\right)^{1/2} - F}{2}\right]^{1/2} z} \tag{98}$$

The second term in equation (98) represents how much the attenuation of wave in ionospheric plasma.

7. Conclusion

Ionospheric plasma has double-refractive index and generally weak conductivity in every direction, season and local time. Due to these, the ionospheric parameters such as conductivity, dielectric constant and the refractive index are different in every direction and have the complex structure (both the real and imaginary part). This shows that the diagonal elements of conductivity, refractive index and dielectric tensor which the conductivities dominate have generally higher conductivity than other elements; besides any electromagnetic wave propagating in ionospheric plasma could be sustained attenuation in every direction. So this attenuation results from the imaginary part of the refractive index.

8. Symbol list

m_e	: Mass of electron
m_i	: Mass of ion
$\mathbf{V}_e$	: Velocity of electron
t	: Time
-e	: Charge of electron
$\mathbf{E}$	: Electric field intensity
$\mathbf{B}$	: Magnetic induction
ν_e	: Electron collision frequency
ν_{ei}	: Electron-ion collision frequency
ν_{en}	: Electron-neutral particle collision frequency
$\mathbf{J}$	: Current density
N_e	: Electron density
σ	: Conductivity
ω_{ce}	: Electron cyclotron frequency
ω	: Wave angular frequency
ω_{pe}	: Electron plasma frequency
ε_0	: Dielectric constant of free space
I	: Dip angle
D	: Declination angle
∇	: Dell operator

ρ : Charge density
μ_0 : Magnetic permeability of free space
c : Speed of light
ε : Dielectric constant of medium
$\underset{\sim}{1}$: Unit tensor
k : Electromagnetic wave vector
r : Displacement vector
n : Refractive index
μ : Real part of refractive index
μ_p : Real part of refractive index of polarized wave
μ_o : Real part of refractive index of ordinary wave
μ_{ex} : Real part of refractive index of extra-ordinary wave
D : Displacement current
H : Magnetic field intensity
S : Surface
V : Volume
τ : Collision time
τ_{rel} : Relaxation time
α : The measure of the rate of collisions per unit time
f : Linear frequency
χ : Imaginary part of refractive index

9. References

[1] H. Rishbeth, A reviev of ionospheric F region theory, *Proc. of the IEEE.*, Vol.55(1), pp.16-35, 1967.

[2] H. Rishbeth and M. Mendillo, Patterns of F2-layer variability, *J. Atmos. Solar-Terr. Phys.*, Vol.63(15), pp.1661-1680, 2001.

[3] M. Grabner and V. Kvicera, Refractive index measurements in the lowest troposphere in the Czech Republic, *J. Atmos. Solar-Terr. Phys.*, Vol.68(12), pp.1334-1339, 2006.

[4] H. Rishbeth, Review paper:The ionospheric E layer and F layer dynamos –a tutorial review, *J. Atmos. Solar-Terr. Phys.*, Vol.59(15), pp.1873-1880, 1997.

[5] R. Della and J. Devin, A graphical interpretation of the electrical conductivity tensor, *J. Atmos. Solar-Terr. Phys.*, Vol.67(4), pp.337-343, 2005.

[6] H. G. Booker, Cold Plasma Waves. Martinus Nijhoff Publishers, Dordrecht-Netherlands, 1984.

[7] R. O. Dendy, Plasma Dynamics. Clarendon Press, Oxford, 1990.

[8] B. S. Tanenbaum, Plasma Physics. McGraw-Hill Book Company, New York 1967.

[9] M. Aydoğdu, A. Yeşil and E. Güzel, The group refractive indices of HF waves in the ionosphere and departure from the magnitude without collisions, *J. Atmos. and Solar-Terr. Phys.* Vol. 66(5), pp.343-348, 2004.

[10] M. Aydoğdu, E. Güzel, A. Yeşil, O. Özcan and M. Canyılmaz, Comparison of the calculated absorption and the measured field strength of HF waves reflected from the ionosphere, *Il Nouovo Cimento,* Vol.30C, No.3, pp.243-253, 2007.

[11] A. Yeşil, and İ. Ünal, The effect of altitude and season dielectrical relaxation mechanism of ionospheric plasma, *Il Nouovo Cimento*, Vol.124B, No.7, pp.777-784, 2009.
[12] J. A. Ratcliffe, The Magneto-Ionic Theory and its Applications to the Ionosphere. Cambridge University Press, 1959.

11

Exposing to EMF

Mahmoud Moghavvemi[1], Farhang Alijani[1], Hossein Ameri Mahabadi[1]
and Maryam Ashayer Soltani[2]
[1]*Department of Electrical Engineering, University of Malaya (UM),*
[2]*Department of Bioprocess Engineering, Universiti Teknologi Malaysia (UTM)*
Malaysia

1. Introduction

1.1 Electromagnetic Field Radiation

In the recent years, by developing the usage of new popular Electronic-Communication gadgets and home appliances like Mobile Phones and Microwave Ovens which are mostly sources of Electromagnetic Wave Radiation, a severe public concern regarding the side-effects "whether positive or negative" on human health and environment has been arisen. It means that there are new challenging issues rather than old one "the exposing to low radio-frequency emissions" which have been studied for past decades. Indeed, for more than 100 years, human has exposed to the Radio-waves radiation from Radio & TV broadcasting transmitters besides the Radar and point to point V/UHF/microwave links. However due to the low power density of those high frequency emissions which were far enough, the only low frequency was considered as main and important subject of EMF to human.

To set the limitations on all types of EMF Radiations the "International Non-Ionizing Radiation Committee" INIRC, in cooperation with "World Health Organization" WHO, had started their activities for 4 decades. Their first report to "United Nations Environment Program" UNEP has been approved in 1977. Later in 1992, to professionally study and investigate the detriments of Non-Ionized radiation and its effects on environment, INIRC changed to "International Commission on Non-Ionizing Radiation Protection" ICNIRP. In spite of world activities that have been done so far, it can be easily understood that all kinds of EMF radiations, which we regularly use in industry and/ or in our personal life has become a vital part that we cannot ignore. Therefore, sometimes we expect and also suffer from and its side effects. [1]

In the following sections, the different types of EMF and their characteristics besides of definitions, would be introduced and investigated separately. There are different fields that we use nowadays, i.e.; Electrostatic Field, Magnetic Field and Radio-waves which are our part of interest. It would be showed that the effect of near-field radiation is quite different comparing to RF effect of far-field as we used to know. It means that the closer to the source of radiation there are the new side-effects that we are exposed to. That is important to know that, distinguishing between near and far-field regions for all sources due to their frequency, dimensions and structure are not clarified well and consequently, EMF measuring techniques should be chosen carefully. The engineering techniques for measuring the intensity of radiation is named EMF Dosimetry which would be discussed in the next section and some practical measurement ideas would be given. Further, the Biologic effects on live tissues and human body will be investigated for concerned fields. Finally we will discuss on safety techniques and by introducing techniques will give remedies to keep environment safer.

1.2 EM fields characteristics

Though the radiation generally refers to all kinds of energy transmission but there is a distinguishing difference between energy levels that cause breaking in molecules boundaries. If the energy is enough to make a change in molecules, we would have an ionized radiation. This can be seen in particle based radiations like alpha, beta, gamma, and X-ray as well as, high levels of frequency and energy. The quantized energy level for such radiation could be calculated from the following equation:[2]

$$E=h\nu \qquad (1)$$

In which "ν" is the frequency and "$h = 6.6260693\times10^{-34}$ Js" is the Planck constant.
As for Radio-Wave Radiation, since the concerned spectrum is lower than 300GHz, normally we don't have those high levels of ionizing energy. So a new term as non-ionizing radiation is used instead. This type of energy based on its components which are Electric and Magnetic fields, presents different phenomenon in terms of frequency, level and direction of field vectors, electric and magnetic characteristics of substance, angles of incident fields and the distance between source and measuring point or the type of radiation-zone.
The Microwave ovens, RF diathermy, RF welding or RF-brazing are good common examples of non-ionizing but effective Radio-Wave radiation sources in our life.
Since the ionizing radiation is not considered here, its definitions, units and scales are not part of our interest consequently. Therefore we prefer to use more tangible scales which are derived from Electromagnetic concepts rather than high energy particles in ionizing radiation. It is important to mention that though the energy levels can be converted between the two types of radiations but for instance the results and effects on human tissues are not necessarily the same. It can be interpreted by effecting mechanism on tissues under test.
According to electromagnetic equations, for the transverse electromagnetic mode wave TEM, when we are far from the source, but in the main axes of radiation, the power density can be calculated from electromagnetic fields' components (Poynting vector):

$$S_{(W/m^2)} = |\mathbf{E}|_{(V/m)} . |\mathbf{H}|_{(A/m)} \qquad (2)$$

1.3 Basic concepts

There are 3 distinguished regions around an EMF radiator which have different characteristics. In the vicinity of source (antenna) where we are not far from $\lambda/2\pi$ or 0.159λ the fields are reactive and stored in the volume around the source. We name such a region as near-field where E and H fields have complex relation and all types of polarizations as Vertical, Horizontal and Circular exists but traveling wave is not observed. In each point, the E or H is dominant. Power density cannot be measured since both fields and their phases needed. In this region E and H are stored around antenna and propagate as component. The stored energy in shape of self-capacitance or inductance transact to current source by the moving back and forth of energy. In this region any additional conductor absorbs this energy and the energy doesn't return back again. In this condition the energy of transmitter is pulled out by outer conductor but as soon as outer circuit closes, power consumes and transmitter loads, like primary and secondary winding of a transformer. Therefore, any simple field measurement in this region is invalid.

Far beyond the reactive near-field zone, up to about 1-wavelength from the radiator, the radiative near-field or Fresnel zone could be distinguished. In this region the power density doesn't follow the $1/r^2$ law since the E and H are not contributed in plane-wave yet. This region is very dangerous for human since far-field measuring equipments do not work correctly. Since we are far from energy storing region, all kinds of radiation could be seen in form of spherical waves. This not-uniform radiation propagates in high speed but mixture of E and H components are not same as far-field instead they have phase shifts that don't vary linearly with distance from phase center. In this region re-radiating from metals which acts as parasitic antennas happens and such new antennas also makes its own near-field zone. As exposing mostly occurs by fields components (E and H), to measure the field we need E and H probes. Obviously no pattern and polarization is considered for a source in near-field region. This region extends up to about 2λ as transition zone to reach to far-field (Fraunhofer) region. By considering the maximum overall dimension of an antenna as "D" there is a criterion for distinguishing between these regions as:[4, 5]

$$\begin{aligned} &\text{Near-field:} && R < 0.62(D^3/\lambda)^{1/2} \\ &\text{Fresnel Region:} && 0.62(D^3/\lambda)^{1/2} < R < 2D^2/\lambda \end{aligned} \tag{3}$$

Further than the Fresnel region, we have far-field (Fraunhofer) region which extends up to infinity. The plane-wave with its full characteristics like: radiation impedance ($E/H=\sqrt{\mu/\varepsilon}=120\pi\ \Omega$), pattern, polarization and mode is valid:

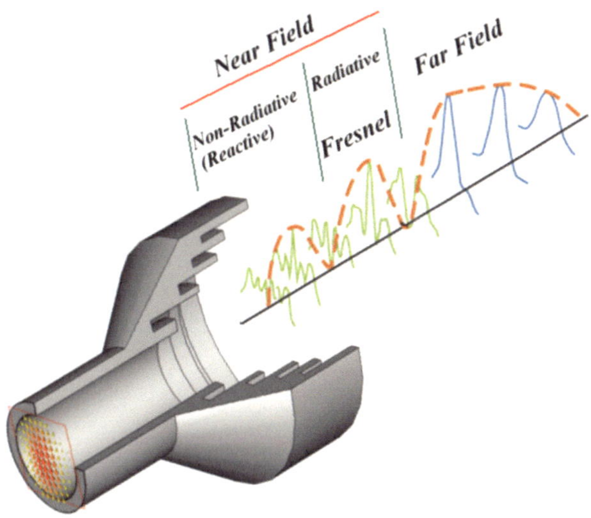

Fig. 1. Near-Field and Far-Field regions for a microwave radiation source sample

1.4 Polarization

The Electrical plane for the far field TEM radiation is called polarization plane and the orientation of the electrical component is polarization's direction. As an instance, for the following TEM wave, the wave propagation is in x-axes and we consider this wave as a linear vertical polarization the E component is in parallel to y-axes.

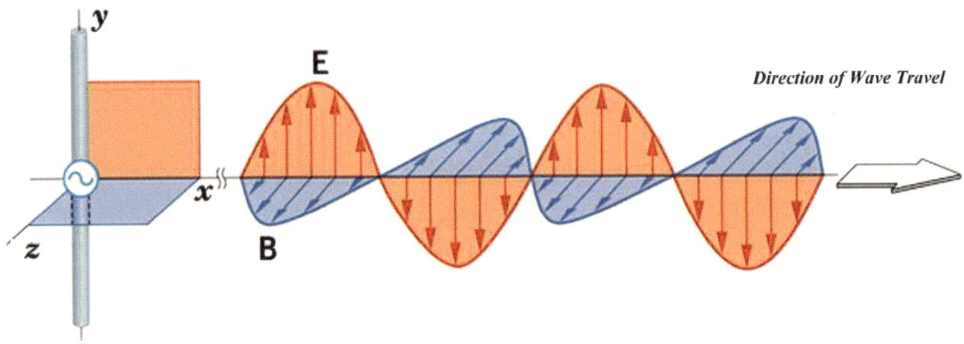

Fig. 2. Plane wave and its components

In reception mode, since the E-field, easily couples an induced current in a conductive material, it is very important to know what the polarization of the incident wave is. Such waves can be stopped easily by putting a conductor in parallel to polarization axes as it would absorb E field energy and eliminating the TEM wave consequently. If the polarization's direction changes by time and for example the E-field vector passes through a circle contour, the circle polarization occurred. It would also goes through an elliptic contour to make an elliptic polarization. This kind of polarization is seen when a plane wave passes through a $\lambda/4$ dielectric or reflects from a metallic plate in non-perpendicular angle. This is because of dividing each linear polarization into two Right Hand Circular (RHC) and Left Hand Circular (LHC) waves. Anyway, since such polarization cannot be stopped easily by parallel E-field plates, they can easily penetrate into any substances and even our bodies.

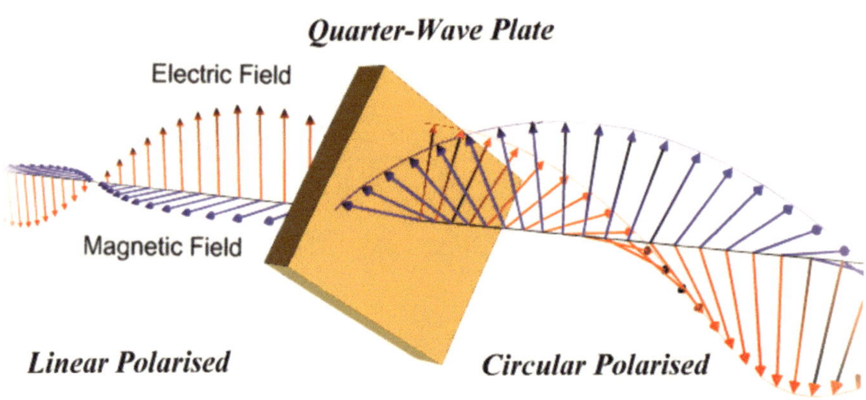

Fig. 3. Generating a circular polarized wave from linear polarized wave

It is so important to know that, if we don't have any idea about the polarization of incident wave, we cannot measure its power and its important parameters. Even the linear sensing dipole, receives half of energy of circularly EMF radiation in any direction of transverse plane.

2. Dosimetry

According to a definition the amount of a substance that is in contact with or exposed to radiation is called as "Dose" and for EMF case we use the expression "EMF radiation dose". The EMF radiation dose represent the different aspects like: induced current density for lower frequency than 100 kHz, Thermal effect by using the specific absorption rate SAR (W/kg) and other non-thermal effects mostly measured by electric or magnetic fields.
For a standing person, the body can couple to a vertical electric field which the induced current is observed in orientation from head to feet and vice versa. However for a properly circulated induction to a standing body, a horizontal magnetic field needed.[6, 7, 14]

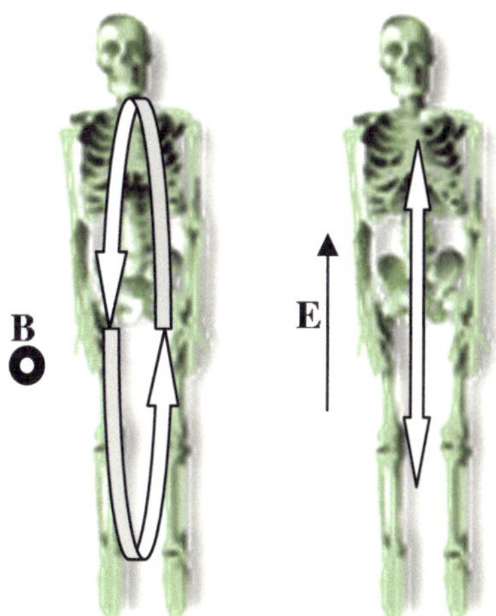

Fig. 4. Coupling of E and B with body

2.1 SAR

Specific Absorption Rate (SAR) is the safety criterion in terms of absorbed energy in human tissue. SAR is measured by Watt per unit mass of tissue and shown by W/kg and used between 100 kHz and 10GHz. If we know the tissue's electric field E, the conductivity σ and the mass density ρ then average SAR can be calculated as:[5 -7]

$$\mathrm{SAR} = \int_{Sample} \frac{\sigma(\mathrm{r})\left|E(\mathrm{r})\right|^2}{\rho(\mathrm{r})} d\mathrm{r} \tag{4}$$

Usually SAR is determined over a certain mass of tissues, like 1 or 10 g and for each given values for SAR such condition should be mentioned. As an example, the maximum SAR of a mobile handset determined by FCC, is 1.6 (W/kg) over a volume containing 1 gram of tissue but, CENELEC the European standard gives an average of 2 (W/kg) over 10gram of tissue.
As a general view, we should know that SAR gives an absolute value and sometimes it could be misleading. So in order to be able to compare the parameters, it is better to issue a standard model for human body. To study the heating effect of EMF on human body, "standard human" has been introduced. The standard human is defined as (height =175cm, weight= 70kg, total surface= 1.85m^2). So exposing to 14W radio-wave radiation, causes an

average SAR equals to 14/70=0.2 W/kg. The absorption for each tissue depends on its specific SAR and we cannot use an absolute value to describe total radiation effect.
The specific absorption of energy or "SA", by knowing the specific absorption rate SAR in each tissue can be defined by:[6]

$$SA_{(J/kg)}=SAR_{(W/kg)} \times \text{Exposure time}_{(Sec)} \quad (5)$$

Relative heath capacity is determined as:

$$C=\Delta Q/\Delta T \quad (6)$$

ΔQ is the amount of heat due to a ΔT rise in temperature. For unit of mass the relative heath capacity is describe as:

$$C\rho = C/m \; (J/kg.K^{o}) \quad (7)$$

As an example, $C\rho$ of the water (liquid) at 15 °C and P=101.325 kPa is:

$$C\rho(\text{water}) = 4.1855 \; (J/kg.K^{o}) = 4185.5 (J/g.K^{o}) \quad (8)$$

Therefore, by knowing the absorbed energy and exposure time, increasing the temperature of tissue by relative heath capacity of C_ρ for unit of tissue mass due to such energy is derived as:

$$\Delta T_{(Deg)}= SA_{(J/kg)}/C\rho \quad (9)$$

So exposing to SAR=4(W/Kg) at 15 minutes can cause an increasing 1^{oC} in human body when there is no cooling system like blood circulation. [6]
Considering a $\Delta T_{(Deg)}= 1^{oC}$ temperature increase in human tissue can be considered as the first restriction criterion for exposure as a rule of thumb.

Tissue	C (J/kg.C°)	Density (kg/m³)
Skeletal muscle	3470	70
Fat	2260	940
Bone, cortical	1260	1790
Bone, Spongy	2970	1250
Blood	3890	1060

Table 1. Specific capacity and density [19]

For exposing to EMF, It should be mentioned that like what we use in microwave ovens there are 2 major factors pair of Time and Power density.
Unfortunately, measuring the SAR in real is not practical so in simulations, the models of human body or plastic dolls (phantom) are used. Since the body is not electrically homogeneous usually the models are filled by same dielectric characteristic absorbing materials and heat can be measured by infra-red camera analyzers.

2.2 Induced current density

According to Maxwell's equations the conductivity and permittivity are in relation together. The induced current based on electric field is expressed as:

$$\vec{J} = (j\omega\varepsilon + \sigma)\vec{E} \quad (\mathrm{A/m^2}) \tag{10}$$

By expanding the permittivity in complex form of real part as dielectric constant, and imaginary part as dielectric losses:

$$\varepsilon_r = \varepsilon_r^{'} - j\varepsilon_r^{''} \quad (\mathrm{F/m}) \tag{11}$$

Indeed $\varepsilon_r^{''}$ is a measure of both the friction of changing polarization and drift of conduction charges.

While their ratio is the tangent of loss angle:

$$\tan(\delta_e) = \frac{\varepsilon''}{\varepsilon'} \tag{12}$$

By combining these equations, we have:

$$\vec{J} = (j\omega(\varepsilon_r^{'} - j\varepsilon_r^{''}) + \sigma)\vec{E} = j\omega\varepsilon_r^{'}\vec{E} + (\omega\varepsilon_r^{''} + \sigma)\vec{E} \tag{13}$$

The real part of equation is the relation between electric field and current:

$$\sigma_{eff} = (\omega\varepsilon_r^{''} + \sigma) \quad (\mathrm{S/m}) \tag{14}$$

The equation shows the frequency dependence of conductivity. Though the best way for calculating the SAR based on induced current is measuring the current in tissues but in real situation, such a measurement is very difficult. So simulating the current by knowing the electrical characteristic is the way that is used. [10]

$$SAR = \frac{\sigma|\mathbf{E}|^2}{\rho_m} = \frac{\omega\varepsilon_o\varepsilon''|\mathbf{E}|^2}{\rho_m} \tag{15}$$

The geometry of human body determines the type of induction. In case of isolated feet from ground we consider an un-grounded case. And in case of bare foot the image of human contributes to make a double size of body that cause a half resonance frequency in compare to un-grounded human. In case of grounded, the area around the ankles might be burned due to maximum current.

The following values are used for modeling a human body to calculate the SAR [7] :

Tissue	σ(S/m)	ε_r	ρ(g/cm³)
muscle(high water content)	46	2.3	1.1
fat (low water content)	5.5	0.15	0.9

Table 2. Characteristics of main tissues for modeling

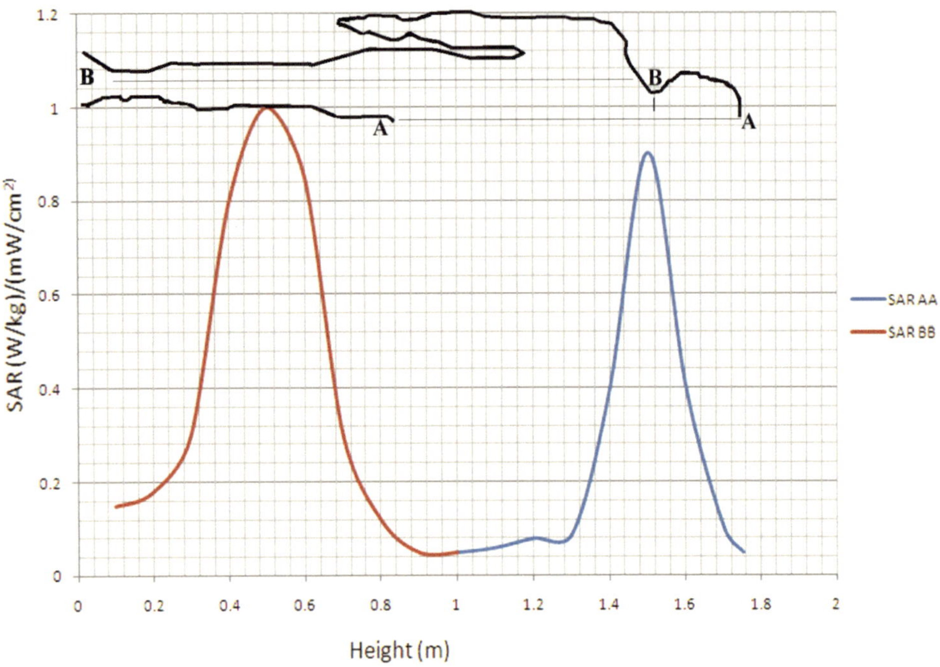

Fig. 5. Calculated SAR along a model in two sections of standard human body. [10] The incident wave is $1mW/cm^2$ on the surface, frequency = 350MHz

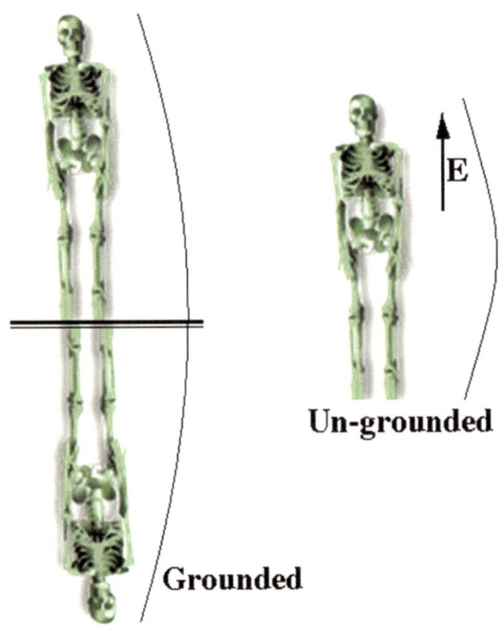

Fig. 6. Electric field induced on grounded and un-grounded human [14, 6]

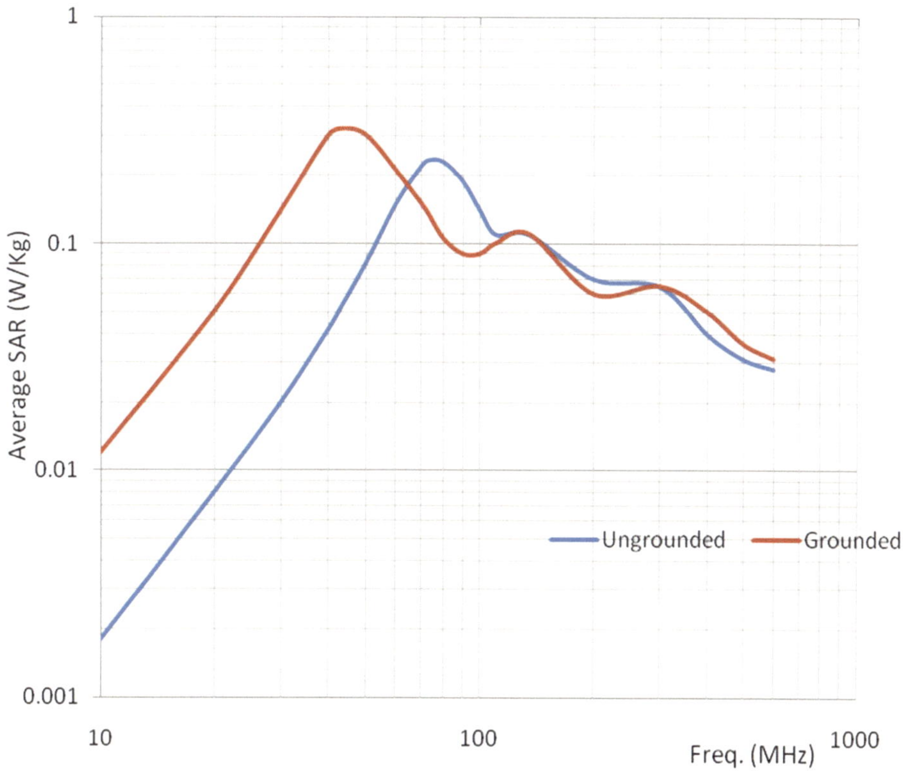

Fig. 7. Resonance of body and SAR. Below resonance the SAR varies by f^2 and beyond resonance by approximately 1/f [6]

2.3 Interaction characteristic

When we are in near-field, the body (tissue) which is under dosimetry's test interacts to fields which perturb the fields consequently. In near-field we have a coupling model between body (tissue) and source that cause changes in the source impedance. The high impedance sources tend to give a high electric coupling to electric fields and low impedance sources tend to couple to magnetic fields.

In real, some EMF sources can be considered as near-field, like mobile handset phones which we keep them on our ear/ head that directly couples its energy to our head and brain. By considering their maximum power when they working in far regions to compensate the link budget losses we cannot ignore its role in EMF near-field exposure (up to about 5W).

Leaky wave cables, RF loose connections, BTS Tx/Rx on roofs, high order modes or Non-TEM ducted waves in buildings, transmitters in tunnels are all can be considered as samples of Near-Field Sources.

2.4 Exposing to VLF and low frequency EMF

Electric fields can induce surface charges on body and cause a surface current consequently. The position of body and its size, conductivity of skin tissues and direction of Electric vector for evaluating the exposure effects, are all important.

The low frequency magnetic fields cause an induced circular current on human body. The magnitude of field and current density depends on circle's radius of current's contour, tissues conductivity and magnetic flux density.
For Magnetic Resonance Imaging, usually high power magnetic field used to penetrate body. Exposing to such fields also limited and in standards like IEC 60601-2-33, the suggested values are given. As a brief result, the maximum 2(W/kg) for whole body at 6 minute exposing is recommended. [12, 6]

2.5 Field measuring techniques

The best measuring method is the one that makes minimum level of perturbation on the existed fields. It means that we should use the smallest probe or sampler inside of the field to measure it. In addition we know that small antennas provide a broadband response. But its gain would be limited due to small size. For this reasons, using the normal dipole or LPDA as broad-band antenna has serious restrictions in measuring the fields and especially dosimetry applications.
Instead, very small electric probe by electrical gap can be used as an element of E-Field sensor. In real, isotropic E-field probe with 3 orthogonal small dipoles usually used and give an rms values to readout device. This E-probe has an easy operation. In order to keep sensors away of contacting to illuminators or resources, the spherical foam-cap used to cover the head of probe.

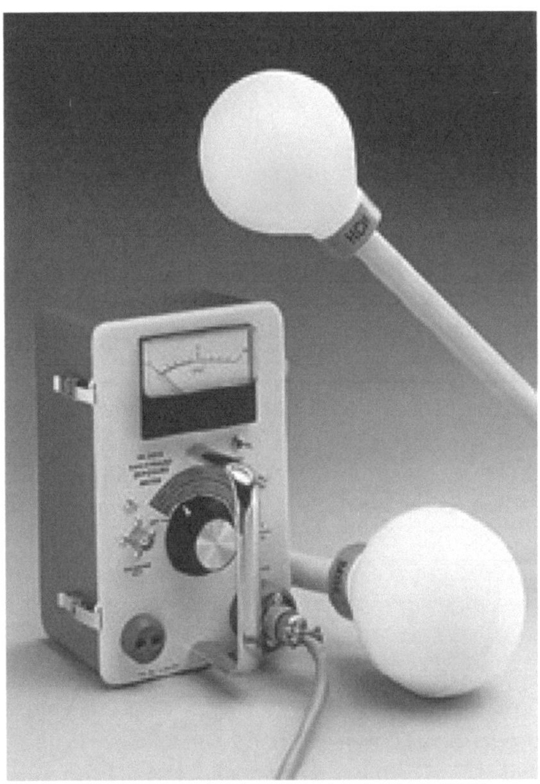

Fig. 8. Electric and Magnetic Field meter. Readout device is same for E and M probes

The B-field sensors are made by using the very small magnetic dipole by small loop. It is usually constructed by 3 orthogonal small loops (for 3 dimensions) as isotropic probe with common readout unit with E-field instrument. This sensor can be used up to 1 GHz and H can be obtained by $H=E/120\pi$ if we have a plane wave metering.
Another type of sensors is synthesized by resonant antenna. Such antenna is narrow-band in frequency response but has smaller size and higher efficiency in reception. Such meters are good enough for especial applications but as general metering tool, they are not useful since their polarization is usually fixed and k-factor of antenna is not valid for near-field measurement.

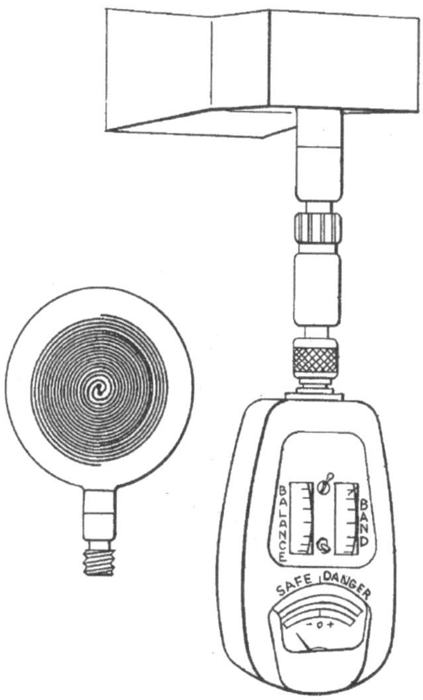

Fig. 9. Two samples of E-probe that are based on wide-band antennas. Any wide-band and narrow-band probes can be used with same readout meter if their size is small enough. Such antennas are directive (whether wide-band spiral or horn) in pattern and are not suitable for isotropic EMF exposure metering

There are other sensors used for low frequency applications based on Pockels-effect (E-field sensor) and Hall-effect (B-field sensor). The size of such sensors is small enough to use them in many applications easily.
For pulsed field measurement and dosimetry applications especially in near-field where electric and magnetic fields have high variations that cause big problem in metering, as we have seen before, it is better to use thermocouple detector rather than diode. This method easily works up to 18 GHz and more. The thermocouple directly converts existed fields that make the heat in matters. There is not an estimation or prediction in metering as we have in E or M-probes cases, while we directly readout the correct values of both fields (mixed) that we needed. [14]

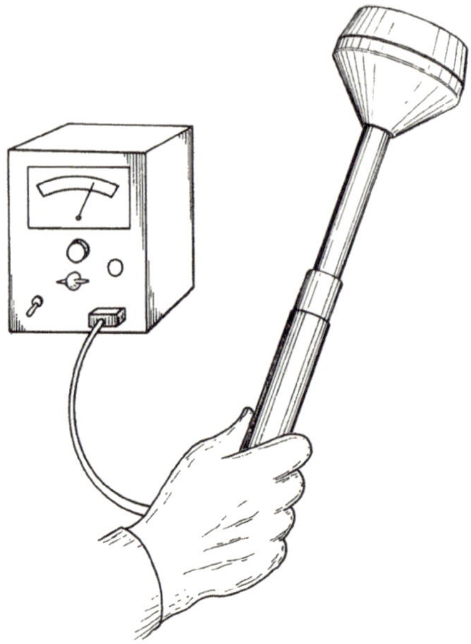

Fig. 10. Common Thermocouple Dosimeter. This design has been used for more than 60 years. The wideband thermocouple in probe's head makes it the best solution for Radiation dosimetry

3. Biologic effects

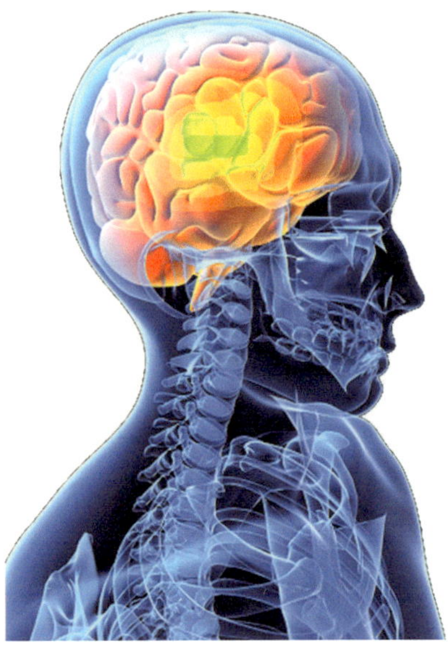

Fig. 11. Brain and nervous system are easily threatened by EMF exposure

Exposing to Radio-Waves radiation does not make the same effects in children and adults. The size and electromagnetic characteristics of children's body is completely different to adults and cause multiple risk ratios in EMF exposure. In addition, we also can add the interfering effects risk by non-heating but biochemistry side effects to growth hormones of glands, which as we expected are in high level for children. That is why the maximum amount and frequency of EMF, in different age groups are the case of study for standard organizations. As we know, EMF lowers our melatonin levels, which control our immune system response to cancer and other diseases. By considering all side-effects, therefore, we are expected to have a practical and handy table for such limitations in terms of age, frequency and other parameters. But unfortunately such table is not available easily.

In order to better realization and study of such effects on body we can start by known symptoms. It means that such effects should be analyzed from different points of view. Thereafter derivation of charts and analyzing reports can be done easily by merging those results together.

3.1 Heating

Absorbed EMF radiation simply deposits its energy by heating the material. Tissues heating due to absorbing the electromagnetic fields can be seen in full range of radio-waves. But in spite of tangent of the loss or relative permittivity which is important in high frequencies, the exceeded current density due to its conductivity in tissues causes a problem in the low frequencies too.

The heating by absorption of EMF completely depends on frequency. For low frequencies (100 kHz up to 20MHz) the neck and feet are the most absorbed parts.

In the frequency band of (20-300MHz) we have a higher absorption in full body and much more in resonated parts like the head. In the (300MHz up to some GHz) the absorption is based on non-uniform and mostly considered as regional. In case of (10-300GHz) we can expect the most absorption is in skin and outer small tissues like eyes and gonads which can show the resonance effects as well. [8-10] What we expect to know is that intense radio waves can thermally burn living tissue same way as we can cook food in microwave ovens!

As we know, human's body fulfilled by a high percentage of water. Water molecules can be polarized due to its molecular structure (104.5^{o} between H-O-H). So the human tissues which have high amount of water are able to be affected by magnetic fields too. The water's molecules forced to follow the magnetic field of incident EMF but due to its lag in vibration the excess energy converted into heat.

There is an adjusting system in human body to fix the body's temperature. Over-exposing to EMF can increase the human's body temperature and hyperthermia occurred consequently. Even increasing the 2.2 degrees in human's body destroy the tissues cell and needs the special treatment and sever cure. Fever is the common symptom that everybody is familiar with and can give a tangible and good idea for understanding the body temperature increasing. Usually the maximum SAR levels depends on many factors like: weight, age, equivalent EM characteristics of tissues and body dimensions but as for standard human mostly the 0.4 (W/kg) is the safety limit introduced for healthy people and for occupational exposure case while an additional safety factor of 5 suggested for public case as 0.08(W/kg) [12]. It also means that for an un-healthy people, the standard is much different. Noticing to this point is very important in using the standard tables.

Just like any exposed material to EMF, an effective penetration depth can be considered for human tissues too. It is where each of field levels degrades by $1/e$ (Euler's constant). Therefore by degrading the e^{-2} we have drop down of power density to 13.5%. This depth depends on wavelength and decreased by increasing the frequency. By considering the saturation level of water in tissues and environment humidity, such following curve derived over the radio frequency band [13].

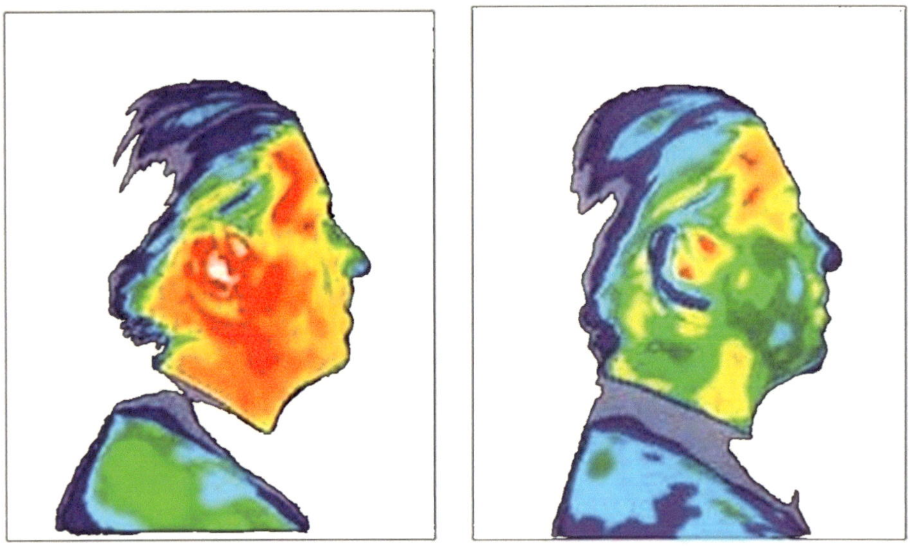

Fig. 12. Thermo-graphic image of the head, Right: no exposure. Left: A 15-minute phone call

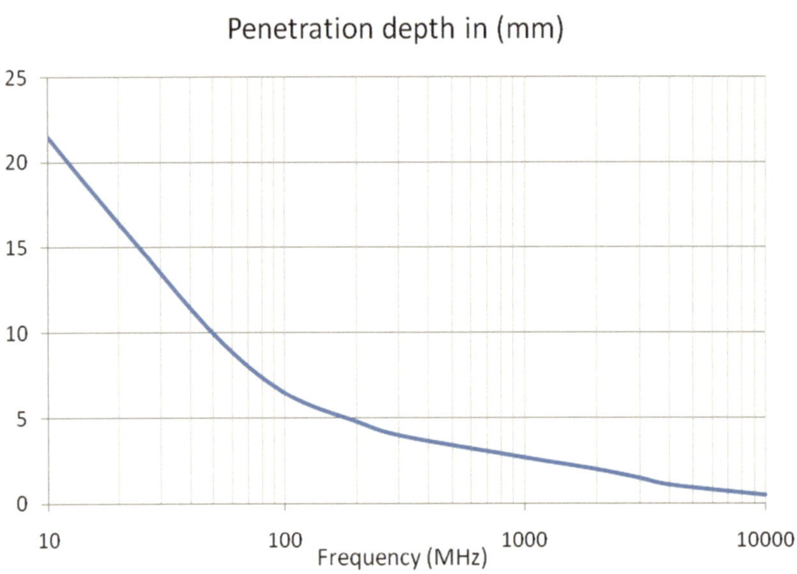

Fig. 13. Sample of penetration depth and its dependency to frequency

This penetration depth can lead us to this fact that 86.5% of incident energy should be dissipate and absorbed in such layer of skin and not affecting on whole body. In other words, for high frequencies we don't involve whole body's weight and meanwhile, skin tissues can burn easily even by not a high power density [11].
The human body like any mixed-dielectric structure can show the resonance frequency. It means that in certain frequencies and in that specific body's position the amount of SAR can raise to a maximum. In other words standing, sitting, lying and so on, have different response to SAR. In resonance of radio-waves the maximum size of body which is parallel to electric field's direction (polarization) is important. So for standing case, by considering the average relative dielectric value as we have seen before, the resonance occurred when the height is about $\lambda_r/2$. Therefore, in case of standard human we have:

$$f_r=c/\ \lambda_r=c/\ \lambda.\varepsilon^{0.5}=3/(2x1.75x(2^{\ 0.5}))=60\text{ MHz} \tag{16}$$

This amount is just for estimation and in real case by considering the different tissues and organs; we should consider each case one by one. Obviously, this frequency is going higher for shorter people and kids. As for instance the resonance frequency for 1-meter tall boy would be around 120 MHz and for 0.5 meter infant is around 240 MHz.
In addition to mentioned effects we have seen the hot spots regions in EMF exposed body too. The hot spots usually occurred due to non-uniformity of EM characteristics of tissues. Because of such mixed structure, the power density is not distributed uniformly between tissues, and in some parts that tangent of loss or SAR is higher it could have maximum temperature. The organic parts that cannot be cooled by blood circulation are the weakest parts in terms of radio-wave radiation. The eyes and gonads are the most sensitive parts since there are not enough vessels for cooling down by blood circulation.
The major effect of EMF on eyes could be the cataract in which the lens clouded. This effect is especially very important when the microwave radiation is considered (10 - 300 GHz) but the destruction completely depends on how incident waves enter eyes in terms of polarization, power, duration, and human personal factors. As ICNIRP recommended, by limiting the power density to lower than $50W/m^2$ eyes defects could be controlled. The SAR for some animal eyes has been measured. As an example the measured SAR is about 100 up to 140 W/kg for rabbits. Therefore EMF exposing in long duration (more than an hour) would increase the eye's temperature up to some degrees which is very harmful. The experiments on rats proved that there is a direct relation between EMF exposing and disability in fertility due to gonads' temperature which has risen by EMF. It is interesting that the effects on sleeping groups are more than awaken groups. In other words the threshold SAR level is lower for sleeping people. This could be explained by temperature auto adjusting system by blood circulation which is in rest during sleeping.
It may be seems strange, but the ears are also able to feel (hear) some radar's high pulsed power. By different experiments on volunteers the power density threshold of ($4\text{-}20W/m^2$) for frequency range of (200MHz-3GHz) by PRF of 200 and 400Hz could be heard. But it was cleared that peak pulsed power was more important than average power and some perception by brain was involved in hearing the microwave pulses [6]. In following table, some clinical symptoms are given. It shows the direct relation between symptoms caused by exposing to EMF.

Source	Ave. Power	Exposing duration	Power density Wm^{-2}	Clinical/Biological Symptoms
Radar 3GHz	"high"	5h	300~700	Heat, Headache, Vertigo
Radar 9-10GHz		80 sec	600~900	Increasing in Creatine kinase
microwave oven 2.45GHz	600W	5 sec		Neuropathy
wood heater 20kW	20kW	2 sec		Burn metallic material
wood heater 20kW	20kW	26sec		Oedema, paresthesia
UHF TV antenna 1.75kW	1.75kW	1 to 2.5 min	>200	Diarrhoea, dysesthesia

Table 3. Some reported EMF exposing clinical effects [17, 18]

3.2 Burning and shocking

Burning or shocking can be seen by contacting to conductive materials inside of the EMF. This effect can be separated from indirect effect of EMF on implanted circuits like pace maker or ear-amplifiers. Indeed, the induced current in body due to exposing in magnetic field occurred. Such induction especially in lower part of body which contacted to ground has higher amount. Since the ankle cross section is lower than other parts of leg, therefore the high current density may cause a burning in tissues there.
For frequency lower than 27 MHz an empirical formula used for evaluating such induced current [6]

$$I_{(mA)}=0.108 \times h^2_{(m)} \times f_{(MHz)} \times E_{(V/m)} \quad (17)$$

Further by simulation techniques this burning effect has been modeled. The shoes have a great role in controlling the maximum current in feet that reduce it to 0.6 ~0.8 of its peak [9]. Other phenomenon that causes burning is happened when we touch metallic parts which have HF, high induced currents on. As soon as the connection occurred the high density of current passes through junction and it could cause a sever injury. Usually we named such effects as shocking and it considered as side effects of EMF on our body. There are lots of instructions and recommendations like the ones issued by WHO/IRPA or ICNIRP98 for reducing such effects in sites.

3.3 Non-thermal effects

As we have informed the Non-ionizing radiation is considered to be essentially harmless below the levels that cause heating since absorbed EMF radiation simply deposits its energy by heating the material. But how much this is true?

Such non-thermal effects can be seen by direct effect of EMF to biological tissues without making a significant heat effect. There are lots of researches done so far on finding out the relation between EMF and cancer tumors. The reports presented contradiction results by different research groups and seems need to work more. What we would expect is that EMF can affect on DNA, Genes, Nervous system and totally Biochemistry of body which is under control of hormones. Therefore even the unknown illness symptoms, may have a root in our EMF polluted environment. There are some experiments on ELF magnetic exposure and its relations to brain cancer and leukemia as reported [13]. Based on epidemiologic studies for field magnitude of 0.3 - 0.4T it was shown the possible carcinogenic relation to childhood leukemia.

3.4 Treatment by EMF

All discussions till now were about un-wanted EMF radiations. But, if we can raise the temperature of a tissue by EMF, we can use it for medical treatment as well. This is the basic idea for lots of methods that we have done so far in medical world. Nowadays we use EMF to make a faster healing in broken bone. In addition, EMF can kill the unwanted cells like cancer cells. This treatment named radio-therapy and it can be used as adjunctive therapy besides of regular methods like chemical-therapy. Heating the prostate gland by EMF also used for keeping it small as a treatment and considered as safe and fast treatment without a need to surgery.

Today, EMF treatment generally is used for fast bone repairing, nerve and immune system stimulation, blood circulation and wound treatment, osteoarthritis, tissue regeneration, electro-acupuncturing as well as using the pulsed EMF and low frequency exposure for soft-tissue injuries [16, 17].

4. Safety techniques

Safety techniques derived from simple rules. In real, safety refers to a number of predictions and preparation to prevent exposing or reducing the exposure to as low as possible. So we can start from main issues that have well worth to remember [10].

1. Wetter materials (muscle) are lossier than drier (fat, bone) materials.
2. For parallel E-field to the body's longitudinal axes, SAR is higher.
3. Sharp edges, concentrate E-fields.
4. Parallel conductor has a high perturbation on E-field rather than perpendicular conductors.
5. Penetration depth decreases by increasing frequency and conductivity.
6. Below resonance frequency the SAR varies by f^2
7. Small objects compared to wavelength cause minimum perturbation on fields

By noting to those items and referring to charts and equations given in last sections, here are the most important safety issues that we could list: [7]

- The safety level for whole-body average SAR should be less than (4 W/kg). This value is measured by average power absorbed over a 6 minute interval per total body weight.
- The maximum partial-body SAR could be up to 20 times of the whole-body averaged SAR (80W/kg). This parameter is defined by strongest radiation over the specified volume like $1cm^3$.

- The product of power density and exposure time should remain below the safe values for sensitive organs such as eyes and testes. The key point is increasing the temperature in organs.

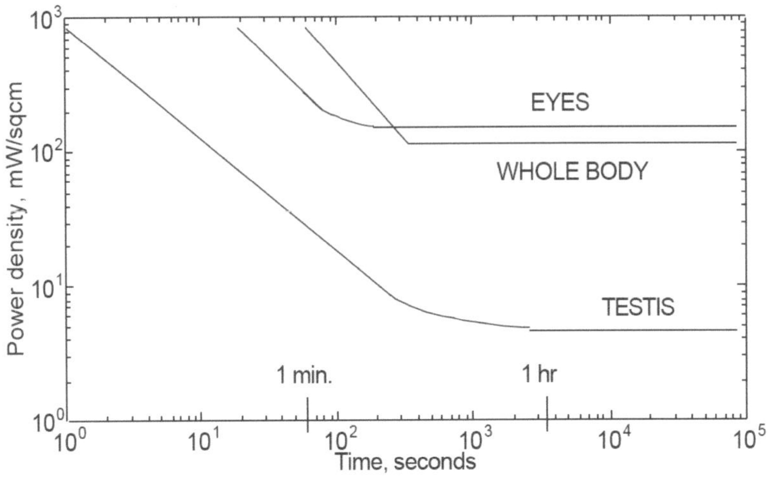

Fig. 14. Threshold levels versus time for sensitive organs [7]

- The current induced in the body due to radiation power density should be low enough to prevent any shock or burning by contacting to metals or conducting materials.

4.1 New technologies

The technology of making light weight electromagnetic absorbers developed to making them popular in our used equipment or devices. By using a number of small resonant antennas that loaded to their matched load, and incident waves damped consequently, we can have a high EMF isolation. Though, these absorbers are designed for far-field, but they can also be used in near-field too. The various spiral and crossed slot antenna were used in past, but fractal antenna are recently introduced to determine the characteristics precisely. Meanwhile there are good progresses in designing of antenna. Such development helps us to keep the unwanted EMF radiations as low as possible. As an example, the main antenna beam of some cell-phones is outward of head. And they use some RF dampers to keep head as cool as possible. In addition, by new technologies, we use the least energy and power for devices. Therefore unwanted emissions are lower than before.

4.2 People duties

Since safety has different aspects, it is impossible to have a safe environment without noticing to people acts and duties as they are main users of such facilities. Here, some ideas to control such unwanted exposure are given:

- Leaving the Hot area: Since, feeling the heat in body, mostly happened when temperature rose due to EMF exposure for more than a minute, and the skin was able to sense the EMF only in UHF/SHF. So in case of feeling a heat, leaving the site is strongly recommended for operators and technicians even the power density is lower than standard threshold. As it was showed, the standard limitations are not considered as guarantee levels for not feeling the heat, and if someone feels heat, it means that it has passed far above standard of safety. Remember that, standards are based on statistical analysis but, replying to EMF is not same for all people. Everyone can have its own personal standards based on his/hers body's electrical characteristics.

- Escaping from resonances: exposing to frequencies near to resonance of our body has higher effect. Even the source power is not too much, the body can absorb and store as much as energy in vicinity of resonance frequency. Therefore, by knowing the emitted frequency, the threshold of hazardous understood. It can also be investigated by Spectrum monitoring of places of concern.
- Don't forget to use suitable shoes to reduce induced currents through feet. This recommendation is important especially for technicians of high power RF equipments, like Radio and TV broadcasting transmitters, Microwave wood drier, RF technicians in Diathermy Clinics, Radar technicians and so on.
- Don't forget that standards limitations are for healthy people. If you are not healthy or have a medical problem, the threshold values for you are very lower than others.
- Restrict using the cell-phones. Use a corded device that allows you to talk on your phone without holding it next to your head. There is some evidence that cell-phone use has caused an increase in brain tumors.
- Don't forget that EMF radiation inhibits melatonin production, which is most active during sleep.
- Don't keep your cell-phones near your bed. Use a corded telephone rather than wireless type if possible. Locate the base of wireless telephone away from beds.
- Reducing the cell-phone's call duration. Standards are for 3-6 minute duration call. Talking in longer time means higher EMF exposure.
- Avoid calling cell-phone when the radio coverage is not fulfilled. In this case, BTS increases its power and send a command to hand sets to increase their power too. It is to keep the link budget between BTS and handset in proper margin. Now, the mobile handsets consume higher power and emit more RF. This can also be checked by battery's discharge speed in such areas.
- Use EMF-protected mobile handsets that used beam forming techniques towards out of head.
- Hug the babies in RF zones. This item may seem to be strange, but as mentioned, the infants and kids have smaller dimensions and have higher resonance frequency. They are so sensitive to EMF exposure than adults. In addition, since they are in their growing age, their growth hormones are in high level, so any even low interfering to their organic system of adjusting hormones, remains the un-returnable side effects that its symptoms can be seen many years later. Besides of mentioned issues, there are some un-proved evidence between EMF, cancer and tumors in children. Therefore, in order to get rid of such worry, we recommend keep babies out of hazardous zone and hug them. It can help to protect from high resonance frequency by uniting the bodies altogether, prevent from direct expose to EMF as supporter buffers a plenty of radiation. Reducing the induced current as they are in higher altitude from ground and supporter's body divide the induced current between kid and itself as well as keeping the frequency resonance as low as possible to reduce intensity of the current.

- Try to live away from high-power lines or cell phone BTS masts. They are installing in different places even on our home's roof. They may be the highest EMF hazard in near to our living place.
- Use a shielded glass for windows if they are exposing from the EMF transmitters. Using a metallic net by maximum $\lambda/10$ in mesh size can make a good shield for windows.
- Try to damp the EMF inside of our room by RF absorbers. Using the metal shield without having the idea about RF and its incident direction, doesn't stop the waves and may differ them toward a more sensitive place.
- Stand away from microwave ovens while they're in use. Restrict using of them as low as possible. They can be used only for warming the food and not cooking. They have two effects. First, on our body by RF leakage. Second, on our food by direct effect on its molecules. So try to put these ovens inaccessibility of children since they are curious to watch inside.
- Having your own EMF dosimeter or detector to find out any unwanted radiation before being too late.

4.3 Standards

Each standard is developed for certain usage. According to our usual case, in our life style we need to know the boundaries of what we call as Public levels of safety. The ICNIRP [1] has defined the limitations for two groups of "Occupational" and "general public" in different reports for certain short exposing duration. Its reports cover the full frequency band from some kHz up to 300GHz. Though other organizations like FCC, IEEE and NRPB93 have different definitions but we believe that the one which has stronger care to health is preferred. Therefore, the standard which based on clinical experiments has been chosen rather than the one which based on compatibility to existed equipments.

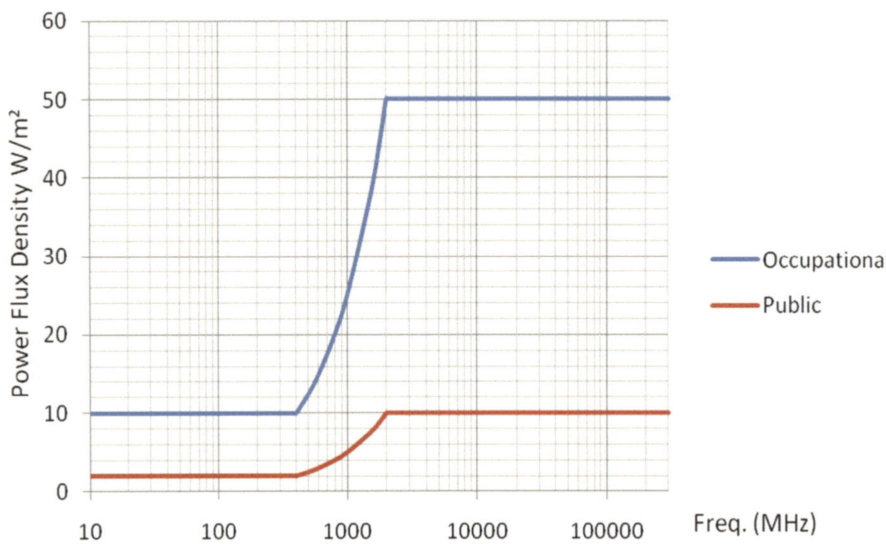

Fig. 15. Power Flux density curves for occupational and public, based on ICNIRP recommendation

	Power flux density	(Wm^{-2})
Frequency (MHz)	Occupational	Public
10-400	10	2
400-2000	f/200	f/40
2000-300000	10	50

Table 4. Power Flux density for occupational and public, based on ICNIRP recommendation

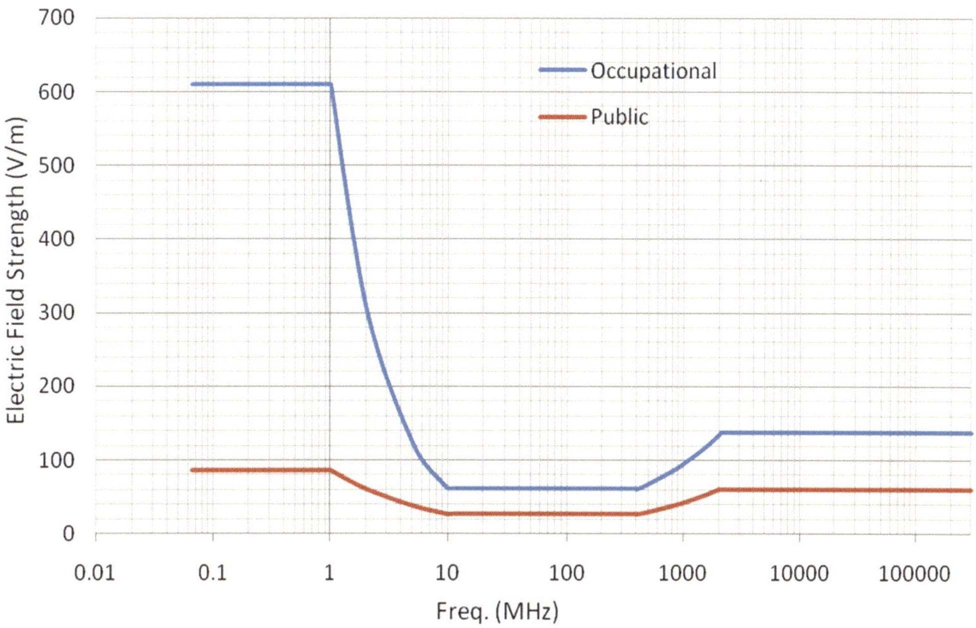

Fig. 16. Electric field strength curves for occupational and public, based on ICNIRP recommendation

	Electric Field Strength	(Vm^{-1})
Frequency (MHz)	Occupational	Public
0.065-1	610	87
1-10	610/f	$87/f^{0.5}$
10-400	61	28
400-2000	$3f^{0.5}$	$1.375f^{0.5}$
2000-300000	137	61

Table 5. Electric field strength for occupational and public, based on ICNIRP recommendation

	Magnetic Field Strength	(Am-1)
Frequency (MHz)	Occupational	Public
0.065-1	1.6/f	0.73/f
1-10	1.6/f	0.73/f
10-400	0.16	0.073
400-2000	$0.008f^{0.5}$	$0.0037f^{0.5}$
2000-300000	0.36	0.16

Table 6. Magnetic field strength for occupational and public, based on ICNIRP recommendation

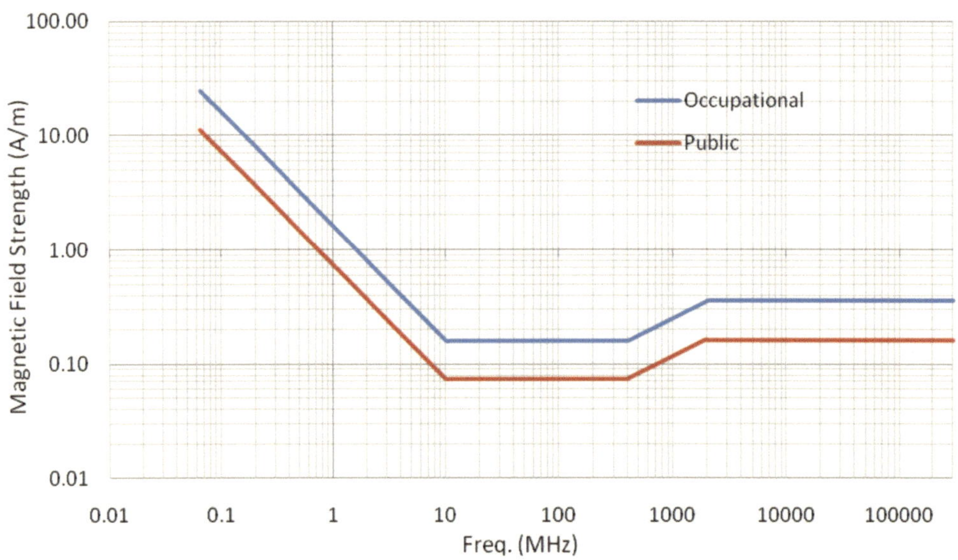

Fig. 17. Magnetic field strength curves for occupational and public, based on ICNIRP recommendation

5. Conclusion

A brief introductory to Radio-wave radiation is given and measuring techniques for each region of radiation is also discussed. There are different definitions regarding safety standards. The main problem for using such safety recommendations is misusing the equipments due to un-aware of EMF exposure effects, lack of general knowledge and society culture. As an example even a standard mobile handset that we called safe, can be harmful if user call duration exceeds its recommended talking time (max 3min/resting time for rehabilitation). This long calling time on mobile handsets is what we have seen in some countries since people use the mobile phone same as fixed line telephone. Therefore, this misusing can be the main problem to any radiation sources but there are some good progresses in terms of EMF exposure prevention methods like radiation absorbers to reduce such EMF exposure on human's body. Using the various absorbers, shielding, and controlling the leakage of cables are some of those methods. Effects of EMF radiation on children and adults are not the same so taking more care for children is the good advisory to keep in mind.

6. References

[1] International Commission on Non-Ionizing Radiation Protection, ICNIRP, *EMF guidelines*, http://www.icnirp.de

[2] Karttunen, H., et al., (2007), Fundamental Astronomy, 5th edition, Springer, New York, ISBN 978-3-540-34143-7

[3] Milligan, T.A., (2005), Modern Antenna Design, 2nd edition, John Wiley & Sons, Inc., New Jersey, ISBN-13978-0-471-45776-3

[4] Balanis, C.A., (2005), Antenna Theory Analysis and Design, 3rd edition, John Wiely & Sons, Inc., New Jersey, pp. 34-36, ISBN-0-471-66782-X

[5] Barnes, F.S., Greenebaum, B., (2007), Bioengineering and Biophysical Aspects of Electromagnetic Fields, *Handbook of Biological effects of Electromagnetic Fields*, 3rd edition, CRC Taylor & Francis press, ISBN-0-8493-9539-9

[6] Kitchen, R.,(2001). RF & Microwave Radiation Safety Handbook, pp. 68, ISBN: 0-7506-4-3552, Newnes, London

[7] Jenn, D. (2009). Electromagnetic Radiation Hazards, EC3630 radiowave propagation, pp.11, Naval Postgraduate school, department of Electrical & Computer Engineering, Monterey, California

[8] Adir, E.R, (1987). Thermophysical Effects of Electromagnetic Radiation, *IEEE Engineering in Medicine and Biology Magazine*, pp.37-41

[9] Chen, J. & Gandhi, O.P.(1989). RF Currents induced in an anatomically based model of a human for plane-wave exposures (20 to 100MHz), *Health Physics*, Vol. 57, No. 1, pp. 89-98

[10] Durney, C.H. et al.,(2002), Radiofrequency Radiation Dosimetry Handbook, 4th edition, The University of Utah, Salt Lake city, USA

[11] Gandhi O.P. et al (1980). State of the knowledge for Electromagnetic Absorbed Dose in Man and Animals; *Proc. IEEE*, Vol.68, No.1, Jan 1980, pp. 24-32

[12] ICNIRP (1998), Guidelines for limiting exposure to time-varying electric, magnetic and electromagnetic fields (up to 300GHz), *Health Physics*, Vol. 74, No. 4, pp. 494-522.

[13] Kheifets, L. (2001). EMF Epidemiology: State of science, *WHO Meeting on EMF Biological Effects & Standards Harmonization in Asia and Oceania* 22 - 24 October, 2001, Seoul, Korea

[14] Masao Taki, (2001). CHARACTERISTICS, DOSIMETRY & MEASUREMENT OF EMF, *WHO Meeting on EMF Biological Effects & Standards Harmonization in Asia and Oceania* 22 - 24 October, 2001, Seoul, Korea

[15] NRPB, (1992). Electromagnetic Fields and the Risk of Cancer; report of an ADVISORY GROUP ON Non-Ionizing Radiation; NRPB Document Vol. 3 , No.1, HMSO Books, London ISBN-0-85951-346-7

[16] Pirogova,E., Vojisavljevic,V., Cosic, I., (2010). *New developments in Biomedical Engineering*, Ch 5., InTech, ISBN 978-953-7610-57-2

[17] Pirogova, E., Vojisavljevic,V., Cosic, I.,(2009) Biological effects of electromagnetic radiation, *Recent Advances in Biomedical Engineering*, ISBN 978-953-7619-X-X, (accepted for publication), In-Tech Vienna, Austria.

[18] Schilling, C.J.(2000). Case report: Effects of exposure to very high frequency radio frequency radiation on six antenna engineers in two incidents; *Occup. Med.*,Vol. 50, No. 1, pp.49-56

[19] Vorst, A.V., Rosen, A., Kotsuka, Y., (2006). *RF/Microwave Interaction with biological Tissue*. Wiley-interscience, IEEE Press, pp. 93-140

12

Low Frequency Electromagnetic Waves Observation During Magnetotail Reconnection Event

X. H. Wei, J. B. Cao and G. C. Zhou

X. H. Wei Key Laboratory of Space Weather, Center for Space Science and Applied Research, Beijing University of Aeronautics and Astronautics, Beijing, China

1. Introduction

Magnetic reconnection is a very important physical process in astrophysical and laboratory plasmas, which enables reconfiguration of the magnetic field topology and converts the magnetic field energy to plasma kinetic and thermal energy. The diffusion region is a crucial region of reconnection where magnetic field and plasma decouple from each other and strong wave activity and complex wave particle interactions occur. In general, the regions where the energy conversion takes place, e.g. substorm, ionosphere, shocks, produce wave emissions or wave turbulence covering a wide frequency range. Reconnection sites are not an exception. Understanding the role of waves and wave turbulence in the energy conversion, energy transport, and structure formation of the reconnection sites is an important and challenging task. When the reconnection takes place at ion inertial length, the wave-particle interaction plays an important role in reconnection process. Both the whistler dynamics and kinetic Alfvén waves can strongly influence the structure of the dissipation region during magnetic reconnection. Hall term in the generalized Ohm's law brings the dynamics of whistler waves into the fluid equations [1]. The only place where a reconnection site can be studied in great detail is laboratory and the Earth magnetosphere (or other environments in our solar system that have been visited by spacecraft, e.g. solar wind, other planets, comets). The spacecraft observations give much more detailed picture of the plasma dynamics at the smallest electron scales than the laboratory experiments, mainly due to the possibility to resolve particle distribution functions and fields at small scales. As the reconnection involves many processes at different spatial and temporal scales, numerical simulations serve as a superior tool for understanding the environment and physical processes near reconnection sites. The subsolar magnetopause and magnetotail are the two main regions in the Earth magnetosphere where the reconnection process has been observed by spacecrafts. The magnetotail reconnection is generally symmetric. In a sense plasmas on both sides of the current sheet have very similar properties. The opposite situation is observed at the magnetopause where the reconnection is mainly asymmetric. Another important difference between the magnetopause and mangetotail is that the typical spatial scales, e.g. ion inertial length, are usually a factor of ten smaller at the magnetopause. This is important for in situ studies where the instrument resolution becomes a limiting factor. Although there is significant amount of studies dealing with low frequency electromagnetic waves at the magentopause and in the magnetotaill, but only in few cases was accompanied by reconnection processes. In many cases it has been speculated about such relationship. In this paper we summarize in situ observations of low frequency electromagnetic waves where reconnection signatures are well defined as well as those observations where one only speculates about such relationship.

2. Whistler observed by spacecrafts in the magnetosphere

Whistler is often observed in the magnetosphere. As early as 1960s, plasma wave was observed in the plasma sheet and neutral sheet region of the distant magnetotail by Ogo1, 3, and 5 satellites, which provide measurements only in the near-earth regions of plasma sheet (at radial distances ~ 17Re). Brody et al [2] have reported observations of brief bursts of whistler mode magnetic noise near the neutral sheet. Scarf et al. [3], using measurements from the Imp7 spacecraft at a radial distance of about 30 Re , have reported observation of moderately intense electric field oscillations in the region immediately outside the plasma sheet an very intense low-frequency magnetic noise in the high-density region of the plasma sheet. Imp8 observed the whistler waves in the region near the neural sheet in the magnetotail at radial distances ranging from -46.3 to -23.1 Re [4]. Three principal types of plasma waves are detected: broad band electrostatic noise, whistler mode magnetic noise bursts, and electrostatic electron cyclotron waves. Gurnett et al. [4] suggested that the whistler waves are most likely produced by current-driven plasma instabilities. The whistler waves are also observed by ISEE3 in the plasma sheet [5] and magnetotail flux ropes [6].

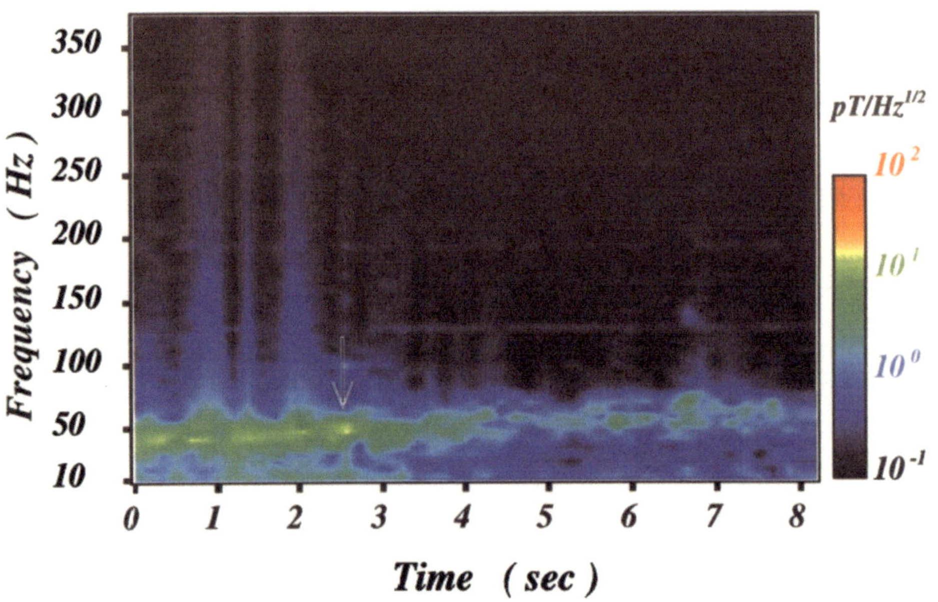

Plate 1. Dynamic spectra of magnetic field from wave form data around 1138.53UT on December 23, 1994. The Geotail was located in the near magnetotail at (-46.94, -6.62, -5.37) Re. The electron cyclotron frequency is 135Hz during the 8s period

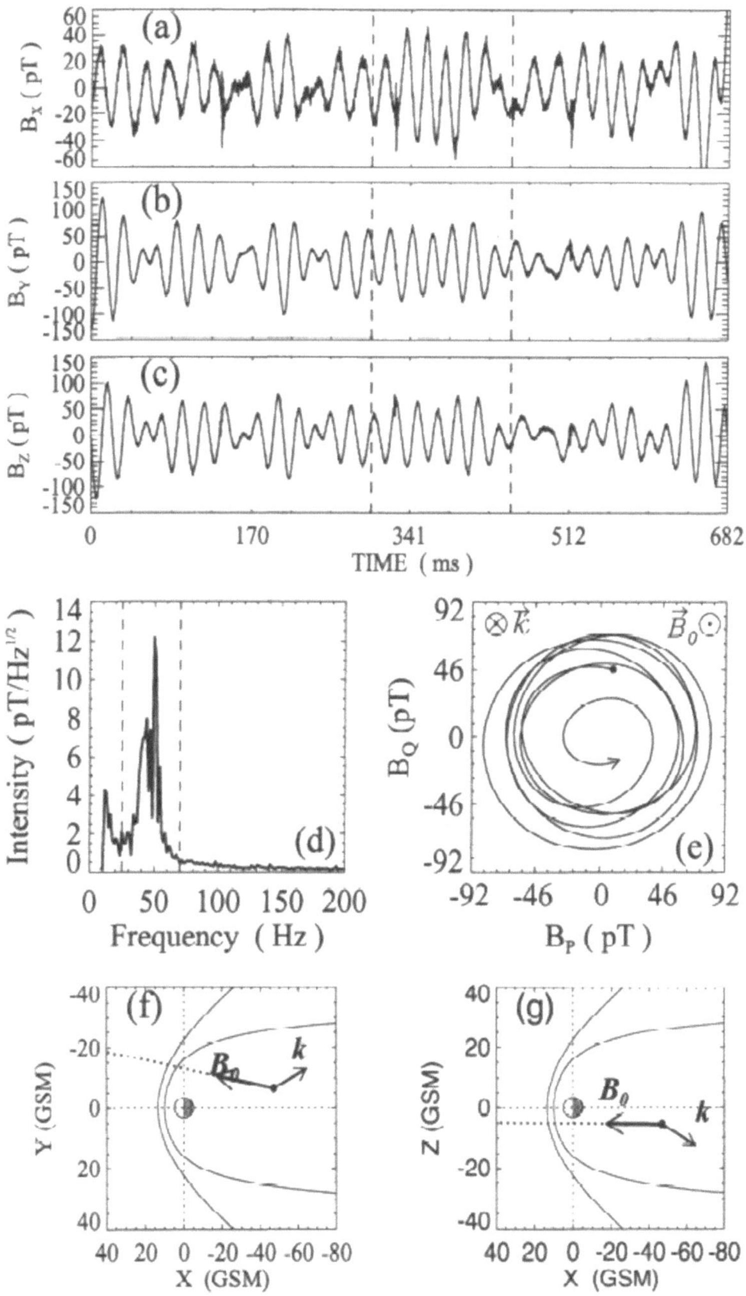

Fig. 1. (a)-(c) Magnetic field waveforms, (d) spectrum, (e) hodograph, and (f) and (g) k vector plots. Bx, By and Bz are the magnetic field components in directions from Earth to the Sun from dawn to dusk, and parallel to the Geotail spin axis, respectively. BP and BQ are two arbitrary orthogonal axes which are perpendicular to k vectors of the whistler mode waves. The wave was recorded around 1138:54UT on December 23, 1994 (see the white arrow in Plate 1)

It is suggested that superthermal electrons with highly anisotropic pitch angle distributions generate the whistler waves. Zhang et al. [7] analyze whistler waves observed by Geotail in the magnetotail at radial distance ranging from -210 Re to -10 Re, and they found that whistler waves can exist in both plasma sheet and plasma sheet boundary layer, and propagate quasi-parallel to the ambient magnetic field with an average propagation angle of 23 degrees. They thought that it is the energetic electron beams that generate the whistler waves. Plate 1 shows the dynamic spectra of magnetic field from wave form data around 1138.53UT on December 23, 1994. The Geotail was located in the near magnetotail at (-46.94, -6.62, -5.37) Re. The electron cyclotron frequency is 135Hz during the 8s period. Details of the bursts marked by a white arrow in Plate 1 are shown in Figure 1. Figure 1a,1b and 1c are magnetic waveforms for a period of 0.7s. They are quasi-monochromatic waves with amplitude around 70pT. Figure 1d confirms that the central frequency (ω) of bursts is 50Hz. The hodograph in Figure 1e (obtained from the data between 300 and 450 ms in Figure 1a, 1b and 1c by using the minimum variance method [8] indicates the wave is right-hand circularly polarized with respect to the ambient magnetic field. The electron cyclotron frequency was 180Hz at this time, the central frequency (50Hz) is 0.28 Ωe. This frequency is well between the electron cyclotron frequency and the lower hybrid frequency. All the wave character shows the magnetic bursts are whistler mode waves.

The above observations discussed the whistler wave activities in the magnetotail. No reconnection events and whistler wave activities were observed by spacecrafts simultaneously in current sheets.

3. Whistler and magnetic reconnection observed by spacecrafts in the magnetostail simultaneously

Whistler waves associated with reconnection in the Earth's magnetopause were also observed by Geotail [9]. Wind observed low frequency Alfvén /whistler waves associated with the LHDI (Low Hybrid Drift Instability) in the near-vicinity of the X-line of reconnection far from the Earth at about 57 Re in the magnetotail [10]-[11]. Cluster II STAFF instrument provides the good opportunity to study the low frequency electromagnetic waves. In succession, the whistler wave and reconnection event observed by Cluster will be significantly discussed.

3.1 Magnetic reconnection event and whistler wave on August 21, 2002

Cluster crossed the magnetotail plasma sheet from 07:00 UT to 09:00 UT on August 21, 2002. Figure 2 gives the ion flow velocity components (Vx), magnetic field components (Bx, By, Bz), plasma density, total magnetic field (B), and plasma Beta, which are observed by C1(black), C3 (green) and C4 (blue) during the interval of 07:50UT - 08:00UT on August 21, 2002. All Cluster spacecrafts were on the dawn side and in plasma sheet. It can be seen that a high-speed tailward ion flow (Vx <0) accompanied by southward magnetic field component was observed by three satellites and it lasted about 5 min. The tailward ion flow with southward magnetic field component appeared at 07:53:50 UT on C1 and C3, and at 07:54:05 UT on C4. The velocity of tailward ion flow was very large and its maximum value even exceeded 1500 km/s. The maximum southward magnetic field component reached 25 nT. Generally, such a high speed tailward ion flow with a large southward magnetic field component is produced by magnetic reconnection. The tailward ion flow with southward magnetic field component disappeared at 07:58:30UT. Figures 1e and 1f give the ion density and plasma beta (β) observed by C1 and C4. From 07:50UT -07:56UT, the plasma β was larger than 0.7, with a peak value 3.17 at 07:54:45UT, where the ion density was about 0.3/cm^3, the total magnetic field was about 15nT, and the proton temperature was about 5897 eV. Thus according to general identification criterion of plasma sheet [12], the Cluster satellites were located in the plasma sheet at least between 07:50UT -07:56UT. The reconnection event details see the reference [13].

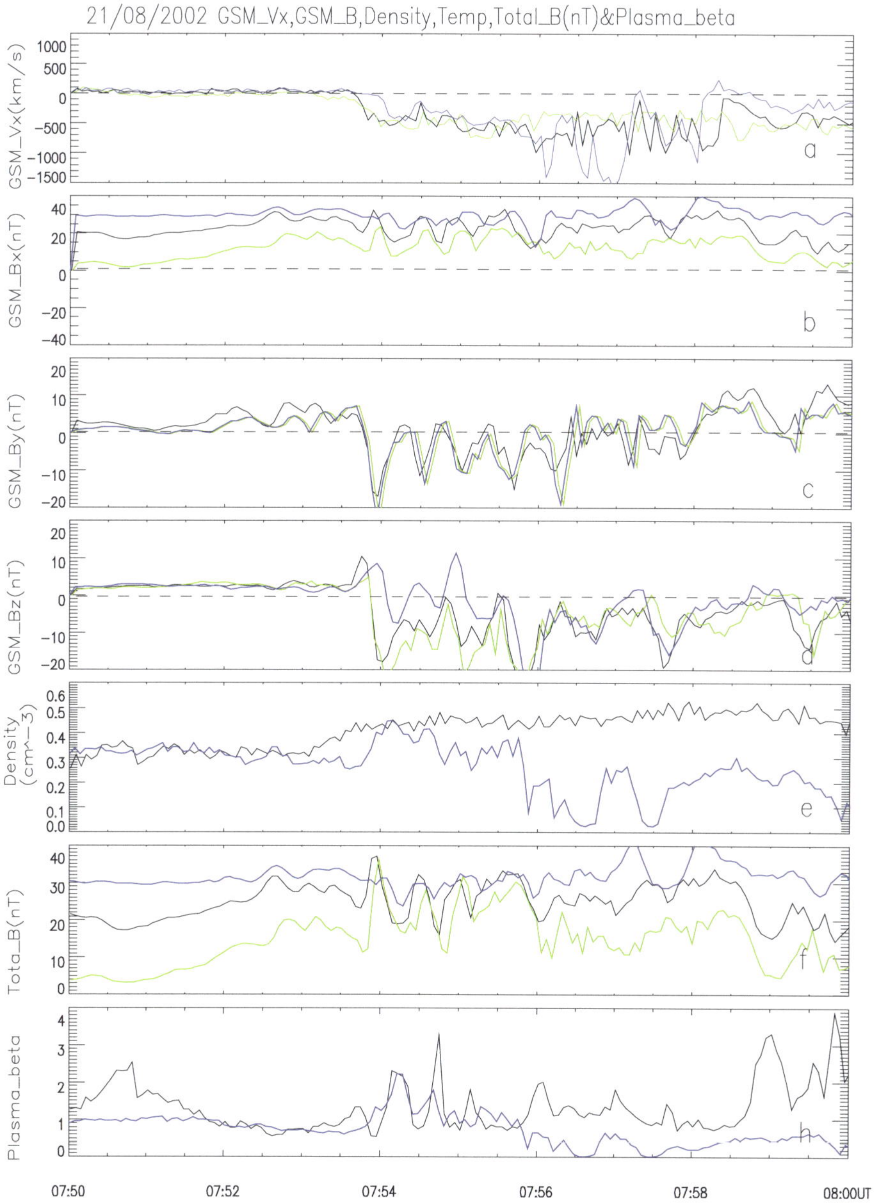

Fig. 2. Plasma parameters observed by C1 (black), C3 (green) and C4 (blue) during the interval of 07:50UT - 08:00UT on August 21, 2002. From top to bottom: plasma flow (Vx), magnetic field components (Bx, By, Bz), ion density

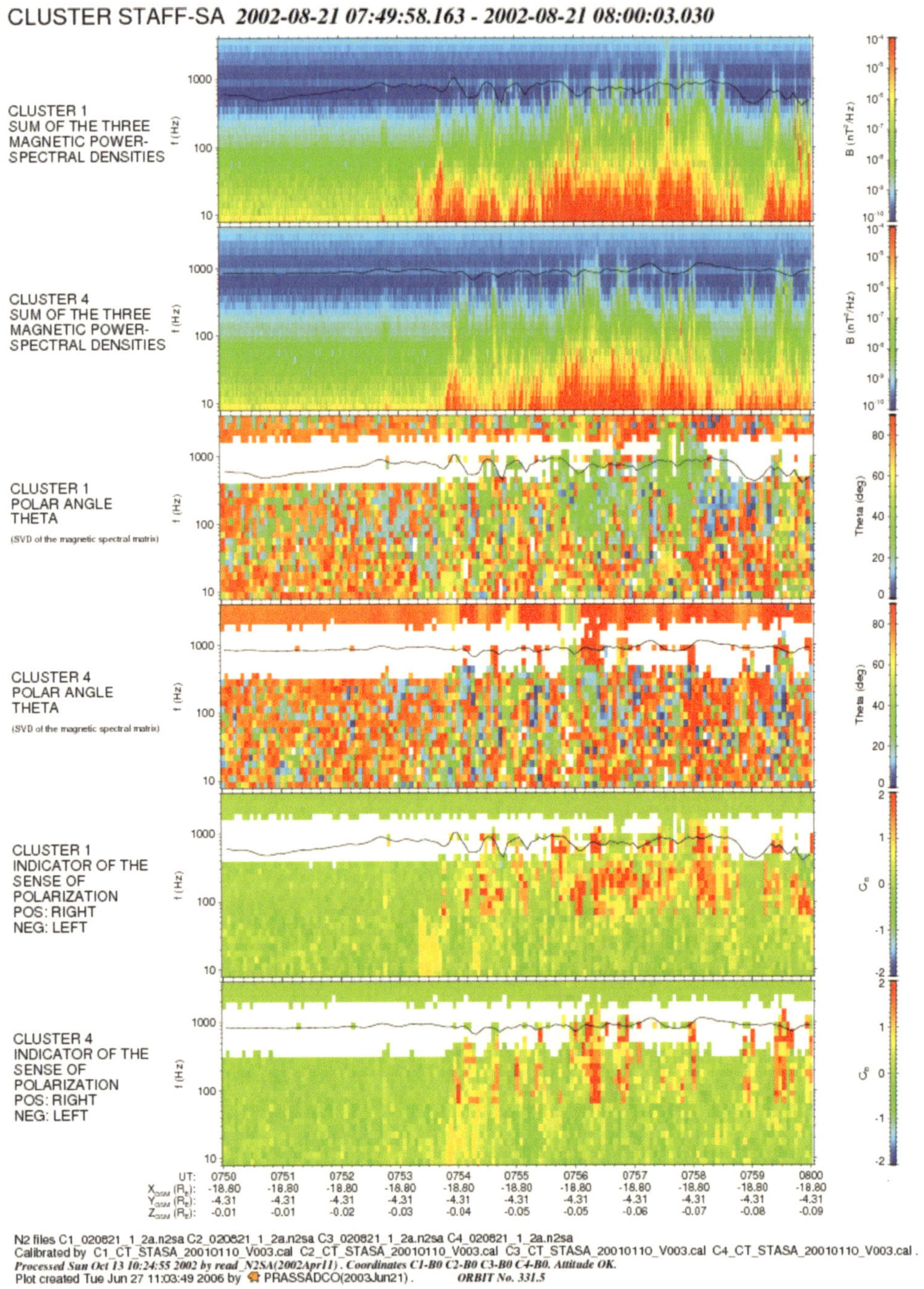

Fig. 3. Wave characteristics observed by C1 and C4 during the period of 07:50UT-08:00UT on August 21, 2002. From top to bottom, Panels 1-2: the dynamic spectra of total field turbulence B-power; Panels 3-4: the polar angles (THETA) of the wave normal direction with respect to ambient magnetic field; Panels 5-6: the sense of polarization. Black curves on the dynamic spectra are the electron cyclotron frequency

The waves in the frequency range 10Hz-4kHz observed by STAFF are analyzed by means of PRASSADCO tool (Propagation Analysis of STAFF-SA Data with Coherency tests)[14]. All three satellites (C1, C3 and C4) observed the whistler waves prior to southward turning of Bz component. The wave characteristics observed by C1, C2 and C3 were nearly identical. However the wave characteristics observed by C4 were different from those of other three satellites. Thus only the wave characteristics observed by C1 and C4 are displayed here. Figure 3 shows the wave characteristics observed by C1 and C4 during the interval of 07:50 UT-08:00 UT on August 21, 2002. The black curves represent the electron cyclotron frequency. The first and second panels show the power-spectral densities of magnetic field. The third and fourth panels represent the polar angles (Theta) of the wave normal direction with respect to the ambient magnetic field. These angles are obtained from the magnetic power spectral with the method of SVD [14]. The fifth and sixth panels represent the sense of polarization, in which the values of $c_B < 0$ indicate a right-hand polarized wave, and the values of $c_B > 0$ indicate a left-hand polarized wave. During the period of 07:50UT-08:00UT, weak (yellow) and enhanced (red) wave activities were observed. The frequency range of waves was between ion cyclotron frequency and electron cyclotron frequency.

total magnetic field (B) and plasma beta. Since the waves prior to the southward turning of Bz component and the waves in the higher frequency range during the magnetic reconnection event were quasi-parallel ($\theta \leq 40$ degree) propagating and right-hand polarized waves, they are typical whistler modes. The waves in the lower frequency range during the magnetic reconnection event were quasi-perpendicular propagating and linearly polarized wave, and they may result from a superposition of several linearly polarized waves [15].

The above analysis shows that the intense whistler activities appeared about 30s prior to the southward turning of magnetic field. The whistler waves were greatly enhanced after the southward turning of magnetic field. In addition, as reconnection proceeded, the wave frequency became higher and higher. When tailward flows reached the maximum, the wave frequency got closer to the electron cyclotron frequency.

3.2 Magnetic reconnection event and whistler wave on September 17, 2003

Figure 4 gives the ion flow velocity components (Vx), magnetic field components (Bx, By, Bz), and total magnetic field (B), respectively, which are observed by C1 (line), C3 (dot line) and C4 (broken line) during the interval of 13:00UT - 13:30UT on 17-09-2003. At 13:11:30UT, high-speed tailward ion flow was observed by C1, C3 and C4 at same time (see figure 4, a). It lasted to 13:19:30UT. At 13:11:30UT, C1, C3 and C4 observed southward field component (see figure5, d). It lasted to 13:15:00UT and then changed northward. The northward magnetic field component lasted to 13:16:50UT, and changed southward again. It lasted to 13:19:30UT. During the period of 13:00:00UT~13:19:30UT, C1, C3 and C4 observed the earthward

During the period of 13:11:30UT~13:19:30UT, C1, C3 and C4 observed almostly dawnward magnetic filed (see figure 4, c). Sometime, this field changed duskward. These field characters coincide with Hall magnetic field polarity of collisionless reconnection. Thus this event was a collisionless reconnection event, and it was observed at 13:11:30UT. Figure 5 shows the wave characteristics observed by C1 and C4 during the interval of 13:00-13:30UT on 17-09-2003. The first and second panels from top to bottom show the power-spectral densities of magnetic field observed by C1 and C4, respectively. At 13:10:40UT, the magnetic field power-spectral have enhanced already. The third and fourth panels show the power-spectral densities of electric field observed by C1 and C4, respectively. The electric field power-spectral enhanced earlier than the magnetic field power-spectral. The fifth and sixth panels represent the polar angles (Theta) of the wave normal direction with respect to the ambient magnetic field observed by C1 and C4, respectively. These electromagnetic waves propagated in the quasi-parallel direction are right-hand polarized (The seventh and eighth panels represent the sense of polarization, in which the values of c_B >0 indicate a right-hand polarized wave, and the values of c_B <0 indicate a left-hand polarized wave. See the red regions). During the period of 13:00-13:30UT, weak (yellow) and enhanced (red) wave activities were observed. The frequency range of waves was between ion cyclotron frequency and electron cyclotron frequency. They may be a whistler mode. The enhanced wave activities were observed earlier than reconnection event 30s. These enhanced waves included the superposition of many linearly polarized waves.

4. Conclusion and discussion

The above observation and analyses discussed two reconnection events and whistler wave activities event. The observation shows the low frequency electromagnetic waves is very strong during the reconnection event period. Before the reconnection event was observed, the strong whistler wave activities are observed in the current sheet. According to above analyses for the character and relationship of reconnection and wave activities, we can get the flowing conclusions. Hall magnetic field and whistler waves were observed in the plasma sheet, and whistler activities prior to reconnection. These wave activities in plasma sheet maybe play an important role on the generation of reconnection, and they maybe excite the reconnection. After reconnection event took place, with the reconnection on going, the frequency of enhanced wave activities increased and the polarization enhanced. During this kind of events, Cluster located in the Hall magnetic field region of reconnection.
The above studies show that the whistler waves can be excited prior to the reconnection event. The analysis of whistler group velocity and ion flow velocity indicates that this phenomenon could not be caused by the propagation time difference between whistler waves and ion flow velocity. The whistler group velocity can be estimated from the whistler dispersion equation:

$$V_g=\frac{d\omega}{dk}=2c^2k/[2\omega+\frac{\omega_{pe}^2\Omega_e}{(\omega-\Omega_e)^2}] \quad (1)$$

where $\omega_{pe}=\left(4\pi n_e e^2/m_e\right)^{1/2}$ and $\Omega_e=eB_0/m_e$ are the electron plasma frequency and cyclotron frequency, respectively. If the wave frequency $\omega << \Omega_e$, the V_g can become as

$$V_g=\frac{2c}{\omega_{pe}}\sqrt{\Omega_e\omega}=\frac{2c}{f_{pe}}\sqrt{f_e f} \quad (2)$$

At the beginning of reconnection observed by C1 (on 17-09-2003 13:00-13:30UT), n_e=0.2/cm^3, $f_{pe}=8.98\times10^3 n_e^{1/2}\approx4.0\times10^3$Hz, $f\approx40$Hz, and $f_e\approx700\text{Hz}<f_{pe}$. Thus the group velocity is about 25000km/s. The ion flow velocity is about 600km/s. The group velocity is about 23 times the ion flow velocity. The ion inertial length (di) in the reconnection layer in the magnetotail near ~18Re is about 500 km. Since the length of reconnection layer is proportional to the square root of di and about several times the ion inertial length in the magnetotail, the length of reconnection layer in the magnetotail is about 1-2 Re[9][16]-[17]. So the propagation time difference of whistler and ion flow from

the neutral line of reconnection to the Cluster satellite position is about 10s. Thus even the propagation time difference is considered, the observed whistler appears at least 30s earlier than the reconnection event. The whistler waves were excited in the central plasma sheet prior to the reconnection event. Both theoretical and experimental studies also showed that there are whistler waves and kinetic Alfvén waves in the reconnection layers. Recently theoretical studies showed that whistler and Alfvén wave instabilities can even be excited in the current sheet prior to the reconnection [18]-[21].

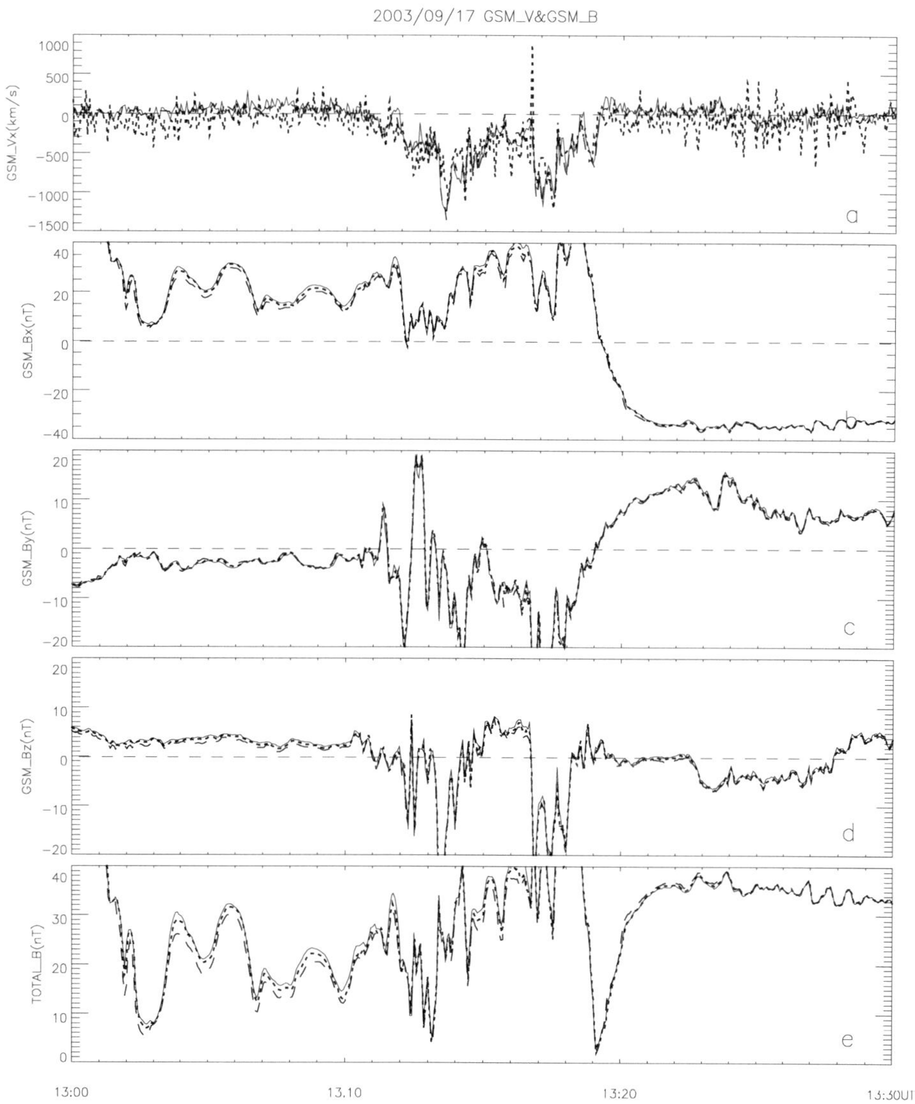

Fig. 4.

Fig 4 shows the flow and magnetic field character on 09-17-2003. a, b, c, d, e give x- component of the ion flow; x-, y-, z- components of magnetic field, and total magnetic field, respectively. The line, dot, broken lines show the data observed by C1, C3 and C4 respectively.

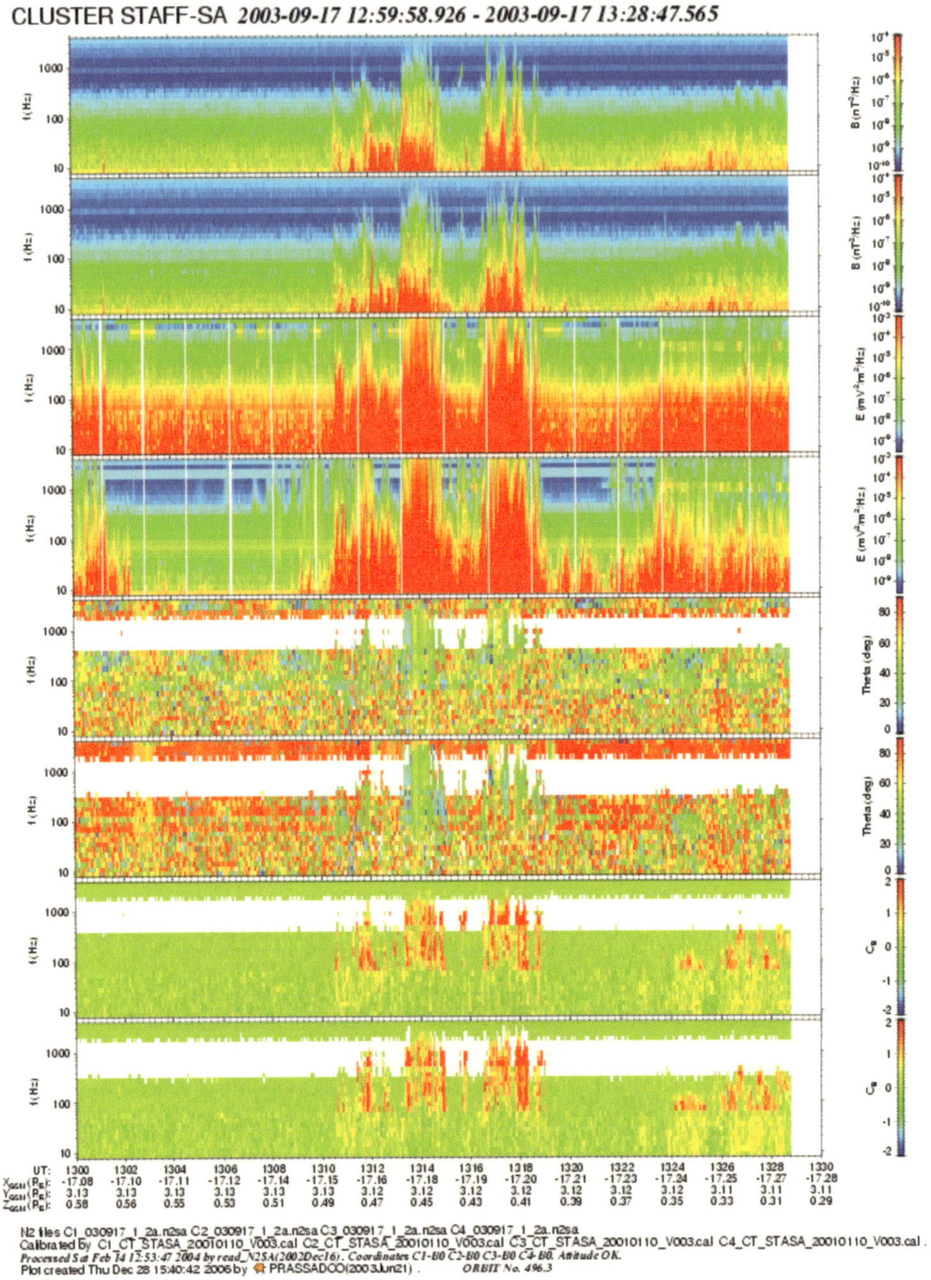

Fig. 5.

Fig. 5 show the characters of waves observed by C1 and C4 on 17-09-2003. From top to bottom, Panels 1-2: the dynamic spectra of total field turbulence B-power; Panels 3-4: the dynamic spectra of total field turbulence E-power; Panels 5-6 the polar angles (THETA) of the wave normal direction with respect to ambient magnetic field; Panels 7-8: the sense of polarization. Black curves on the dynamic spectra are the electron cyclotron frequency. The spectrograms intensities， polar angles size and sense of polarization indicated by color-coded according to the scale on the right.

Up to present, the problem of triggering of magnetic reconnection in the tail is still not fully understood. The MHD theory of collisionless reconnection requires that anomalous resistivity exist in the reconnection layer. The whistler waves prior to the reconnection may play an important role in the triggering of reconnection. The satellite observations show that besides whistler waves, there are LHWs [22]-[23] and solitary waves [24]-[26][10]-[11] in magnetopause and magnetotail reconnections. Also during dayside magnetopause reconnection, whistler modes could be converted to LHW modes [27]-[29]. However it is still an open question that what role these waves play exactly in the reconnection process and triggering of reconnection.

5. Acknowledgment

This work was supported by NSFC grants 40804045.

6. References

[1] Drake, J. F., Magnetic Reconnection: A Kinetic Treatment, in *Physics of the Magnetopause,* edited by P. Song, pp155-165,.AGU, Washington, D.C. 1995

[2] Brody, K.I. R.E. Holzer, and E.J.Smith, Magetic field fluctuations between 2 and 1000 Hz in the magetotail (abstract), *Eos Trans.* AGU, 49, 280, 1968.

[3] Scarf, F., Frank, K. Ackerson and R, Lepping, Plasma wave turbulence at distant crossings of the plasma sheet boundaries and neutral sheet, *Geophys. Res. Lett.*, 1, 189, 1974

[4] Gurnett, D. H., L. A. Frank., and R. P. Lepping, Plasma waves in the distant magnetotail, *J. Geophys. Res.*, 81(34), 6059-6071, 1976

[5] Scarf, F.L., F.V. Cornoniti, C.F. Kennel, R.W. Fredricks, D.A. Curnett, and E.J. Smisth, ISEE 3 wave measurements in the distant geomagnetic tail and boundary layer, *Geophys. Res. Lett.*, 11, 335-339, 1984.

[6] Kennel, C.F., F.V. Coroniti, and F.L. Scarf, Plasma waves in the magnetotail flux ropes, *J. Geophys. Res.*, 91, 1424. 1986.

[7] Zhang, Y., H. Matsumoto, and H. Kojima, Whistler mode waves in the magnetotail, *J. Geophys. Res.*, 104(A12), 28633-28644, 1999.

[8] Sonnerup, B. U. Ö, I. Papamastorakis, G. Paschmann, and H. Lühr, Magnetopause properties from AMPTE/IRM observations of the convection electric field: Method development, *J. Geophys. Res.*, 92(11), 12137-12159, 1987.

[9] Deng, X. H., and H. Matsumoto, Rapid magnetic reconnection in the Earth's magnetosphere mediated by whistler waves, *Nature*, 410, 557–560, 2001.

[10] Farrell, W. M., M. D. Desch, M. L. Kaiser, and K. Goetz, The dominance of electron plasma waves near a reconnection X-line, *Geophys. Res. Lett.*, 29(19), 1902, doi: 10. 1029/2002GL014662, 2002.

[11] Farrell, W. M., M. D. Desch, K.W. Ogilvie, M. L. Kaiser, and K. Goetz, The role of upper hybrid waves in magnetic reconnection, *Geophys. Res. Lett.*, 30(24), 2259, doi: 10. 1029/2003GL01549, 2003.

[12] Hughes, W. J., The magnetopause, magnetotail, and magnetic reconnection,, in *Introduction to Space Physics,* edit by M. G. Kivelson and C. T. Russel, PP227-287, Camb. Univ. Press, London, UK, 1995.

[13] X. H. Wei, J. B. Cao, G. C. Zhou, O. Santolík, H. Réme, I. Dandouras, N. Cornilleau-Wehrlin, E. Lucek, C. M. Carr, A. Fazakerley, Cluster observations of waves in the whistler frequency range associated with magnetic reconnection in the Earth's magnetotail, *J. Geophys. Res.*, 112, A10225, doi:10.1029/2006JA011771, 2007

[14] Santolík, O., M. Parrot, and F. Lefeuvre, Singular value decomposition methods for wave propagation analysis, *Radio Sci.*, 38(1), 1010, doi:10.1029/2000rs002523,2003.

[15] Santolík, O., J.S. Pichett, D.A. Gurnett, L. R. O. Storey, Magnetic component of narrowband ion cyclotron waves in th auroral zone, *J. Geophys. Res.*, 107(12), 1444, doi:10.1029/2001JA00146, 2002

[16] Wang, X. G, A. Bhttacharjee, and Z. W. Ma, Scaling of Collisionless Forced Reconnection. *Phys. Res. Lett.*, 87(26), doi:10.1103/PhysRevLett.87.265003, 2001.

[17] Shay M A., Drake J F, and Rogers B N. The scaling of collisionless, magnetic reconnection for large systems. *Geophys. Res. Lett.*, 26, 2163~2166, 1999.

[18] Wang, X.G, A. Bhttacharjee, and Z. W. Ma, Collisionless reconnection: Effects of Hall current and electron pressure gradient. *J. Geophys. Res.*, 105(12), 27633-27648, 2000.

[19] Zhou, G. C., J. B. Cao, D. J. Wang, and C. L. Cai, Low-frequency waves in collisionless plasma current sheet, *Acta Physica Sinica*, 53(8) (in Chinese), 2644-2653, 2004.

[20] Zhou, G. C., X. H. Wei, J. B. Cao, D. J. Wang, and C. L. Cai, Whistler wave instabilities in the collisionless current sheet, *Chinese Astronomy and Astrophysics*, 29, 404-412, 2005.

[21] Wei, X. H., G. C. Zhou, J. B. Cao, and L. Y. Li, Low-frequency electromagnetic instabilities in a collisionless current sheet: MHD model (in Chinese), *Acta Phys. Sin.*, 54(7), 3228- 3235, 2005.

[22] Bale, S. D., F. S. Mozer, and T. Phan, Observation of low hybrid drift instability in the diffusion region at a reconnecting magnetopause, *Geophys. Res. Lett.*, 29(24), 2180, doi:10.1029/2002GL016113, 2002.

[23] Øieroset, M., R. P. Lin, T. D. Phan, D. E. Larson, and S. D. Bale, Evidence for electron acceleration up to 300 keV in the magnetic reconnection reconnection diffusion region of Earth's magnetotail, *Phys. Rev. Lett.*, 89(19), doi:10.1103/PhysRevlett.89195001, 2002.

[24] Matsumoto, H., X. H. Deng, H. Kojima, and R. R. Anderson, Observation of electrostatic solitary waves associated with reconnection on the dayside magnetopause boundary, *Geophys. Res. Lett.*, 30(6), 1326, doi:10.1029/2002GL016319, 2003.

[25] Cattell, C., J. Crumley, J. Dombeck, J. Wygant, and F. S. Mozer, Polar observations of solitary waves at the Earth's magnetopause, *Geophys. Res. Lett.*, 29(5), 1065, doi:10.1029/2001GL014046, 2002.

[26] Cattell, C., et al., Cluster observations of electron holes in association with magnetotail reconnection and comparison to simulations, *J. Geophys. Res.*, 110, doi:10.1029/2204JA010519, 2005.

[27] Carter, T. A. et al., Measurement of Lower-Hybrid Drift Turbulence in a Reconnecting Current Sheet, *Phys. Res. Lett.*, 88, DOI: 10.1103/PhysRevLett.88.015001,2005.

[28] Carter, T. A. et al., Experimental study of lower-hybrid drift turbulence in a reconnecting current sheet, *Phys. Plasmas* 9, 3272-3288, 2002.

[29] Daughton, W., Electromagnetic properties of the lower-hybrid drift instability in a thin current sheet, *Physics of Plasma*, 10(8), 3103-3119, 2003.

13

Solitary Electromagnetic Waves Generated by the Switching Mode Circuit

Hirokazu Tohya and Noritaka Toya
ICAST, Inc.
Japan

1. Introduction

The switching mode circuit (SMC) performs by the switching transistor or the other switching devices. Usually, unlike the distorted continuous signal voltage on the linear circuit, the signal voltage of the SMC has two stationary states. The SMC has been applied widely to the power circuit and the signal circuit including the digital circuit now. The performance of the SMC has been supported by the improvement of the semiconductors. For example, the International Technology Roadmap for Semiconductors (ITRS) is showing the details of the technology trend of the worldwide semiconductors [1]. According to the ITRS, the technology about the high performance system on a chip (SoC) or the VLSI has been improved to 45*nm* from 78*nm* in the last five years. However the gate delay of the PMOS FET stays between 1.65*ps* and 1.73*ps*, and is increasing slightly. In addition, the improvement of the on-chip clock frequency is staying in 5-6*GHz* approximately. Many results of the research about the on-chip interconnect were presented [2, 3]. The themes of these studies are the influence of the resistance, the parasitic capacitances, the power noise, the substrate noise, and others to the performance of the SoC [4, 5]. Meanwhile, the on-chip transmission line interconnect [6, 7] and the optical interconnect [8] were proposed for improving the performance of the SoC. Many results of the study of the power distribution network (PDN) about the EMI also were presented [9-14]. The relevant papers which include above ones and the conventional theories were discussed from the point of view of the stagnation of the performance of the SoC and the suppression of the EMI of the SMC. Concurrently, the development of the theory and the technologies suitable to the SMC was started. As a result, the solitary electromagnetic wave (SEMW) theory and the technologies of the low impedance lossy line (LILL) and the matched impedance lossy line (MILL) were successfully developed. These are presented below together with the result of the analysis, experiments, and the review of the conventional theory and engineering.

2. Review of the conventional theories and engineering

2.1 EMW theory

The EMW theory is the most trusted fundamental theory for the AC circuit including the SMC. It was presented by James Clerk Maxwell in 1873 and the existence of the EMW was first confirmed experimentally by H. R. Hertz in 1888. This theory was based on the vector EMW equation. The vector EMW equation has been developed by fusing the Maxwell's theory and the undulation equation. The undulation equation handles the continuous wave and it was presented by Jean Le Rond d'Alembert in 1750. In 1881, Oliver Heaviside replaced the electromagnetic potential field by the force field and simplified the complexity of Maxwell's theory to four differential equations which are known now as Maxwell's laws or Maxwell's equations.

Maxwell's vector EMW equations in vacuum are

$$\nabla^2 \dot{E} - \mu_0 \varepsilon_0 \frac{\partial^2 \dot{E}}{\partial t^2} = 0 \, , \; \nabla^2 \dot{H} - \mu_0 \varepsilon_0 \frac{\partial^2 \dot{H}}{\partial t^2} = 0 \tag{1}$$

The traveling speed of the EMW is

$$c = 1 \big/ \sqrt{\mu_0 \varepsilon_0} \tag{2}$$

In (2), c corresponds to the velocity of the light in vacuum.

When the EMW is vibrating at a constant angular velocity, the modified (1) is

$$\dot{E} = i\sqrt{2}E_o \cos\left\{\omega\left(t \mp z\sqrt{\mu_o \varepsilon_o}\right) + \theta\right\}, \; \dot{H} = \pm j\sqrt{2}H_0 \sqrt{\frac{\varepsilon_o}{\mu_o}} \cos\left\{\omega\left(t \mp z\sqrt{\mu_o \varepsilon_o}\right) + \theta\right\} \tag{3}$$

where i is the unit vector of the standard transverse direction against the traveling direction, j is the unit vector of the perpendicular direction to the standard transverse direction.

According to (3), the EMW consists of the electric field wave and the magnetic field wave. These waves are at right angles to each other and both wave shapes are same except its magnitude.

The alternate current (AC) circuit is defined by the electromagnetism as the circuit of the EMW because the electric field and the magnetic field are vibrating on it. According to the Ampère's circuital law, the AC is the integrated magnetic field around a closed loop to the conductor. The vibrating magnetic field forms the magnetic wave which constitutes the EMW. The EMW can travel at near speed to the light through the insulator of the transmission line. The EMW cannot travel through the conductor. The skin depth causes the attenuation when the EMW travels on the transmission line. It shows the sinking degree of the electric field or the magnetic field into the conductor.

The quasi-stationary state is one of the definitions in the electromagnetism. The EMI is negligible when the length of all wires in the circuit is quite shorter than the wave length. Such AC circuit can be handled as the quasi-stationary closed circuit (QSCC).

The EMW theory handles the continuous wave as shown in (3). Therefore, the Fourier transform is necessary to analyze the distorted EMW. It is believed that the signal voltage wave of the SMC consists of many harmonic waves by the idea of the Fourier transform. It's very convincing mathematically. However, the following serious problems will be caused when it is applied to the SMC. That is, though the wave shape of the signal voltage of the SMC has two stationary states obviously, the stationary state got from the Fourier transform is only one. Furthermore, the timing information of the intermittent signal wave on the SMC is lost.

2.2 AC circuit theory

The Kirchhoff's circuit law is one of the most important laws in the AC circuit theory. It was presented in 1845. The principle of superposition which is based on this law is quite convenient for analyzing the complex AC circuit. The AC circuit theory is the basic theory of the simulation program with integrated circuit emphasis (SPICE) which is used for the design and analysis of the AC circuit all over the world. It includes HSPICE, PSPICE, and XSPICE. The SPICE has been used also for the design and the analysis of the SMC for a long time.

The lumped element model is used in this theory. The lumped element model consists of 4 kinds of elements which are the inductance (L), capacitance (C), resistance (R), and conductance (G). Each element is formed as the component usually and functions independently from the electromagnetic phenomenon in the circuit.

Fig. 1 shows the transmission coefficient (S_{21}) when the capacitor and the short circuit are connected to the transmission line. They were measured by the network analyzer.

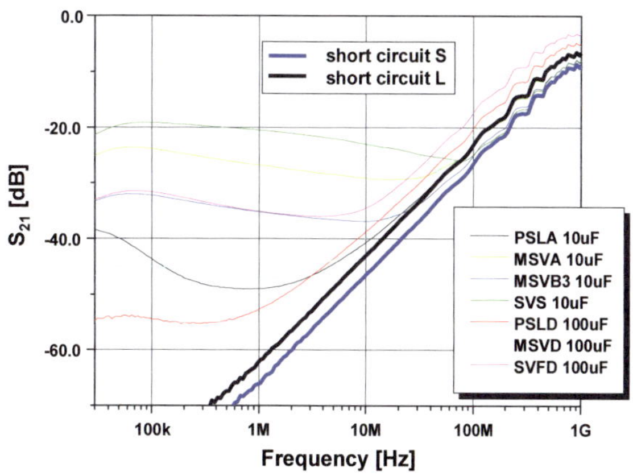

Fig. 1. Measured S_{21} of the transmission line

The small tantalum capacitors of $10\mu F$ and the large tantalum capacitors of $100\mu F$ were mounted on the coplanar line S and the coplanar line L each, the characteristic impedance of each coplanar line was 50Ω, each short circuit S and short circuit L were formed by shorting the pads for the capacitor on the coplanar line S and the coplanar line L.

In Fig. 1, each S_{21} of the small tantalum capacitor and large tantalum capacitor are approaching to S_{21} of the short circuits S and short circuits L at the frequency which is more than 10 MHz.

The transmission coefficient of the transmission line when the capacitor connected in parallel with it is

$$S_{21} = \frac{2Z_C}{2Z_C + Z_0} \tag{4}$$

where $Z_C=(2\pi fC)^{-1}$, C is the capacitance of the capacitor, Z_0 is the characteristic impedance of the transmission cable equipped to the network analyzer.

The idea that the impedance of the capacitor can be got from (4) approximately has been believed by almost all circuit engineers as well as the capacitor manufacturers. However, this idea is not correct because the impedance of the capacitor is got from $(2\pi fC)^{-1}$ theoretically and the capacitor is used also in series to the transmission line.

S_{21} depends on the materials and the form of the transmission line. However the capacitor has not the structure of the transmission line. Therefore the capacitor which is connected to the transmission line in parallel disturbs slightly the traveling of the EMW as the short circuit in Fig. 1. The curves of S_{21} in Fig. 1 will move to the low frequency side when the capacitors are connected by the through holes to the power plane and the ground plate as conventional instead of the coplanar. From above, it can be said that Fig. 1 and (4) show the application limitation of the Kirchhoff's circuit law as well as the principle of superposition. According to the AC circuit theory, the current in the circuit is defined as the average speed of the electron (dq/dt) in the wire which is called the electron current hereinafter. According to the physics, dq/dt is quite small. For example, it becomes $3.68\times10^{-4} m/s$ in the case of the copper wire which has the cross-sectional area of $1mm^2$. Therefore, the electron current is valid only on the direct current (DC) circuit or the stationary circuit. The idea that one of the functions of the capacitor is charging the electron does not conform to the electromagnetism which defines it as the electrostatic energy consisting of the electric field. Nevertheless, above definition and idea will be permitted for the actual design and analysis when the AC circuit is the QSCC.

From above, it should be understood that the AC circuit theory is the theory for the QSCC. Therefore, the design and analysis by the SPICE will be available for the QSCC part of the AC circuit including the SMC without troubles. However, it has been the difficult problem how to be formed to the QSCC because the power lines and the signal lines are included in them.

2.3 Telecommunication engineering (TELE)

The TELE has been used for design of the transmission line of the SMC as well as the high-frequency AC circuit. The idea of the characteristic impedance is useful for designing of the transmission line of the SMC. The theory of the TELE is based on the telegrapher's equation and it was presented by Oliver Heaviside in 1880. The telegrapher's equation was developed by fusing the undulation equation and the AC circuit theory instead of the Maxwell's equation, because the AC circuit theory was the only theory for the electric engineers at this times. Therefore, the line current in the telegrapher's equation is the electron current. In addition, some breaches in the TELE to the EMW theory are found at the boundary of the conductor and the insulator.

Nonetheless the existence value for the design and analysis of the transmission line is not spoiled. The concept of the characteristic impedance is effective to only the useful signal wave travelling on the transmission line.

3. SEMW theory

We imagined that the SMC generates the SEMW instead of the distorted EMW. This idea will be effective for finding way out of the above mentioned stagnation of the semiconductor technologies because it will solve above mentioned serious problems caused by the application of the Fourier transform. The SEMW theory was developed by fusing the EMW theory and the NLU theory. It is also the fruits of the discussion with the co-author who graduated from the department of the physics of the university.

3.1 NLU theory

The solitary wave was found out as the great wave by John Scott Russell who was making an experiment for smoothing the run of the boat in a canal in 1834. However this discovery was not recognized by the scientist at this times. D. J. Korteweg and G. de Vries (KdV) presented the equation about the wave traveling one direction on the shallow water in 1895. The KdV equation is

$$\frac{\partial \eta}{\partial t}+\frac{3c_0}{2h}\eta\frac{\partial \eta}{\partial \xi}+\frac{c_0 h^2}{6}\frac{\partial^3 \eta}{\partial \xi^3}=0 \tag{5}$$

where η is the altitude from the average surface of the water, ξ is the coordinate which is moving at the velocity of $c_0 = (gh)^{1/2}$ and, $\xi = x - c_0 t$, g is the constant of gravitation, and h is the average depth of the water.
The special solution of (5) by KdV is

$$\eta = \eta_0 \operatorname{sec}h^2\left(\frac{1}{2}\sqrt{\frac{3\eta_0}{h^3}}(x-ct)\right) \tag{6}$$

$$c = c_0\left(1+\frac{\eta_0}{2h}\right) = \sqrt{gh}\left(1+\frac{\eta_0}{2h}\right) \tag{7}$$

where η_0 is the magnitude of the wave.
N. J. Zabusky and M. D. Kruskal confirmed the existence of the solitary wave at the computer simulation based on the KdV equation about the heat conducting system in 1965. They discovered the following characteristics similar to the particle to the solitary wave and named it soliton.

a. Soliton travels without changing its speed and wave shape.
b. The magnitude and the speed of the soliton are not influenced by crossing.
c. When the magnitude becomes larger, the wave length becomes smaller and the traveling speed becomes faster.

3.2 Generation mechanism of the SEMW

The SoC consists of the on-chip inverters which are formed by the NMOS FET and the PMOSFET. The PMOS FET leads the switching motion of the CMOS circuit when it turns on. On the other hand, the NMOS FET leads the switching motion of the on-chip inverter when it turns off.
According to the semiconductor physics [15] and the ITRS, the saturation drain current of the MOS FET at the voltage higher than V_T is

$$I_{dsat} = A \cdot I_{dsat,p} \cdot (V_g - V_{T,sat}) / z(V_{dd} - V_{T,sat}) \tag{8}$$

where, A is the conversion coefficient, V_{dd} is the supply voltage, $I_{dsat,p}$ is the peak saturation drain current of the MOS FET, $V_{T,sat}$ is the saturation threshold voltage, z is the gate width, V_T is the threshold voltage, and V_g is the gate voltage.
The drain current of the MOS FET at up to the saturation threshold voltage is

$$I_d = (I_{sd \cdot leak}/z) \cdot V_g / V_{dd} \tag{9}$$

where $I_{sd\ leak}$ is the subthreshold leakage current.
The drain current increases in response to the half of VDD (1.1V) from zero according to the curve line got from (8) and (9). The drain current decreases in response to zero from 0.55V according to the inverse curve line formed by (8) and (9).
The 2009 technology node of the ITRS 2008update is used at the following calculation. The calculation condition in (8) and (9) about the PMOS FET is as follows; V_{dd} is 1.1V, $I_{dsat,p}$ is 1.317$mA/\mu m$ at V_{dd}, $V_{T,sat}$ is 196mV, $I_{sd\ leak}$ is 0.17$\mu A/\mu m$ at V_{dd}, and z is 108nm.
According to the ITRS, the voltage between drain and source (V_{ds}) is

$$V_{ds} = \frac{1}{C_{ox}}\int_0^{2T_1} I_{dsat1a} dt \tag{10}$$

T_1 is 1.7*ps* which is got from (10) when V_{ds} is 0.55V, C_{ox} is 0.875*fF*.
Fig. 2 shows the calculated waveform of the PMOS FET. Fig. 2. a shows V_{ds1} and Fig. 2. b shows I_{dsat2a}.

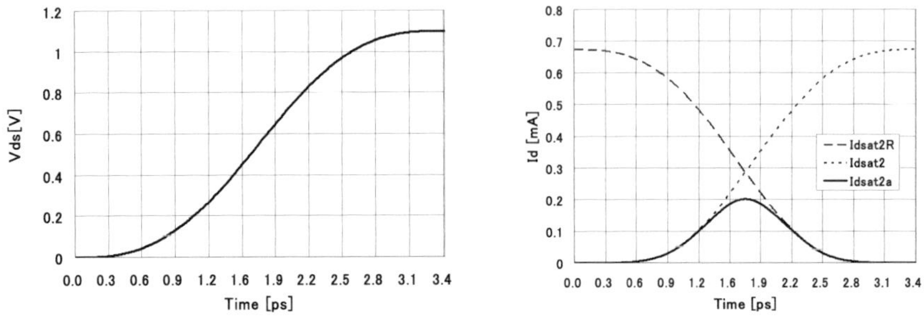

a) V_{ds1} b) I_{dsat2a}

Fig. 2. Calculated waveform of the PMOS FET

In Fig. 2. a, V_{ds1} was got from (10). In Fig. 2. b, each I_{dsat2} and I_{dsat2R} was got from (8) and (9) by using the wave shape of V_{ds1} in Fig. 2. a as Vg, and the drain current I_{dsat2a} was got from merging I_{dsat2R} and I_{dsat2}. At all calculations for getting the wave shape, the vector is ignored.
Fig. 3 shows the calculated waveform of the PMOS FET on the second stage of the on-chip inverter. Fig. 3. a shows the drain-source voltage V_{ds}. Fig. 3. b shows the electric field of between the drain and the source. Fig. 3. c shows the magnetic field.

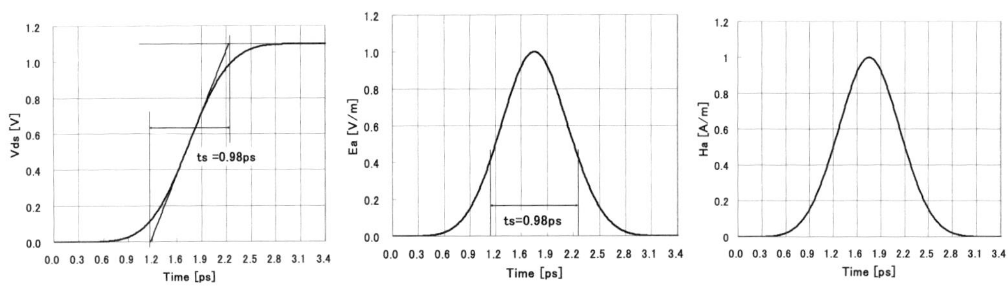

a) V_{ds} b) E_a c) H_a

Fig. 3. Calculated waveform of the PMOS FET on the second stage of the on-chip inverter

In Fig. 3. b, E_a is the differential value of the drain-source voltage V_{ds} according to the definition of the voltage potential in the electromagnetism. In Fig. 3. c, and H_a is got from the drain current I_{dsat2a} by applying the Ampère's law. Each peak value of E_a and H_a was set to 1 and the actual peak values of them depend on the parameters of the formation and the materials of the PMOS FET.
According to (1), the changing electric field and the magnetic field form the EMW and traveling. Therefore both E_a and H_a should be form to a kind of the EMW.
The EMW is the field free from the mass. When the EMW is applied to (6), the equation of the soliton is

$$u(t) = A \cdot \operatorname{sec}h^2\left(B\left(t - T_1\right)\right) \tag{11}$$

where A is the magnitude of the wave, B is the constant of the wave shape.

Fig. 4. shows the calculated result about the soliton. Fig. 4. a shows the wave shape of the soliton calculated by (11) when $T_1 = 1.7ps$, $A = 1$, and $B = 20.55$. Fig. 4. b shows the integral wave shape of the soliton in Fig. 4. a.

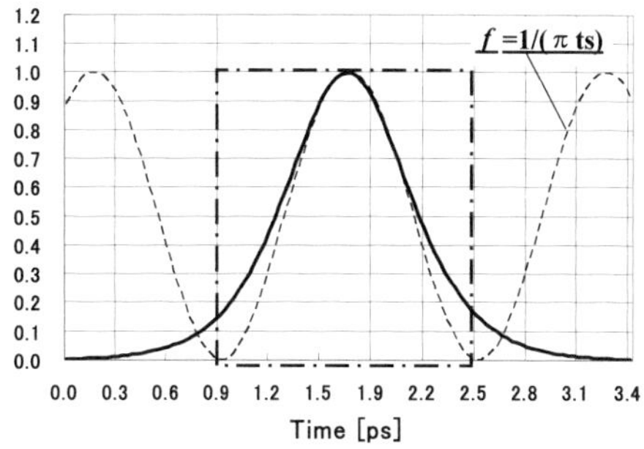

a) Wave shape of the Soliton

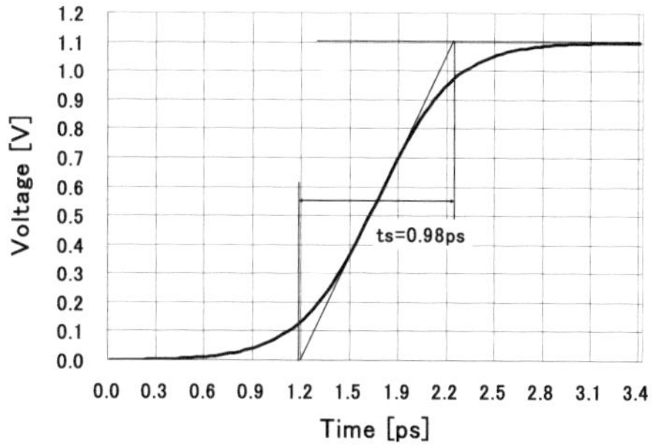

b) Integral wave shape of the soliton

Fig. 4. Calculated result about the soliton

In Fig. 4. a, the calculated soliton wave is similar to E_a and H_a in Fig. 3. In the calculation, t was set to $3.4ps$ from $3.7fs$. In Fig. 4. b, the maximum magnitude was set to $1.1V$ and the calculated soliton wave is similar to the wave shape in Fig. 3. a. The wave shapes in Fig. 4. a and Fig. 4. b are formed by one switching action by the PMOS FET. Therefore, unlike the Gaussian wave, no harmonic waves can be included in these wave shapes. Therefore, we named them the SEMW. The generation mechanism of the SEMW is similar to the great wave which was observed by John Scott Russell. In addition, it is also similar to the generation mechanism of the tsunami which was recognized as a kind of the soliton.
According to (3), the vector wave equations of the SEMW on the transmission line is

$$\dot{E} = i\sqrt{2}E_o \operatorname{sec}h^2\left(B\left(t \mp z\sqrt{\mu\varepsilon}\right) + T_1\right),\ \dot{H} = \pm j\sqrt{2}\sqrt{\frac{\varepsilon}{\mu}} \operatorname{sec}h^2\left(B\left(t \mp z\sqrt{\mu\varepsilon}\right) + T_1\right) \tag{12}$$

where i is the unit vector of the standard transverse direction against the traveling direction, j is the unit vector of the perpendicular direction to the standard transverse direction, μ is the dielectric constant, ε is the permeability, z is the travelling distance, and T_1 is the initial value of time.
According to (12), the SEMW consists of the solitary electric field wave (SEW) and the solitary magnetic field wave (SMW) and these waves are at right angles to each other and both wave shapes are same except its magnitude.
The definition of the wave length (λ_S) of the SEMW in the SEMW theory is

$$\lambda_S = \sqrt{\mu\varepsilon}/t_S \tag{13}$$

where t_S is the specified rise time of the V_{DS} of the PMOS FET in Fig. 3. a and Fig. 3. b.
Hundreds of millions of transistors are formed in the semiconductor layer, the on-chip interconnect layers are placed on the semiconductor layer, the power supply wiring layer (PSWL) which consist of the power mesh and the ground mesh are placed on the on-chip interconnect layer in the typical SoC.
The distributed element model is not necessary for the design and the analysis of the SMC when the rise time (t_r) of the signal voltage is larger than 2.5 times of the traveling time (t_f)

on the transmission line [16]. t_r is defined the time between 10 percent and 90 percent of the final voltage. This means the SMC can be handled as the QSCC when $t_r > 2.5t_f$,. λ_S on the on-chip interconnect is $179\mu m$ at the 2009 technology node of the ITRS 2008update because t_S is $0.98ps$ and the relative dielectric constant (ε_r) is 2.7. When all of the on-chip interconnect including the power supply wiring are shorter than $72\mu m$, the closed SMC at this technology can be handled as the QSCC.

In Fig. 4, the sine-wave having the frequency of $(\pi t_S)^{-1}$ is shown. The idea about the significant frequency (SF) was presented [17]. This idea cannot be directly applied to the SEMW theory because it has been discussed about the harmonic waves.

The properties of the modified significant frequency (MSF) which is defined according to the SEMW theory are as follows;

a. The correlation of the time domain wave shape of the SEMW and half wave shape of the sine-wave having the frequency of $(\pi t_S)^{-1}$ is about 85%.
b. No spectra exist except the MSF.

As the result, both frequency analysis and time analysis of the SMC become easy by the idea of the MSF.

The calculation time to get the wave shapes of Fig. 3 was about ten hours by using the desktop computer. The calculation to get the wave shapes of the next stage of the inverter was not finished during one week. On the other hand, the calculation to get the wave shapes of Fig. 4 was finished at once. Therefore, the design and analysis will quicken when the SEMW theory is applied to the SMC.

By the way, conventionally, I_{dsat2a} corresponding to H_a in Fig. 3. c has been recognized as the penetrating current. This current has been hated conventionally because this induces a brief spike in power consumption and becomes a serious issue at high-frequencies. However, according to the SEMW theory, this current or magnetic field is necessary to form the signal and its magnitude should be designed suitably.

3.3 Behaviours of the SEMW on the transmission line

Fig. 5 shows the equivalent circuit of the SMC including the on-chip inverter.

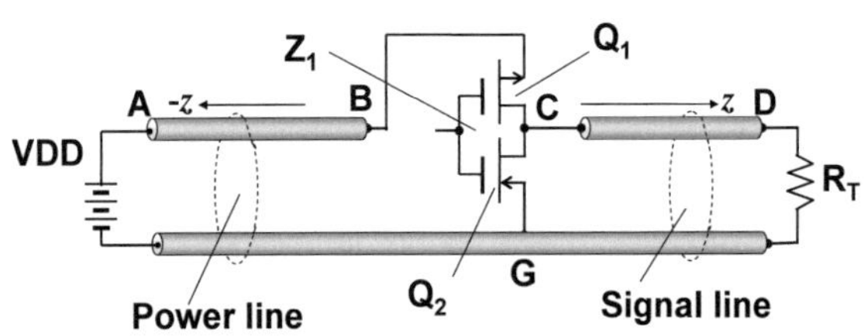

Fig. 5. Equivalent circuit of the SMC including the on-chip inverter

In Fig. 5, VDD is the ideal DC voltage source, Z_1 is the on-chip inverter formed by one stage, however it is formed by several stages in actual use, Q_1 is the PMOS FET, Q_2 is the NMOS FET, R_T is the terminating resistor, "0" is the state that Q_1 is OFF and Q_2 is ON, "1" is the inverse state of it, and VDD supplies VDDV.

A couple of the SEMW is generated when the inverter Z_1 changes to "1" from "0" or the inverter Z_1 changes to "0" from "1". The SEMW travels in the insulator of the transmission line at $\varepsilon_r^{-1/2}$ of the light speed. The normal SMC is a kind of the voltage source circuit. The SEW acts as the leading player in such circuit.

Fig. 6 shows the wave shapes of the SEW and the voltage on the transmission line.

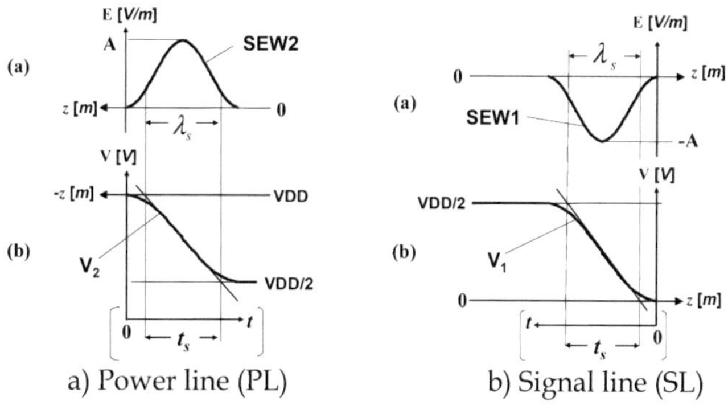

Fig. 6. Wave shapes of the SEW and the voltage on the line

In Fig. 6, z shows the traveling direction of the SEW, t shows the time direction which is opposite direction to z, a. (a) shows the SEW2 on the power line (PL), a. (b) shows the falling part of the voltage on the PL, b. (a) shows the SEW1 on the signal line (SL), b. (b) shows the rising part of the voltage on the SL. The SEW1 travels on the SL with charging to 0.5VDDV from 0V and the SEW2 travels on the PL with discharging to 0.5VDDV from VDDV. When the PL has the infinite length or the matched termination, the voltage on the SL and PL remains the half of the source voltage as shown in (b) of Fig. 6. However, the voltage on the SL and PL should reach VDDV finally because the output impedance of VDD is zero.
Fig. 7 shows the calculated wave shapes of the SEW and the rising part of the signal voltage at the point D in Fig.5.

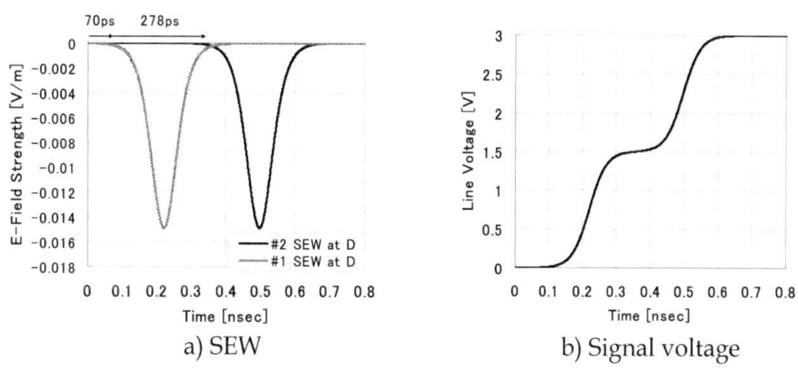

Fig. 7. Calculated wave shapes

In Fig. 5 and Fig. 7, the calculation condition is as follows; the gate delay of the inverter is 200*ps*, ε_r is 4.35, each length of the SL and the PL is 10*mm* and 20*mm*, the calculated delay time till the point D from the point C is 70*ps*, and the calculated delay time till the point A from the point B is 139*ps*. The generated #1 SEW reaches to the point D after the delay of 70*ps*. The generated #2 SEW reaches to the point D after the delay of 348*ps*, because the #2 SEW travels on the PL and it reflects at the point A. The signal voltage is got by the integral of the #1 SEW and #2 SEW because it is formed by charging of the SL by the #1 SEW and #2 SEW. From above, it is clarified that the signal voltage reaches to VDDV from the 0.5VDDV by the reflected SEW generated on the PL and the rise time of the signal depends on length of the PL.
It has been considered conventionally that the low output impedance of the voltage source is effective for only keeping the magnitude of the signal. However, by the SEMW theory, it

is clarified that both low impedance of the voltage source and the short length of the PL are necessary to keep the signal integrity of the SMC.
The SEMW theory first clarified the existence of three kinds of the current on the SMC. The current observed on the transmission line is the first current to follow the Ampère's law. This current has the same wave shape as the SEMW. When the electrostatic energy is taken out from the power source, the current which follows the Ampère's law flows and it is the second current. The magnitude of this current is decided by the source voltage and the characteristic impedance of the transmission line. The electron current for the matched termination resistance flows whenever the SL is being charged. This is the third current.
Te reflection or the bounce on the transmission line of the SMC has been analyzed by using the method of the lattice diagram currently, which has the other names such as the bounce diagram, the echo diagram, or the reflection diagram. This method has been used widely for the design and analysis of the SMC [18]. However this method cannot handle the electromagnetic phenomenon. The SEMW theory enables the analysis of the reflection or the bounce from the point of view of the electromagnetism easily.

3.4 Simulation of the influence of the length of the PL

Fig. 8 shows the equivalent circuit for simulation by HSPICE.

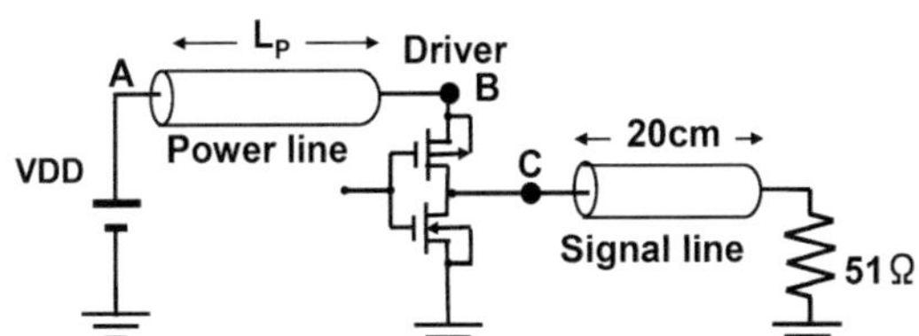

Fig. 8. Equivalent circuit for simulation

In Fig. 8, VDD is the ideal DC voltage source and the out-put voltage is 2.5V. The driver was described by the sub-circuit depending on the design parameter. The calculated characteristic impedance of the PL and the SL were 51.28Ω, which was got from ApsimRLGC®. The simulation result will be reliable in spite of the HSPICE because the lossless transmission line is used.
Fig. 9 shows the simulated voltage of the point B and C at the varied length of the PL.

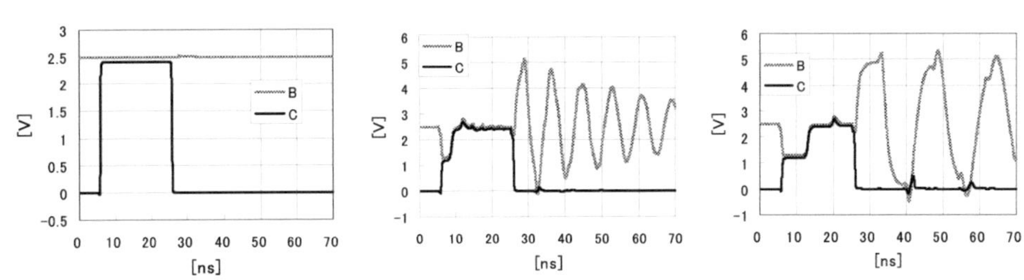

a) $0cm$ b) $20cm$ c. $50cm$

Fig. 9. Simulated signal voltage depending on each length of the PL

In Fig 9, the rise time does not increase from the gate delay when the length of the PL is zero and it increases in proportion to the length of the PL, the voltage at the point B in Fig. 8 is vibrating after the turn-off of the driver when the length of the PL is not zero.

When the PL has the length and the electron current exists on the SL, the static magnetic energy exists on the SL and the PL. In this situation, the SEMW is generated on the PL and the SL at the turn-off motion of the driver. The generation of the SEMW on the PL is caused to the electron current on the PL. The SEMW generated on the SL travels to the matched termination and is consumed. However, the SEMW generated on the PL shuttles between the point A and the point B. As the result, the voltage at the point B vibrates.

4. LILL technologies

When the ideal voltage source is located near the on-chip inverter on the PL, it is expected that the rise time of the signal will become close to the gate delay of the on-chip inverter. The vibration on the PL will not occur because the termination resistance does not exist in the SoC. The capacitor cannot function as ideal voltage source. It is caused by the structure of the capacitor which is not the transmission line. When the transmission line has the low impedance and the large loss, this transmission line which was named LILL will be able to function as the ideal voltage source. To get the cooperation from the semiconductor industries is necessary to actualize the on-chip LILL. However, unfortunately the EMW theory and the transmission line technologies are not being applied to their current design rule of the SoC. Therefore, the first development of the LILL was started from the on-board LILL.

4.1 On-board LILL technologies

Many spectra are observed on the board as well as on the chip. They are not caused by the harmonics based on the Fourier transform but are caused by many reflections and many repetitions of the SEMW. The frequency of these many spectra is lower than the MSF.
When the LILL is connected to the power terminal of the SoC closely, the LILL provides the ideal DC source to the SoC and it reflects the EMW including the SEMW at the power terminal. As the result the stability of the SoC will be improved and the EMI on the board will be suppressed. The electromagnetic susceptibility also will be improved by it because it is well known that many troubles of the SoC are caused by the EMW which comes through the PDN. However the rise time of the on-chip inverter will not be shortened enough.
The on-board LILL is most useful when the on-chip LILL technologies are not applied to the SoC and other ICs.

4.1.1 Necessary decoupling performance

All on-chip inverters are connected to the PL in parallel. Therefore, it can say that the PDN causes the EMI. Fig. 10 shows an example of the power current of the DDR2 dual-in-line memory module (DDR2 DIMM).
In Fig. 10, the x-axis shows the frequency allocated to $1GHz$ from $10MHz$ on the log scale, the y-axis shows the S_{21} allocated to $120dB\mu A$ from $-40dB\mu A$ on the linear scale. The power current of the DDR2 DIMM was measured by the committee members by using the setup for the kit-module in the rule of VCCI [19]. The magnetic probe which was standardized by IEC in "magnetic probe method" was used for this measurement.
The measured maximum current was $78dB\mu A$ or $7.9mA$ at $140MHz$. The power of it is $0.31mW$ because the measured input impedance was 5Ω at $140MHz$.
The electric field strength at the distance r from the antenna when the EMW of P watt is radiated from the antenna [20] is

$$E = \frac{7\sqrt{P}}{r} \quad [V/m] \tag{14}$$

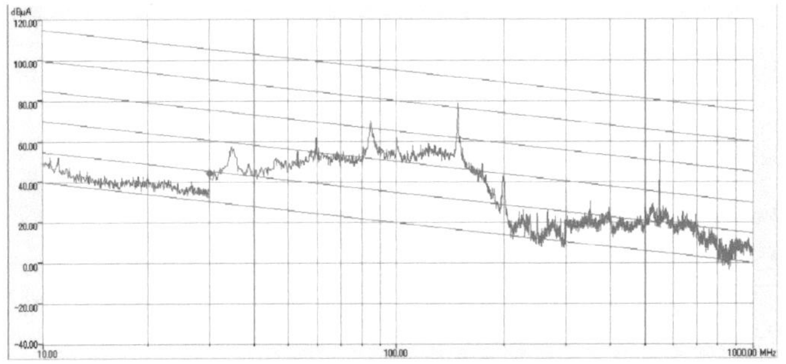

Fig. 10. Measured power current of the DDR2 DIMM

According to the IEC CISPR22, the limit of the electric field at 140MHz from the class B information technology equipment (ITE) is 30dB$\mu V/m$ at the distance of 10m. From (15), E is 12.4mV/m or 82dB$\mu V/m$ when p is 0.31mW. The DRAM module including the DDR2 DIMM is known as one of the devices which cause the interference. Therefore, the attenuation of the power decoupling is hoped to be more than 50dB at 140MHz.

4.1.2 Necessary value of the terminal impedance

The characteristic impedance of the on-board LILL should be small enough than it of the on-chip interconnect.
Fig. 11 shows the analyzing model of the on-chip interconnect as the transmission line.

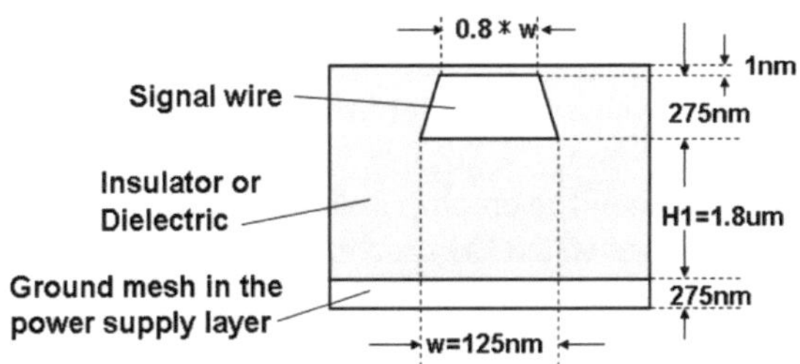

Fig. 11. Analyzing model of the on-chip interconnect

In Fig. 11, this model was formed by quoting the data of the 2006TN in the ITRS 2005. Each ε_r and tanδ of the insulator is 3.2 and 0.001. The simulated characteristic impedance by ApsimRLGC® was 148.5Ω at 200GHz and 143Ω at 1THz. According to the MSF, 200GHz corresponds to the switching time of 1.59ps and 1THz corresponds to the switching time of 0.32ps. The simulated characteristic impedance by MW STUDIO® was 10Ω approximately. When characteristic impedance was calculated by the conventional microstrip equation in TELE, it was 177Ω. When ε_r is 3.2, the intrinsic impedance of the insulator is 210Ω. The characteristic impedance is not defined by the conventional EMW theory. From above, we decided that the conventional equation is most reliable.
As the result, the minimum characteristic impedance of the on-chip interconnect and the PSWL was estimated to 177Ω.

4.1.3 Development of the equation for getting the characteristics

Fig. 12 shows the cross-section of the chip of the on-board LILL.

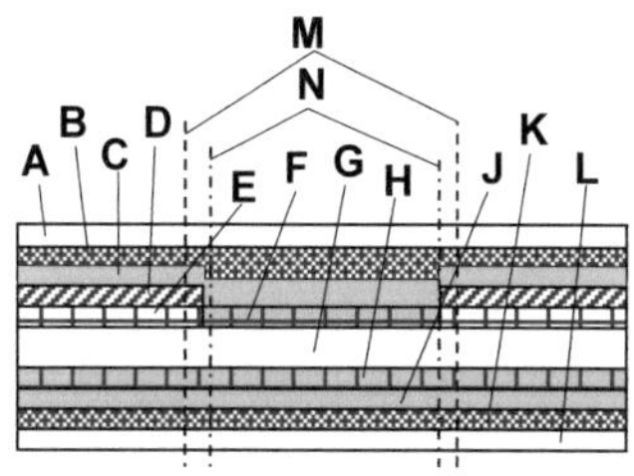

Fig. 12. Cross-section of the chip of the on-board LILL

In Fig. 12, the cross-section view is common to the two sides of the chip. A and L are the silver coating layers, A is the cathode and L is the anode, B and K are the carbon graphite layers, C and J are the conductive polymer layers, E is the etched layer of the aluminum without the conductive polymer, F and H are the etched layer including the conductive polymer, and G is the aluminum layer, N is the boundary of the available chip, and M is the line for cut. D is the masking layer to keep the insulation between the aluminum layer and the conductive polymer layer at the cut surface.

The equations have been improved through prototyping of many numbers of times till now. The extension ratio of the etching layer is

$$k = \frac{C_1 \cdot a}{\varepsilon_0 \varepsilon_r \cdot 2 \cdot 10^{-4}} \tag{15}$$

where C_1 is the capacitance of the capacitor which is made of the etched aluminum foil of $1cm^2$, a is the thickness of the alumina layer on the etching surface.

The impedance of the capacitance of the chip is

$$Z_C = \frac{10^{-4}}{2\pi f C_1 z w} \tag{16}$$

where z is the effective chip length, w is the effective chip width.

The transmission coefficient of the capacitor when it is connected to a point on the way of the transmission line is

$$S_{21C} = \frac{2Z_C}{2Z_C + Z_0} \tag{17}$$

where Z_0 is the characteristic impedance of the transmission line.

The effective thickness of the insulator layer is

$$d_e = \frac{a \cdot Z_I + b_S \cdot Z_S + b_C \cdot Z_C + b_V \cdot Z_V}{Z_I} \tag{18}$$

where each a, b_S, and b_C is the thickness of the insulator layer, the conductive polymer layer, and the effective carbon graphite layer, b_V is the equivalent thickness of the void, and each Z_I, Z_S, Z_C, and Z_V is the intrinsic impedance of the insulator layer, the conductive polymer layer, the carbon graphite layer, and the air.

The characteristic impedance of the LILL chip is

$$Z_{CI} = \sqrt{\frac{\mu_0}{\varepsilon_r \varepsilon_0}} \cdot \frac{d_e}{w \cdot R_a \cdot \sqrt{k}} \tag{19}$$

where R_a is the appearance ratio of the capacitance.
The transmission coefficient of the on-chip LILL caused by the reflection at each edge is

$$S_{21R} = \sqrt{1 - \left(\frac{Z_{CI} - Z_0}{Z_{CI} + Z_0} \right)^2} \tag{20}$$

The transmission coefficient of the on-chip LILL having the finite line length is

$$S_{21RA} = S_{21C} + S_{21R}{}^2 \tag{21}$$

The rate of the absorption loss in each material of the layer is

$$A_M = 1 - e^{\frac{-b_M}{\delta_M}} \tag{22}$$

where b_M is the thickness of each layer, and δ_M is the skin depth of each material of the layer.
The effective attenuation constant at each material of the layer is

$$\alpha_M = \frac{A_M}{2 \cdot Z_{CI} \cdot w \cdot R_a \cdot \sqrt{k} \cdot \delta_M \cdot \sigma_M} \tag{23}$$

The transmission coefficient of the LILL chip caused by the absorption loss is

$$S_{21\alpha} = e^{-\alpha_S \cdot z \cdot R_a \cdot \sqrt{k} \cdot R_S} \tag{24}$$

where a_S is the sum of a_M, and R_S is the shortening ratio by the void.
The transmission coefficient between the terminals by the effect of the surface wave coupling is

$$S_{21T} = \frac{2}{(Z_{CT}/Z_{CI}) + 2} \tag{25}$$

where Z_{CT} is the impedance of the capacitance between the terminals of the LILL.
The transmission coefficient of the LILL with the effect of the surface wave coupling is

$$S_{21} = \sqrt{(S_{21RA} \cdot S_{21\alpha})^2 + S_{21T}{}^2} \tag{26}$$

The terminal impedance of the on-board LILL is

$$Z_L = Z_C + Z_{CI} \tag{27}$$

The terminal impedance of the on-chip LILL with the effect of the surface wave coupling is

$$Z_a = Z_L \left(1 + \frac{120\pi}{1 + Z_{CT}} \right) \tag{28}$$

4.1.4 Prototyping of the on-bard LILL

The on-board LILL has been prototyped for 5 times since 2008.
Fig. 13 shows the examples of the prototype of the on-board LILL

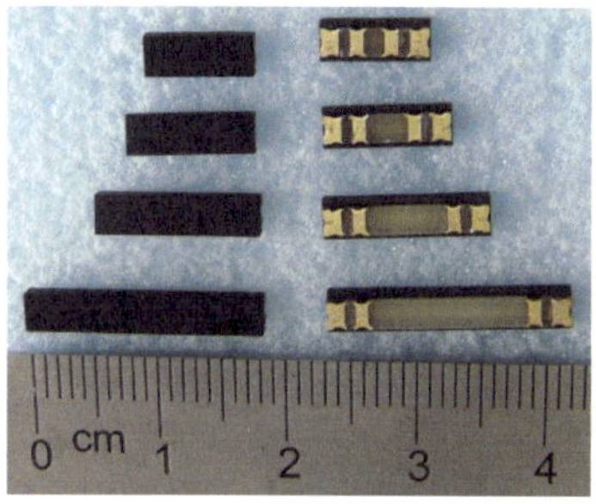

Fig. 13. Prototypes of the on-board LILL

In Fig. 13, each LILL03, LILL05, LILL08, and LILL14 has the line length of 3.5*mm*, 5*mm*, 8*mm*, and 14*mm*. The conductive polymer layer is formed in the water solution of the microscopic particles of the conductive polymer. The thickness of the chip is 200μm approximately.
Fig. 14 shows an example of the transmission coefficient (S_{21}) of the latest prototype of the LILL14. Fig. 14. a shows the measured S_{21} and Fig. 14. b shows the calculated S_{21}.

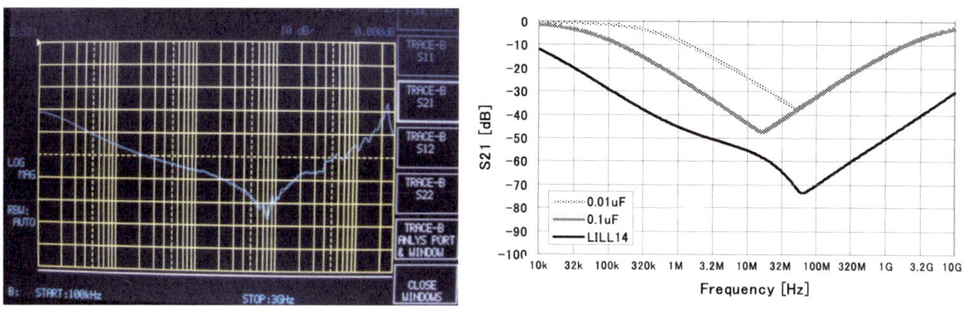

a) Measured S_{21} b) Calculated S_{21}

Fig. 14. An example of the transmission coefficient (S_{21}) of the latest prototype of the LILL14

In Fig. 14. b, the calculation condition is as follows; C_1 is 33.9μF, ε_r of the alumina is 8.5, σ_S is 12,000, σ_C is 72,727, R_a is 0.8, R_S is 0.45, w is 1*mm* at the calculation of (16) and Z_{CI} in (23), w is 1×2*mm* at the calculation of Z_{CI} in (20), z is 14*mm*, a is 22.8*nm*, b_S is 1μm, b_C is 0.92μm, and b_V is 0.8μm, the capacitance for Z_{CT} is 7×10^{-17} F/m.
The calculated S_{21} well matches to the measured value.
Fig. 15 shows the calculated characteristics of the examples of the improved on-board LILL. In Fig. 15, the calculation condition is same as the improved LILL14 in Fig. 14 except that σ_S is 12,000, b_S is 20μm, b_C is 0.46μm, the numeric following after LILL means the chip length (z). The calculation condition of the characteristics as follows; the numeric following after LILL means the chip length (z), the capacitance for Z_{CT} is 4×10^{-18} F/m, each LILL is used together with the 1*mF* capacitor and which are connected to the power traces of 40*mm*×40*mm* and 10*mm*×200*mm*, C01 consists of the 0.1μF capacitor and 1*mF* capacitor connected to the power trace of 100*mm*×200*mm*, and C02 consists of the 0.1μF capacitor and 1mF capacitor connected to the power trace of 10*mm*×200*mm*.

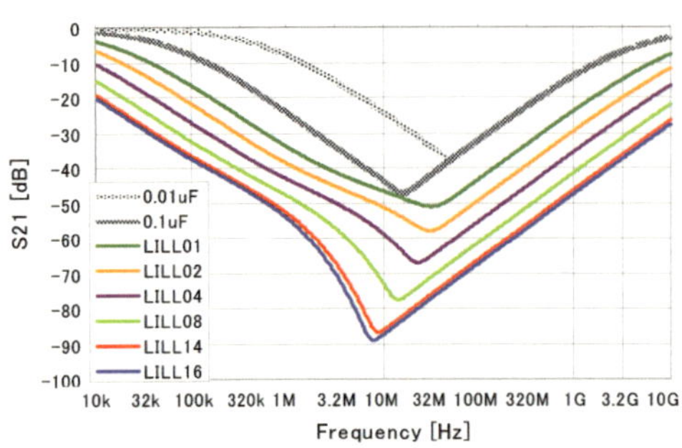

a) Transmission coefficient (S_{21})

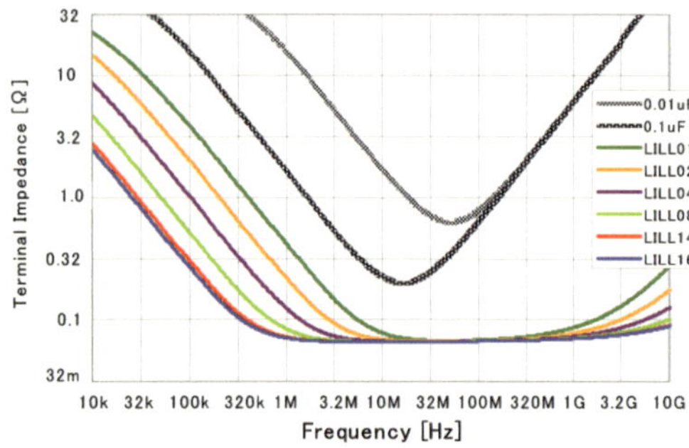

b) Terminal impedance (Z_a)

Fig. 15. Calculated characteristics of the examples of the improved on-board LILL

Fig. 16 shows the calculated characteristics of the improved on-board LILL on the PCB.

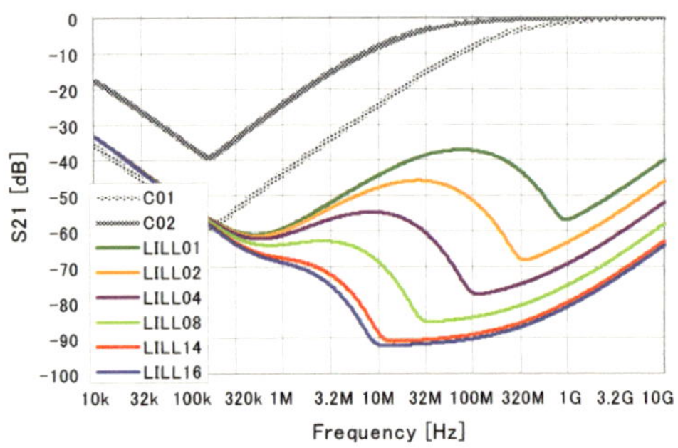

a) Transmission coefficient (S_{21})

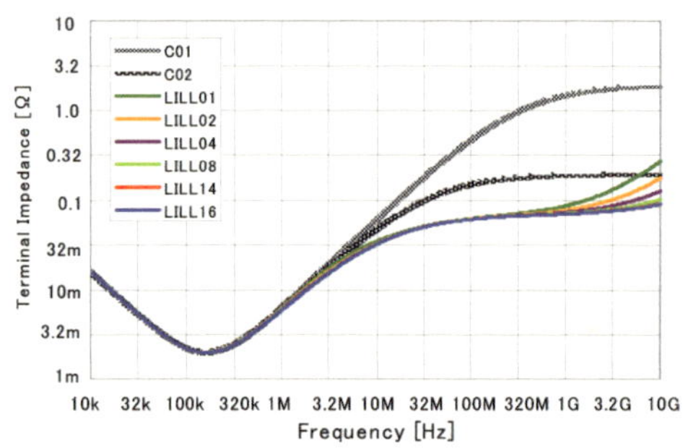

b) Terminal impedance (Z_a)

Fig. 16. Calculated characteristics of the improved on-board LILL on the actual PCB

In Fig. 16, at the frequency of lower than 10*MHz*, the decoupling performance of C01 is decuple of C02 approximately and the terminal impedance of C01 and C02 are equal. On the other hand, at the higher frequency than 1*GHz*, decoupling performance of both is zero and the terminal impedance of C01 is decuple of C02. Thus it is difficult to find the optimum solution about both decoupling performance and low impedance on the board when the capacitors are used for the decoupling circuit in the PDN. S_{21} of the LILL02 is -56dB at 140*MHz* and this value satisfies enough the hoped value which is -45dB. Z_a at 140*MHz* is small enough compared with the characteristic impedance of the on-chip interconnect or it of the PSWL.

The improved on-board LILL can decouple effectively the power source and the switching device including the SoC. The power source only provides the electrostatic energy under this situation.

Fig. 17 shows an example of the appearance of the on-board LILL for the commercialization.

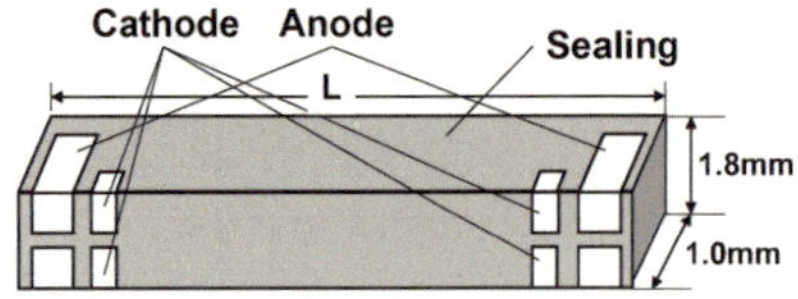

Fig. 17. Example of the appearance

In Fig.17, the improved on-board LILL is sealed by the transfer molding or other equivalent method. Each L means the sum of the chip length and 3.4mm. The rated voltage is 3.2V. The rated current is more than 10A and it depends on the thickness of the lead frame of the anode. The specification of the chip is corresponding to the calculation condition of Fig. 15. Fig. 18 shows an example of the application of the on-board LILL.

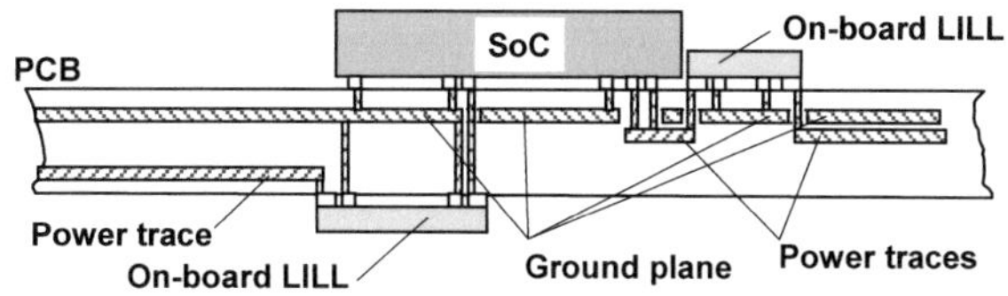

Fig. 18. Example of the application

In Fig. 18, the power traces of the supply-side should be slender because the through holes of the signal traces exist on the power traces when it is formed widely. The embedded LILL has advantage which minimizes the electromagnetic coupling between the terminals and reduces both mounting space and the manufacturing cost.

4.1.5 Comparison of the conventional decoupling components and LILL

Fig. 19 shows the appearance of the chip of the low impedance line component (LILC). The LILC was the first prototyped at NEC in 2000.

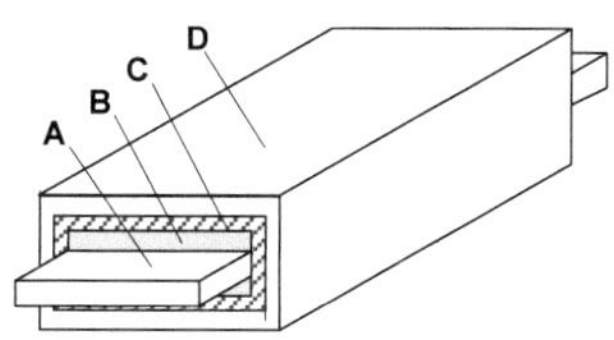

Fig. 19. Outline of the chip of the LILC

In Fig. 19, A is the anode which is formed by scraping the both exposed part of the etched aluminum foil, B is the conductive polymer layer, C is the carbon paste layer, and D is the cathode which is coated by the silver paste. All surfaces except the anode of the etched aluminum foil are covered by the insulation coating formed by the chemical conversion. The chip width is 1.5mm. Each LILC04, LILC08, LILC16, and LILC24 has the line length of 4mm, 8mm, 16mm, and 24mm.

Fig. 20 shows the calculated transmission coefficient (S_{21}) of the LILC. Fig. 20. a shows the specified S_{21} corresponding to the measuring value of the network analyzer. Fig. 20. b shows S_{21} on the actual PCB.

In Fig.20, the calculation condition is as follows; σ_S is 12,000, C_1 is 66μF, R_a is 1, k is 51.3, R_S is $1/\sqrt{k}$, w is 1.5×2mm at the calculation of the capacitance, w is 0.5×2mm at the calculation of Z_{CI}, a is 20.3nm, b_S is 3μm, b_C is 0 and b_V is 13nm for LILC04, b_C is 2.1μm and b_V is 10nm for LILC08, b_C is 12μm and b_V is 0 for LILC16, b_C is 2.9μm and b_V is 0 for LILC24, and the capacitance (C_T) for Z_{CT} in relation to the distance between the terminals is 4×10^{-17} F/m. In Fig .20. b, the calculation condition is same as the improved on-board LILL on the actual PCB in Fig. 16.

In Fig. 20. a well matches to measured S_{21} by using the network analyzer.

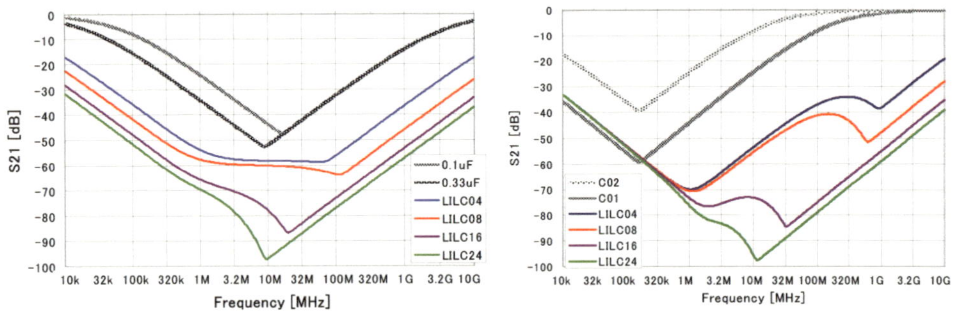

a) Specified S_{21} b) S_{21} on the actual PCB

Fig. 20. Calculated transmission coefficient (S_{21}) of the LILC

From the calculation result it was clarified that the structure of Fig. 19 has some following defects; the electromagnetic wave attenuates only at the cut surface of the aluminum foil, the formation of the carbon graphite layer on the cut surface is very difficult from the point of view of the quality. In addition, the thick etched aluminum foil is necessary to increase the rated current.

The multilayer structure of the LILC (Multi LILC) can improve the above mentioned defects. However the great side effects develop in addition to the cost increases. The attenuation coefficient is not improved because the escape channel of the EMW increases. In contrast, the terminal-impedance reduces. As the result, S_{21} on the actual PCB will decay from the specified S_{21} because the characteristic impedance of the PL is very smaller than 50Ω.

Table 1 shows the comparison of the decoupling components.

	Capacitor	LILC	Multi LILC	Prototype LILL
Insulation of Cut Surface	Chemical coating			Without treatment
Material of Anode	Etched aluminum			Lead frame
Formation process of Conductive Polymer Layer	Chemical reaction by Conductive Monomer			Coating by Conductive Polymer
Numbers of Anode	Single/ Multiple	Single	multiple	Single
Absorption loss	Zero	Large	Small	Large
Reflection loss	Small	Large	Very Large	Large
Limit factor of Electron current	No limit	Thickness of Aluminum foil	Numbers of Aluminum foil	Thickness of Lead frame
Improvement of Reduction of Decoupling on the PCB	Impossible	To increase the Chip Length		Possible without increasing Chip Length

Table 1. Comparison of the decoupling components

4.2 On-chip LILL technologies

The on-chip LILL should be connected to all of the on-chip inverters on the SoC. Therefore, for example, several tens of thousands of on-chip LILL will be used and each on-chip LILL will be connected to several tens of thousands of the on-chip inverter.

4.2.1 Validation of the function of the LILL and the influence of the length of the PL

Fig. 21 shows the circuit diagram of the test board for the validation.

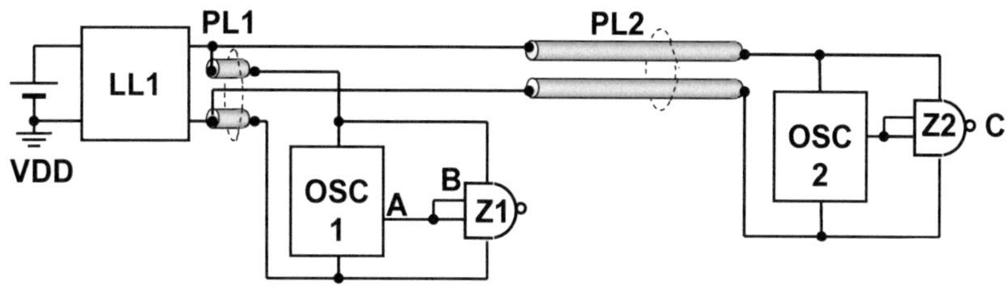

Fig. 21. Circuit diagram of the test board

In Fig. 21, VDD was 5.06*V*, which was supplied from the commercial power unit, CMX309HWC50MHz was used for the oscillator (OSC1, OSC2), SN 74AS00N was used for the gates (Z1, Z2), no SL were connected to the drivers, the prototype of the on-board LILL was used for LL1, the rated typical rise time of SN74AS00N is 0.5*ns*, each length of the PL1 and PL2 was 1*cm* and 36*cm*, the characteristic impedance of the PL was designed to be 73Ω, each PL was mounted on the insulator layer of the single sided board of FR-4, and each calculated round-trip time of the PL of 1*cm*-long and 36cm-long is 0.14*ns* and 5*ns*.

Fig. 22 shows the measured signal voltage at the point of A, B, and C in Fig. 21.

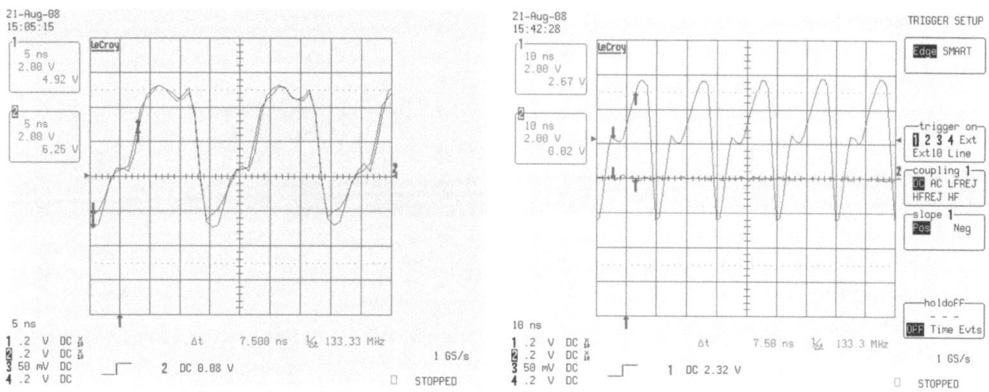

a) Point A and B b) Point C

Fig. 22. Measured signal voltage

In Fig. 22, the *y*-axis shows the voltage of 2*V* step, and the *x*-axis shows the time which is each 5*ns*/div. in Fig. 22. a and 10*ns*/div. in Fig. 22. b.

It were confirmed that the rise time increases to the calculated round-trip time approximately and that the LILL acts effectively as the ideal voltage source. From this experiment, it was clarified that the SEMW generated by each Z1 and Z2 returns to each certainly even though the LILL is shared.

4.2.2 Formation of the on-chip LILL

Fig. 23 shows an example of the layer formation of the on-chip LILL

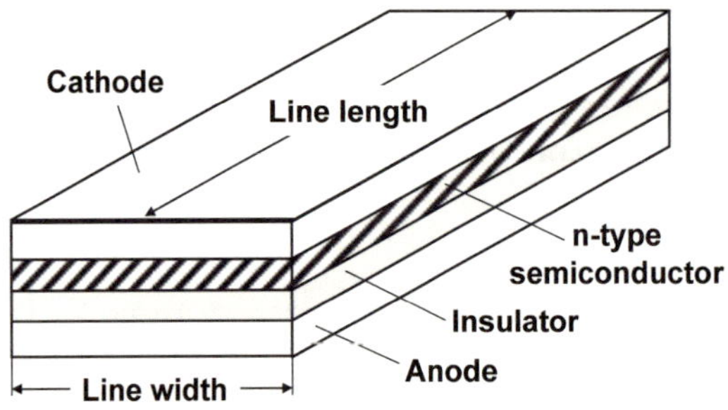

Fig. 23. Layer formation of the on-chip LILL

In Fig. 23, the attenuation appears caused by the n-type semiconductor layer which is made of the organic material or the inorganic material, and the n-type semiconductor and the cathode form the Schottky junction diode. The p-type semiconductor is also permitted. In this case, the p-type semiconductor is located between the anode and the insulator.

The low impedance is got by the thin insulator film of which the relative dielectric constant is large. The attenuation depends on the line length and the conductance of the n-type semiconductor.

4.2.3 Design of the on-chip LILL

The calculation equations for the on-board LILL were used for the on-chip LILL.

Fig. 24 shows the calculated characteristics of the on-chip LILL of line length.

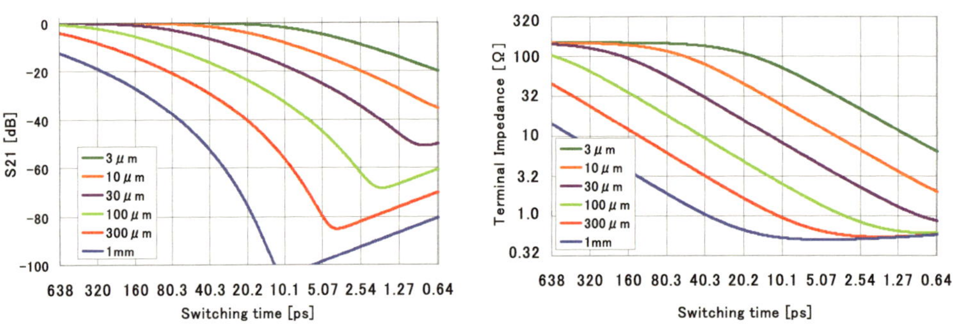

a) Transmission coefficient (S_{21}) b) Terminal impedance (Z_a)

Fig. 24. Calculated characteristics of the on-chip LILL

In Fig. 24, the calculation condition is as follows; σ_S is 10^5 S/m, the thickness of the insulator layer a is $10nm$, b_S is $1\mu m$, ε_r is 8.5, w is $3\mu m$, C_T is 10^{-22} F/m, Z_0 which is the minimum characteristic impedance of the on-chip interconnect is 177Ω, and the switching time is got from the idea of the MSF after calculating to $500GHz$ from $500MHz$.

When the suitable line length is selected, the calculated Z_a is small enough than Z_0 and the transmission coefficient is smaller than -40dB.

5. MILL technologies

The crosstalk is one of the hardest problems for transmitting the signal at high-speeds. The crosstalk does not occur on the DC circuit or the QSCC. Therefore, the crosstalk is a kind of the EMI. According to the SEMW theory, the SEMW forms the signal voltage wave by charging and discharging the transmission line. Nothing except the SEMW can charge and discharge rapidly on the SMC.

The magnitude of the signal voltage should not reduce when the electron current does not exist on the SL. Even if the SL is the lossy line, the SEMW should travel without changing its speed and wave shape except reduction of the magnitude. In addition, as being shown in Fig. 6, the rise time (t_S) of the signal voltage depends on the wave length (λ_S) of the SEMW. Therefore, the wave-shape of the signal voltage should be maintained in the lossy line when the characteristic impedance of the lossy line is matched to it of the signal line. From above discussions, the concept of the MILL was born.

The MILL is the lossy line and the characteristic impedance of it matches the SL. The MILL is connected to the SL near to the driver. The MILL decouples the SEMW between the driver and the receiver on the SL effectively.

5.1 Function of the MILL

Fig. 25 shows the improved SMC by applying the technologies of the LILL and MILL.

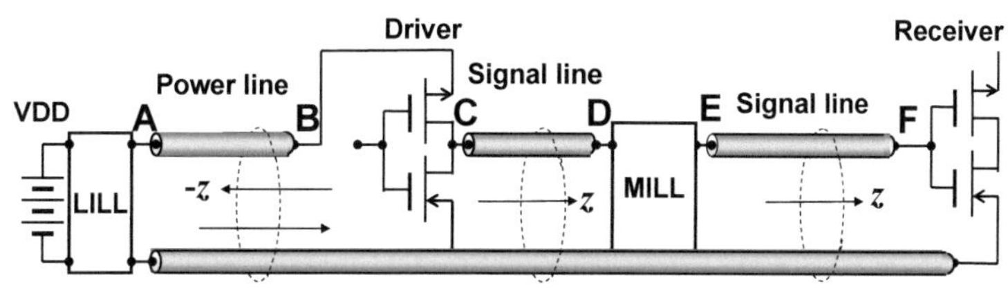

Fig. 25. Improved SMC

In Fig. 25, In Fig. 14, the LILL is connected to VDD and the point A on the PL in series and it forms the ideal power source, the MILL is connected to the point D and the point E on the SL in series, and the matched termination is not used on the SL.

Fig. 26 shows the SEW on the MILL.

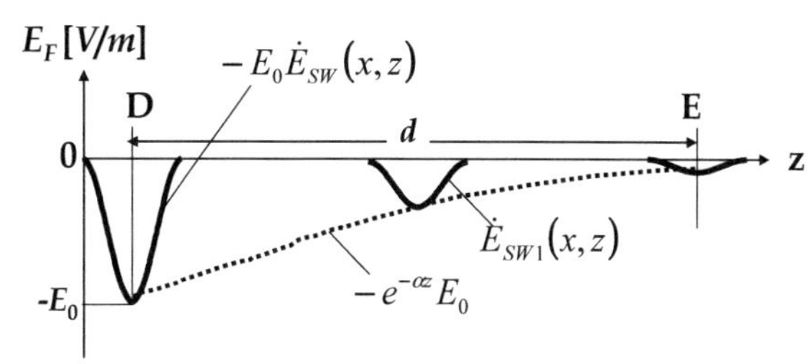

Fig. 26. SEW on the MILL

In Fig. 26, the electric field strength of the SEW on the MILL is

$$\dot{E}_{SW1}(x,z) = -e^{-\alpha z}E_0\dot{E}_{SW}(x,z) \tag{29}$$

where each x is the direction of the thickness, z is the traveling direction of the SL, and a is the attenuation constant.
According to the definition of the electromagnetism, the signal voltage on the MILL is

$$V_1(z) = \int_0^{\lambda_S}\int_0^{d} \dot{E}_{SW1}(x,z) = e^{-\alpha z} E_0 \int_0^{\lambda_S}\int_0^{d} \dot{E}_{SW}(x,z)\partial x \cdot \partial z \tag{30}$$

where, λ_S is the wave length of the SEW, d is the thickness of the insulator layer of the MILL. When $x=d$ and $z=\lambda_S$, (30) is

$$V_1(z) = e^{-\alpha z} V_S \tag{31}$$

The charge voltage of the SL at the exponential change of the magnitude of the SEW is

$$V_2(z) = \left(1 - e^{-\alpha z}\right)\ V_S \tag{32}$$

From (31), (32), a total of the signal voltage is

$$V_1(z) + V_2(z) = V_S \tag{33}$$

From above, it is clarified that the wave shape of the signal voltage is kept on the MILL even though the SEMW attenuate on it.
Meanwhile, the crosstalk and bounce on the SL between the point E and F in Fig. 25 will be suppressed because the magnitude of the SEMW is attenuated by the MILL.

5.2 Validation of the function of the MILL

Fig. 27 shows the measured S_{21} and S_{11} of the lab sample of the MILL.

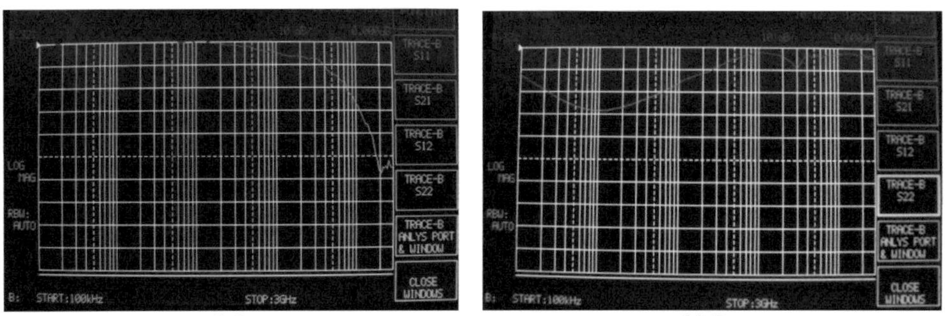

a) S_{21} b) S_{11}

Fig. 27. Measured characteristics of the lab sample of MILL

In Fig. 27, the x-axis shows the frequency allocated to 3*GHz* from 100*kHz* on the log scale, the y-axis is allocated to -100dB from 0dB on the linear scale.
The lab sample of the MILL was made of the carbon graphite, the silver paste, and the polyurethane enamel wire (UEW). The thickness of the carbon graphite layer was 0.1*mm* approximately, the diameter of the UEW was 0.1*mm*, the thickness of the insulator of the UEW was 5*μm*, and the line length was 100*mm*. Z_a of the lab sample was designed to 6.8Ω which is very lower than 50Ω, because there was no choice except using the carbon paste for getting the attenuation.

Fig. 28 shows the circuit diagram of the test board.

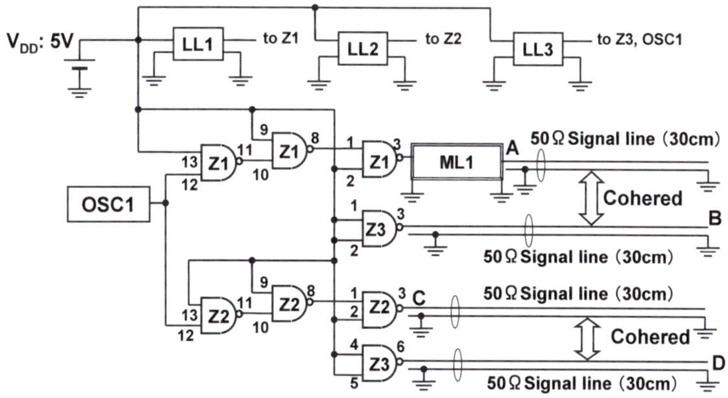

Fig. 28. Circuit diagram of the test board

Fig. 29 shows the measured voltage wave forms on the test board at the point A, B, C, and D in Fig. 28.

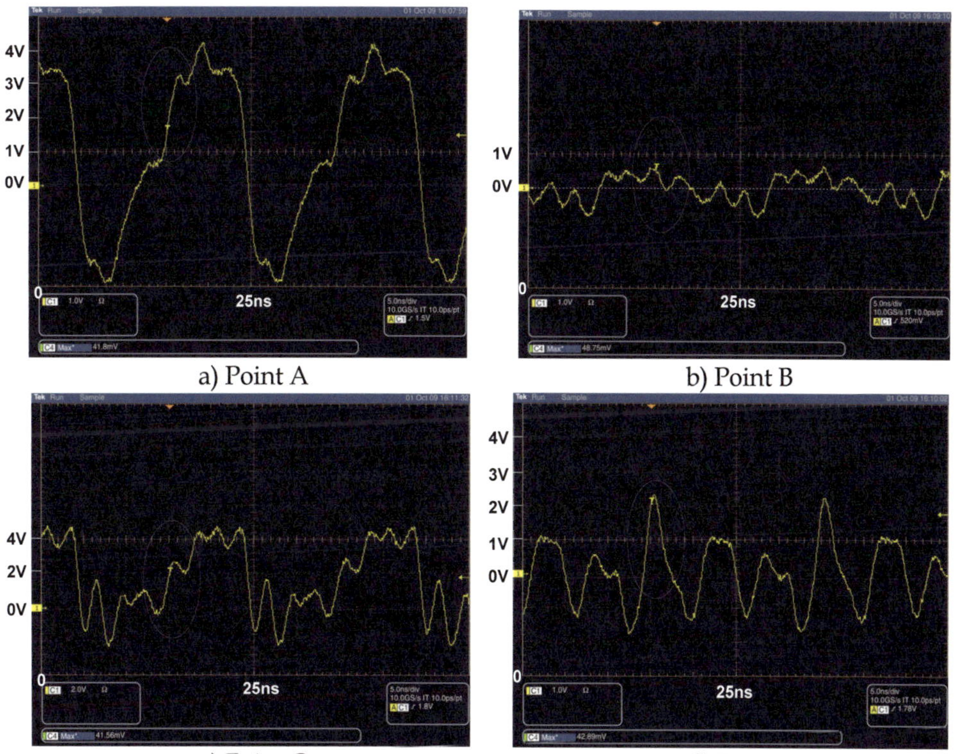

a) Point A
b) Point B
c) Point C
d) Point D

Fig. 29. Measured voltage wave forms on the SL

In Fig. 28, CMX309HWC50MHz was used for the oscillator (OSC1), SN 74AS00N was used for the gates (Z1, Z2, and Z3), LL1, LL2, and LL3 are the prototypes of the LILL14 shown in

Fig.13, the lab sample of MILL was used as ML1, the SL on the test board was formed by the PVC wire and the copper plane, ε_r of the PVC is 4, and the designed characteristic impedance was 50.1Ω. LL1, LL2, and LL3 were connected to all gates by the short wires.
In Fig. 29, the rising time shown in the circle of Fig. 29. a is almost equal to the rising time shown in the circle of Fig. 29. c. However, the crosstalk amplitude in the circle of Fig. 29. b is one quarter of it in the circle of Fig. 29. d. The reduction ratio of the crosstalk is almost equal to measured S_{21} in Fig. 27. a at 637MHz which is the MSF of the rise time of 0.5*ns*. The vibration is observed in Fig. 29. c, and its cycle time is 4*ns* approximately. This cycle time is equal to the calculated round trip time of the SEMW on the SL of 30*cm* long. However the vibration is not exists on the signal voltage in Fig. 29. a.
From above, the basic function of the MILL was validated.

5.3 Design example of the MILL

The layer formation of the MILL is same as the formation of the design example of the on-chip LILL in Fig. 23.
Fig. 30 shows the calculated characteristics of the design example of the MILL depending on each line length.

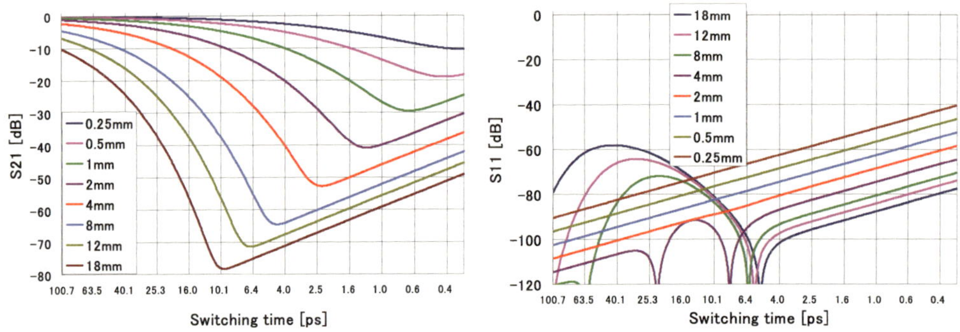

a) S_{21} b) S_{11}

Fig. 30. Calculated characteristics of the design example of the MILL

In Fig. 30, the equations for the calculation are same as these of the on-chip LILL. The thickness of the insulator was minimized for getting the large attenuation. The line width of the MILL was designed in the way that S_{11} becomes smaller than -40dB. The calculation condition is as follows; the n-type semiconductor is $700S/m$, a is $13\mu m$, b_S is $30\mu m$, ε_r is 3.5, w is $55\mu m$, and C_T is 10^{-19} F/m, and Z_0 is 50Ω. When the suitable line length is selected, S_{21} is smaller than -20dB.
In Fig. 25, when above mentioned design examples of the LILL and the MILL are used, and the wire length in the driver and the receiver is shorter than 2.5λs, the improved SMC shown in Fig. 25 will be able to be handled as the QSCC.
When the MILL designed by above way is applied to the SMC, the super-high-speed data transmission by parallel bit will be achieved relatively easily because the crosstalk is suppressed. At the same time, the multiple-value logic and the high-speed serialization/de-serialization (SerDes) will become unnecessary for the present. The MILL technologies are actualized to the on-board component or the on-chip component as with the LILL technologies. The on-board MILL will consist of several MILL chips. Each chip should be

connected to the on-chip drivers by as short wire as possible. The embedded MILL into the circuit board will be useful for increasing the performance and reducing both mounting space and the manufacturing cost.

6. Conclusions

The SEMW theory and both technologies of the LILL and the MILL were presented. The SEMW theory was developed by fusing the EMW theory and the NLU theory for the design and analysis of the SMC. The SMC generates the SEMW. The SEMW has no harmonic waves. Many spectra in the signal voltage are caused by the reflections and the repetitions of the SEMW on SMC. The wave shape of the signal voltage is got by integral of the wave shape of the SEW because the signal voltage is formed by the charging action of the SEW. The timing analysis with the high-accuracy will become possible when the SEMW theory is applied to the tools for the design or analysis. When the MILL is used together with the on-chip LILL, the signal integrity will be improved excellently and the almost all parts of the SMC will be able to be formed as the QSCC. This means that the design of all of the SMC become easy and free from the EMI. However, about the transmission line, more strict design of both characteristic impedance and skew or timing will be required.
The SEMW theory will be supposed to cause the innovation to the method of the manufacturing, design and analysis of the SMC. Improving the completeness of the SEMW theory is the future problem. The technologies of the LILL and MILL will create the emerging industries. The challenge for the commercialization of these technologies by getting the patron is the immediate issues for us. MathCAD Professional® and Excel® were used for all calculations.

7. Acknowledgment

The LILC was developed in NEC by being subsidized from NEDO from 2000 to 2003. The circuit simulation by using the HSPICE and the simulation for getting the characteristic impedance were done from 2005 to 2008 by the cooperation of the server platform division of NEC System Technologies, Ltd. The on-board LILL has been prototyped since 2008 by using the Clevios offered by Heraeus in Germany, the etched aluminum foil offered by JAPAN CAPACITOR INDUSTRIAL CO., LTD., and some production equipment offered by Kohzan Corporation in Japan.

8. References

[1] ITRS, *ITRS Reports*, Available from http://www.itrs.net/reports.html, 2011.

[2] Bendik Kleveland, et al., Exploiting CMOS Reverse Interconnect Scaling in Multigigahertz Amplifier and Oscillator Design, *IEEE JOURNAL OF SOLID-STATE CIRCUITS*, vol. 36, no. 10, pp.1480-1488, 2001.

[3] Bamal, M.; List, et al, , Performance Comparison of Interconnect Technology and Architecture Options for Deep Submicron Technology Nodes, *IEEE Interconnect Technology Conference*, pp. 202-204, 2006.

[4] Mark Shane Peng, et al., Study of Substrate Noise and Techniques for Minimization, *IEEE JOURNAL OF SOLID-STATE CIRCUITS*, vol. 39, no. 11, pp. 2080-2086, 2004.

[5] Makoto Nagata, et al., A Built-in Technique for Probing Power Supply and Ground Noise Distribution Within Large-Scale Digital Integrated Circuits. *IEEE JOURNAL OF SOLID-STATE CIRCUITS*, vol. 40, no. 4, pp. 813-819, 2005.

[6] Kazuya Masu, et al., On-Chip Transmission Line Interconnect for Si CMOS LSI. *IEEE Silicon Monolithic Integrated Circuits in RF Systems, 2006 Topical Meeting*, pp. 353- 356, 2006.

[7] Akiko Mineyama, et al., LVDS-type On-Chip Transmission Line Interconnect with Passive Equalizers in 90nm CMOS Process, *IEEE ASPDAC 2008 Asia and South Pacific*, pp.97- 98, 2008.

[8] Hoyeol Cho , et al., Power comparison between high-speed electrical and optical interconnects for interchip communication, *IEEE Journal of Lightwave Technology*, vol. 22, Issue 9, pp. 2021-2033, 2004.

[9] Venkataramanan, G, Characterization of capacitors for power circuit decoupling applications, *IEEE, Industry Applications Conference, Thirty-Third IAS Annual Meeting*, vol.2, pp.1142 - 1148, 1998.

[10] Larry D. Smith, et al., Power Distribution System Design Methodology and Capacitor Selection for Modern CMOS Technology, *IEEE Trans. on AP*, Vol. 22, No. 3, pp. 284-291, 1999.

[11] Keng L. Wong, et al., Enhancing microprocessor immunity to power supply noise with clock-data compensation, *IEEE Journal of SSC, Vol. 41*, No. 4, pp. 749-758, 2006.

[12] Istvan Novak, et al., Distributed Matched Bypassing for Board-Level Power Distribution Networks, *IEEE Tran. Advanced Packaging*, vol. 25, no. 2, pp. 230-243, 2002.

[13] Theodore M. Zeeff, Todd H. Hubing, Reducing Power Bus Impedance at Resonance with Lossy Components, *IEEE, Tran. Advanced Packaging*, vol. 25, no. 2, pp. 307-310, 2002.

[14] Shiho Hagiwara, et al., Linear Time Calculation of On-Chip Power Distribution Network Capacitance Considering State-Dependence. *IEICE Trans. Fundamentals*, vol. E93-A, no. 12, pp. 2409-2416, 2010.

[15] S. M. SZE & KWOK K. NG , *Physics of Semiconductor Devices, Third Edition*, John Wiley & Sons, pp. 293-373, New Jersey, USA, 2007.

[16] H. B. Bakoglu, *Circuits, Interconnections, and Packaging for VLSI*, Addison-Wesley Pub, pp. 239-244, 1990.

[17] Chuhg- Kuan Cheng, et al., *Interconnect Analysis and Synthesis*, John Wiley & Sons, INC. pp.105-108, New York, USA, 2000.

[18] Tsai, C.-T. Signal integrity analysis of high-speed, high-pin-count digital packages, *IEEE, Electronic Components and Technology Conference*, vol. 2, pp. 1098–1107, 1990.

[19] VCCI (2010). *Activities, Kit module Program*. Available from http://www.vcci.jp/

[20] IEC, Specification for radio disturbance and immunity measuring apparatus and methods, *CISPR 16-2-3*, pp. 79, 2003.

14

Effect of Magnetic Field on Nonlinear Absorption of a Strong Electromagnetic Wave in Low-dimensional Systems

Nguyen Quang Bau, Le Thai Hung and Hoang Dinh Trien
Hanoi University of Science, Vietnam National University
Vietnam

1. Introduction

Recently, there are more and more interests in studying and discovering the behavior of low-dimensional system, such as compositional superlattices, doped superlattices, quantum wells, quantum wires and quantum dots. The confinement of electrons and phonons in low-dimensional systems considerably enhances the electron mobility and leads to unusual behaviors under external stimuli. Many attempts have conducted dealing with these behaviors, for examples, electron-phonon interaction effects in two-dimensional electron gases (graphene, surfaces, quantum wells) (Ruker et al., 1992; Richter et al., 2009; Butscher et al., 2006). The dc electrical conductivity (Vasilopoulos et al., 1987; Suzuki, 1992), the electronic structure (Sager et al., 2007), the wavefunction distribution (Samuel et al., 2008) and the electron subband (Flores, 2008) in quantum wells have been calculated and analyzed. The problems of the absorption coefficient for a weak electromagnetic wave in quantum wells (Bau&Phong, 1998), in doped superlattices (Bau et al., 2002) and in quantum wires (Bau et al., 2007) have also been investigated and resulted by using Kubo-Mori method. The nonlinear absorption of a strong electromagnetic wave in low-dimensional systems have been studied by using the quantum transport equation for electrons (Bau&Trien, 2011). However, the nonlinear absorption of a strong electromagnetic wave in low-dimensional systems in the presence of an external magnetic field with influences of confined phonons is stills open to study. In this chapter, we consider quantum theories of the nonlinear absorption of a strong electromagnetic wave caused by confined electrons in the presence of an external magnetic field in low dimensional systems which considered the effect of confined phonons.

The problem is considered for the case of electron-optical phonon scattering. Analytic expressions of the nonlinear absorption coefficient of a strong electromagnetic wave caused by confined electrons in the presence of an external magnetic field in low-dimensional systems are obtained. The analytic expressions are numerically calculated and discussed to show the differences in comparison with the case of absence of an external magnetic with a specific AlAs/GaAs/AlAs quantum well, a compensated n-p n-GaAs/p-GaAs doped superlattices and a specific GaAs/GaAsAl quantum wire.

This book chapter is organized as follows: In section 2, effect of magnetic field on nonlinear absorption of a strong electromagnetic wave in a quantum well. Section 3 presents the effect of magnetic field on nonlinear absorption of a strong electromagnetic wave in a doped superlattice. The effect of magnetic field on nonlinear absorption of a strong electromagnetic wave in a cylindrical quantum wire is presented in section 4. Conclusions are given in the section 5.

2. Effect of magnetic field on nonlinear absorption of a strong electromagnetic wave in a quantum well

2.1 The electron distribution function in a quantum well in the presence of a magnetic field with case of confined phonons

It is well known that in quantum wells, the motion of electrons is restricted in one dimension, so that they can flow freely in two dimensions. In this article, we assume that the quantization direction is in z direction and only consider intersubband transitions (n≠n′) and intrasubband transitions (n=n'). As well as, we consider a quantum well with a magnetic field $\vec{B}$ applied perpendicular to its barriers. The Hamiltonian of the confined electron-confined optical phonon system in a quantum well in the presence of an external magnetic field $\vec{B}$ in the second quantization representation can be written as (Mori & Ando, 1989; Bau & Phong, 1998; Bau et al, 2009):

$$H=\sum_{n,N,\vec{k}_\perp}\varepsilon^H{}_{n,N}\left(\vec{k}_\perp-\frac{e}{\hbar c}\vec{A}(t)\right).a^+_{n,N,\vec{k}_\perp}a_{n,N,\vec{k}_\perp}+\sum_{m,\vec{q}_\perp}\hbar\omega_{m,\vec{q}_\perp}.b^+_{m,\vec{q}_\perp}.b_{m,\vec{q}_\perp}+ \\ +\sum_{n,N,\vec{k}_\perp}\sum_{m,\vec{q}_\perp}C^m_{\vec{q}_\perp}I^m_{n,n'}J_{N,N'}(u)a^+_{n',N',\vec{k}_\perp+\vec{q}_\perp}a_{n,N,\vec{k}_\perp}\left(b_{m,\vec{q}_\perp}+b^+_{m,\vec{q}_\perp}\right) \quad (1)$$

where N is the Landau level index (N = 0, 1, 2 …); n (n = 1, 2, 3, ...) denotes the quantization of the energy spectrum in the z direction, $(n,N,\vec{k}_\perp)$and$(n',N',\vec{k}_\perp+\vec{q}_\perp)$ are electron states before and after scattering, $\vec{k}_\perp(\vec{q}_\perp)$ is the in plane (x, y) wave vector of the electron (phonon), $a^+_{n,N,\vec{k}_\perp}$, $a_{n,N,\vec{k}_\perp}$ ($b^+_{m,\vec{q}_\perp}$,$b_{m,\vec{q}_\perp}$) are the creation and the annihilation operators of the confined electron (phonon), respectively, $\vec{A}(t)$ is the vector potential of an external electromagnetic wave $\vec{A}(t)=e\vec{E}_o\sin(\Omega t)/\Omega$ and $\hbar\omega_{m,\vec{q}_\perp}$ is the energy of a confined optical phonon. The electron energy $\varepsilon^H{}_{n,N}(\vec{k}_\perp)$ in quantum wells takes the simple form (Bau et al., 2009):

$$\varepsilon^H{}_{n,N}(\vec{k}_\perp)=(N+\frac{1}{2})\hbar\Omega_B+\frac{n^2\pi^2\hbar^2}{2m^*L^2} \quad (2)$$

where $\Omega_B = eB/m^*$ is the cyclotron frequency, m* is the effective mass of electron; L is the width of quantum wells and $C^m_{\vec{q}_\perp}$ is the electron-phonon interaction factor. In the case of the confined electron- confined optical phonon interaction, we assume that the quantization direction is in z direction, $C^m_{\vec{q}_\perp}$ is:

$$\left|C^m_{\vec{q}_\perp}\right|^2=\frac{2\pi e^2\hbar\omega_o}{\varepsilon_o V}\left(\frac{1}{\chi_\infty}-\frac{1}{\chi_o}\right)\frac{1}{q_\perp^2+\left(\frac{m\pi}{L}\right)^2} \quad (3)$$

where V, e and ε_o are the normalization volume, the effective charge and the electronic constant (often V=1); $\hbar\omega_o$ is the energy of a optical phonon ($\hbar\omega_{m,\vec{q}_\perp}\approx\hbar\omega_o$); m (m=1, 2, …), is the quantum number m characterizing confined phonons, L is well's width, χ_∞ and χ_o are the static and the high-frequency dielectric constant, respectively. The electron form factor in case of confined phonons is written as (Rucker et al., 1992):

$$I_{nn'}^{m} = \frac{2}{L}\int_0^L \left[\eta(m)\cos\frac{m\pi z}{L} + \eta(m+1)\sin\frac{m\pi z}{L} \right] \sin\frac{n'\pi z}{L}\sin\frac{n\pi z}{L}dz \tag{4}$$

with $\eta(m)=1$ if m is even number and $\eta(m)=0$ if m is odd number, and $J_{N,N'}(u)$ takes the simple form:

$$J_{N,N'}(u) = \int_{-\infty}^{+\infty} \phi_{N'}\left(\vec{r}_\perp - a_c^2\left(\vec{k}_\perp - \vec{q}_\perp\right)\right) e^{i\vec{k}_\perp \vec{q}_\perp} \phi_N\left(\vec{r}_\perp - a_c^2\vec{k}_\perp\right)dr. \tag{5}$$

$\vec{r}_\perp$ and $a_c = c/eB$ is position and radius of electron in the (x, y) plane, e is the electron charge, c is the light velocity, $u = a_c^2 q_\perp^2/2$, $\phi_N(x)$ represents the harmonic wave function.
Starting from the Hamiltonian (Eqs. (1-5)) and realizing operator algebraic calculations, we obtain the quantum kinetic equation for confined electrons in the presence of an external magnetic field with case confined phonons:

$$\begin{aligned}
&\frac{\partial n_{n,N,\vec{k}_\perp}(t)}{\partial t} = \left\langle \left[a^+_{n,\ell,N,\vec{k}_\perp} a_{n,\ell,N,\vec{k}_\perp}, H \right] \right\rangle_t = \\
&= -\frac{1}{\hbar^2} \sum_{n,N,\vec{k}_\perp} \sum_{m,\vec{q}_\perp} \left|C_{m,\vec{q}_\perp}\right|^2 . \left|I^m_{n,n'}\right|^2 . \left|J_{N,N'}(a_c^2 q_\perp^2/2)\right|^2 . \sum_{g,s=-\infty}^{+\infty} J_g\left(\frac{e\vec{E}_o\vec{q}}{m^*\Omega^2}\right) . J_s\left(\frac{e\vec{E}_o\vec{q}}{m^*\Omega^2}\right) . \exp\left[i(g-s)\Omega t\right] \int_{-\infty}^{t} dt' \times \\
&\quad \times \Big\{ \left[n_{n,N,\vec{k}_\perp}(t').\left(N_{m,\vec{q}_\perp}+1\right) - n_{n',N',\vec{k}_\perp+\vec{q}_\perp}(t').N_{m,\vec{q}_\perp} \right] \\
&\quad \times \exp\left\{ i\left(\varepsilon^H_{n',N'}(\vec{k}_\perp+\vec{q}_\perp) - \varepsilon^H_{n,N}(\vec{k}_\perp) + \hbar\omega_{m,\vec{q}_\perp} - g\hbar\Omega + i\hbar\delta \right)(t-t') \right\} + \\
&+ \left[n_{n,N,\vec{k}_\perp}(t').N_{m,\vec{q}_\perp} - n_{n',N',\vec{k}_\perp+\vec{q}_\perp}(t').\left(N_{m,\vec{q}_\perp}+1\right) \right] \times \exp\left\{ i\left(\varepsilon^H_{n',N'}(\vec{k}_\perp+\vec{q}_\perp) - \varepsilon^H_{n,N}(\vec{k}_\perp) - \hbar\omega_{m,\vec{q}_\perp} - g\hbar\Omega + i\hbar\delta \right)(t-t') \right\} - \\
&- \left[n_{n',N',\vec{k}_\perp-\vec{q}_\perp}(t').\left(N_{m,\vec{q}_\perp}+1\right) - n_{n,N,\vec{k}_\perp}(t').N_{m,\vec{q}_\perp} \right] \times \exp\left\{ i\left(\varepsilon^H_{n,N}(\vec{k}_\perp) - \varepsilon^H_{n',N'}(\vec{k}_\perp-\vec{q}_\perp) + \hbar\omega_{m,\vec{q}_\perp} - g\hbar\Omega + i\hbar\delta \right)(t-t') \right\} - \\
&- \left[n_{n',N',\vec{k}_\perp-\vec{q}_\perp}(t').N_{m,\vec{q}_\perp} - n_{n,N,\vec{k}_\perp}(t').\left(N_{m,\vec{q}_\perp}+1\right) \right] \times \exp\left\{ i\left(\varepsilon^H_{n,N}(\vec{k}_\perp) - \varepsilon^H_{n',N'}(\vec{k}_\perp-\vec{q}_\perp) - \hbar\omega_{m,\vec{q}_\perp} - g\hbar\Omega + i\hbar\delta \right)(t-t') \right\} \Big\}
\end{aligned} \tag{6}$$

where $\langle\psi\rangle_t$ is the statistical average value at the moment t and $\langle\psi\rangle_t = Tr(\hat{W}\hat{\psi})$ ($\hat{W}$ being the density matrix operator); $J_g(x)$ is the Bessel function, m* is the effective mass of the electron, the quantity δ is infinitesimal and appears due to the assumption of an adiabatic interaction of the electromagnetic wave; $n_{n,N,\vec{k}_\perp}(t) = a^+_{n,\ell,N,\vec{k}_\perp} a_{n,\ell,N,\vec{k}_\perp}$ is electron distribution function in quantum well; $N_{m,\vec{q}_\perp}$ which comply with Bose-Einstein statistics, is the time-independent component of the phonon distribution function. In the case of the confined electron-confined optical phonon interaction, $N_{m,\vec{q}_\perp}$ can be written as (Abouelaoualim, 1992):

$$N_{m,\vec{q}_\perp} = \frac{1}{e^{\frac{\hbar\omega_{m,\vec{q}_\perp}}{k_B T}} - 1} \tag{7}$$

After using the first order tautology approximation method (Malevich&Epstein, 1974) to solve this equation, the expression of electron distribution function can be written as:

$$n_{n,N,\vec{k}_\perp}(t) = -\sum_{n,N,\vec{k}_\perp}\sum_{m,\vec{q}_\perp}\left|C_{m,\vec{q}_\perp}\right|^2 . \left|I_{n,n'}^{m}\right|^2 . \left|J_{N,N'}\right|^2 \sum_{g,s=-\infty}^{+\infty} J_g\left(\frac{e\vec{E}_o\vec{q}}{m^*\Omega^2}\right).J_{g+s}\left(\frac{e\vec{E}_o\vec{q}}{m^*\Omega^2}\right).\frac{\exp(-is\Omega t)}{s\Omega}\times$$

$$\times\left\{-\frac{\bar{n}_{n,N,\vec{k}_\perp}.N_{m,\vec{q}_\perp} - \bar{n}_{n',N',\vec{k}_\perp+\vec{q}_\perp}.(N_{m,\vec{q}_\perp}+1)}{\varepsilon^H_{n',N'}(\vec{k}_\perp+\vec{q}_\perp) - \varepsilon^H_{n,N}(\vec{k}_\perp) - \hbar\omega_{m,\vec{q}_\perp} - g\hbar\Omega + i\hbar\delta}\right.$$

$$-\frac{\bar{n}_{n,N,\vec{k}_\perp}.(N_{m,\vec{q}_\perp}+1) - \bar{n}_{n',N',\vec{k}_\perp+\vec{q}_\perp}.N_{m,\vec{q}_\perp}}{\varepsilon^H_{n',N'}(\vec{k}_\perp+\vec{q}_\perp) - \varepsilon^H_{n,N}(\vec{k}_\perp) + \hbar\omega_{m,\vec{q}_\perp} - g\hbar\Omega + i\hbar\delta} + \tag{8}$$

$$+\frac{\bar{n}_{n',N',\vec{k}_\perp-\vec{q}_\perp}.N_{m,\vec{q}_\perp} - \bar{n}_{n,N,\vec{k}_\perp}.(N_{m,\vec{q}_\perp}+1)}{\varepsilon^H_{n,N}(\vec{k}_\perp) - \varepsilon^H_{n',N'}(\vec{k}_\perp-\vec{q}_\perp) - \hbar\omega_{m,\vec{q}_\perp} - g\hbar\Omega + i\hbar\delta}$$

$$\left.+\frac{\bar{n}_{n',N',\vec{k}_\perp-\vec{q}_\perp}.(N_{m,\vec{q}_\perp}+1) - \bar{n}_{n,N,\vec{k}_\perp}.N_{m,\vec{q}_\perp}}{\varepsilon^H_{n,N}(\vec{k}_\perp) - \varepsilon^H_{n',N'}(\vec{k}_\perp-\vec{q}_\perp) + \hbar\omega_{m,\vec{q}_\perp} - g\hbar\Omega + i\hbar\delta}\right\}$$

where $\bar{n}_{n,N,\vec{k}_\perp}$ is the time - independent component of the electron distribution fuction; $J_g(x)$ is the Bessel function. Eq.(8) also can be considerd a general expression of the electron distribution function in two dimensional systems with the electron form factor and the electron energy spectrum of each system.

2.2 Calculations of the nonlinear absorption coefficient of a strong electromagnetic wave by confined electrons in a quantum well in the presence of a magnetic field with case of confined phonons

The nonlinear absorption coefficient of a strong electromagnetic wave by confined electrons in the two-dimensional systems takes the simple form (Shmelev, 1978):

$$\alpha = \frac{8\pi}{c\sqrt{\chi_\infty}E_o^2}\left\langle \vec{j}_\perp(t)\vec{E}_o \sin\Omega t\right\rangle_t \tag{9}$$

Because the motion of electrons is confined along z direction in quantum wells, we only consider the in plane (x, y) current density vector of electrons so the carrier current density formula in quantum wells is taken the form(Shmelev, et al., 1978):

$$\vec{j}_\perp(t) = \frac{e\hbar}{m^*}\sum_{n,N,\vec{k}_\perp}(\vec{k}_\perp - \frac{e}{\hbar c}\vec{A}(t))n_{n,N,\vec{k}_\perp} = -\frac{e^2\hbar n_o^*\vec{E}_o}{m^*\Omega}\cos(\Omega t) + \frac{e\hbar}{m^*}\sum_{n,N,\vec{k}_\perp}\vec{k}_\perp n_{n,N,\vec{k}_\perp}(t) \tag{10}$$

with $n_o^* = \frac{n_o(\pi e)^{3/2}}{V.(m_o k_B T)^{3/2}}$, n_o is the electron density in quantum well, m_o is the mass of free electron, k_B is Boltzmann constant.

By using Eq.(10), the confined electron-confined optical phonon interaction factor $C_{m,\vec{q}_\perp}$ in Eq.(3) and the Bessel function, from the expression of current density vector in Eq.(10) and the relation between the nonlinear absorption coefficient of a strong electromagnetic wave with $\vec{j}_{\perp}(t)$ in Eq.(9), we established the nonlinear absorption coefficient of a strong electromagnetic wave in a quantum well in the presence of an external magnetic field under influnece of confined phonons:

$$\alpha = \frac{e^4 n_0^* k_B T \Omega_B^2 \hbar^2}{4\varepsilon_o c \Omega^3 \pi a_c^2 \sqrt{\chi_\infty}} . (\frac{1}{\chi_\infty} - \frac{1}{\chi_0}) . \sum_{n \neq n', N, N'} \sum_m | I^m_{n,n'} |^2 (1 + \frac{3e^2 E_0^2}{16 a_c^2 m^{*2} \Omega^4}(N + N' + 1)) \times$$

$$\times \left\{ \exp\left[\frac{(N + \frac{1}{2})\hbar\Omega_B + \frac{\pi^2\hbar^2 n^2}{2m^{*2}L^2}}{k_B T} \right] - \exp\left[- \frac{(N' + \frac{1}{2})\hbar\Omega_B + \frac{\pi^2\hbar^2 n'^2}{2m^{*2}L^2}}{k_B T} \right] \right\} \times \tag{11}$$

$$\times \frac{\hbar\sqrt{A_Q |N - N'|}}{|N - N'| [(N' - N)\hbar\Omega_B + \frac{\pi^2\hbar^2 (n'^2 - n)}{2m^{*2}L^2} + \hbar\omega_0 - \hbar\Omega + A_Q \hbar^2}$$

here, $M = N - N'$; $A_Q = N_o \frac{|C_o|^2}{4\pi\hbar^2} \sum_{n,n'} |I^m{}_{n,n'}|^2$; $N_o = \frac{k_B T}{\hbar\omega_o}$; $|C_o|^2 = \frac{2\pi e^2 \hbar\omega_o}{\varepsilon_o V}\left(\frac{1}{\chi_\infty} - \frac{1}{\chi_o}\right)$

In Eq. (11), it's noted that we only consider the absorption close to its threshold because in the rest case (the absorption far away from its threshold) α is very smaller. In the case, the condition $|g\Omega - \omega_0| << \bar{\varepsilon}$ ($\bar{\varepsilon}$ is the average energy of electron) must be satisfied (Pavlovich & Epshtein, 1977). The formula of the nonlinear absorption coefficient contains the quantum number m characterizing confined phonons. When quantum number m characterizing confined phonons reaches to zero, the expression of the nonlinear absorption coefficient for the case of absorption in quantum wells without influences of confined phonons can be written as:

$$\alpha = \frac{e^4 n_o^* k_B T \Omega_B^3}{2\pi c L \sqrt{\chi_\infty} a_c^2 . \varepsilon_o . \Omega^3 \hbar^2} \left(\frac{1}{\chi_\infty} - \frac{1}{\chi_o} \right) \sum_{n,n'} \sum_{N,N'} \left[1 + \frac{3e^2 E_o^2}{8m^{*2} a_c^2 \Omega^4} \right] \times$$

$$\times \left\{ \exp\left[-\frac{1}{k_B T}\left(N + \frac{1}{2} \right)\hbar\Omega_B - \frac{\pi^2 n^2 \hbar^2}{2m^* L^2} \right] - \exp\left[-\frac{1}{k_B T}\left(N' + \frac{1}{2} \right)\hbar\Omega_B - \frac{\pi^2 n'^2 \hbar^2}{2m^* L^2} \right] \right\} \times \tag{12}$$

$$\times \frac{\hbar\sqrt{A|M|}}{|M|\left(\hbar\Omega - \hbar\omega_o + M\hbar\Omega_B + \frac{\pi^2\hbar^2}{2m^* L^2}\left(n^2 - n'^2 \right)^2 + \hbar^2 A \right)}$$

with $A = N_o \frac{e^2 \omega_o}{2\pi\hbar^2 L} (\frac{1}{\chi_\infty} - \frac{1}{\chi_o})$

In Eqs.(12) we can see that the formula of the nonlinear absorption coefficient easy to come back to the case of linear absorption when the intensity (E_o) of external electromagnetic wave reaches to zero which was calculated Kubo - Mori method (Bau & Phong, 1998).

2.3 Numerical results and discussion

In order to clarify the mechanism for the nonlinear absorption coefficient of a strong electromagnetic wave in a quantum well with case of confined, in this section, we will evaluate, plot and discuss the expression of the nonlinear absorption coefficient for a specific quantum well: AlAs/GaAs/AlAs. We use some results for linear absorption in (Bau&Phong, 1998) to make the comparison. The parameters used in the calculations are as follows (Pavlovich, 1978): ε_0=12.5; $\chi_o = 12.9$, $\chi_\infty = 10.9$, $n_o = 10^{23} m^{-3}$, $L = 100 A^0$, $m^* = 0.067 m_o$, m_o =9.1.10^{-31}Kg, k_B=1.3807×10^{-23}, $\Omega = 2.10^{14} s^{-1}$, $\hbar\omega_{m,\vec{q}_\perp} \approx \hbar\omega_o = 36.25 meV$. Fig. (1-2) show the nonlinear and the linear absorption coeffcients in quantum wells in the presence of an external magnetic field for the case of confined electron - optical phonon scattering. The dependence of the absorption coeffcient on the frequency Ω of an external, strong electromagnetic wave and on the cyclotron frequency Ω_B for the case of an external magnetic field has one main maximum and several neighboring secondary maxima. The further away from the main maximum, the secondary one is the smaller. But in case of absence of an external magnetic field, there are only two maxima of nonlinear absorption coeffcient (Bau et al., 2010). When we consider case E_o = 0 and m reaches to zero, the nonlinear result with case confined phonon, will turn back to the linear result with case of unconfined phonons which was calculated by using another method the Kubo - Mori method (Bau&Phong, 1998).

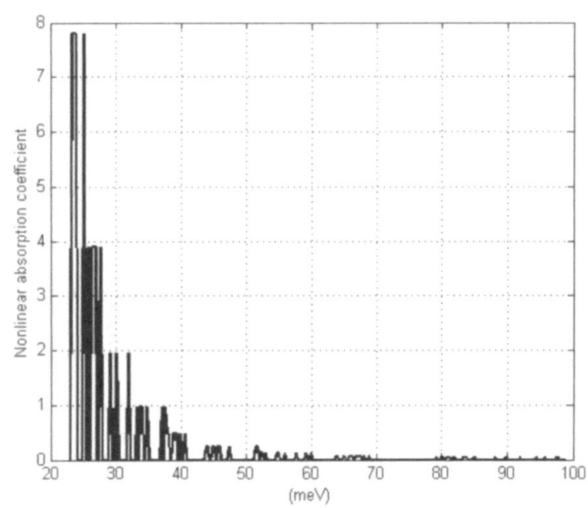

Fig. 1. The dependence of nonlinear absorption coefficient on $\hbar\Omega$ in case of confined phonons

Another point is that the absorption coeffcient in the presence of an external magnetic field is smaller than that without a field (Bau et al., 2010) because in this case, the number of electrons joining in the absorption process is limited. In addition, the Landau level that electrons can reach must be defined. In other word, the index of the Landau level that an electron can reach to after the absorption process must satisfy the condition:

$$(n'^2 - n^2)\frac{\pi^2\hbar^2}{2m^* L^2} + M\hbar\Omega_B - \hbar\omega_o + \hbar\Omega = 0$$

This is different from that for normal bulk semiconductors (index of the Landau level that electrons can reach after the absorption process is arbitrary), therefore, the dependence of the absorption coeffcient on Ω_B and Ω is not continuous.

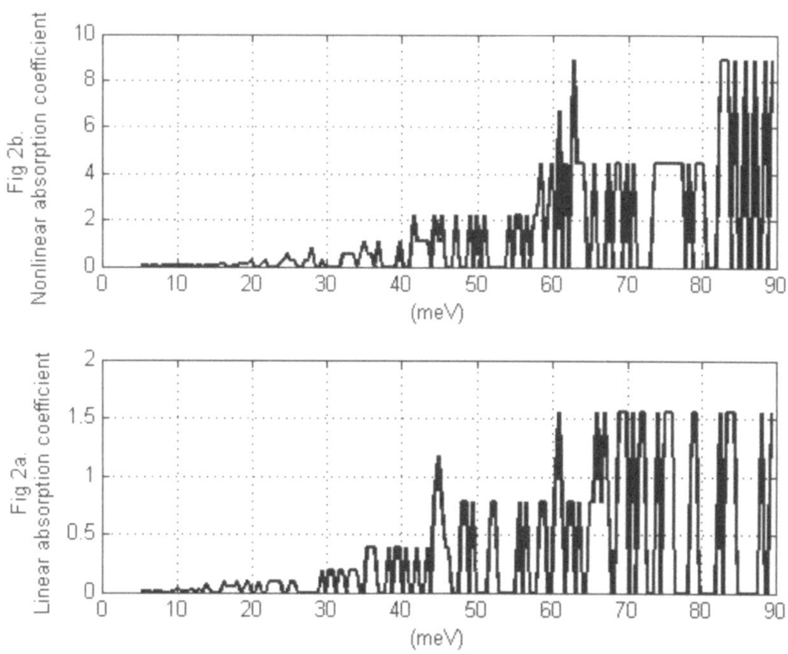

Fig. 2. The dependence of absorption coefficient on $\hbar\Omega_B$ in case of confined phonons

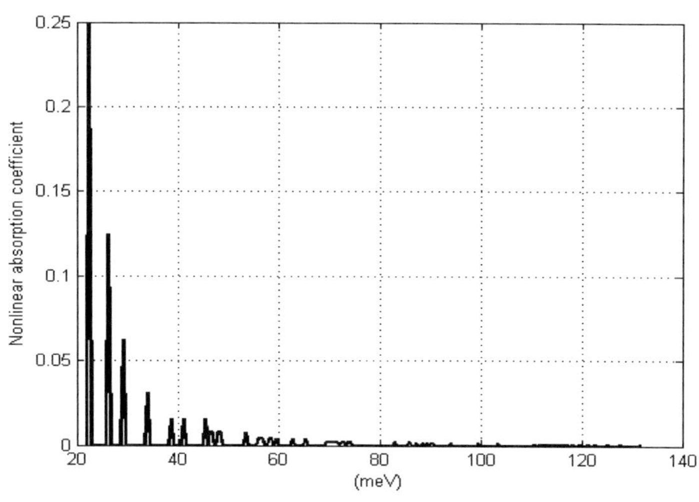

Fig. 3. The dependence of nonlinear absorption coefficient on $\hbar\Omega$ in case of unconfined phonons

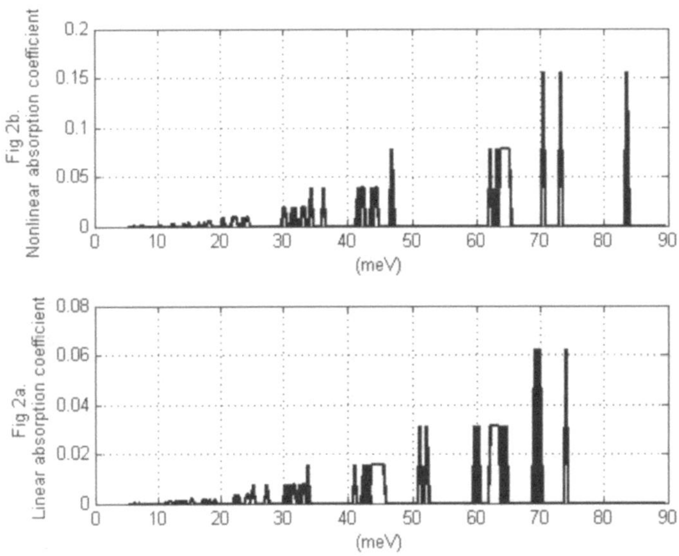

Fig. 4. The dependence of absorption coefficient on $\hbar\Omega_B$ in case of unconfined phonons

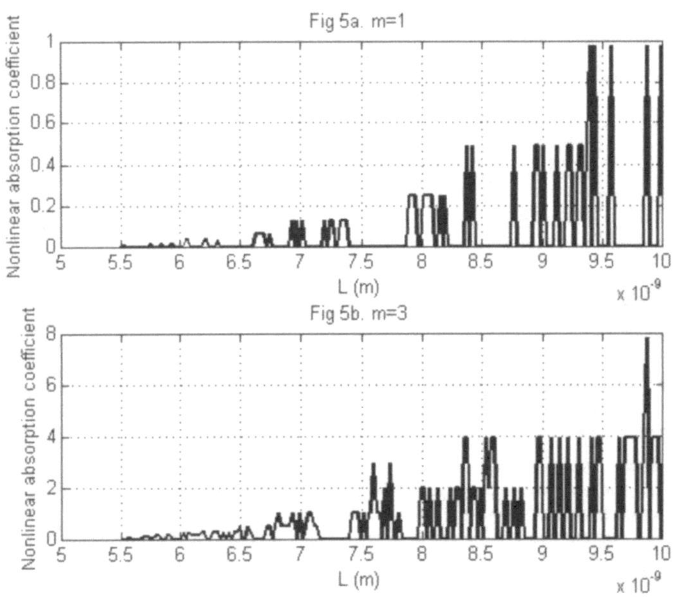

Fig. 5. The dependence of nonlinear absorption coefficient on L in case of confined phonons (m=1, m=3)

Fig. (3-4) show the nonlinear and the linear absorption coeffcients in quantum wells in the presence of an external magnetic field for the case of unconfined phonons (Bau et al., 2009). All figs show that the confinement of optical phonons have effected much strongly on absorption coeffcients. There is difference in the spectrum of absorption coeffcients in case confined phonons from its in case unconfined phonons. In fig.(1), the density of resonancepeaks is greater and the value of absorption coeffcients is higher than its with case unconfined phonons (fig.3) in both of the nonlinear and the linear absorption. In fig.4, some of resonance peaks have changed these position. These points are quite similar to case of confined phonons with out influence of an external magnetic field (Bau et al., 2010).

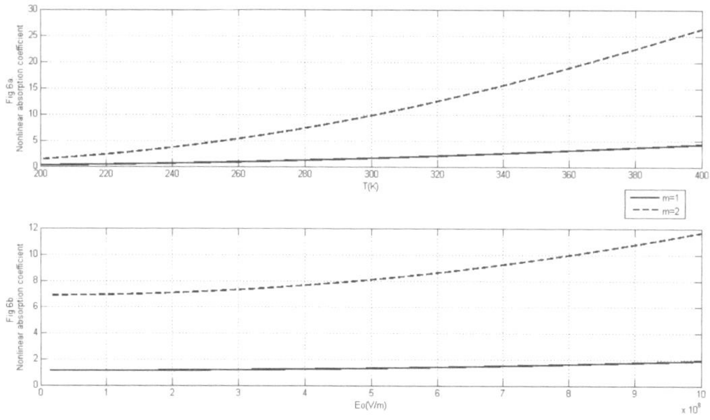

Fig. 6. The dependence of nonlinear absorption coefficient on T and E_o in case of confined phonons (m=1, m=2)

In fig.5, the dependence of the nonlinear absorption coeffcient on the quantum well's width has several maxima while each curve has one maximum peak in absence of a magnetic field (Bau et al., 2010). Fig.6 shows the dependence of the nonlinear absorption coeffcient on T and E_o. Both of fig.(5,6) show that the absorption coeffcient depends much strongly on quantum number m characterizing confined phonons, it gets greater when m increases.

3. Effect of magnetic field on nonlinear absorption of a strong electromagnetic wave in a doped superlattice

3.1 Calculations of the nonlinear absorption coefficient of a strong electromagnetic wave by confined electrons in a doped superlattice in the presence of a magnetic field with case of confined phonons

In doped superlattices, in the presence of an external magnetic field, the confined electron energy takes the simple form:

$$\varepsilon_{n,N}^{H}\left(\vec{k}_{\perp}\right)=\hbar\omega_{p}\left(n+\frac{1}{2}\right)+\hbar\Omega_{B}\left(N+\frac{1}{2}\right) \tag{13}$$

with, $\omega_p=\left(\dfrac{4\pi e^2 n_D}{\varepsilon_o m^*}\right)^{1/2}$ is the frequency plasma caused by donor doping concentration, n_D is the doping concentration. The electron form factor $I^{m}{}_{n,n'}$ can be written as:

$$I^{m}{}_{n,n'}=\sum_{m}^{N_1}\int_{0}^{d}\phi_{n'}(z-md)e^{iq_z md}\phi_n(z-md)dz \tag{14}$$

Here $\phi_n(z)$ is the eigenfunction for a single potential which compose the doped superlattices potential, d is the doped superlattices period, N_1 is the number of doped superlattices period.

The electron distribution function in doped superlattices can be written as Eq.(8) with the electron energy spectrum and the electron form factor in Eq. (13) and Eq.(14). We insert the expression for $n_{n,N,\vec{k}_\perp}(t)$ for a doped superlattice into the expression for $\vec{j}_\perp(t)$ and then insert the expression for $\vec{j}_\perp(t)$ into the expression for α in Eq.(9). Using the properties of Bessel function and realizing the calculations, the time - independent component of the electron distribution function and the confined electron-confined optical phonon interaction constants, we can calculate to obtain expression of the carrier current density and the nonlinear absorption coefficient of a strong electromagnetic wave by confined electrons in a doped superlattice. We obtain the explicit expression of α in a doped superlattice for the case electron-optical phonon scattering:

$$\alpha = \frac{e^4 n_o^* k_B T \Omega_B}{4\pi e_o m^* c\sqrt{x_\infty} a_c^2 \Omega^3}\left(\frac{1}{\chi_\infty}-\frac{1}{\chi_\infty}\right)\sum_{n,n',N,N'}\left|I^m{}_{n,n'}\right|^2\left[1+(N+N'+1)\frac{3e^2E_o^2}{16m^{*2}a_c^2\Omega^4}\right]\times n\sqrt{A_{DSL}\left|N-N'\right|}$$
$$\times\frac{\left\{\exp\left[-\frac{1}{k_BT}\left(\hbar\omega_p(n+\frac{1}{2})+\hbar\Omega_B(N+\frac{1}{2})\right)\right]-\exp\left[-\frac{1}{k_BT}\left(\hbar\Omega_B(N'+\frac{1}{2})+\hbar\omega_p(n'+\frac{1}{2})\right)\right]\right\}}{\left|N-N'\right|\left[n(\Omega-\omega_o)+n\Omega_B(N-N')+n\omega_p(n-n')\right]^2+\hbar^2A_{DSL}} \quad (15)$$

Here,

$$A_{DSL} = N_o\frac{|C_o|^2}{4\pi\hbar^2}\sum_{m,n,n'}\left|I^m{}_{n,n'}\right|^2;$$

In Eq. (15), it's noted that we only consider the absorption close to its threshold. In the case, the condition $|g\Omega-\omega_0| << \bar{\varepsilon}$ must be satisfied (Pavlovich & Epshtein, 1977). The formula of the nonlinear absorption coefficient contains the quantum number m characterizing confined phonons. When quantum number m characterizing confined phonons reaches to zero, the expression of the nonlinear absorption coefficient for the case of absorption close its threshold in a doped superlattice in case of unconfined phonons can be written as:

$$\alpha = \frac{e^4 n_o^* k_B T \Omega_B}{4\pi e_o m^* c\sqrt{x_\infty} a_c^2 \Omega^3}\left(\frac{1}{\chi_\infty}-\frac{1}{\chi_\infty}\right)\sum_{n,n',N,N'}\left|I^m{}_{n,n'}\right|^2\left[1+(N+N'+1)\frac{3e^2E_o^2}{16m^{*2}a_c^2\Omega^4}\right]$$
$$\times\frac{\left\{\exp\left[-\frac{1}{k_BT}\left(\hbar\omega_p(n+\frac{1}{2})+\hbar\Omega_B(N+\frac{1}{2})\right)\right]-\exp\left[-\frac{1}{k_BT}\left(\hbar\Omega_B(N'+\frac{1}{2})+\hbar\omega_p(n'+\frac{1}{2})\right)\right]\right\}}{\left|N-N'\right|\left[n(\Omega-\omega_0)+n\Omega_B(N-N')+n\omega_p(n-n')\right]^2+\hbar^2A^*} \quad (16)$$
$$\times\frac{1}{\pi}e^{-2\left(\frac{N_1d}{a_1}\right)^2}\left(\frac{N_1d}{a_1}\right)^2\left[1+\frac{2}{3}\left(\frac{N_1d}{a_1}\right)^2\right]^2 n\sqrt{A^*\left|N-N'\right|}$$

Here,

$$A^* = \frac{N_0 e^2 \omega_0}{2\hbar}\left(\frac{1}{\chi_\infty} - \frac{1}{\chi_0}\right) e^{-2\left(\frac{N_1 d}{a_1}\right)^2} \left(\frac{N_1 d}{a_1}\right)^2 \left[1 + \frac{2}{3}\left(\frac{N_1 d}{a_1}\right)^2\right]^2$$

$$a_1 = \frac{\hbar}{m^* \omega_p};$$

The term in proportion to quadratic intensity of a strong electromagnetic wave tend toward zero, the nonlinear result in Eqs.(16), will turn back to the linear case which was calculated by another method the Kubo - Mori (Bau et al., 2002).

3.2 Numerical results and discussion

In order to clarify the mechanism for the nonlinear absorption of a strong electromagnetic wave in a doped superlattices, in this section, we will evaluate, plot and discuss the expression of the nonlinear absorption coefficient for the case of a specific doped superlattices n-GaAs/p-GaAs. We use some results for nonlinear absorption in case absence of external magnetic field (Bau et al., 2010) to make the comparison. The parameters used in the calculations are as follows: $\chi_\infty = 10{,}9$; $\chi_o = 12{,}9$; $n_o = 10^{23}m^{-3}$; $\varepsilon_o=12.5$; $m^* = 0{,}067m_o$, d = 800A^0 being the doped superlattices period, $\hbar\omega_{m,\vec{q}_\perp} \approx \hbar\omega_o = 36.25meV$; Ω= 5.5.1013 s^{-1}, T = 300 K, $e = 2.07e_0$, $e_o = 1.6.10^{-19}$C.

Fig.7 and fig.8 show the dependence of nonlinear absorption coefficient α in doped superlattices on the intensity, E_o, of the electromagnetic wave is strong for the two cases present and absent magnetic field. Graph shows, the absorption coefficient dependency strong and nonlinear on the intensity E_o of electromagnetic waves. When the intensity of E_o increases, the absorption coefficient also increased rapidly. However, for the case present external magnetic field, the value of the nonlinear coefficient absorption larger than when absent external magnetic field.

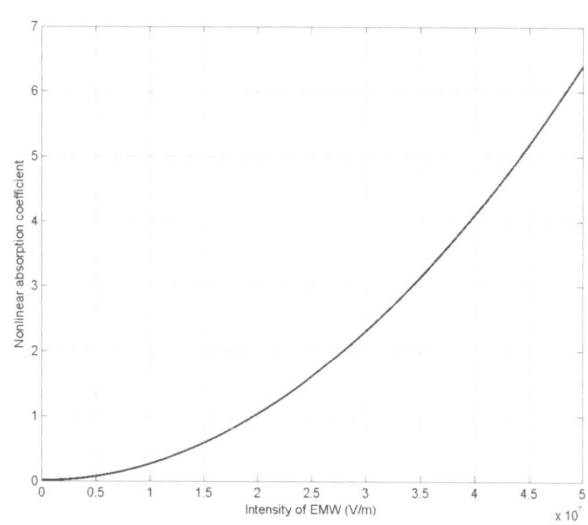

Fig. 7. The dependence of α on the E_o (Absence of an external magnetic field)

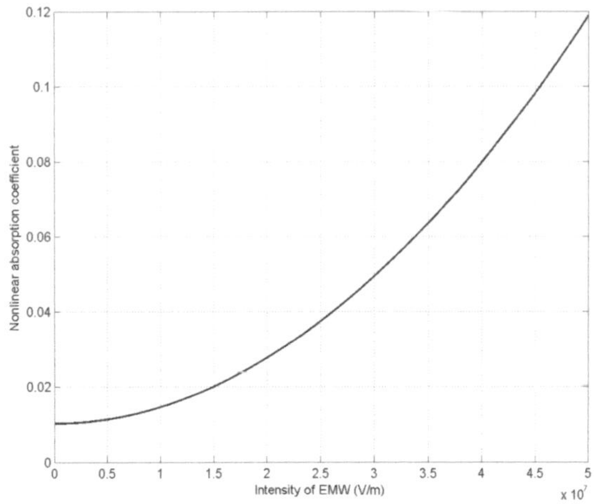

Fig. 8. The dependence of α on the E_o (Presence of an external magnetic field)

Fig.9 and fig.10 are the results from the survey of absorption coefficient on temperature and concentration of concentration doped. Graph shows absorption coefficient is also dependence strong and nonlinear on temperature T and concentration doped n_D. For both cases present and absent magnetic field, nonlinear absorption coefficient increased, when the temperature and the concentration doped of the system increases. Causes of this phenomenon: when the temperature and the concentration doped of system increases, leading to increased particle density, increased stimulation resulting vibrations in the material particles so that absorption coefficient of electromagnetic waves increased.

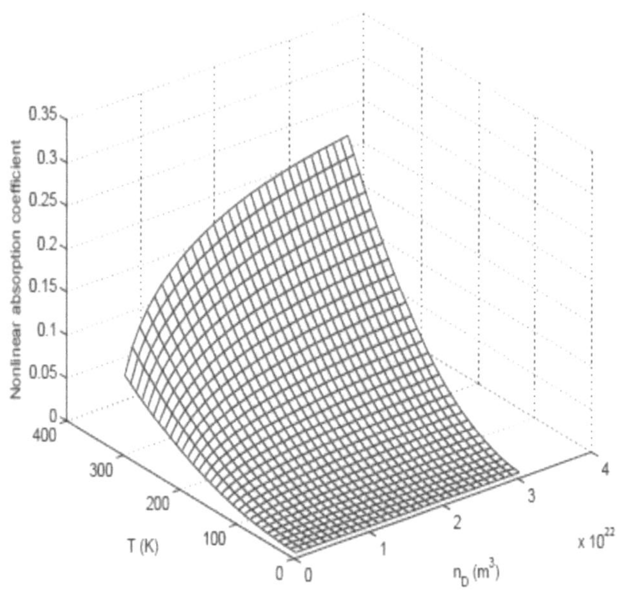

Fig. 9. The dependence of α on the n_D and T (Absence of an external magnetic field)

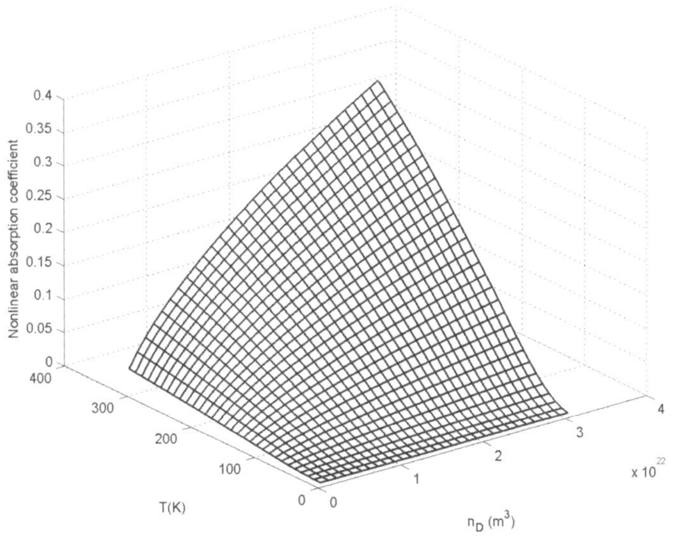

Fig. 10. The dependence of α on the n_D and T (Presence of an external magnetic field)

Fig.11 and fig. 12 show the dependence of the nonlinear absorption coefficient α on the energy $\hbar\Omega$ in doped superlattices of an external strong electromagnetic wave for the cases two absent and present magnetic field. Graphs show clearly the difference between them.

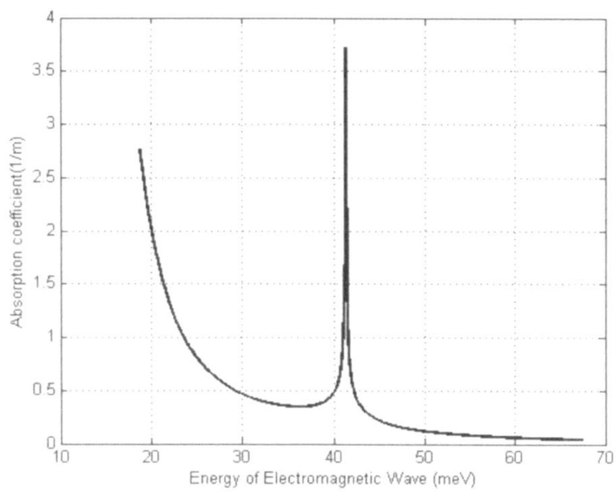

Fig. 11. The dependence of α on the $\hbar\Omega$ (Absence of an external magnetic field)

In fig.11, energy absorption spectrum have one resonance peak in case absence of magnetic field, but in fig.12, energy absorption spectrum is interrupted in the case presence of magnetic field. The absorption coefficient strong increases more in the case absence of magnetic field. The line spectrum formation of absorption coefficient in fig.12 clearly shows the influence of magnetic field on the absorption coefficient of strong external electromagnetic waves. This difference can be explained: in the presence of a magnetic field, the energy spectrum of electronic is interruptions. In addition, the Landau level that electrons can reach must be defined. In other word, the index of the absorption process must satisfy the condition:

$$\hbar\omega_p(n'-n)+\hbar\Omega_B(N'-N)+\hbar\omega_o-\hbar\Omega=0 \tag{17}$$

in function Delta - Dirac. This is different from that for case absent a magnetic field (index of the landau level that electrons can reach after the absorption process is arbitrary), therefore, the dependence of the absorption coefficient α on $\hbar\Omega$ is not continuous.

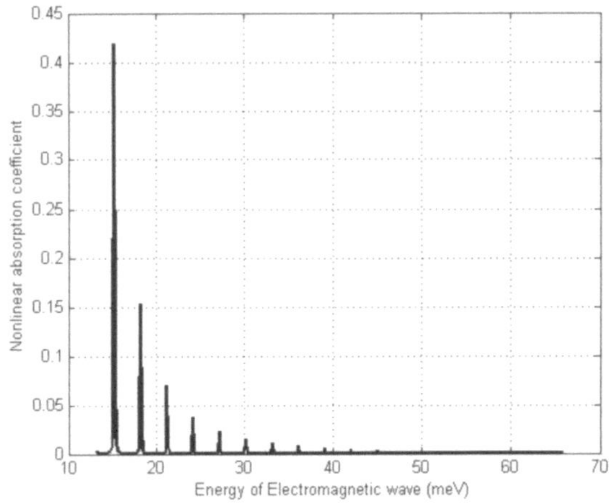

Fig. 12. The dependence of α on the $\hbar\Omega$ (Presence of an external magnetic field)

4. Effect of magnetic field on nonlinear absorption of a strong electromagnetic wave in a cylindrical quantum wire

4.1 The electron distribution function in a cylindrical quantum wire in the presence of a magnetic field with case of confined phonons

We consider a wire of GaAs with a circular cross section with radius R and length L_z embedded in AlAs. The carries (confined electrons) are assumed to be confined by infinite potential barriers and free along the wire's axis (Oz). A constant magnetic field with the magnitude $\vec{B}$ is applied parallel to the axis of wire. In the case, the Hamiltonian is given by (Bau & Trien, 2010):

$$H=\sum_{n,\ell,N,\vec{k}_z}\varepsilon_{n,\ell,N}\left(\vec{k}_z-\frac{e}{\hbar c}\vec{A}(t)\right).a^+_{n,\ell,N,\vec{k}_z}a_{n,\ell,N,\vec{k}_z}+\sum_{m,k,\vec{q}_z}\hbar\omega_{m,k,\vec{q}_z}.b^+_{m,k,\vec{q}_z}.b_{m,k,\vec{q}_z}+$$
$$\sum_{\substack{n',\ell',N',\vec{k}_z\\ n,\ell,N,\vec{q}_z}}\sum_{m,k}C^{m,k}_{\vec{q}_z}I^{m,k}_{n\ell,n'\ell'}J_{N,N'}(u)a^+_{n',\ell',N',\vec{k}_z+\vec{q}_z}a_{n,\ell,N,\vec{k}_z}\left(b_{m,k,\vec{q}_z}+b^+_{m,k,\vec{q}_z}\right) \tag{18}$$

Where the sets of quantum numbers (n,ℓ,N) and (n',ℓ',N'), characterizing the states of electron in the quantum wire before and after scattering with phonon; $a^+_{n,\ell,N,\vec{k}_z}\left(a_{n,\ell,N,\vec{k}_z}\right)$ is the creation (annihilation) operator of a confined electron; $\vec{k}_z$ is the electron wave vector (along the wire's z axis); $b^{+}_{m,k,\vec{q}_z}\left(b_{m,k,\vec{q}_z}\right)$ is the creation operator (annihilation operator) of a confined optical phonon for state have wave vector $\vec{q}_z$; $\omega_{m,k,\vec{q}_z}$ is the frequency of confined optical phonon, which was written as (Yu et al., 2005; Wang et al., 1994): $\omega^2_{m,k,\vec{q}} = \omega_0^2 - \beta^2\left(q_z^2 + q^2_{m,k}\right)$, with β is the velocity in cylindrical quantum wire and ω_0 is the frequency of optical phonon; (m, k) are quantum numbers characterizing confined phonons; the electron form factor $I^{m,k}_{n\ell,n'\ell'}$ can be written from (Li et al., 1992):

$$I^{m,k}_{n\ell,n'\ell'}(q_{m,k}) = \frac{2}{R^2}\int_0^R J_{|n-n'|}(q_{m,k}R).\psi^*_{n',\ell'}(r).\psi_{n,\ell}(r).rdr \tag{19}$$

The electron-optical phonon interaction constants can be taken as:

$$\left|C^{m,k}_{\vec{q}_z}\right|^2 = \pi e^2\hbar\omega_o.(1/\chi_\infty - 1/\chi_o)/2V\varepsilon_o(q_z{}^2 + q^2_{m,k}) \tag{20}$$

Here V is the normalization volume, ε_o is the permittivity of free space, χ_∞ and χ_o are the high and low-frequency dielectric constants, $q^2_{m,k}$ can be written from (Yu et al., 2005; Wang et al., 1994):

$$q_{m,k} = \begin{cases} x_k^1/R & \text{when } m = 0 \\ h_k^m/R & \text{when } |m| = 2s+1;\ s = 0,1,2... \\ g_k^m/R & \text{when } |m| = 2s\ ;\ s = 1,2,3... \end{cases}, \tag{21}$$

and $J_{N,N'}(u)$ takes the form (Suzuki et al., 1992; Generazio et .al, 1979; Ryu et al., 1993; Chaubey et al., 1986):

$$J_{N,N'}(u) = \int_{-\infty}^{+\infty} dr\phi_{N'}\left(\vec{r}_z - a_c^2\left(\vec{k}_z - \vec{q}_z\right)\right).e^{i\vec{q}_z.\vec{k}_z}\phi_N\left(\vec{r}_z - a_c^2\vec{k}_z\right) \tag{22}$$

Here $\phi_N(x)$ represents the harmonic wave function.
When the magnetic field is strong and the radius R of wires is very bigger than cyclotron radius a_c, the electron energy spectra have the form:

$$\varepsilon_{n,\ell,N,\vec{k}_z} = \frac{\hbar^2 k_z{}^2}{2m^*} + \hbar\Omega_B\left(N + \frac{n}{2} + \frac{\ell}{2} + \frac{1}{2}\right) \text{ with } N = 0,1,2,...\ , \tag{23}$$

In order to establish analytical expressions for the nonlinear absorption coefficient of a strong electromagnetic wave by confined electrons in cylindrical quantum wire, we use the quantum kinetic equation for particle number operator of electron $n_{n,\ell,N,\vec{k}_z}(t) = \left\langle a^+_{n,\ell,N,\vec{k}_z}a_{n,\ell,N,\vec{k}_z}\right\rangle_t$:

$$\frac{i\partial_{n_{n,\ell,N,\vec{k}_z}(t)}}{\partial t} = \left\langle\left[a^+_{n,\ell,N,\vec{k}_z}a_{n,\ell,N,\vec{k}_z}, H\right]\right\rangle_t \tag{24}$$

Where $\langle\psi\rangle_t$ is the statistical average value at the moment t and $\langle\psi\rangle_t = Tr(\hat{W}\hat{\psi})$ ($\hat{W}$ being the density matrix operator). Starting from the Hamiltonian Eq. (18) and using the commutative relations of the creation and the annihilation operators, we obtain the quantum kinetic equation for electrons in cylindrical quantum wire in the presence of a magnetic field with case of confined phonons:

$$\begin{aligned}
\frac{\partial n_{\gamma,\vec{k}_z}(t)}{\partial t} = & -\sum_{\gamma',\vec{q}_z}\sum_{\lambda}\left|C^{\lambda}_{\vec{q}_z}\right|^2.\left|I^{\lambda}_{n\ell,n'\ell'}\right|^2.\left|J_{N,N'}\right|^2.\sum_{g,s=-\infty}^{+\infty} J_g\left(\frac{e\vec{E}_0\vec{q}}{m^*\Omega^2}\right).J_s\left(\frac{e\vec{E}_0\vec{q}}{m^*\Omega^2}\right).\exp\left[i(g-s)\Omega t\right]\int_{-\infty}^{t}dt'\times \\
& \times\Big\{\left[n_{\gamma,\vec{k}_z}(t').\left(N_{\lambda,\vec{q}_z}+1\right)-n_{\gamma',\vec{k}_z+\vec{q}_z}(t').N_{\lambda,\vec{q}_z}\right]\times \\
& \times\exp\left\{i\left(\varepsilon^H_{\gamma'}(\vec{k}_z+\vec{q}_z)-\varepsilon^H_{\gamma}(\vec{k}_z)+\hbar\omega_{\lambda,\vec{k}_z}-g\hbar\Omega+i\hbar\delta\right)(t-t')\right\}+ \\
& +\left[n_{\gamma,\vec{k}_z}(t').N_{\lambda,\vec{q}_z}-n_{\gamma',\vec{k}_z+\vec{q}_z}(t').\left(N_{\lambda,\vec{q}_z}+1\right)\right]\times \\
& \times\exp\left\{i\left(\varepsilon^H_{\gamma'}(\vec{k}_z+\vec{q}_z)-\varepsilon^H_{\gamma}(\vec{k}_z)-\hbar\omega_{\lambda,\vec{q}_z}-g\hbar\Omega+i\hbar\delta\right)(t-t')\right\}- \\
& -\left[n_{\gamma',\vec{k}_z-\vec{q}_z}(t').\left(N_{\lambda,\vec{q}_z}+1\right)-n_{\gamma,\vec{k}_z}(t').N_{\lambda,\vec{q}_z}\right]\times \\
& \times\exp\left\{i\left(\varepsilon^H_{\gamma}(\vec{k}_z)-\varepsilon^H_{\gamma'}(\vec{k}_z-\vec{q}_z)+\hbar\omega_{\lambda,\vec{q}_z}-g\hbar\Omega+i\hbar\delta\right)(t-t')\right\}- \\
& -\left[n_{\gamma',\vec{k}_z-\vec{q}_z}(t').N_{\lambda,\vec{q}_z}-n_{\gamma,\vec{k}_z}(t').\left(N_{\lambda,\vec{q}_z}+1\right)\right]\times \\
& \times\exp\left\{i\left(\varepsilon^H_{\gamma}(\vec{k}_z)-\varepsilon^H_{\gamma'}(\vec{k}_z-\vec{q}_z)-\hbar\omega_{\lambda,\vec{q}_z}-g\hbar\Omega+i\hbar\delta\right)(t-t')\right\}\Big\}
\end{aligned} \tag{25}$$

where $J_g(x)$ is the Bessel function, m is the effective mass of the electron, $N_{\lambda,\vec{q}_z}$ is the time - independent component of the phonon distribution function, $n_{\gamma,\vec{k}_z}(t)$ is electron distribution function in cylindrical quantum wire and the quantity δ is infinitesimal and appears due to the assumption of an adiabatic interaction of the electromagnetic wave; $\gamma = n,\ell,N;\ \gamma' = n',\ell',N';\ \lambda = m,k$.

It is well known that to obtain the explicit solutions from Eq. (25) is very difficult. In this paper, we use the first - order tautology approximation method (Pavlovich & Epshtein, 1977; Malevich & Epstein, 1974; Epstein, 1975) to solve this equation. In detail, in Eq. (25), we use the approximation: $n_{\gamma,\vec{k}_z}(t') \approx \bar{n}_{\gamma,\vec{k}_z}$; $n_{\gamma',\vec{k}_z\pm\vec{q}_z}(t') \approx \bar{n}_{\gamma',\vec{k}_z\pm\vec{q}_z}$.

where $\bar{n}_{\gamma,\vec{k}_z}$ is the time - independent component of the electron distribution function in cylindrical quantum wire. The approximation is also applied for a similar exercise in bulk semiconductors (Pavlovich & Epshtein, 1977; Malevich & Epstein, 1974). We perform the integral with respect to t. Next, we perform the integral with respect to t of Eq. (25). The expression of electron distribution function can be written as:

$$\begin{aligned}
n_{\gamma,\vec{k}_z}(t) = & -\sum_{\gamma',\vec{q}_z}\sum_{\lambda}\left|C^{\lambda}_{\vec{q}_z}\right|^2.\left|I^{\lambda}_{n\ell,n'\ell'}\right|^2.\left|J_{N,N'}\right|^2\sum_{g,s=-\infty}^{+\infty} J_g\left(\frac{e\vec{E}_0\vec{q}}{m^*\Omega^2}\right).J_{g+s}\left(\frac{e\vec{E}_0\vec{q}}{m^*\Omega^2}\right).\frac{\exp(-is\Omega t)}{s\Omega}\times \\
& \times\left\{-\frac{\bar{n}_{\gamma,\vec{k}_z}.N_{\lambda,\vec{q}_z}-\bar{n}_{\gamma',\vec{k}_z}.\left(N_{\lambda,\vec{q}_z}+1\right)}{\varepsilon^H_{\gamma'}(\vec{k}_z+\vec{q}_z)-\varepsilon^H_{\gamma}(\vec{k}_z)-\hbar\omega_{\lambda,\vec{q}_z}-g\hbar\Omega+i\hbar\delta}-\frac{\bar{n}_{\gamma,\vec{k}_z}.\left(N_{\lambda,\vec{q}_z}+1\right)-\bar{n}_{\gamma',\vec{k}_z+\vec{q}_z}.N_{\lambda,\vec{q}_z}}{\varepsilon^H_{\gamma'}(\vec{k}_z+\vec{q}_z)-\varepsilon^H_{\gamma}(\vec{k}_z)+\hbar\omega_{\lambda,\vec{q}_z}-g\hbar\Omega+i\hbar\delta}+\right. \\
& \left.+\frac{\bar{n}_{\gamma'\vec{k}_z-\vec{q}_z}.N_{\lambda,\vec{q}_z}-\bar{n}_{\gamma,\vec{k}_z}.\left(N_{\lambda,\vec{q}_z}+1\right)}{\varepsilon^H_{\gamma}(\vec{k}_z)-\varepsilon^H_{\gamma'}(\vec{k}_z-\vec{q}_z)-\hbar\omega_{\lambda,\vec{q}_z}-g\hbar\Omega+i\hbar\delta}+\frac{\bar{n}_{\gamma',\vec{k}_z-\vec{q}_z}.\left(N_{\lambda,\vec{q}_z}+1\right)-\bar{n}_{\gamma,\vec{k}_z}.N_{\lambda,\vec{q}_z}}{\varepsilon^H_{\gamma}(\vec{k}_z)-\varepsilon^H_{\gamma'}(\vec{k}_z-\vec{q}_z)+\hbar\omega_{\lambda,\vec{q}_z}-g\hbar\Omega+i\hbar\delta}\right\}
\end{aligned} \tag{26}$$

From Eq.(26) we see that the electron distribution function depends on the constant in the case of confined electron - confined phonon interaction, the electron form factor and the electron energy spectrum in cylindrical quantum wire. Eq.(26) also can be considered a general expression of the electron distribution function in cylindrical quantum wire with the electron form factor and the electron energy spectrum of each systems.

4.2 Calculations of the nonlinear absorption coefficient of a strong electromagnetic wave by confined electrons in a cylindrical quantum wire in the presence of a magnetic field with case of confined phonons

The nonlinear absorption coefficient of a strong electromagnetic wave α in a cylindrical quantum wire take the form similary to Eq.(9):

$$\alpha = \frac{8\pi}{c\sqrt{\chi_\infty}E_o^2}\left\langle \vec{j}_z(t)\vec{E}_o \sin\Omega t\right\rangle_t \tag{27}$$

where $\langle X\rangle_t$ means the usual thermodynamic average of X at moment t, $\vec{A}(t)$ is the vector potential, E_o and Ω is the intensity and frequency of electromagnetic wave. The carrier current density formula in a cylindrical quantum wire takes the form similary to in (Pavlovich & Epshtein, 1977):

$$\vec{j}_z(t) = \frac{e\hbar}{m^*}\sum_{\gamma,\vec{k}_z}\left(\vec{k}_z - \frac{e}{c}\vec{A}(t)\right).n_{\gamma,\vec{k}_z}(t) \tag{28}$$

Because the motion of electrons is confined along the (x, y) direction in a cylindrical quantum wire, we only consider the in - plane z current density vector of electrons, $\vec{j}_z(t)$. Using Eq. (28), we find the expression for current density vector:

$$\vec{j}_z(t) = -\frac{e^2\hbar n_o^*.\vec{E}_o}{m^*\Omega}\cos(\Omega t) + \frac{e\hbar}{m^*}\sum_{\gamma,\vec{k}_z}\vec{k}_z.n_{\gamma,\vec{k}_z}(t) \tag{29}$$

We insert the expression of $n_{\gamma,\vec{k}_z}(t)$ into the expression of $\vec{j}_z(t)$ and then insert the expression of $\vec{j}_z(t)$ into the expression of α in Eq.(27). Using property of Bessel function $J_{k+1}(x)+J_{k-1}(x)=2kJ_k(x)/x$, and realizing calculations, we obtain the nonlinear absorption coeffcient of a strong electromagnetic wave by confined electrons in cylindrical quantum wire with case of confined phonons:

$$\begin{aligned}\alpha = &\frac{4\pi^3.e^2.\Omega.k_B.T}{\varepsilon_o.c\sqrt{\chi_\infty}E_o^2.V}\left(\frac{1}{\chi_\infty}-\frac{1}{\chi_0}\right)\sum_{\gamma,\gamma',\lambda}\sum_{\vec{q}_z,\vec{k}_z}\sum_{g=-\infty}^{+\infty}\left|C_{\vec{q}_z}^{\lambda}\right|^2.\left|I_{n\ell,n'\ell'}^{\lambda}\right|^2.\left|J_{N,N'}\right|^2.k.J_g^2\left(\frac{e\vec{E}_o\vec{q}}{m^*\Omega^2}\right)\times\\ &\times\frac{1}{q_z^{\ 2}+q_\lambda^2}\times\left(\bar{n}_{\gamma,\vec{k}_z}-\bar{n}_{\gamma',\vec{k}_z+\vec{q}_z}\right).\delta\left(\varepsilon_{\gamma'}^H(\vec{k}_z+\vec{q}_z)-\varepsilon_\gamma^H(\vec{k}_z)+\omega_o-g\Omega\right)\end{aligned} \tag{30}$$

We only consider the absorption close to its threshold because in the rest case (the absorption far away from its threshold) α is very smaller. In the case, the condition $|g\Omega-\omega_o| << \bar{\varepsilon}$ must be satisfied (Pavlovich & Epshtein, 1977). We restrict the problem to the case of one photon absorption and consider the electron gas to be non-degenerate:

$$\bar{n}_{\gamma,\vec{k}_z} = n_o^*.\exp\left\{-\frac{\varepsilon_{\gamma,\vec{k}_z}}{k_B T}\right\}, \text{ with } n_o^* = \frac{n_o(\pi e)^{3/2}}{V(m_o k_B T)^{3/2}}$$

By using the electron - optical phonon interaction factor, the Bessel function and the electron distribution function, from the general expression for the nonlinear absorption coefficient of a strong electromagnetic wave in a cylindrical quantum wire Eq.(30), we obtain the explicit expression of the nonlinear absorption coefficient α in cylindrical quantum wire for the case confined electron-confined optical phonon scattering:

$$\alpha = \frac{e^4.n_0^*.\Omega_B.k_B T}{4\sqrt{2}.\varepsilon_o.c\sqrt{\chi_\infty}m^*.a_c.\Omega^3.V}\times\left(\frac{1}{\chi_\infty}-\frac{1}{\chi_o}\right)\sum_{\gamma,\gamma',\lambda}\left|I^{\lambda}_{n,\ell,n',\ell'}\right|^2.\left(\frac{N_n!}{N_m!}\right)\times$$
$$\times\left[\exp\left\{-\frac{\Omega_B}{k_B T}\left(N+\frac{n}{2}+\frac{1}{2}+\frac{|\ell|}{2}\right)\right\}-\exp\left\{-\frac{\Omega_B}{k_B T}\left(N'+\frac{n'}{2}+\frac{1}{2}+\frac{|\ell'|}{2}\right)\right\}\right]\times \quad (31)$$
$$\times\sum_{M=0}^{+\infty}(-1)^M\left(\frac{1}{2}a_c^2 q_\lambda^2\right)^M\times\left\{C_1(M)+\frac{1}{a_c^2}\left(\frac{eE_0}{2m^*\Omega^2}\right)^2.C_2(M)\right\}\times\frac{\sqrt{A_{QR}}|Q|}{Q(\Omega-\omega_0+M\Omega_B)^2+A_{QR}}$$

Where $C_1(M)$, $C_2(M)$ are functions of M:

$$C_1(M) = \frac{\Gamma\left(N_m-N_n-M+\frac{1}{2}\right)\Gamma(1+N_m)}{(N_n!)^2\Gamma(1+N_m-N_n)}\times$$
$$\times\left\{\frac{d^{N_n}}{dh^{N_n}}\left[\frac{F\left(\frac{N_m-N_n-M+\frac{1}{2}}{2};\frac{N_m-N_n-M+\frac{3}{2}}{2};1+N_m-N_n;\frac{4h}{(1-h)^2}\right)}{(1-h)^{M+\frac{1}{2}}.(1+h)^{N_m-N_n-M+\frac{1}{2}}}\right]\right\}_{h=0}$$

$$C_2(M) = \frac{\Gamma\left(N_m-N_n-M+\frac{3}{2}\right).\Gamma(1+N_m)}{(N_n!)^2.\Gamma(1+N_m-N_n)}\times$$
$$\times\left\{\frac{d^{N_n}}{dh^{N_n}}\left[\frac{F\left(\frac{N_m-N_n-M+\frac{3}{2}}{2};\frac{N_m-N_n-M+\frac{5}{2}}{2};1+N_m-N_n;\frac{4h}{(1-h)^2}\right)}{(1-h)^{M-\frac{1}{2}}.(1+h)^{N_m-N_n-M+\frac{3}{2}}}\right]\right\}_{h=0}$$

and A_{QR} is written as:

$$A_{QR} = \frac{e^2 k_B T}{8\pi^2 \varepsilon_o V}\left|I^{\lambda}_{n\ell,n'\ell'}\right|^2\left(\frac{1}{\chi_\infty}-\frac{1}{\chi_o}\right);\ Q=\left[N-N'+\frac{n-n'}{2}+\frac{|\ell|-|\ell'|}{2}\right];$$

$$N_n := \min\{N,N'\};\ N_m := \max\{N,N'\}$$

From analytic expressions of the nonlinear absorption coefficient of a strong electromagnetic wave by confined electrons in cylindrical quantum wire with infinite potential in the presence of a magnetic field (Eq.31), we can see that quantum numbers (m, k) characterizing confined phonons reaches to zero, the result of nonlinear absorption coefficient will turn back to the case of unconfined phonons (Bau & Trien, 2010):

$$\alpha = \frac{e^4.n_0^*.\Omega_B.k_BT}{2\varepsilon_o.c.\sqrt{\chi_\infty}.V_o.m^*.a_c^2.\Omega^3}\left(\frac{1}{\chi_\infty}-\frac{1}{\chi_0}\right)\sum_{\substack{n,\ell,N\\ n',\ell',N'}}\frac{N_m!}{N_n!}\times$$

$$\times\left[\exp\left\{-\frac{\Omega_B}{k_BT}\left(N+\frac{n}{2}+\frac{1}{2}+\frac{|\ell|}{2}\right)\right\}-\exp\left\{-\frac{\Omega_B}{k_BT}\left(N'+\frac{n'}{2}+\frac{1}{2}+\frac{|\ell'|}{2}\right)\right\}\right]\times$$

$$\times\left[1+(N_m+N_n+1)\frac{3}{4a_c^2}\left(\frac{e.E_o}{2m^*\Omega^2}\right)^2\right]\frac{\sqrt{A|M|}}{M(\Omega-\omega_o+M\Omega_B)^2+A} \quad (32)$$

with $A = \frac{e^2 k_B T}{8\pi^2\varepsilon_o V}\left|I_{n\ell,n'\ell'}\right|^2\left(\frac{1}{\chi_\infty}-\frac{1}{\chi_o}\right)$; which $I_{n\ell,n'\ell'}$ is written as in (Bau & Trien, 2010)

4.3. Numerical results and discussion

In order to clarify the results that have been obtained, in this section, we numerically calculate the nonlinear absorption coefficient of a strong electromagnetic wave for a GaAs=GaAsAl cylindrical quantum wire. The nonlinear absorption coefficient is considered as a function of the intensity E_o and energy of strong electronmagnet wave, the temperature T of the system, and the parameters of cylindrical quantum wire. The parameters used in the numerical calculations (Bau&Trien, 2010) are ε_0=12.5,χ_∞= 10.9, χ_o= 13.1, m = 0.066m_o, m_o being the mass of free electron, $\hbar\omega_o \approx \hbar\omega_{m,k,\vec{q}_z} = 36.25\text{meV}$, k_B = 1.3807×10-23j/K, $n_o = 10^{23}\text{m}^{-3}$, e = 1.60219 ×10^{-19}C, ħ = 1.05459 ×10^{-34}j/s.

Fig. (13,14) shows the dependence of nonlinear absorption coefficient of a strong electromagnetic wave on the radius of wire. It can be seen from this figure that α depends strongly and nonlinear on the radius of wire but it does not have the maximum value (peak), the absorption increases when R is reduced. This is different from the case of the absence of a magnetic field. Fig. (13) show clearly the strong effect of confined phonons on the nonlinear absorption coefficient, It decreases faster in case of confined phonons.

Fig. (15) presents the dependence of nonlinear absorption coefficient on the electromagnetic wave energy at different values of the temperature T of the system. It is shown that nonlinear absorption coefficient depends much strongly on photon energy but the spectrum quite different from case of unconfined phonons (Bau&Trien, 2010). Namely, there are more resonant peaks appearing than in case of unconfined phonons and the values of resonant peaks are higher. These sharp peaks are demonstrated that the nonlinear absorption coefficient only significant when there is the condition. This means that α depends strongly on the frequency Ω of the electromagnetic wave.

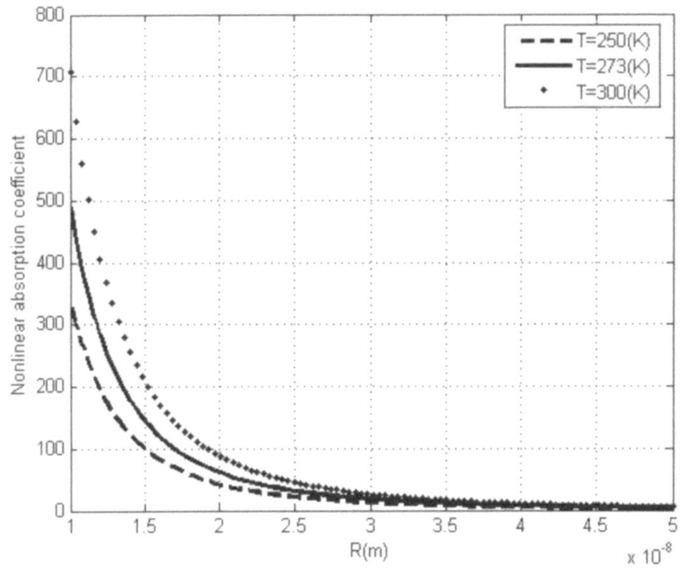

Fig. 13. The dependence of nonlinear absorption coefficient on R and T in case of confined phonons

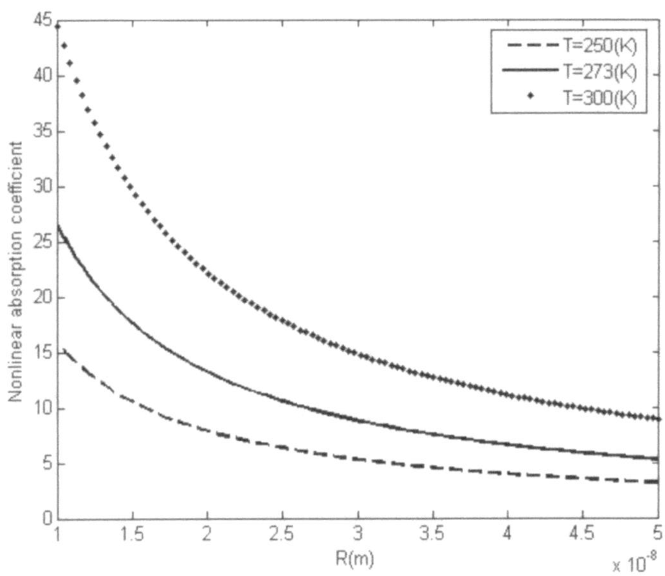

Fig. 14. The dependence of nonlinear absorption coefficient on R and T in case of unconfined phonons

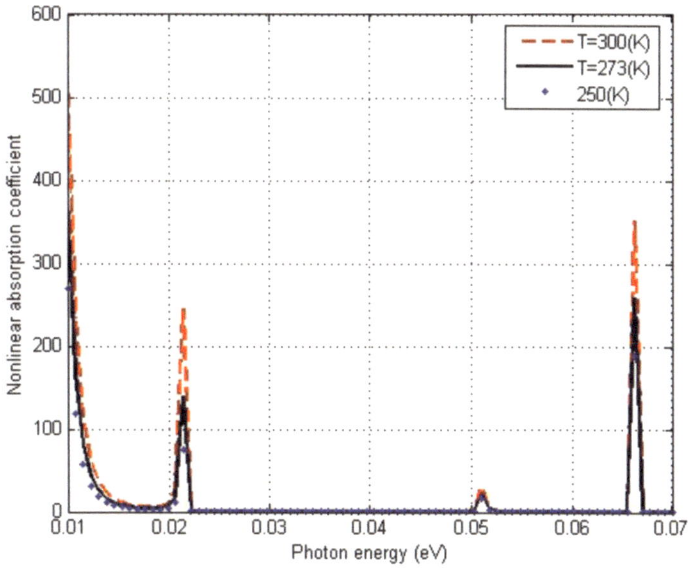

Fig. 15. The dependence of nonlinear absorption coefficient on $\hbar\Omega$ and T in case confined phonons

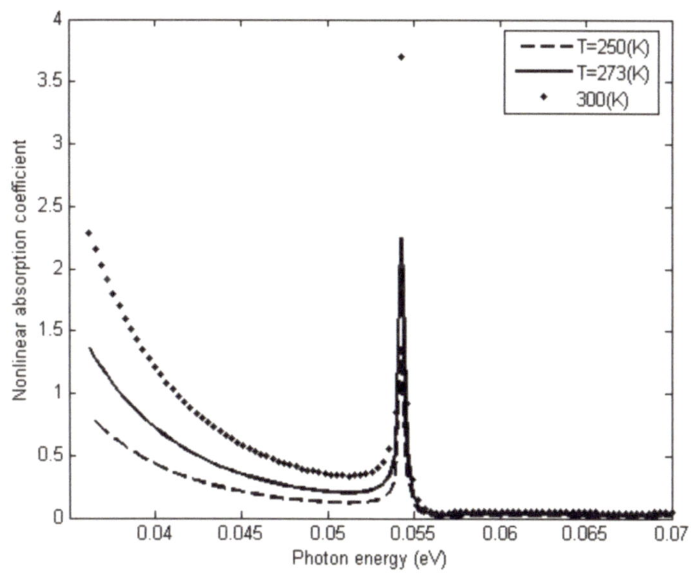

Fig. 16. The dependence of nonlinear absorption coefficient on $\hbar\Omega$ and T in case unconfined phonons

It can be seen from this figure that nonlinear absorption coefficient depends strongly and nonlinearly on T, α is stronger at large values of the temperature T.

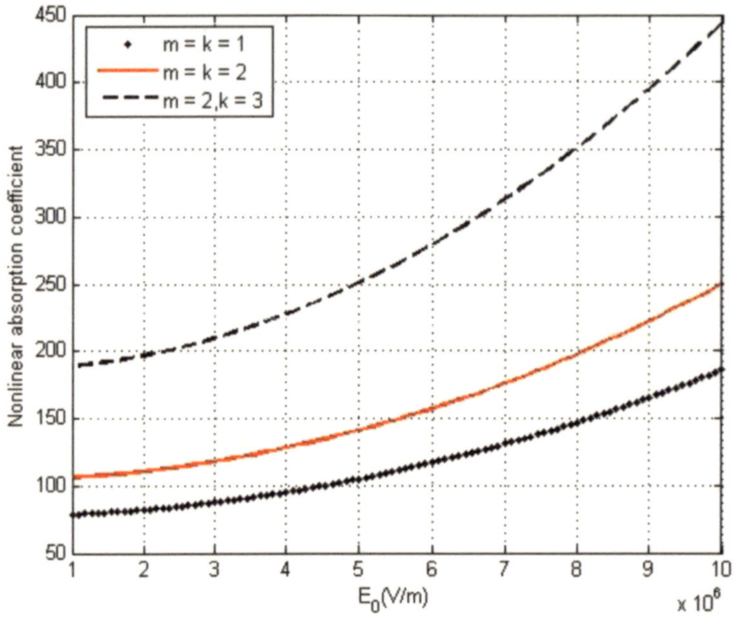

Fig. 17. The dependence of nonlinear absorption coefficient on R and m,k in case confined phonons

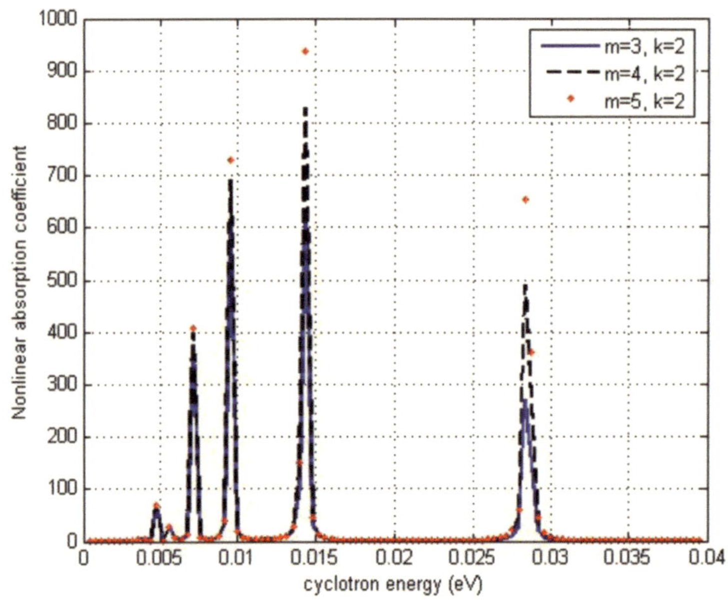

Fig. 18. The dependence of nonlinear absorption coefficient on $\hbar\Omega_B$ and m,k in case confined phonons

Fig. 17 shows the dependence of nonlinear absorption coefficient on intensity E_0 of electromagnetic wave in case confined phonons. It can be seen from this figure that α depends strongly and nonlinearly on E_0. α is stronger at large values of the intensity E_0 of electromagnetic wave.
Fig. 18 presents the dependence of α on the cyclotron energy ($\hbar\Omega_B$) of the magnetic field. It can be seen from this figure that there are same resonance peaks at different values of cyclotron frequency $\hbar\Omega_B$. The nonlinear absorption coefficient only significant at these resonance peaks. Based on this result we make the following remarks: The index of Landau level N′ which electrons can move to after absorption. Only at these peaks, strong electromagnetic wave is absorbed strongly. In addition, the density of resonance peaks is very high in the region where $\Omega_B < \Omega$, corresponding to the weak magnetic field $\vec{B}$, but this density is low when $\vec{B}$ increases. These resonance peaks, reflect the effect of quantum magnetic field on the quantum wire. When the magnetic field is stronger, the peaks is more discrete, the influence of the magnetic field is shown more clearly.
Fig. (17,18) show that nonlinear absorption coefficient depends much strongly on quantum numbers characterizing confined phonons (m, k), it increases fllowing (m,k). It mean that the confined phonons have huge effect on nonlinear absorption coefficient of a strong electromagnetic wave in cylindrical quantum wire.

5. Conclusion

In this chapter, the nonlinear absorption of a strong electromagnetic wave by confined electrons in low-dimensional systems in the presence of an external magnetic field is investigated. By using the method of the quantum kinetic equation for electrons, the expressions for the electron distribution function and the nonlinear absorption coefficient in quantum wells, doped superlattics, cylindrical quantum under the influence of an external magnetic field are obtained. The analytic results show that the nonlinear absorption coefficient depends on the intensity E_0 and the frequency Ω of the external strong electromagnetic wave, the temperature T of the system, the cyclotron frequency, the quantum number characterizing confined phonons and the parameters of the low-dimensional systems as the width L of quantum well, the doping concentration n_D in doped superlattices, the radius R of cylindrical quantum wires. This dependence are complex and has difference from those obtained in case unconfined phonon and absence of an external magnetic field (Pavlovich & Epshtein, 1977), the expressions for the nonlinear absorption coefficient has the sum over the quantum number of confined electron-confined optical phonon and contain the cyclotron frequency. All expressions for the nonlinear absorption coefficient turn back to case of unconfined phonon and absence of an external magnetic field if the quantum number and the cyclotron frequency reaches to zero.
The numerical results obtained for a AlAs/GaAs/AlAs quantum well, a n-GaAs/p-GaAs doped superlattice, a GaAs/GaAsAl cylindrical quantum show that α depends strongly and nonlinearly on the intensity E_0 and the frequency Ω of the external strong electromagnetic wave, the temperature T of the system, the cyclotron frequency, the quantum number characterizing confined phonons and the parameters of the low-dimensional systems. The numerical results shows that the confinement effect and the presence of an external magnetic field in low dimensional systems has changed significantly the nonlinear absorption coefficient. The spectrums of the nonlinear absorption coefficient have changed

in value, densty of resononlinear absorption coefficiente peaks, position of resononlinear absorption coefficiente peaks. Namely, the values of nonlinear absorption coefficient is much higher, densty of resononlinear absorption coefficiente peaks is bigger.

In addition, from the analytic results, we see that when the term in proportion to a quadratic in the intensity of the electromagnetic wave (E_o)(in the expressions for the nonlinear absorption coefficient of a strong electromagnetic wave) tend toward zero, the nonlinear result will turn back to a linear result (Bau & Phong, 1998; Bau et al., 2002; 2007).

The nonlinear absorption in each low-dimensional systems in the presence of an external magnetic field is also quite different, for example, the nonlinear absorption coefficient in quantum wires is bigger than those in quantum wells and doped superlattices.

6. Acknowledgment

This work is completed with financial support from the Vietnam National Foundation for Science and Technology Development (Vietnam NAFOSTED).

7. References

Abouelaoualim,D.(1992). *Electron–confined LO-phonon scattering in GaAs-$Al_{0.45}Ga_{0.55}As$ superlattice*, Pramana Journal of physics, Vol.66, pp. 455-465, ISSN 0304-4289, Available from www.ias.ac.in/pramana/v66/p455/fulltext.pdf

Bau, N.Q. & Trien, H.D. (2011). The Nonlinear Absorption of a Strong Electromagnetic Wave in Low-dimensional Systems, *Intech: Wave Propagation*, chaper 22, pp. 461-482, ISBN 978-953-307-275-3, Available from www.intechopen.com/download/pdf/pdfs_id/14174

Bau, N. Q.; Hung, L. T. & Nam, N. D. (2010). *The Nonlinear Absorption Coefficient of a Strong Electromagnetic Wave by Confined Electrons in Quantum Wells under the Influences of Confined Phonons*, J. of Electromagn. Waves and Appl., Vol.24, pp. 1751-1761, ISSN 0920-5071, ISSN(Online) 1569-3937, Available from http://www.ingentaconnect.com/content/vsp/jew/2010/00000024/00000013/art00006

Bau, N.Q. & Trien, H.D. (2010). *The Nonlinear Absorption Coefficient of Strong Electromagnetic Waves Caused by Electrons Confined in Quantum Wires*, J. Korean Phys. Soc, Vol. 56, pp. 120-127, ISSN 0374-4884, Available from http://www.kps.or.kr/home/kor/journal/library/downloadPdf.asp?articleuid=%7B26781ED0-F7CB-42A7-8C64-152F89D300B5%7D

Bau, N. Q. & Trien,H.D. (2010). *The nonlinear absorption of a strong electromagnetic wave by confined electrons in rectangular quantum wires*, PIERS Proceedings (March 22-26), pp. 336-341, ISSN 1559-9450, Available from http://piers.org/piersproceedings/piers2010XianProc.php?searchname=Bau

Bau,N.Q.; Hung,M.D.&Hung,.L.T. (2010). *The influences of confined phonons on the nonlinear absorption coefficient of a strong electromagnetic wave by confined electrons in doping superlattices*, Pier Letters, Vol. 15, pp. 175-185, ISSN 1937-6480, Available from www.jpier.org/PIERL/pierl15/20.10030911.pdf

Bau N.Q.; Hung, N.M. & Ngoc ,N.B. (2009). *The Nonlinear Absorption Coeffcient of a Strong Electromagnetic Wave Caused by Conffined Electrons in Quantum Wells*, J. Korean. Phys. Soc, Vol. 42, No.2, pp. 765-773, ISSN 0374-4884

Bau,N.Q.; Dinh,L.& Phong,T.C(2007). *Absorption coefficient of weak electromagnetic waves caused by confined electrons in quantum wires*, J. Korean.Phys. Soc., Vol. 51, pp. 1325-1330, ISSN 0374-4884, Available from http://www.kps.or.kr/jkps/downloadPdf.asp?articleuid=%7B9BAB8518-F80B-41F5-A49D-5407B4614DD9%7D

Bau, N. Q. and Phong ,T. C. (1998). *Calculations of the Absorption Coefficient of a Weak Electromagnetic Wave by free Carriers in Quantum Wells by the Kubo-Mori Method*, J.Phys. Soc. Japan, Vol. 67, pp. 3875-3880, , ISSN 0031-9015, ISSN(Online) 1347-4073

Bau, N.Q.; Nhan, N.V. &Phong, T.C. (2002). *Calculations of the absorption coefficient of a weak electromagnetic wave by free carriers in doped superlattices by using the Kubo - Mori method*, J. Korean. Phys. Soc., Vol. 41, pp. 149-154, ISSN 0374-4884, Available from http://www.kps.or.kr/home/kor/journal/library/downloadPdf.asp?articleuid=%7BD05E09B1-B3AA-4AFD-9748-1D4055F885DB%7D

Butscher.S.& Knorr,A.(2006). *Occurrence of Intersubband Polaronic Repellons in a Two-Dimensional Electron Gas*, Phys. Rev. L, Vol. 97, pp. 197401-197405, ISSN 0031-9007, ISSN(Online) 1079-7114, Available from http://prl.aps.org/abstract/PRL/v97/i19/e197401

Chaubey, M.P.& Carolyn, M. (1986). *Transverse magnetoconductivity of quasi-two-dimensional semiconductor layers in the presence of phonon scattering*, Phys. Rev. B, Vol.33, pp. 5617-5622, ISSN 1098-0121, ISSN(Online) 1550-235X, Available from http://prb.aps.org/abstract/PRB/v33/i8/p5617_1

Flores, A. D. A.(2008). *Electron subband structure and mobility trends in p-n delta-doped quantum wells in Si*, Pier Letters, Vol. 1, pp. 159-165, ISSN: 1937-6480, Available from www.jpier.org/PIERL/pierl01/19.07120607.pdf

Generazio, E.R.& Spector.H.N. (1979). *Free-carrier absorption in quantizing magnetic fields*, Phys. Rev. B, Vol. 20, pp. 5162-5167, ISSN 1098-0121, ISSN(Online) 1550-235X, Available from http://prb.aps.org/abstract/PRB/v20/i12/p5162_1

Richter, M.; Carmele, A.; Butscher, S.; Bücking, N.; F. Milde.; Kratzer P.; Scheffler M. & Knorr, A. (2009). *Two-dimensional electron gases: Theory of ultrafast dynamics of electron-phonon interactions in graphene, surfaces, and quantum wells*, J. Appl. Phys., Vol. 105, pp. 122409-122416, Print: ISSN 0021-8979, ISSN(Online) 1089-7550, Available from http://jap.aip.org/japiau/v105/i12/p122409_s1?bypassSSO=1

Li ,W. S.; Gu, S.W.; Au-Yeung, T. C.& Yeung,Y. Y. (1992). *Effects of the parabolic potential and confined phonons on the polaron in a quantum wire*, Phys. Rev. B, Vol.46,pp. 4630-4637, Available from http://prb.aps.org/abstract/PRB/v46/i8/p4630_1

Malevich, V. L. & Epstein, E. M. (1974). *Nonlinear optical properties of conduction electrons in semiconductors*, Sov. Quantum Electronic, Vol. 4, p. 816, ISSN: 0049-1748, Available from http://iopscience.iop.org/0049-1748/4/6/L27

Sager, L.M.G.; Martine, N. M.; Vargas I. R.; Alvarez, R. Pe.; Grimalsky,V. V.& Mora - Ramos,M. E.(2007). *Electronic structure as a function of temperature for Si doped quantum wells in GaAs*, PIERS Online, Vol. 3, No. 6, pp. 851-854, ISSN 1931-7360, Available from www.jpier.org/PIERL/pierl01/19.07120607.pdf

Mori, N. and Ando, T. (1989). *Electron-optical-phonon interaction in single and double heterostructures*, Phys. Rev.B, Vol.40, pp. 6175-6188, ISSN: 1937-6480, Available from http://prb.aps.org/abstract/PRB/v40/i9/p6175_1

Pavlovich, V. V. & Epshtein, E. M. (1977). *Quantum theory of absorption of electronmagnetic wave by free carries in simiconductors*, Sov. Phys. Solid State., Vol.19, pp. 1760, ISSN: 0038-5654

Ryu, J.Y.& O'Connell, R. F. (1993). *Magnetophonon resonances in quasi-one-dimensional quantum wires*, Phys. Rev. B, Vol. 48, pp. 9126-9129, ISSN 1098-0121, ISSN(Online) 1550-235X, Available from http://prb.aps.org/abstract/PRB/v48/i12/p9126_1

Rucker ,H.; Molinari ,E. & Lugli, P.(1992). *Microscopic calculation of the electron-phonon interaction in quantum wells*, Phys. Rev. B, Vol. 45, pp. 6747-6756, ISSN 1098-0121, ISSN(Online) 1550-235X, Available from http://prb.aps.org/abstract/PRB/v45/i12/p6747_1

Samuel, E. P.,& Patil, D. S. (2008). *Analysis of wavefunction distribution in quantum well biased laser diode using transfer matrix method,* Pier Letters, Vol. 1, pp. 119-128, ISSN: 1937-6480, Available from www.jpier.org/PIERL/pierl01/15.07111902.pdf

Suzuki, A.(1992). *Theory of hot-electron magneto phonon resonance in quasi-two-dimensional quantum-well structures,* Phys. Rev. B, Vol.45, pp. 6731-6741, ISSN 1098-0121, ISSN(Online) 1550-235X, Available from http://prb.aps.org/abstract/PRB/v45/i12/p6731_1

Shmelev,G. M. ; Chaikovskii, L. A. & Bau ,N. Q. (1978). *HF conduction in semiconductors superlattices,* Soc. Phys. Tech. Semicond, Vol. 12, No.10, p. 1932, ISSN 0018-9383

Vasilopoulos,P. M.; Charbonneau& Vliet,C.M.Van(1987). *Linear and nonlinear electrical conduction in quasi-two-dimensional quantum wells,* Phys. Rev. B, Vol. 35, pp. 1334-1344, ISSN 1098-0121, ISSN(Online) 1550-235X, Available from http://prb.aps.org/abstract/PRB/v45/i12/p6731_1

Yu, Y.B.; Guo, K.X. & Zhu. S.N.(2005). *Polaron influence on the third-order nonlinear optical susceptibility in cylindrical quantum wires,* Physica E, Vol. 27, pp. 62-66, ISSN: 1386-9477, Available from http://adsabs.harvard.edu/abs/2005PhyE...27...62Y

Wang,X. F. & Lei, X. L. (1994). *The polar-optic phonons and high field electron transport in cylindrical GaAs/AlAs quantum wires,* Phys. Rev. B, Vol. 49, pp. 4780-4789, ISSN 1098-0121, ISSN(Online) 1550-235X, http://prb.aps.org/abstract/PRB/v49/i7/p4780_1

15

Chiral Transverse Electromagnetic Standing Waves with E||H in the Dirac Equation and the Spectra of the Hydrogen Atom

H. Torres-Silva
*Instituto de Alta Investigación, Universidad de Tarapacá, Arica
Chile*

1. Introduction

This article examines the conditions under which transverse electromagnetic (TEM) waves exist in a sourceless medium. It shows that TEM waves can be classified according to whether their Poynting vector is identically zero or nonzero. The former are non propagating TEM standing waves with $\mathbf{E} \parallel \mathbf{H}$ and the latter are TEM traveling and standing waves with $\mathbf{E} \perp \mathbf{H}$. The Chiral general condition under which TEM standing waves with $\mathbf{E} \parallel \mathbf{H}$ exist is derived. The behavior of these waves under Lorentz transformations is discussed and it is shown that these waves are lightlike and that their fields lead to well-defined Lorentz invariants. Two physical examples of these standing waves are given.

The first example is about the Dirac Equation for a free electron, which is obtained from the Maxwell Equations under the Born Fedorov approach, ($\mathbf{D} = \varepsilon(1 + T\nabla\times)\mathbf{E}$), and ($\mathbf{B} = \mu(1 + T\nabla\times)\mathbf{H}$). Here, it is hypothesized that an elementary particle is simply a standing enclosed electromagnetic wave with a half or whole number of wavelengths (λ). For each half number of λ the wave will twist 180° around its travel path, thereby giving rise to chirality. As for photons, the Planck constant (h) can be applied to determine the total energy (E): $E = nhc/\lambda$, where n = 1/2, 1, 3/2, 2, etc., and c is the speed of light in vacuum. The mass m can be expressed as a function of λ, since $E = mc^2$ gives $m = nh/c\lambda$, from the formula above. This result is obtained from the resulting wave equation which is reduced to a Beltrami equation $\nabla \times \mathbf{E} = -(1/2T)\mathbf{E}$ when the chiral factor T is given by $T = n\hbar/mc$. The chiral Pauli matrices are used to obtain the Dirac Equation.

The second example is on a new interpretation of the atomic spectra of the Hydrogen atom. Here we study the energy conversion laws of the macroscopic harmonic LC oscillator, the electromagnetic wave (chiral photon) and the hydrogen atom. As our analysis indicates that the energies of these apparently different systems obey exactly the same energy conversion law. Based on our results and the wave- particle duality of electron, we find that the atom in fact is a natural microscopic LC oscillator.

In the framework of classical electromagnetic field theory we analytically obtain, for the hydrogen atom, the quantized electron orbit radius $r_n = a_o n^2$, and quantized energy $E_n = -R_H hc/n^2$, (n = 1, 2, 3, · · ·), where a_0 is the Bohr radius and R_H is the Rydberg constant. Without the adaptation of any other fundamental principles of quantum mechanics, we present a reasonable explanation of the polarization of photon,selection rules and Pauli exclusion principle. Our results also reveal an essential connection between electron spin and the intrinsic helical movement of electron and indicate that the spin itself is the effect of quantum confinement.

2. Chiral Transverse Electromagnetic standing waves with $\mathbf{E} \parallel \mathbf{H}$

This section provides the basis for reexamining the electromagnetic model of the electron, which is developed in others sections appearing in the present paper.
The solutions of Maxwell's equations of common interest are concerned with the propagation of electromagnetic (EM) energy in the form of transverse electromagnetic (TEM) waves in free space, material media, and transmission lines as well as transverse electric and magnetic waves in waveguides. Such solutions are characterized by the orthogonality of the electric field $\mathbf{E}$ and the magnetic flux density $\mathbf{H}$: i.e., $\mathbf{E} \perp \mathbf{H}$. An examination of many senior undergraduate and graduate electrodynamics textbooks [Jackson, 1975; Marion & Herald, 1980; Stratton, 1941; Panofsky & Phillips, 1955; Cook, 1975; Reitz & Milford, 1967; Lorrain & Corson, 1970; Portis, 1978; Smythe, 1950; Purcell, 1965] reveals that there is very little discussion of the conditions under which TEM standing waves exist. Textbooks with more of an engineering emphasis [Ramo et al., 1965.; Jordan & and Balmain, 1968.; Bekefi. & Barrett, 1977.; Shadowitz, 1975.; Rao, 1977], generally discuss standing waves in the context of the measurement of the voltage standing-wave ratio for transmission lines [Rao, 1977]. At the first time a work by Chu and Ohkawa [Chu & and Ohkawa, 1982] shows that a class of TEM waves with $\mathbf{E} \parallel \mathbf{H}$ exists, also, Zaghloul et al. [Zaghloul, et al.,1988] have derived the general conditions under which we have TEM waves with $\mathbf{E} \parallel \mathbf{H}$. From here, that it is useful to classify TEM waves according to whether their Poynting vector is identically zero or nonzero. The former are $\mathbf{E} \parallel \mathbf{H}$ standing waves and the latter are traveling and standing TEM waves with $\mathbf{E} \perp \mathbf{H}$ and $\mathbf{E} \nparallel \mathbf{H}$. Using the Born Fedorov formalism [Torres-Silva, 2008], this article provides a discussion of the physical properties of TEM standing waves with $\mathbf{E} \parallel \mathbf{H}$ in connection with the interaction matter waves from its basic principles. It also suggests that $\mathbf{E} \parallel \mathbf{H}$ standing waves can be generated by the superposition of waves traveling in opposite directions and that the electromagnetic density of these standing waves somehow plays the same role in electrodynamics that the rest mass plays in relativistic dynamics. The background assumed in this article is that of a senior undergraduate physics or electrical student. It is hoped that this article will encourage instructors of electrodynamics courses to include a discussion of $\mathbf{E} \parallel \mathbf{H}$ TEM standing waves.
In this article, SI units are used. Vectors are denoted by bold letters. Other notational definitions are given as required.

2.1 Solutions of Maxwell's equations in a sourceless medium

Under the Born-Fedorov, constitutive relations between $\mathbf{D}$, $\mathbf{E}$, $\mathbf{B}$, $\mathbf{H}$ are [Torres-Silva, 2008]

$$\mathbf{D} = \varepsilon_0(1 + T\nabla\times)\mathbf{E}\,,\ \mathbf{B} = \mu_0(1 + T\nabla\times)\mathbf{H} \tag{1a}$$

In such a representation (BF), rotation terms are added to the basic constitutive relation whose chiral coefficients εT or μT can be either positive or negative for two stereoisomer structures. Solving the constitutive relation together with Maxwell's equations, we can easily get two eigenwaves, which are left and right circularly polarized with different wavevectors. The Maxwell's equation are

$$\nabla\times\mathbf{E}+\frac{\partial(1+T\nabla\times)\mathbf{H}}{c\partial t}=0,\ \nabla\times\mathbf{H}-\frac{\partial(1+T\nabla\times)\mathbf{E}}{c\partial t}=0 \tag{1b}$$

$$\nabla\bullet\mathbf{D}=0,\ \nabla\bullet\mathbf{B}=0 \tag{1c}$$

since $1/c^2=\varepsilon_0\mu_0$, where c is the speed of light in vacuum and ε_0, μ_0 and T (meter) are the vacuum permittivity, permeability, and the scalar chiral factor respectively.
Considering time variation as $\mathbf{E}\sim\mathbf{E}e^{i\omega_0 t}$ etc, the solution of Eqs. (1a)-(1c) can be obtained such that

$$k\nabla^2\mathbf{E}+k_0^2(1-k_0^2T^2)^{-1}\mathbf{E}+2\omega_0\mu_0\varepsilon_0T((1-k_0^2T^2)^{-1}\nabla\times\mathbf{E}=0, \tag{2a}$$

$$k\nabla^2\mathbf{H}+k_0^2(1-k_0^2T^2)^{-1}\mathbf{H}+2\omega_0\mu_0\varepsilon_0T((1-k_0^2T^2)^{-1}\nabla\times\mathbf{H}=0 \tag{2b}$$

Due to the chiral scalar $T\geq 0$ or $T\leq 0$, a general TEM solution of Eq. (2a) can be expressed in the form due to d'Alembert [Smythe, 1950],

$$\boldsymbol{E}(\mathrm{r},t)=\boldsymbol{E}_+(\eta)+\boldsymbol{E}_-(\xi), \tag{3}$$

where $\boldsymbol{E}_+(\eta)=\sum_{i=1}^{3}E_+(\eta)_i\mathbf{l}_i$, $\boldsymbol{E}_-(\xi)=\sum_{i=1}^{3}E_-(\xi)_i\mathbf{l}_i$, $\eta\equiv\mathrm{k}\bullet\mathrm{r}-\omega_0 t$, $\xi\equiv\mathrm{k}\bullet\mathrm{r}+\omega_0 t$, and l_i are mutually orthogonal unit vectors in an arbitrary orthogonal coordinate system. Notice that the most general TEM solution of Eq. (2a or 2b) is a sum of solutions of the form of Eq. (3a) over all possible propagation vectors, k. Here, k and ω, the angular frequency, are related by $k_0^2=\omega_0^2/c^2$. The two solutions of Eq. (2a or 2b) given by Eq. (3a) can be considered as the vector electromagnetic field of two TEM waves traveling in opposite directions, i.e., energy propagation takes place in two opposite directions. In general, there is more energy flow in one direction than in the other so there is net energy flow in one direction. Such a TEM wave is called a traveling wave. In the special case where the energy propagated in one direction is equal to that propagated in the opposite direction, there is no net energy flow in the medium and the sum of the two TEM waves form what is generally known as a standing wave. Mathematically, the amount of energy density propagated is proportional to the magnitude of the Poynting vector [Marion & Herald, 1980] $\mathbf{S}$, where

$$\mathbf{S}\propto\mathbf{E}\times\mathbf{H}. \tag{4}$$

The condition for a standing wave is that the time average of $\mathbf{S}$ vanishes . This can be achieved if (i) $\mathbf{S}$ is zero all the time everywhere in the region of space under consideration, i.e., $\mathbf{S}(\mathrm{r},t)=0$, or (ii) $\mathbf{S}$ changes sign with time and has a zero time average, i.e., $\mathbf{S}(\mathrm{r},t)_{\mathrm{av}}=0$. Examination of Eq. (4) shows that $\mathbf{S}=0$ if either $\mathbf{E}$ or $\mathbf{H}$ is zero (the cases of electro- and magnetostatics) or if $\mathbf{E}\parallel\mathbf{H}$. In this last case, a particular solution of Eq. (2a, 2b)

is when $k_0^2T^2 = 1$, where we have the condition $\mathbf{E} \parallel \mathbf{H}$, and $\mathbf{E} = i\eta\mathbf{H}$, so we find the Beltrami force free equation $\mathbf{E} + 2T\nabla \times \mathbf{E} = 0$ and the vector Poynting vanishes.
This means that for time-varying fields (i) leads to TEM standing waves with $\mathbf{E} \parallel \mathbf{H}$ whereas (ii) leads to the more familiar TEM standing waves with $\mathbf{E} \perp \mathbf{H}$.The fact that $\boldsymbol{S}(\mathrm{r},t) = 0$, for time-varying fields leads to $\mathbf{E} \parallel \mathbf{H}$ and TEM standing waves will be used to derive the general conditions and connection to derive the Dirac equation.
The derivation of the general conditions for the existence of TEM standing waves with $\mathbf{E} \parallel \mathbf{H}$ was given first in a condensed form by [Zaghloul, et al.,1988]. Our derivation can be simplified by assuming that $\nabla \cdot (1 + k_0^2T^2)\boldsymbol{A} + \partial\phi/\partial t = 0$,. If $k_0^2T^2 = 1$, $\partial\phi/\partial t = 0$, and $\nabla\phi = 0$. This is equivalent and consistent with the choice of the Lorentz gauge.
In our approach if only if $kT = \pm 1$, then we have TEM standing waves with $\mathbf{E} \parallel \mathbf{H}$. Here, T is a chiral scalar factor. This approach defines a class of $\mathbf{E} \parallel \mathbf{H}$ TEM standing waves that may be characterized by the fact that the chiral vector potential $\mathbf{F}$ satisfies the equation $\nabla \times \mathbf{F} = k\mathbf{F}$ as has been shown by [Torres-Silva, 2008]. An example of a vector potential satisfying $\nabla \times \mathbf{F} = k\mathbf{F}$ with $\mathbf{E} \parallel \mathbf{H}$ fields

$$\boldsymbol{F}(\mathrm{r},t) = F_0(\sin kz, \cos kz, 0)\cos\omega t\,, \tag{5}$$

$$\mathbf{E}(\mathrm{r},t) = \omega F_0(\sin kz, \cos kz, 0)\sin\omega t\,, \tag{6}$$

$$\mathbf{H}(\mathrm{r},t) = \eta F_0(\sin kz, \cos kz, 0)\cos\omega t\,. \tag{7}$$

This solution corresponds to two circularly polarized waves [Chu & and Ohkawa, 1982] propagating opposite to each other in such a way that their Poynting vectors are cancelled out, so $\mathbf{E} = \mathrm{i}\eta\mathbf{H}$. Therefore, a single helical photon with energy $\hbar\omega$ carries a magnetic helicity of $\hbar c$.

2.2 Physical properties of $\mathbf{E} \parallel \mathbf{H}$ TEM standing waves

The TEM waves with $\mathbf{E} \parallel \mathbf{H}$ do not propagate, and hence are standing waves. It has been shown earlier that they are characterized by $\mathbf{S}(\mathrm{r},t) = 0$. This means that there is no energy transfer between any two points in space at any time. How such waves can be generated and maintained in some region of space is an interesting question. One answer is that they are due to the superposition os waves traveling in opposite directions. It should be noted that for TEM waves with $\mathbf{E} \parallel \mathbf{H}$, investigation of the direction of $\mathbf{E}$ and $\mathbf{H}$ at any one point in space cannot uniquely determine the direction of the propagation vector $\mathbf{k}$. The direction of k can be inferred from a consideration of the directions of the fields at different points. The two fields $\mathbf{E}$ and $\mathbf{H}$ remain parallel to each other at each point but their direction will change from point topoint such that all of these directions lie in parallel planes, the transverse planes. This provides a unique definition of the direction of the propagation vector. It might be more appropriate to use the term wave vector in place of propagation vector for these standing waves since they do not propagate.
If $\mathbf{E} \parallel \mathbf{H}$ in one Lorentz frame K, then in any other frame K', $\mathbf{E}'$ may not be parallel to $\mathbf{H}'$. Moreover, no frame exists for which $\mathbf{E}' \perp \mathbf{H}'$. This is a direct consequence of the invariance

of $\mathbf{E}\cdot\mathbf{H}$, which will be discussed in next Sec. The TEM standing waves with $\mathbf{E}\parallel\mathbf{H}$, like all other TEM waves with $\mathbf{E}\perp\mathbf{H}$, keep $k_1=\mathbf{E}^2-\eta^2\mathbf{H}^2$ and $k_2=2\mathbf{E}\cdot\eta\mathbf{H}$ invariant in all Lorentz frames ($\eta=\sqrt{\mu/\varepsilon}$). The TEM waves with $k_1=k_2=0$ are known as null or pure radiation fields .

Notice that the invariants k_1 and k_2 are also gauge invariant. The EM invariants k_1 and k_2 can be combined to form another invariant [Zaghloul, et al; Torres-Silva, 2008],

$$\begin{aligned} k_1^2+k_2^2 &= \left(\mathbf{E}^2-\eta^2\mathbf{H}^2\right)^2+4\left(\mathbf{E}\cdot\eta\mathbf{H}\right)^2 \\ &= \left(\mathbf{E}^2+\eta^2\mathbf{H}^2\right)^2-4\left|\mathbf{E}\times\eta\mathbf{H}\right|^2, \\ &= 4\left(\varepsilon^2-S^2/c^2\right) \end{aligned} \tag{8}$$

where ε is the EM energy density and $\mathbf{S}$ is the Poynting vector. In a reference frame with $\mathbf{E}\parallel\mathbf{H}$ and $\mathbf{S}=0$,

$$k_1^2+k_2^2=4\varepsilon_{\rm sw}^2, \tag{9}$$

where $\varepsilon_{\rm sw}$ is the EM standing-wave energy density. In any other frame, if $\mathbf{S}'\neq 0$, and $\varepsilon'\neq\varepsilon_{\rm sw}$, then

$$k_1'^2+k_2'^2=4\left(\varepsilon'^2-S'^2/c^2\right)=k_1^2+k_2^2=4\varepsilon_{\rm sw}^2 \tag{10}$$

or

$$\varepsilon'^2-S'^2/c^2=\varepsilon_{\rm sw}^2\,. \tag{11}$$

Equation (11) states that for any TEM wave, the difference between the square of the EM energy density (ε') and the square of the magnitude of the Poynting vector (S') in any frame is equal to the square of the EM energy density ($\varepsilon_{\rm sw}$) in the frame for which $\mathbf{E}\parallel\mathbf{H}$. This is true of all EM waves since $\varepsilon_{\rm sw}=0$ for classical EM traveling waves with $\mathbf{E}\perp\mathbf{H}$ (null fields) whilst $\varepsilon_{\rm sw}\neq 0$ for parallelizable fields (generic fields) and for classical standing waves with $\mathbf{E}\perp\mathbf{H}$. It should be emphasized that $\varepsilon_{\rm sw}$ is an invariant and that ε is not an invariant [Zaghloul, et al.,1988].

Equation (11) can be compared with the expression for the invariant squared magnitude of the energy-momentum four-vector p_μ of a particle that is given by [Lorrain & Corson, 1970; Torres-Silva, 2008]

$$c^2p_\mu p^\mu=E^2-c^2p^2=m_0^2c^4, \tag{12}$$

where $E=mc^2$ is the self-energy, p is the magnitude of the momentum, m_0 is the rest mass, and m is the mass of the particle. This similarity suggests that somehow $\varepsilon_{\rm sw}$ plays the same role in electrodynamics that the rest energy of a particle plays in relativistic dynamics. In other words, the generation of $\mathbf{E}\parallel\mathbf{H}$ TEM standing waves allows the localization of electromagnetic energy. This result for $\mathbf{E}\parallel\mathbf{H}$ standing waves differs from that of $\mathbf{E}\perp\mathbf{H}$ standing waves since the energy oscillates back and forth, for the latter waves, and there exists no Lorentz frame where the energy is static, which is the situation for the former waves. This localization of EM energy could open the door for some very interesting possibilities such as the storage ("bottling") of EM energy using, for example, a laser beam to make what could be thought of as a laser battery as well as verifying experimentally if the storing of EM energy E in a medium can be associated with an increase of the mass of the medium by m_0, where $m_0 = E/2c^2 = \hbar / 2cT$. "Electromagnetic mass" where gravitational mass and other physical quantities originate from the electromagnetic field alone. We can say that the electromagnetic mass models which are the sources of purely electromagnetic origin have not only connection associated with the conjecture of Lorentz but even a physics having novel features.

3. Chiral Dirac equation from electromagnetic standing waves with $\mathbf{E} \parallel \mathbf{H}$

The purpose of this section is to (a) derive a two-component force-free Dirac particle equation that incorporates all of the physically meaningful information about the particle contained in the force-free, four-component Dirac equation, (b) solve that two-component equation and, from the resulting two-component wave function, obtain the internal spin of the particle, (c) show that the spin and rest mass of the particle are the result of the phenomenon described by Schrodinger [Schrodinger, 1930] as *Zitterbewegung*, and (d) demonstrate that the relativistic increase in mass with velocity of the force-free particle is due to an increase in the rate of *Zitterbewegung* with velocity.

We introduce a new unitary transformation under which, for force-free motion, we obtain *uncoupled two-component* equations *which we can identify separately as a particle equation and an antiparticle equation.* In uncoupling the two equations and solving each equation as separate from and unrelated to the other, although that information is still there, hidden in the Hamiltonian, since the two equations can be put back together and transformed once again into the usual four-component Dirac equation, with a particle solution and an antiparticle solution. Our particle and antiparticle spinors are not Weyl or Majorana spinors; nevertheless and state that two of our spinors equal one pair of particle and antiparticle Dirac spinors.

The Foldy-Wouthuysen transformation [Foldy & Wouthuysen, 1950] decomposes the Dirac equation into two-component states of positive and negative energy. However, as the Foldy-Wouthuysen transformation does not provide a pair of particle and antiparticle equations. For the two-component equation produced by our separation process to be an equation for a particle (or for the companion equation to be an equation for an antiparticle), we must be able of demonstrating that, like the four-component equation, each of the two-component equations obtained leads to a relativistically correct four-vector, the fourth component of which is a scalar probability density, with a three-vector describing a probability current. For definiteness, throughout most of this section we will refer to the particle under consideration as an electron.

Based on the results of Fues and Hellmann [Fues & Hellmann, 1930], Schrödinger had come to the conclusion that it was not practical to obtain general information about the electron by solving the four-component Dirac equation directly, and set out in [Schrodinger, 1930] to see what general information could be obtained without direct solutions. Schrodinger focused on that aspect of the Dirac equation and introduced the concept of Zitterbewegung in an attempt to deal with this problem when obtaining general information about solutions to the Dirac equation. In addition to being discussed by Schrödinger in incisive detail in his original article [Schrodinger, 1930], *Zitterbewegung* in the four-component Dirac basis has been extensively addressed in books devoted to the the Dirac equation, [Rose, 1961.] and [Thaller, 1992] (who summarize Schrödinger's approach), as well as by Hanl and Papapetrou [Hanl & Papapetrou, 1940] in a paper that first analyzes Zitterbewegung in detail and then attempts to mimic the effects of *Zitterbewegung* with a mechanical model based on a classical point charge.

The relativistic particle equation in our paper presents an interpretation of Zitterbewegung similar to, yet quite different from, the interpretation for the force-free, four-component Dirac equation. For the Dirac Hamiltonian *Zitterbewegung* results from an interference between two positive and two negative energy components of the Dirac spinor [Lock, 1979]. Solutions of our two-component, force-free equation will be shown to have one positive energy component interfering with one negative energy component. In contrast to the solutions in the Dirac basis, the concept of the motion of a point has disappeared in our two-component case, and has been replaced by a density function and an internal motion (rotation) of that density function.

The difference between the interpretation of *Zitterbewegung* for our two-component particle Hamiltonian and for the Dirac Hamiltonian is to be expected, since it has been demonstrated that *Zitterbewegung* has different interpretations after different unitary transformations of the Dirac equation [Lock, 1979]. From another perspective, the different interpretation in our paper is also to be expected from comparing our result to equation-reduction techniques due to Feshbach [Feshbach, 1958]. In particular, whereas the Hamiltonian for the Dirac equation is linear in the components of the momentum operator, the Hamiltonian of our two-component particle equation is nonlocal.

The central difference between the two-component relativistic particle equation developed in the present section and the four-component Dirac equation is that for the Dirac equation the *Zitterbewegung* is explicit, as demonstrated by Schrödinger. In the case of the two-component particle equation obtained here, the effect of *Zitterbewegung* is implicit. Also, as Schrödinger noted before demonstrating the utility and importance of the concept of *Zitterbewegung*, information about a Dirac particle described in the usual Dirac basis can be difficult to extract. As an example consider Huang's insight (using the Dirac basis) [Huang, 2006] into the fact that internal rotation of the Dirac electron is associated with *Zitterbewegung*.

In many respects, the present section is a direct extension of papers written by Weyl in 1929 [Weyl, 1929 a;. Weyl, 1929 b; Weyl, 1929 c]. The importance of Weyl in the early interpretation of the Dirac equation has been succinctly described by Miller [Miller,1994]. In the first of the three papers Weyl asserts that, since the wave function of the electron "can only involve two components", two components of the solution of the Dirac equation shou1d be ascribed to the electron and two to the proton [Weyl, 1952]. In light of Weyl's later proof [Dirac, 1931] that negative and positive energy electrons must have the same mass, Dirac was led to the concept of the second solution of the Dirac equation being not a proton but antimatter, "having the same mass and opposite charge to an electron" [Dirac,1928]. However, in our case we have the condition +m, -e, T>0 to the electron and -m, +e, T<0 to the positron because for a particle in a rest frame $T = \hbar / 2mc$.

As Dirac well understood [Dirac, 1928], a two-component theory linear in the derivatives with respect to both the time and spatial coordinates is not possible. However, as we will demonstrate, by relaxing the restriction of linearity with respect to derivatives of the spatial coordinates, a two-component equation for the force-free electron is possible, and the solutions we obtain to that equation provide the insights that Schrödinger was seeking [Torres-silva, 2011]. Moreover, when we obtain the complete solution of the two-component equation for the electron, we find significant additional characteristics not apparent in solutions of the four-component equation of the Dirac theory of the electron.

Finally, because in the Dirac basis the chirality operator does not commute with the Hamiltonian, it might appear that chirality (handedness, as opposed to the kinematic concept of helicity, associated with the labels ψ_L and ψ_R) cannot be a characteristic of a Dirac particle with mass. We will use our two-component solutions to demonstrate that this would be a false conclusion; for a Dirac particle with mass the usual definition of chirality can be replaced by a concept specific to a particle with mass, a concept that we will refer to a intrinsic electromagnetic chirality, $T = \hbar / 2mc$ with $\psi_L = \psi_{\vec{E}} = \vec{\sigma} \cdot \mathbf{E}$ and $\psi_R = \psi_{\vec{H}^\bullet} = \vec{\sigma} \cdot \boldsymbol{H}^\bullet$ explained below.

3.1 Two-component equations for the electron and anti-electron

The usual choice of an orthogonal set of four plane-wave solutions of the free-particle Dirac equation does not lend itself readily to direct and complete physical interpretation except in low energy approximation. A different choice of solutions can be made which yields a direct physical interpretation at all energies. Besides the separation of positive and negative energy states there is a further separation of states for which the spin is respectively parallel or antiparallel to the direction of the momentum vector. This can be obtained from the Maxwell's equation without charges and current in the $\mathbf{E} \| \boldsymbol{H}$ configuration. Dirac's four-component equation for the relativistic electron is.

$$i\hbar \frac{\partial}{\partial t} \psi = \hat{H}^D \psi , \tag{13}$$

where:

$$\hat{H}^D = c(\bar{\alpha} \cdot \hat{\bar{p}}) + mc^2 \beta , \tag{14}$$

$$\alpha_k = \begin{pmatrix} 0 & \sigma_k \\ \sigma_k & 0 \end{pmatrix}, \quad k = 1,2,3, \tag{15}$$

$$\beta = \begin{pmatrix} I & 0 \\ 0 & -I \end{pmatrix} \tag{16}$$

and I is the two-by-two identity matrix. We now reduce the force-free Dirac equation.

This Hamiltonian commutes with the momentum vector $\hat{\mathrm{p}}$, and the usual procedure is to seek simultaneous eigenfunctions of H and p. These eigenfunctions are, however, not uniquely determined, and for given eigenvalues of H and p, there remains a twofold degeneracy. In order to resolve this degeneracy we seek a dynamical variable which commutes with both H and $\hat{\mathrm{p}}$. Such a variable is $\hat{\sigma} \cdot \hat{\mathrm{p}}$, where $\hat{\sigma}$ is the matrix Pauli. It is obvious that this variable commutes with $\hat{\mathrm{p}}$. To verify that it also commutes with H, we write $\hat{\alpha} = \rho_1 \hat{\sigma} = \hat{\sigma} \rho_1$ and recalling that β commutes with operator $\hat{\sigma}$, we have

$$\hat{\sigma}\cdot\hat{\mathrm{p}}H - H\hat{\sigma}\cdot\hat{\mathrm{p}} = \left(c\rho_1\hat{\sigma}\cdot\hat{\mathrm{p}}\right)\left(\hat{\sigma}\cdot\hat{\mathrm{p}}\right) - \left(\hat{\sigma}\cdot\hat{\mathrm{p}}\right)\left(c\rho_1\hat{\sigma}\cdot\hat{\mathrm{p}}\right) = 0\,, \tag{17}$$

since ρ_1 commutes with $\hat{\sigma}$.
We now proceed to find simultaneous eigenfunctions of the commuting variables *H*, p and $\hat{\sigma}\cdot\hat{\mathrm{p}}$. We have, since the components of p commute,

$$\left(\hat{\sigma}\cdot\hat{\mathrm{p}}\right)^2 = p^2\,, \tag{18}$$

where *p* is the magnitude of the momentum vector. Thus for a simultaneous eigenstate of $\hat{\mathrm{p}}$ and $\hat{\sigma}\cdot\hat{\mathrm{p}}$, the value of $\hat{\sigma}\cdot\hat{\mathrm{p}}$ will be *+p* or *–p*, corresponding to states for which the spin is parallel or antiparallel, respectively, to the momentum vector.
A simultaneous eigenfunction of *H* and *p* will have the form of a plane wave

$$\psi_j = u_j \exp\left[i(\mathrm{p}\cdot\mathrm{r} - Wt)/h\right], \qquad j = 1,2,3,4\,, \tag{19}$$

where the ψ_j are the four components of the state function and u_j four numbers to be determined. In the argument of the exponential function, p represents the eigenvalues of the components of the momentum for this state and *E* the corresponding eigenvalue of *H*. Then *E* can have either of the two values.

$$W = \pm\varepsilon = \pm\left(m^2c^4 + c^2p^2\right)^{\frac{1}{2}}. \tag{20}$$

We now demand that ψ_j be also an eigenfunction of $\hat{\sigma}\cdot\hat{\mathrm{p}}$ belonging to one of the eigenvalues $p_{\mathbf{E}}$, say, where $p_{\mathbf{E}} = \pm p$. Employing the usual matrix representation for $\hat{\sigma}$, we have

$$\hat{\sigma}\cdot\hat{p} = \begin{pmatrix} p_z & p_x - ip_y & 0 & 0 \\ p_x + ip_y & -p_z & 0 & 0 \\ 0 & 0 & p_z & p_x - ip_y \\ 0 & 0 & p_x + ip_y & -p_z \end{pmatrix}. \tag{21}$$

In the above matrix, p_x, p_y, and p_z are operators, but since this matrix is to operate on an eigenfunction of p, the operators can be replaced by their eigenvalues. We shall, without risk of confusion, use the same symbols for the eigenvalues as for the corresponding operators.
The eigenvalue equation is

$$\hat{\sigma}\cdot\hat{\mathrm{p}}\psi = p_{\mathbf{E}}\psi\,, \tag{22}$$

Since *W* can be given either of the two values $\pm\varepsilon$ and $p_{\mathbf{E}}$, the two values $\pm p$, we have found for given p four linearly independent plane wave solutions. It is easily verified that they are mutually orthogonal.
The physical interpretation of the solutions is now clear. Each solution represents a homogeneous beam of particles of definite momentum p, of definite energy, either $\pm\varepsilon$, and with the spin polarized either parallel or antiparallel to the direction of propagation.

3.2 Dirac equation deduced from Maxwell's equations with $\mathbf{E} \parallel \mathbf{H}$

Following our first section Transverse electromagnetic standing waves with $\mathbf{E} \parallel \boldsymbol{H}$ the vector potential is

$$\boldsymbol{F}(\mathrm{r},t) = \mathrm{Re}\left(e^{i\omega t}\right)\left[\mathbf{A}_k(\mathrm{r}) + k^{-1}\nabla \times \boldsymbol{A}_k(r)\right] \tag{23}$$

will lead to TEM standing waves with $\mathbf{E} \parallel \boldsymbol{H}$. This approach defines a class of $\mathbf{E} \parallel \boldsymbol{H}$ TEM standing waves that may be characterized by the fact that the vector potential A satisfies the equation $\nabla \times \mathbf{A} = k\mathbf{A}$, with $k = \omega / c$. Thus we have $\nabla \times \boldsymbol{E} = k\boldsymbol{E}$ and $\nabla \times \boldsymbol{H}^{\bullet} = k\boldsymbol{H}^{\bullet}$, where $\boldsymbol{H}^{\bullet}$ is the complex conjugate of $\boldsymbol{H}$. If we make the conjecture that the chiral factor T considered as a hidden scalar satisfies $Tp_{\mathbf{E}} = \pm\hbar / 2$, then we can transform $\nabla \times \boldsymbol{E} = k\boldsymbol{E}$ as

$$i\hat{\sigma}\cdot \nabla \times \boldsymbol{E} = k\hat{\sigma}\cdot i\boldsymbol{E} \tag{24}$$

which can be put as

$$\hat{\sigma}\cdot p(\hat{\sigma}\cdot \boldsymbol{E}) = p_{\mathbf{E}}(\hat{\sigma}\cdot \boldsymbol{E}) \tag{25}$$

where we are considered $\nabla \cdot \boldsymbol{E} = 0$ (no charges), and

$$\hat{\sigma}\cdot \boldsymbol{E} = \begin{pmatrix} E_z & E_x - iE_y & 0 & 0 \\ E_x + iE_y & -E_z & 0 & 0 \\ 0 & 0 & E_z & E_x - iE_y \\ 0 & 0 & E_x + ie_y & -E_z \end{pmatrix} \tag{26}$$

The vector function ψ is given by

$$\psi_{\mathbf{E}} = \begin{pmatrix} E_x - E_y \\ |\boldsymbol{E}| - E_z \\ -\dfrac{|\boldsymbol{E}|c}{W - mc^2}(E_x - E_y) \\ -\dfrac{|\boldsymbol{E}|c}{W - mc^2}(|E| - E_z) \end{pmatrix} \tag{27}$$

with $|\boldsymbol{E}| = \dfrac{Wmc}{eh}$, $E = \hbar\omega$ and $p_{\mathbf{E}}T = \pm n\dfrac{\hbar}{2}$, where $n = 1/2, 1, 3/2, 2$ In this case of an electron n=1/2, the electric and magnetic fields are 90° out of phase, the energy density ε_{sw} is constant and proportional to $4\pi G(\boldsymbol{E}^2 + i\eta\boldsymbol{H}^2) \sim 4\pi k^2 F_0^2$, which numerically, we find in SI units as $4\pi k^2 F_0^2 = (1.3\times 10^{-9} EE_c^{-1} m^{-1})^2$. Here, E_c is the critical field for electron-positron pair production $E_c = m_e{}^2 e^{-1} \sim 1.3\times 10^{18} Vm^{-1}$ (see equations (5,6,7)of this paper) and [Torres-Silva, 2011]. Here, it is hypothesized that an elementary particle is simply a standing enclosed electromagnetic wave with a half or whole number of wavelengths (λ). For each half number of λ the wave will twist 180° around its travel path, thereby giving rise to chirality. As for photons, the Planck constant (h) can be applied to determine the total energy (E):

$E = nhc / \lambda$, where n = 1/2, 1, 3/2, 2, etc., and c is the speed of light in vacuum. The mass (m) can be expressed as a function of λ, since $E = mc^2$ gives $m = nh / c\lambda$, from the formula above.

In this form we have a close connection between the Dirac Equation and the Maxwell' equations, with a direct and complete physical interpretation in the $\mathbf{E} \parallel \boldsymbol{H}$ configuration. Here we are obtained a clear connection between the Planck constant $\hbar$, the quiral factor T, and the electromagnetic mass m.

To find the two uncoupled, two-component equations, one for the electron and one for the anti-electron (positron), we begin this reduction with the Cini-Touschek transformation [Cini & Touschek, 1958].

$$\hat{H}^T = M\hat{H}^D M^{-1}, \tag{28}$$

where:

$$M = \exp[iS], \tag{29}$$

$$S = \frac{\mathrm{i}}{2\hat{p}}\beta\left[\vec{\alpha}\cdot\hat{\vec{p}}\right]\text{arc cot}\left(\frac{\hat{p}}{mc}\right), \tag{30}$$

and $\hat{\vec{p}}$ and $\hat{p}$ are defined by

$$\hat{\vec{p}} = -\mathrm{i}\hbar\left(\hat{\mathbf{i}}\frac{\partial}{\partial x} + \hat{\mathbf{j}}\frac{\partial}{\partial y} + \hat{\mathbf{k}}\frac{\partial}{\partial z}\right);\ \hat{p}^2 = \hat{\vec{p}}\cdot\hat{\vec{p}}\,. \tag{31}$$

Under this transformation

$$\hat{H}^D \to \hat{H}^T = \sqrt{\left(\frac{m^2c^4}{\hat{p}^2} + c^2\right)}\vec{\alpha}\cdot\hat{\vec{p}}\,. \tag{32}$$

Instead of the transformation M in Eq. (29) we introduce the transformation

$$U = \mathrm{N}_{chiral}M, \tag{33}$$

where

$$\mathrm{N}_{chiral} = \frac{1}{\sqrt{2}}\begin{pmatrix} I & I \\ -I & I \end{pmatrix} \tag{34}$$

in which each component I is the two-by-two identity matrix.

Now we operate Eq. (25) with $\mathrm{i}\hbar\partial / \partial t$. This reduces the force-free Dirac equation to the pair of uncoupled two-spinor equations

$$\mathrm{i}\hbar\frac{\partial}{\partial t}(\vec{\sigma}\cdot\vec{E}) = \sqrt{\left(\frac{m^2c^4}{\hat{p}^2} + c^2\right)}\vec{\sigma}\cdot\hat{\vec{p}}(\vec{\sigma}\cdot\vec{E}) \tag{35}$$

and

$$-\mathrm{i}\hbar\frac{\partial}{\partial t}(\vec{\sigma}\cdot\vec{H}^{\bullet}) = \sqrt{\left(\frac{m^2c^4}{\hat{p}^2}+c^2\right)}\vec{\sigma}\cdot\hat{p}(\vec{\sigma}\cdot\vec{H}^{\bullet})\,. \tag{36}$$

In addition to Eqs. (35) and (36), it is instructive to factor the operator $\hat{p}^2$ out of the denominator and write Eqs. (35) and (36), respectively, as

$$\mathrm{i}\hbar\frac{\partial}{\partial t}(\vec{\sigma}\cdot\vec{E}) = \sqrt{(m^2c^4+c^2\hat{p}^2)}\frac{1}{\sqrt{\hat{p}^2}}\vec{\sigma}\cdot\hat{p}(\vec{\sigma}\cdot\vec{E}),\ (\omega_0 T/c\le 1) \tag{37}$$

and

$$-\mathrm{i}\hbar\frac{\partial}{\partial t}(\vec{\sigma}\cdot\vec{H}^{\bullet}) = \sqrt{(m^2c^4+c^2\hat{p}^2)}\frac{1}{\sqrt{\hat{p}^2}}\vec{\sigma}\cdot\hat{p}(\vec{\sigma}\cdot\vec{H}^{\bullet}),\ (\omega_0 T/c\le 1) \tag{38}$$

Where, $\psi_L=\psi_{\mathbf{E}}=\vec{\sigma}\cdot\vec{E}$, and $\psi_R=\psi_{\mathbf{H}^{\bullet}}=\vec{\sigma}\cdot\vec{H}^{\bullet}$. In this form it is easy to see that one can obtain Eq. (38) from Eq. (37) either by replacing *t* in Eq. (37) by *-t*, or by replacing $\sqrt{(m^2c^4+c^2\hat{p}^2)}$, the energy operator for a particle of momentum $\vec{p}$, by the operator $-\sqrt{(m^2c^4+c^2\hat{p}^2)}$, the operator for the negative of the energy in Eq. (37).
For the equation as written by Dirac, the pair of solutions are states of positive and negative energy, interpreted by Dirac and others as the electron and, eventually, the anti-electron (the positron). A later interpretation [Mulligan, 2008], [Feynman, 1949] treats the solutions of the Dirac equation as a pair of different types of solutions-one type moving forward in time and the other type moving backward in time;
the solutions which move forward in time are interpreted as electrons, while those that move backwards in time are interpreted as anti-electrons. Although a study of the literature shows that the physical implications of the two different interpretations are profound, Feynman demonstrated that, *mathematically*, organizing the types of solutions according to his interpretation is equivalent to organizing the solutions according to the Dirac interpretation. From the way in which we rewrite the Dirac equation here as the combination of Eqs. (37) and (38), the difference between the two interpretations can easily be seen as simply a matter of whether to associate a minus sign with the energy *E* or with the time *t* compared with the sign of the chiral scalar T.
That Eq. (38) be a time-reversed or negative energy solution of Eq. (37) is clearly a necessary condition for recognizing the two-component spinor $\vec{\sigma}\cdot\vec{E}$ as a force-free electron.
As required by Dirac, the four-component operator $\hat{H}^D$ in Eq. (13) depends only linearly on the operator $\hat{\vec{p}}$. In contrast, the operator on the right-hand side of Eqs. (35) and (36) also depends on $\hat{\vec{p}}^2$, which appears in the denominator. A series expansion of the right-hand side of Eqs. (37) and (38) involves $\hat{\vec{p}}^2$ to all orders. Thus, whereas the Hamiltonian in Eq. (1) is local, the operator on the right-hand side of Eqs. (37) and (38) is nonlocal. It is well-known that nonlocality can result [13,14] when eliminating explicit reference to channels-in this case, the reference to the coupling of anti-electrons to electrons, which is explicit in Eq. (19) but no longer explicit in Eqs. (37) and (38), since each equation can be solved independently of the other but with the connection $\boldsymbol{E}=i\eta\boldsymbol{H}$.
One of the conditions imposed by Dirac in writing down a relativistic equation for the electron was that the "square" of that equation be the Klein-Gordon equation. Imposing the

additional condition of linearity of $\hat{H}^D$ in the components $\hat{p}_k$ led Dirac to Eqs. (13) and (14). It is trivial to see that squaring either Eqs. (37) and (38) also leads to the Klein-Gordon equation.

3.3 Internal rotation and electron mass

As was the case with the four-component force-free Dirac equation, our two-component force-free equation exhibits Zitterbewegung as an interference between components. In our case, each solution has one positive energy component and one negative energy component. Also, our solutions immediately exhibit a frequency

$$\Omega' = W / \hbar \tag{39}$$

and relate that frequency to an internal rotation of the electron. This frequency agrees with the frequency of rotation determined by Huang from consideration of the characteristics of the *Zitterbewegung* associated with the Dirac equation, but not obtained there so readily; as Huang points out, *Zitterbewegung* as rotation is not a feature which is immediately apparent in the four-component solutions of the Dirac equation.

Here, we consider in more detail the mass of the electron as a property which can be attributed to its internal rotation.

The angular frequency Ω' given by Eq. (39) is the accepted value of the rotational frequency associated with the intrinsic spin of an electron in motion with momentum $\vec{p}$, as discussed by Huang [Huang, 2006]. An additional important aspect of the internal rotation of the electron is the realization and identification that the rest-mass energy of a Dirac particle "appears simply as the energy of the internal motion in the rest frame."

The condition $\omega_0 T / c = 1$, in eq. allow to write the angular frequency Ω for an electron at rest, Eq. (39) becomes

$$\omega_0 T / c = 1 \rightarrow \omega_o = \Omega = \frac{2mc^2}{\hbar} \tag{40}$$

Eq. (40) demonstrates that if *Zitterbewegung* is the mechanism that gives rise to the rest mass of the electron, it also must be the mechanism that gives rise to the relativistic mass as a function of the velocity of the electron.

Moreover, we can attribute the increase in *Zitterbewegung* for an electron in motion directly to the time dilation due to motion of the electron relative to the observer. This can be demonstrated by repeating the well-known fact that if we define a velocity υ_z for the electron by

$$\upsilon_z = \frac{p_z}{M}, \tag{41}$$

where M the relativistic mass

$$M = \frac{m}{\sqrt{1 - \frac{\upsilon_z^2}{c^2}}}, \tag{42}$$

the expression $\sqrt{m^2c^4 + p_z^2c^2}$ for the energy E becomes

$$E = \frac{mc^2}{\sqrt{1 - \frac{\upsilon_z^2}{c^2}}} \tag{43}$$

The time dependence of the angle of the intrinsic rotation of the electron is given by $\frac{2}{\hbar}Et$, or

$$\frac{2}{\hbar}Et = \frac{2}{\hbar}\frac{mc^2}{\sqrt{1 - \frac{\upsilon_z^2}{c^2}}}t \tag{44}$$

We can write Eq. (75) in the alternate form

$$\frac{2}{\hbar}Et = \frac{2}{\hbar}mc^2\tau , \tag{45}$$

where

$$\tau = \frac{1}{\sqrt{1 - \frac{\upsilon_z^2}{c^2}}}t . \tag{46}$$

Consequently, the angular frequency Ω' of the rotation measured by an observer who sees the electron moving with velocity υ_z is related to the rest frame angular frequency Ω by

$$\Omega' = \frac{\Omega}{\sqrt{1 - \frac{\upsilon_z^2}{c^2}}} . \tag{47}$$

The above discussion is presented in detail because Hönl and Papapetrou, in their mechanical model of of the electron, accidentally apply time dilation to *decrease* the rotational frequency, rather than to dilate the time. Accordingly, they conclude that the rate of rotation decreases rather than increases as the velocity of the particle increases. In light of Torres-Silva's [18] insight that the internal motion of the electron is the source of its mass, the relation between increased internal motion and increased mass becomes a critical aspect of the source of mass in a Dirac particle.

Dorling [Dorling, 1970] recognized that Eq. (47) represents an adjustment to a frequency associated with the electron at rest and attempts to relate this frequency to *Zitterbewegung*. We can now readily make the necessary connection based on our results for the force-free electron. To effect the decrease in mass required by the binding process, according to our results the rotational frequency associated with the *Zitterbewegung* would have to decrease by an amount appropriate in energy of the electron. That is exactly what Eq. (47) describes. Therefore, in the case of Eq. (47) the role of time as measured by a clock attached to the electron and a dock held by an external observer is the reverse of that for the free electron. For the bound electron the external observer sees the rate of rotation due to Zitterbewegung correspondingly reduced.

In his paper on *Zitterbewegung* [Schrodinger, 1930], Schrödinger demonstrates that *Zitterbewegung* introduces motion perpendicular to the direction of net motion of the electron.

The need to introduce components p_x and p_y was first recognized by Neamtan [Neamtan, 1952], who did not associate these components with *Zitterbewegung*. Moreover, Neamtan was working with the four-component Dirac equation, and his solutions do not lend themselves to the interpretation possible here as a result of our use of two-component wave functions, to which we can apply Eqs. (37)-(38).

Our results allow us to write a solution for a particle which is intrinsically right-handed, a particle which is intrinsically left-handed, or, by taking a linear combination of solutions, a particle such as the electron that has no intrinsic chirality. In obtaining the solution of Eq. (35) we have effectively solved the Dirac equation for the particle moving in the direction of the positive z-axis and projected out that part of the particle wave function with positive helicity and positive intrinsic chirality; in obtaining one solution, we have effectively solved the Dirac equation for the particle moving in the direction of the positive z-axis and projected out that part of the particle wave function with positive helicity and negative intrinsic chirality. This is of substantial practical importance, since in electron β-decay only that part of the wave function with negative intrinsic chirality takes part.

The general consensus of papers on *Zitterbewegung* is that in the rest frame of the electron *Zitterbewegung* has a characteristic amplitude of $\hbar / 2mc$. This information can be used to obtain an order of magnitude estimate of el. in the rest frame. We set

$$\frac{\int_0^\infty e^{(-2/\hbar)\alpha x} x\, dx}{\int_0^\infty e^{(-2/\hbar)\alpha x} dx} = \frac{\hbar}{2mc} \tag{48}$$

which yields

$$\alpha = mc\,. \tag{49}$$

Before leaving the subject of the relation of *Zitterbewegung* to mass, we return to discussion of Zitterbewegung in the context of the Weyl equations. he treats the mass term in each of the two coupled spinors of the Weyl basis as a source for the other, a coupling proportional to the electron mass. Accordingly, while we already have related the size of the electron perpendicular to its motion to the electron mass. Our analysis allows us also to relate the length of the electron to its mass. Penrose describes the motion of the Dirac point position of the stationary electron along the axis of rotation of the electron (which we have taken to be the z-axis) as motion first with velocity e and negative helicity in the positive z-direction and then with positive helicity in the negative z-direction. The Dirac point performs this oscillatory motion with the de Broglie frequency

$$\upsilon = \frac{mc^2}{2\pi\,\hbar}\,. \tag{50}$$

Multiplying the time for one motion in each direction by the velocity c gives for the length L of the electron

$$L = \frac{\pi \hbar}{mc}. \tag{51}$$

While this calculation results in an order of magnitude estimate of the length of the electron, even when at rest the motion of the electron distribution along its axis of rotation.

4. New interpretation of the atomic spectra of the hydrogen atom: a mixed mechanism of classical LC circuits and quantum wave-particle duality

Following Huang, [Huang, 2006], we study the energy conversion laws of the macroscopic harmonic LC oscillator, the electromagnetic wave (photon) and the hydrogen atom. As our analysis indicates that the energies of these apparently different systems obey exactly the same energy conversion law. Based on our results and the wave- particle duality of electron, we find that the atom of Hydrogen in fact is a natural chiral microscopic LC oscillator.
In the framework of classical electromagnetic field theory we analytically obtain, for the hydrogen atom, the quantized electron orbit radius $r_n = a_0 n^2$, and quantized energy $W_n = -R_H hc / n^2$, (n = 1, 2, 3, ·), where a_o is the Bohr radius and R_H is the Rydberg constant. Without the adaptation of any other fundamental principles of quantum mechanics, we present a reasonable explanation of the polarization of photon, selection rules and Pauli exclusion principle. Our results also reveal an essential connection between electron spin and the intrinsic helical movement of electron and indicate that the spin itself is the effect of quantum confinement.
In fact, at the heart of quantum mechanics lies only the Schrödinger equation, which is the fundamental equation governing the electron. According to quantum theory, it is the electromagnetic interaction (by the exchange of photons) which hold electrons and nuclei together in the atoms [Planck, 1900; Einstein, 1905; Bohr, 1913; Stern,1920; de Broglie, 1924 ; Pauli, 1924 ; Heisenberg, 1927 ; Yang, 1950]. But, up to now, quantum theory never provides a practical scheme that electron and nuclei can absorb and emit photons.
In this section, we investigate the energy relationship of electron in the hydrogen atom. Significantly, we find a process of perfect transformation of two forms of energy (kinetic and field energy) inside the atom and the conservation of energy in the system. By applying the principle of wave-particle duality and comparing to known results of the macroscopic harmonic LC oscillator and microscopic photon, we are assured that electron kinetic energy in fact is a kind of magnetic energy and the atom is a natural microscopic LC oscillator. Moreover, the mixed mechanism (classical LC circuits / quantum wave particle duality) turns out to have remarkably rich and physical properties which can used to describe some important quantum principles and phenomena, for instance, polarization of photon, Zeeman effect, Selection rules, the electron's mass and spin, zero point energy (ZPE) , the Pauli exclusion principle.

4.1 Energy transformation and conversion in hydrogen atom

Classically, as shown in Fig. 1, the hydrogen atom consists of one electron in orbit around one proton with the electron being held in place via the electric Coulomb force. Equation of motion is

$$\frac{e^2}{4\pi\varepsilon_0 r^2} = \frac{m_e u^2}{r}, \tag{52}$$

where m_e is mass of electron. Eq. (52) can be rewritten in the form of kinetic energy W_k and field energy W_f (stored in the capacitor of hydrogen atom) as follows:

$$\frac{e^2}{2C_r} = \frac{1}{2} m_e u^2 , \tag{53}$$

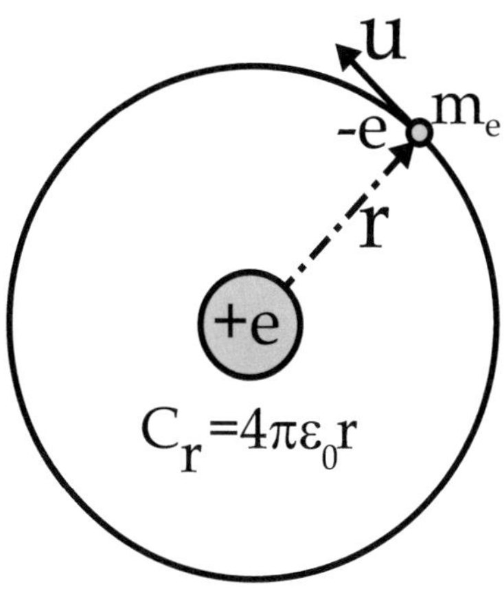

Fig. 1. The diagram illustrating the hydrogen atom

where $C_r = 4\pi\varepsilon_0 r$ is the capacitance of the hydrogen system. Thus the total energy of the hydrogen system is given by

$$W_{total} = \left| \frac{1}{2} m_e u^2 - \frac{e^2}{4\pi\varepsilon_0 r} \right| = \frac{e^2}{2C_r} . \tag{54}$$

It should be pointed out that Eq. (53) and (54) are the foundation of our study. These two equations together indicate a process of perfect periodically transformation of two forms of energy (kinetic energy $W_k = m_e u^2 / 2$ and field energy $W_f = \frac{e^2}{2C_r}$) inside the atom and the conservation of energy in the system

$$W_{total} = W_f = W_k \tag{55}$$

Recall the macroscopic harmonic LC oscillator where two forms of energy, the maximum field energy $W_f = \frac{Q_0^2}{2C}$ of the capacitor C (carrying a charge Q_0) and the maximum magnetic energy $W_m = \frac{L_0^2}{2L}$ of the inductor L, are mutually interchangeable ($W_{total} = W_f = W_k$) with a exchange periodic $T = 2\pi\sqrt{LC}$. And for a microscopic photon (electromagnetic wave), the maximum field energy $W_f = \frac{1}{2}\varepsilon_0 \boldsymbol{E}_0^2$ and the maximum magnetic energy $W_m = \frac{1}{2}\mu_0 \boldsymbol{H}_0^2$ also satisfy $W_{total} = W_f = W_k$.

Based on the above energy relationship for three totally different systems and the requirement of the electromagnetic interaction (by exchanging photon) between electron and nuclei, we assure that the kinetic energy of electron, Eq. (53) is a kind of magnetic energy and the hydrogen atom is a natural microscopic LC oscillator.

Recently, a multinational team of physicists had observed for the first time a process of internal conversion between bound atomic states when the binding energy of the converted electron becomes larger than the nuclear transition energy [Carreyre et al. 2000; Kishimoto et al. 2000]. This observation indicate that energy can pass resonantly between the nuclear and electronic parts of the atom by a resonant process similar to that which operates between an inductor and a capacitor in an LC circuit. These experimental results can be considered a conclusive evidence of reliability of our LC mechanism.

Here raise an important question: how can the electron function as an excellent microscopic inductor?

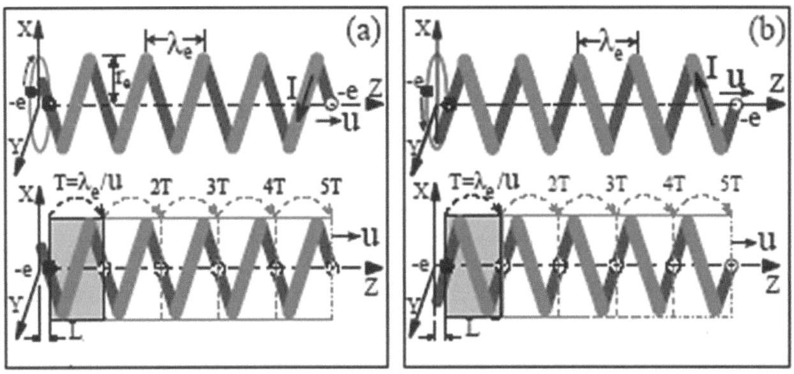

a) Left hand electron (S=1), b) Right hand electron (S=-1)

Fig. 2. A free electron moving along a helical orbit with a helical pitch of de Broglie wavelength λ_e

The answer lies in the intrinsic wave-particle duality nature of electron. In our opinion, the wave-particle nature of electron is only a macroscopic behavior of the intrinsic helical motion of electron within its world.

4.2 Chirality and "inducton" of free electron

De Broglie suggested that all particles, not just photons, have both wave and particle properties [Reines & Sobel, 1974]. The momentum wavelength relationship for any material particles was given by

$$\lambda = h / p, \tag{56}$$

where λ is called de Broglie wavelength, h is Planck's constant and p the momentum of the particle. The subsequent experiments established the wave nature of the electron [9, 10]. Eq. (56) implies that, for a particle moving at high speed, the momentum is large and the wavelength is small. In other words, the faster a particle moves, the shorter is its wavelength. Furthermore, it should be noted that any confinement of the studied particle will shorten than λ and help to enhance the so-called quantum confinement effects.

As shown in Fig. 2 (a) and (b), based on Eq. (56) and the demanding that the electron would be a microscopic inductor, we propose that a free electron can move along a helical orbit (the helical pitch is de Broglie wavelength λ_e) of left-handed or right-handed. In this paper, the corresponding electrons are called "Left-hand" and "Right-hand" electron which are denoted by Chirality Indexes S = 1 and S = −1, respectively. Hence, the electron can now be considered as a periodic-motion quantized inductive particle which is called "inducton" (see Fig. 2). Moreover, the particle-like kinetic energy of electron can be replaced with a dual magnetic energy carried by a "inducton". Therefore, we have

$$W_k = \frac{1}{2} m_e u^2 = \frac{1}{2} L_e I^2, \tag{57}$$

where u is the axial velocity of the helical moving electron and L_e is the inductance of the quantized "inducton".

The above relation indicates that the mass of electron is associated with an amount of magnetic energy. From Fig. 2, the electric current, for one de Broglie wavelength, is given by

$$I = \frac{eu}{\lambda_e}. \tag{58}$$

From Eq. (58), it is important to note that the electric current should be defined within an integral number of de Broglie wavelength. Hence, the electric current $I = \frac{eu}{2\pi r}$ $I = \frac{eu}{2\pi r}$ (where r is the electronic orbital radius in the hydrogen atom), which was widely used in the semiclassical Bohr model, may be physically invalid. Collecting Eq. (57) and (58) together, we have the inductance of single "inducton"

$$L_e = \frac{m_e \lambda_e^2}{e^2}. \tag{59}$$

Then the dual nature of electron can be uniquely determined by L_e, the periodic T (or frequency $f = \frac{1}{T} = \frac{u}{\lambda_e}$), the initial phase φ_0 and the chirality (S = 1 or S = −1).

4.3 Atomic spectra of hydrogen atom

A. Quantized radius and energy by the application of helical electron orbit to the hydrogen atom (Fig. 2), we can explain the stability of the atom but also give a theoretical interpretation of the atomic spectra. Fig. 3 shows four possible kinds of stable helical electron orbits in hydrogen atom, and each subgraph corresponds to a electron of different motion manner within the atom. The electrons can be distinguished by the following two aspects. First consider the chirality of electron orbits, as shown in Fig. 3, the electrons of Fig. 3(a) and (c) are "Left-hand" labelled by S = 1, while electrons of Fig. 3(b) and (d) are "Right-hand" labelled by S = −1. Secondly consider the direction of electron orbital magnetic moment μL, Fig. 3(a) and (b) show that the μL are in the Z direction (Up) while (c) and (d) in the −Z direction (Down), the corresponding electrons are labelled by J = 1 and J = −1, respectively, here J is called Magnetic Index. Hence, the electrons of different physical properties become distinguishable, they are Up "Left-hand" (ULH) electron (J = 1, S = 1), Up "Right-hand" (URH) electron (J = 1, S = −1), Down "Left-hand" (DLH) electron (J = −1, S = 1) and Down "Right-hand" (DRH) electron (J = −1, S = −1).

As shown in Fig. 3(a), the helical moving electron around the orbit mean radius r can now be regarded as a quantized "inducton" of λ_r, thus the hydrogen atom is a natural microscopic LC oscillator.

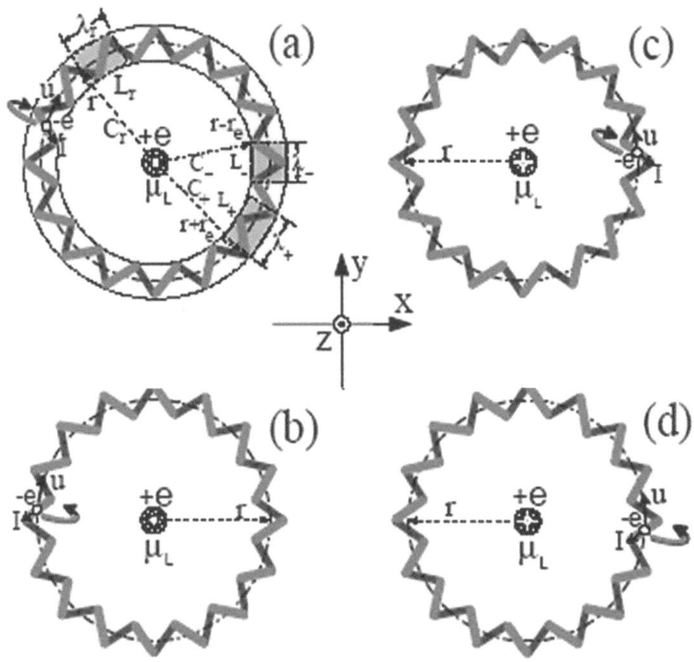

Fig. 3. The quadruple degenerate stable helical electron or-bits in hydrogen atom. a) Up Left-hand electron ULH electron (J=1, S=1); b) Up Right-hand electron URH electron (J=1, S=-1); c) Down Left-hand electron DLH electron (J=-1, S=1); d) Down Right-hand electron DRH electron (J=-1, S=-1)

We consider that the physical properties of the hydrogen atom can be uniquely determined by these natural LC parameters. To prove that our theory is valid in explaining the structure of atomic spectra, we study the quantized orbit radius and the quantized energy of hydrogen atom and make a comparison between our results of LC mechanism and the known results of quantum theory. For the system of λ_r, the LC parameters of the hydrogen atom is illustrated in Fig. 3. Then the LC resonant frequency is

$$\nu_r = \frac{1}{2\pi\sqrt{L_r C_r}} . \quad (60)$$

Recall the well-known relationship $E = h\nu_r$, we have

$$W = E = h\nu_r = \frac{e^2}{8\pi\varepsilon_0 r} . \quad (61)$$

Combining Eq. 60 and Eq. 61 gives

$$\lambda_r = \frac{2h}{e}\sqrt{\pi\varepsilon_0 r / m_e} \,. \tag{62}$$

Then the stable electron orbits are determined by

$$\frac{2\pi r}{\lambda_r} = n \,, \text{ (n = 1, 2, 3 } \cdots\text{)}, \tag{63}$$

where n is called Principal oscillator number. The integer n shows that the orbital allow integer number of "induction" of the de Broglie wavelength λ_r . From Eq. 52 and Eq. 53, the quantized electron orbit mean radius is given by

$$r_n = \frac{\varepsilon_0 h^2}{\pi m_e e^2} n^2 = a_0 n^2 \,, \tag{64}$$

where a_0 is the Bohr radius. And the quantized energy is

$$W_n = -\frac{e^2}{8\pi\varepsilon_0 r_n} = -\frac{m_e e^4}{8\varepsilon_0^2 h^2}\frac{1}{n^2} = -R_H \frac{hc}{n^2} \,, \tag{65}$$

where R_H is the Rydberg constant. Surprisingly, the results of Eq. (64) and (65) are in excellent agreement with Bohr model. Besides, taking Fig. 3 into account, we can conclude that the quantized energies of Eq. (65) are quadruple degenerate.

5. Concluding remarks

In this article we have examined the conditions under which transverse electromagnetic (TEM) waves according to whether their Poynting vector is identically zero or nonzero. We have studied the non propagating TEM standing waves with $\mathbf{E} \parallel \mathbf{H}$. The Chiral general condition under which TEM standing waves with $\mathbf{E} \parallel \mathbf{H}$ exist is derived. Two physical examples of these standing waves are given.

The first example is about the Dirac Equation for a free electron, which is obtained from the Maxwell Equations under the Born Fedorov approach, ($\mathbf{D} = \varepsilon_0(1 + T\nabla\times)\mathbf{E}$), and ($\mathbf{B} = \mu_0(1 + T\nabla\times)\mathbf{H}$). Here, it is hypothesized that an elementary particle is simply a standing enclosed electromagnetic wave with a half or whole number of wavelengths (λ). For each half number of λ the wave will twist 180° around its travel path, thereby giving rise to chirality. As for photons, the Planck constant (h) can be applied to determine the total energy (E): $E = nhc / \lambda$, where n = 1/2, 1, 3/2, 2, etc., and c is the speed of light in vacuum. The mass m can be expressed as a function of λ, since $E = mc^2$ gives $m = nh / c\lambda$, from the formula above. This result is obtained from the resulting wave equation which is reduced to a Beltrami equation $\nabla \times \mathbf{E} = -(1/2T)\mathbf{E}$ when the chiral factor T is given by $T = n\hbar / mc$. The chiral Pauli matrices are used to obtain the Dirac Equation.

In the second example, we have found a perfect transformation of two forms of energy (kinetic and field energy) inside the hydrogen atom and the conservation of energy in the system. Then, we have shown that the helical moving electron can be regarded as an inductive particle ("inducton") while atom as a microscopic LC oscillator, then the indeterministic quantum phenomena can be well explained by the deterministic classical theory. For a microscopic photon (electromagnetic wave), the maximum field energy $W_f = \frac{1}{2}\varepsilon_0 \boldsymbol{E}_0^2$ and the maximum magnetic energy $W_m = \frac{1}{2}\mu_0 \boldsymbol{H}_0^2$ are connected iff $\boldsymbol{E} = i\eta \boldsymbol{H}$. The Poynting vector vanishes and the Hydrogen atom does not radiate and it is stable. In particular, with this approach we can show another phenomena such how a pairing Pauli electron can move closely and steadily in a DNA-like double helical electron orbit. Moreover, we can have pointed out that the mass of electron, the intrinsic "electron spin", the Pauli exclusion principle and the Dirac equation are all really the quantum confinement effects of the intrinsic chirality of particles of helical motion produced by electromagnetic fields..

We have shown that the quantum mechanism is nothing but an electromagnetic theory (with the radius of the helical orbit $r_e \to 0$) of the LC/wave-particle duality mixed mechanism. Our mixed mechanics force us to rethink the nature and the nature of physical world. We believe all elementary particles, similar to photon and electron, are only some different types of energy representation.

From our study, it has been shown that the electron follows a perfectly defined trajectory in its motion, which confirms the de Broglie-Bohm's prediction (Bohm, 1952). Also in our work, it is found that the known wave-particle duality can be best manifested by showing that the wave motion associated with a electron is just the phenomenon of its complex helical motion in real space.

It is hoped that this article will encourage electrodynamics course instructors to include a discussion of $\mathbf{E} \parallel \mathbf{H}$ TEM standing waves in atomic systems.

6. References

Bekefi, G. & Barrett, A. H. (1977). *Electromagnetic Vibrations, Waves, and Radiation,* MIT, Cambridge.

Bohm, D. (1952). *Phys. Rev.* 85 166 .

Bohr, N. (1913). *Phil.Mag.* 26, 576 .

Carreyre, T et al. (2000). *Phys. Rev.* C 62.

Cini, M. & Touschek, B. (1958). *Nuovo Cimento* 7, 422-423.

Chu, C. & and Ohkawa, T. (1982). *Phys. Rev. Lett.* 48, 837.

Cook, D. M. (1975). *The Theory of the Electromagnetic Field* Prentice-Hall, Englewood Cliffs, NJ.

de Broglie, L. (1924). *Phil. Mag.* 47, 446.

Dirac, P. A. M. Proc. (1928). *Royal Soc London A* 117, 610-624.

Dirac, P. A. M. Proc. (1931) *Royal Soc London* A 133, 60-72.

Dorling, J. (1970). *Am. J. Phys.* 38, 510-512.

Einstein, A. (1905). *Ann.Phys.*17, 132.

Feshbach, H. (1958). *Ann. Phys.* 5, 357-390.

Feynman, R. P. (1949). *Phys. Rev.* 76, (1949) 749-759.

Foldy, L. L. & Wouthuysen, S. A. (1950). *Phys. Rev.* 78, 29-36.

Fues, E. & Hellmann, H. (1930). *Physikalisehe Zeitschrift* 31, 465-478
Hanl, H. & Papapetrou, A. (1940). *Zeitschrift fur Physik* 1, 6, 153-183.
Heisenberg, W. (1927). *Z. Phys.* 43, 172 (1927).
Huang, X, (2006). arxiv: *physics*/0601169.
Jackson, J. D. (1975). *Classical Electrodynamics,* Wiley, New York, 2nd ed.
Jordan, E. C. & Balmain, K. G. (1968). *Electromagnetic Waves and Radiating Systems,* Prentice-Hall, Englewood Cliffs, NJ, 2nd ed.
Kishimoto, S et al. (2000). *Phys. Rev. Lett.* 85, 1831.
Lock, J. A. (1979). *Am J. Phys.* 47, 797-802.
Lorrain, P. & Corson, D. (1970). *Electromagnetic Fields and Waves,* Freeman, San Francisco, 2nd ed.
Marion J. B. & Herald, M. H.(1980). *Classical Electromagnetic Radiation,*Academic, New York, 2nd ed.
Miller, A. (1994). *Early Quantum Electrodynamics*: A Source Book, Cambridge University Press, Cambridge, 38-40.
Mulligan, B. (2008). *Annals of Physics* 321, 1865-1891.
Neamtan, S. M. (1952). *Am. J. Phys.* 20, 450-451.
Panofsky, W. K. H. & Phillips, M. (1955). *Classical Electricity and Magnetism,* Addison-Wesley, Reading, MA.
Pauli, W. (1924). *Z. Phys.* 31, 373.
Planck, M. (1900). *Ann. Phys.* 1, 69.
Portis, A. M. (1978). *Electromagnetic Fields-Sources and Media,* Wiley, New York.
Purcell, E. M. (1965). *Electricity and Magnetism,* McGraw-Hill, New York.
Ramo, S.; Whinnery, J. R. & Van Duzer, T. (1965). *Fields and Waves in Communication Electronics,* Wiley, New York.
Rao, N. N. (1977). *Elements of Engineering Electromagnetics,* Prentice-Hall, Englewood Cliffs, NJ.
Reines F. & Sobel, W. H. (1974). *Phys. Rev. Lett.* 32, 954.
Reitz, J. R. & Milford, F. J. (1967).*Foundations of Electromagnetic Theory,* Addison-Wesley, Reading, MA, 2nd ed.
Rose, M. E. (1961). *Relativistic Electron Theory,* Wiley, New York.
Schrodinger, E. (1930). *Sitzber. Preuss. Akad. Wiss. Physik-Math Klasse* 24, 418-428.
Shadowitz, A. (1975). *The Electromagnetic Field,* McGraw-Hill, New York.
Smythe, W. R. (1950). *Static and Dynamic Electricity*, McGraw-Hill, New York, 2nd ed.
Stern, O. (1920). *Z. Phys.*2, 49.
Stratton, J. A. (1941). *Electromagnetic Theory*, McGraw-Hill, New York..
Thaller, B. (1992) *The Dirac Equation*, Springer, New York.
Torres-Silva, H. (2008) *Ingeniare.* vol. 16 pp. 24-30, pp. 36-42, pp. 43-47, pp.119-122.
Torres-Silva, H. (2011). *The new unification of gravitation and electromagnetism* , Andros Press, ISBN 978-956-345-037-8, Santiago-Chile.
Weyl, H. (1929 a). *Proc. Natl. Acad. Sci. USA* 15, 323-334.
Weyl, H. (1929 b). *The Rice Institute Phamphlet* 16, 280-295.
Weyl, H. (1929 c). *Zeitsehrift fur Physik* 56, 330-352.
Weyl, H. (1952). *The Theory of Groups and Quantum Mechanics* (translated by H.P. Robertson), Dover, New York, 193, p. 263.

Yang, C. N. (1950). *Phys. Rev.* 77, 242-245.
Zaghloul, H.; Buckmaster, H. A. & Volk, K.(1988). *Am. J. Phys.* 56, 274.

16

Electromagnetic Response of Extraordinary Transmission Plates Inspired on Babinet's Principle

Miguel Navarro-Cía, Miguel Beruete and Mario Sorolla
Millimetre and Terahertz Waves Laboratory, Universidad Pública de Navarra
Spain

1. Introduction

This chapter is devoted to polarization effects arisen from perforated metallic plates exhibiting extraordinary transmission (ET). Setting aside the state-of-the-art of perforated metallic plates, we show that by applying Babinet's principle, subwavelength hole arrays (SHAs) arranged in rectangular lattice can further enhance its potential polarization response. Different perspectives are brought about to describe and understand the particular behaviour of self-complementariness-based SHAs: Babinet's principle, equivalent circuit analysis, retrieved constitutive parameters, etc. Afterwards, we embark on the numerical analysis of stacked self-complementariness-based perforated plates. It is shown the potential of having a birefringent artificial medium behaving like negative and positive effective refractive index for the vertical and horizontal polarization, respectively. All these findings are experimentally demonstrated at millimetre-waves.

2. Background

Since the ending of the Second World War, the electromagnetic (EM) properties of artificial materials (dielectric and magnetic) have been studied by both physics and engineering communities (Collin, 1991). Unlike traditional materials, these man-made composites achieve their EM properties by directly manipulating EM waves, yielding to frequency selective behaviours. This property has a great practical consequence since it is the operative foundation of a wide range of devices such as filters, signal couplers, radiation elements, etc.

2.1 Classical perforated plates: dichroic filters

The analysis of light transmission through holes or gratings has a long history, tracing back to the beginning of the 20th century (Rayleigh, 1907; Synge, 1928; Wood, 1902). However, perforated plates had their golden years in the 50s, 60s and 70s in microwave engineering (Brown, 1953; Chen 1971, 1973; Robinson, 1960) and infrared (Ulrich 1967), when they were exploited as frequency selective surfaces/filters (FSSs) and their 100% transmittance at wavelengths slightly above the period was discussed. All the FSSs considered at that time had aperture sizes considerably large for those frequencies at which there was transmittance, see Fig. 1(left). The principle of operation of those FSSs called dichroic filters is excellently described in (Goldsmith, 1998), where the analysis is discussed under the transmission line formalism (Fig. 1.1(right)): The waveguide defined by the hole is represented as a series admittance Y and the discontinuities between the guide and free-space are modelled with shunt admittances Y_s at each end of the perforated plate. Needless to say, the free-space is represented by shunt admittances Y_{fs}. Nevertheless, this approach inherently ignores diffraction (thus, the in-plane period must be small enough relative to the operating wavelength). As it was pointed out in (Beruete et al., 2006b), this restriction is of major importance in the description of extraordinary transmission phenomena.

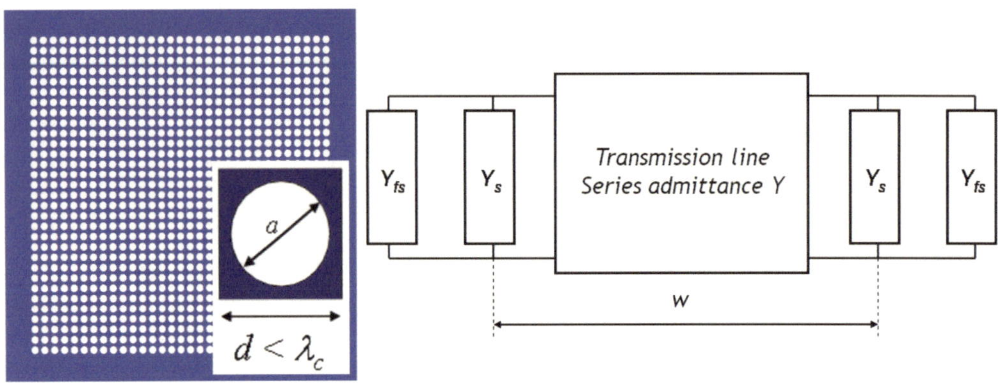

Fig. 1. (left) Schematic of a classical frequency selective surface. Inset: unit cell. (right) Transmission line model to analyse the frequency response of this kind of perforated plates. The length of the waveguide, that is, the plate thickness is denoted as w

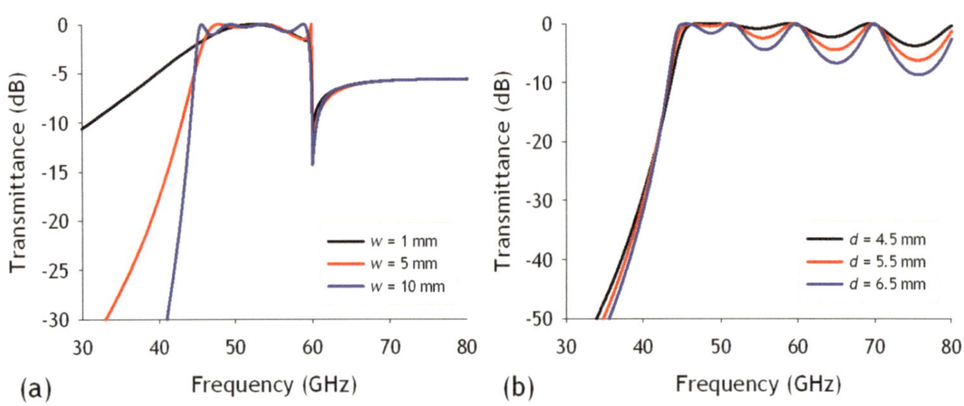

Fig. 2. Frequency response of a dichroic filter with hole diameter a = 4 mm: (a) and in-plane lattice constant d = 5 mm as a function of the waveguide length w = 1, 5 and 10 mm; (b) idem with metal thickness w = 10 mm as a function of the in-plane lattice constant d = 4.5, 5.5 and 6.5 mm applying the equivalent circuit proposed in (Goldsmith, 1998). Note that the Wood's anomaly is not captured with the equivalent circuit

In short, the main features of these dichroic filters are (see Fig. 2):

- The lower limit of the fundamental transmission frequency band is determined by the cut-off frequency of the waveguide defined by the hole, whereas the higher cut-off frequency is governed by the periodicity (Hessel & Oliner, 1965; Rayleigh, 1907; Wood, 1902). At that particular wavelength, a diffracted order that previously contained energy becomes imaginary (goes evanescent), and, for a perfectly conducting grating, the law of conservation of energy demands that the energy previously contained in that order be redistributed among the other remaining diffracted orders.

- The period of the ripples within the passband decreases as the plate thickness increases, because of the increment of electrical length between the guide - free-space discontinuities.
- Also, the longer the plate thickness, the steeper the rise from total reflection to transmission at the lower frequency.
- The larger the in-plane periodicity, the higher the ripple within the passband because the shunt reactance Y_s increases.

Because of the last property, the holes were densely packed at that time (ripple is undesired for technological applications) and this may be one of the reasons of the ignorance of the ET in the electrical engineering community.

2.2 Overcoming Bethe-Bouwkamp's theoretical model

In 1940s Bethe developed a deep study of the field scattered by a single aperture on a thin opaque (perfect electric conductor - PEC) screen by considering it as a magnetic dipole parallel to the screen along with an electric dipole perpendicular to it (the latter one is only required for oblique incidence) (Bethe, 1944). Notice that parallel electric dipoles and perpendicular magnetic dipoles are prohibited by the boundary conditions that impose the parallel electric field and the perpendicular magnetic field to vanish at a perfect electric conductor. He showed that under normal incidence and assuming that the incident intensity is constant over the area of the hole, the transmittance is proportional to $(r/\lambda)^4$ where r is the hole radius. This factor implies extremely low transmission for subwavelength holes. Subsequently, higher-order analytical corrections were introduced by Bouwkamp (Bouwkamp, 1950, 1950b, 1954).

The discovery in 1998 of enhanced light transmission compared to Bethe-Bouwkamp's model in silver plates perforated with periodic subwavelength hole arrays, see Fig. 3(a), launched the field of extraordinary optical transmission (EOT). However, there are signs indicating that the effect was actually discovered during the development of the near-field scanning optical microscopy (NSOM) (Betzig et al., 1986). In 1928 Synge proposed the use of a small aperture close to a sample surface to overcome the diffraction limit imposed by the fact that the propagating waves are limited to $k_x^2 + k_y^2 < (\omega/c)^2$ and thus, the maximum resolution in the image cannot be greater than $\Delta \approx 2\pi/k_{max} = 2\pi c/\omega = \lambda$ (Synge, 1928). This approach was reused by Betzig *et al.* who designed, characterized and implemented hole arrays for the NSOM. During the arrays characterization, transmittance overcoming Bethe's model was observed and tentatively attributed the observation to surface plasmons with no further analysis (Betzig, 1988): *"From the lower graph in Fig. 3.9 it is clear that a significant amount of visible light can be passed through apertures almost as small as λ/10 in diameter, (...) This may once again be a consequence of the finite conductivity and opacity of the screen, or some other phenomenon may be involved which enhances the transport of radiation (e.g. the existence of surface plasmon within the aperture). (...) This provides further circumstantial evidence for hypothesis that some unknown phenomenon is responsible for the high transmission coefficients for the smaller apertures."*

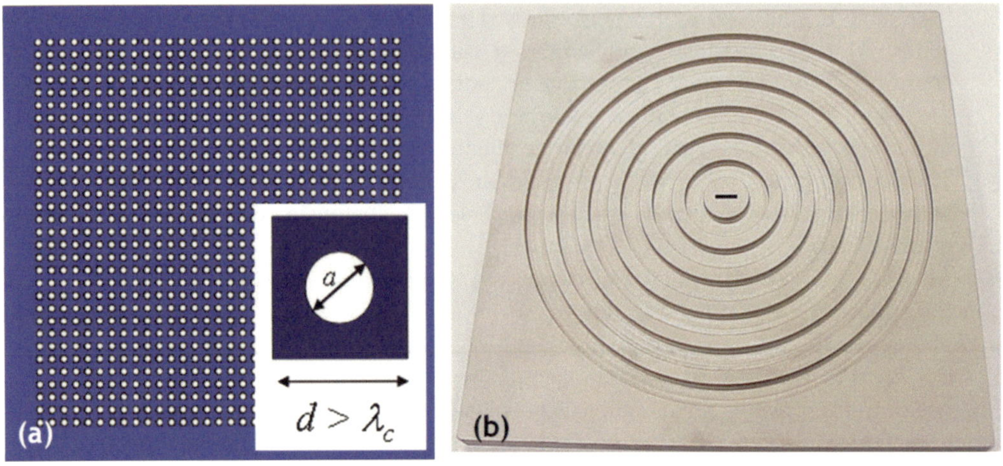

Fig. 3. (a) Schematic of subwavelength hole arrays. Inset: unit cell. (b) Prototype of a slot surrounded by concentric corrugations (Bull's eye configuration)

Nevertheless, it is true that the field took off with Ebbessen and co-workers' deep analyses which brought both theory and experiments to bear on the remarkable conclusion that transmission is mediated via SPPs.

Countless papers followed Ebbesen's seminal paper calling some of them the SPP hypothesis into question or giving other perspectives to the underlying physics of the EOT. Firstly, because of its simplicity, theoretical analyses considered the one-dimensional grating: a periodic array of slits (Porto et al., 1999; Schröter & Heitmann, 1998; Treacy, 1999, 2002). All these works showed that one-dimensional gratings have efficient channels for light transmission because of the non-cut-off slit waveguide that do not exist for hole arrays, so the SPPs theory of one-dimensional slits (although interesting) might not be useful for explaining EOT effect on SHAs.

Subsequently, researchers embarked on the study of the underlying physics of the bigrating. The main works, from the point of view of these authors, are gathered together in the following points according to the different theses proposed.

The first mechanism proposed was the interaction of the apertures with surface plasmons (Barnes et al., 2004; Degiron et al., 2002; Ebbesen et al., 1998; Ghaemi et al. 1998; Krishnan et al., 2001; Martín-Moreno et al., 2001; Salomon et al., 2001). The work of Ebbesen *et al.* was mainly devoted to experimentally study the influence of Wood's anomaly (but not linking it with the resonant peak) (Ghaemi et al. 1998), the dielectric environment (Krishnan et al., 2001), the hole depth (Degiron et al., 2002) and the angle of incidence (Barnes et al., 2004) so as to support their hypothesis based on SPPs. On the other hand, Martín-Moreno, García-Vidal and colleagues (Martín-Moreno et al., 2001) presented a rigorous theory (minimal model) that considered SPPs on infinite grids coupled through an evanescent mode(s) of the hole. Similar line of reasoning was also followed by Salomon *et al.* (Salomon et al., 2001) who presented a near-field picture of the holes to highlight the SPPs Bloch waves as well. Martín-Moreno *et al.* only modelled the response at normal incidence and captured many (but not all) of the prominent features of the experimental results. The minimal model is based on the multiple scattering formalism and assumes the imaginary part of the electric permittivity of the metal $\varepsilon_i = 0$, i.e. no absorption, and one fundamental mode, the least decaying mode inside the subwavelength hole (if the hole is circular, it is the TE_{11}). The main counterintuitive results is that the reflection amplitude is larger than unity at the resonance, but it is explained because the fundamental mode inside the hole is evanescent, for which current conservation only restricts that the imaginary part of the reflection coefficient must be equal or greater than 0, with no constraint on the real part of it. This fact lets the zeroth order transmission coefficient achieve arbitrarily high values. An extension of this model that accounts for the penetration of the EM waves into metals at higher frequencies considers surface impedance boundary conditions (J.D. Jackson, 1999) on the metal-interfaces defining the metal film (but not inside the hole) (Martín-Moreno et al., 2001). An unforeseen consequence of the model of Martín-Moreno *et al.* is that the results may allow EOT in any frequency range because of the PEC boundary conditions (this fact makes all lengths involved in the system scalable), even at frequencies such as microwaves or millimetre-waves (Beruete et al., 2004) where SPPs make no sense. Thus, despite they were still discussing about SPPs, their work was another proof that the physics underlying the mechanism of enhanced transmission through SHAs is more complicated. Needless to say, although all these works are founded on SPPs, they are also considering the periodicity as an important ingredient because the dispersion diagram of SPP has no discontinuities if the interface is smooth. However, when a periodic perforation/corrugation is introduced along the interface, the dispersion relation shows band structure and energy gaps that allow phase-matching between the incident light and the SPPs.

Other mechanism proposed was dynamical light diffraction and Wood's anomaly (García-de-Abajo et al., 2005; Genet et al., 2003; D.R. Jackson et al., 2005; Lezec & Thio, 2004; Lomakin et al., 2004, 2005; Sarrazin et al., 2003, 2005). Genet *et al.* and Sarrazin *et al.* stressed the role of multiple-scattering paths. However, the former emphasized their work on resonant and nonresonant mechanisms from Fano (Fano, 1941) perspective (Genet *et al.*, 2003), whereas the latter devoted their study on Wood's anomalies which are intrinsically related to periodic structures (Sarrazin et al., 2003), and afterwards to Brewster-Zenneck modes (Sarrazin 2005). EOT on nonmetallic systems gave cause for introducing a new model based on interference of diffracted evanescent waves generated by subwavelength marks at the surface of the screen, leading to transmission suppression as well as enhancement (Lezec & Thio, 2004). Babinet's principle was used by García de Abajo and co-workers in order to study light transmission through hole arrays in PEC thin films by relating it to the reflection on planar arrays of metallic disks, which are solved from the multipolar polarizability of single disks (García-de-Abajo et al., 2005). Finally, Jackson *et al.* (D.R. Jackson et al., 2005) and Lomakin *et al.* (Lomakin et al., 2004, 2005) focused their effort on leaky waves formalism and Wood's anomalies (Hessel & Oliner, 1965). Remarkable of this approach is that it gives explanation for any frequency range, which instead of relying on SPPs, it points out the relevance of the periodicity and the surface waves regardless of their exact nature: leaky SPPs (optics) or leaky waves (microwaves and millimetre-waves). Needless to say, it does not account unfortunately for the potential coupling between surface waves of each interface through the metal in layers whose thickness is smaller than the skin depth as it may happen in optics. It is worth recalling that Ulrich already studied hole arrays under the perspective of surface waves, although the meshes were composed of dense-packed holes like the FSSs shown before (Ulrich & Tacke, 1972, 1974).

To bridge the gap between both theories, Pendry *et al.* captured the physics of EOT on SHAs in a unified theoretical framework called spoof surface plasmon (García-Vidal et al., 2005;

Pendry et al., 2004). However, the importance of this theory was not the mimicked response of SPPs by surface waves in PEC screens because this is already implicit in the dynamical diffraction formalism (leaky waves) - and in classical books of microwave engineering (Collin, 1991) -, but to apply the homogenization theory to show that the effective dielectric function of mesoscopic SHAs and SSAs structures resembles a Drude model, whose plasma frequency can be engineered via the geometry of the holes/corrugations alone, allowing the excitation of SPPs-like states in any region of the spectrum where the PEC condition approximates the metallic response.

Despite their differences, the shared line of reasoning of all theories presented before is that the phase-matching of the incident radiation to a surface wave provided by the periodicity is crucial for transmission enhancement via evanescent tunnelling. Therefore, the same process may arise for a single aperture surrounded by a regular array of corrugations (Fig. 3(b)) (Beruete et al., 2004b, 2005; García-Vidal et al., 2003; Grupp et al., 1999; D.R. Jackson, 2005; Lezec et al., 2002; Thio et al., 2001, 2002).

EOT is still a subject of debate, but after clarifying the limits of each theory, all seem to be in perfect agreement with experiment. In addition, equivalent circuit models have emerged that explain the phenomenon from an engineering point of view (Beruete et al., 2011; Medina et al., 2008). This theory is based on the concept of impedance matching model which predicts enhanced transmission in systems that are not periodic at all and where SPPs are not even defined. Remarkable is that it has been experimentally validated in circular waveguide environment with an off-centred diaphragms (Medina et al., 2009). Thus, this is another theory to add to all that have been already proposed to explain the really complex physics of EOT (or ET at frequencies different than visible).

2.3 Stacked subwavelength hole arrays for negative refractive index metamaterials: fishnet-like structure

Some particular relationships among the phenomena electromagnetic/photonic band-gaps, ET and negative refraction were proposed in the past years, i.e., photonic crystals exhibit negative refraction effects (Notomi, 2002) and electromagnetic and photonic band gaps have been reported in SSHAs (Ye & Zhang, 2005). Soon after, the first fabrication and experimental verification of a holey metal-dielectric-metal structure exhibiting a negative refractive index around 2 microns was reported (Zhang et al., 2005b). A numerical demonstration of the negative index of refraction achievable with this structure was presented in (Zhang et al., 2005). Similar results were published in (Dolling et al., 2006).

The full connection among the three phenomena was shown subsequently with periodically stacked metallic plates perforated with SHAs to form a structure showing backward wave propagation within the ET regime (Beruete et al., 2006). It has to be noted that the in-plane periodicity is of the order of free space wavelength, but it is the longitudinal periodicity which determines the kind of propagation. Given the subwavelength stacking essential to achieve backward propagation, the structure can be homogenized along the longitudinal dimension, allowing the use of the metamaterial term in this dimension. This is the key difference between the ET metamaterial (ETM) and the backward propagation observed in standard PBGs (Notomi, 2002) where it takes place above the forbidden frequency gap with a lattice constant to free space wavelength ratio larger than 0.5. On the other hand, the ETM is apparently similar to the realizations of (Zhang et al., 2005b) and (Dolling et al., 2006). Nevertheless, the key difference is that the ETM introduced by Beruete *et al.* (Beruete et al.,

2006) is a real tuneable EBG comprising several periods, whereas the structures presented in (Dolling et al., 2006; Zhang et al., 2005b) consist of an isolated period which cannot account for Floquet-Bloch framework. Moreover, one of the drawbacks of such structures (Dolling et al., 2006; Zhang et al., 2005b) is that they suffer from losses, even for its minimal size structure due to the operating wavelength. The conductor losses are relevant at optical frequencies but not so important at millimetre-waves or even in the infrared. Dielectric losses are less relevant and can be overcome by using air as a dielectric insulator provided membranes are employed as described in (Beruete et al., 2006). Thus, regarding conducting and dielectric losses, not only the free-standing ETM is more advantageous, but also because it is optimized in terms of ET to enhance transmission unlike (Dolling et al., 2006; Zhang et al., 2005b). Recently, magnetic photonic crystals constructed from periodic arrangements of available (possibly anisotropic) homogeneous material layers have been proposed. These structures can exhibit the phenomena of minimal reflection at their interface, large amplitude growth of the harmonic wave within the crystal, and concurrent group velocity slow-down (Mumcu et al., 2006).

3. Babinet's principle, complementariness, and self-complementariness

Prior to any analysis of some curious polarization effects presented here in modified SHAs, we require clarifying some concepts. Let us start with the meaning of complementariness for this work. Consider a free-standing thin metallic screen with air inclusions as sketched on the left-hand side of Fig. 4. The complementary structure is the one shown on the right, where the metallic part has been substituted by air and the apertures by metallic parts. The mathematical description of this concept is fairly simple: if one denotes S_a the set of apertures (and obstructions) on the left, and S_b the one of the right, their union is the completely obstructing surface $S = S_a + S_b$.

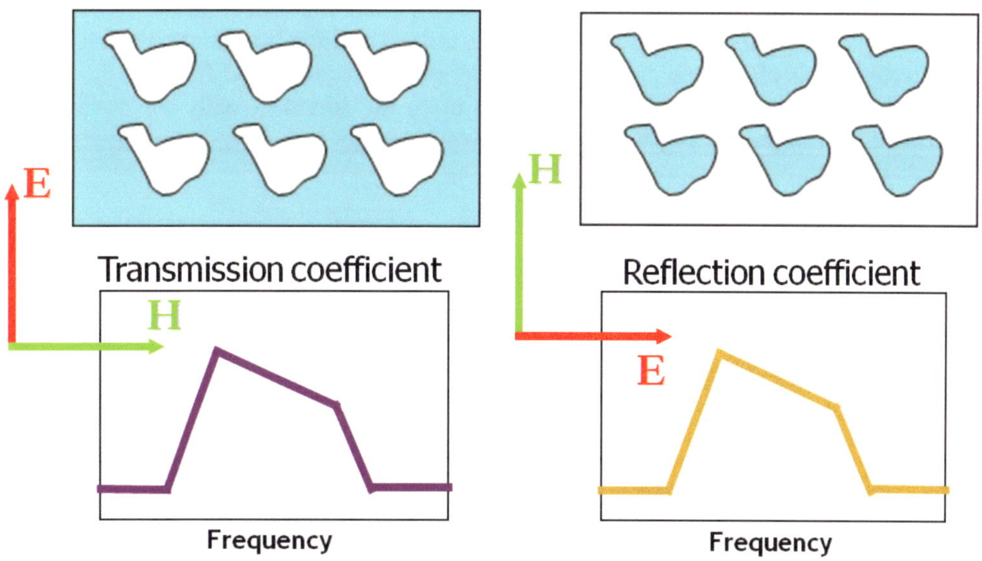

Fig. 4. Example of complementary screens (top sketches) and their electromagnetic response (bottom spectra)

An extension of this idea is the self-complementariness. In this case, the layer has a metal zone congruent to the open area, i.e. the two areas coincide by shift. Thus, the structure is intrinsically composed of its complementary topology, see Fig. 5.

Closely related to the previous concepts, there is an important and common used principle in optics, antenna field and also in metamaterials (Falcone et al., 2004) which establishes the relationship between the diffraction fields of complementary perfectly conducting ($\sigma \to \infty$) infinitely thin ($t \to 0$) screens of infinite extent: the Babinet's principle (Born & Wolf, 1999), (Ishimaru, 1990), (Jackson, 1999). This principle states that the sum of the diffraction fields U_{Sa} and U_{Sb} behind two complementary screens, S_a and S_b, is the field U_0 observed without diffraction objects:

$$U_0 = U_{Sa} + U_{Sb} \tag{1}$$

Fig. 5. Example of self-complementary structure

Rewriting this principle more conveniently for our purposes in terms of electric and magnetic fields, it can be read as follows: if $F = (E, H)$ is the solution of S_a, then $F'(E', H') = (-(\mu/\varepsilon)^{1/2}\cdot H, (\varepsilon/\mu)^{1/2}\cdot E)$ is the solution of S_b. This is indeed nothing but the connection between the reflected fields of S_b for a given polarization with the transmitted fields of S_a, which in plain words says that the reflectance spectra of S_b are identical with the transmittance spectra of S_a, see bottom of Fig. 4, and the other way around.

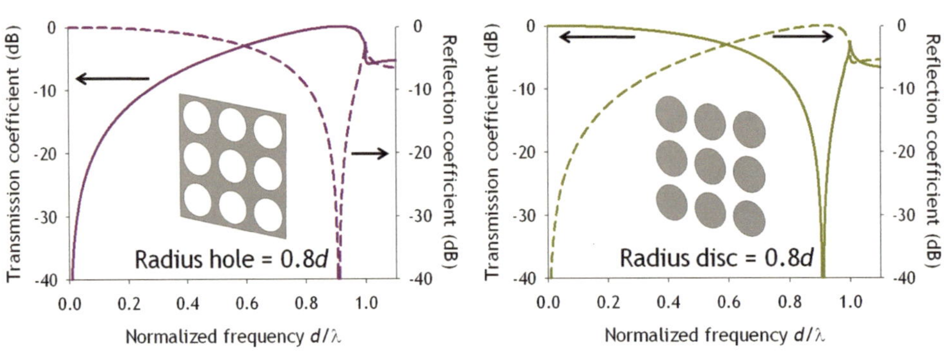

Fig. 6. Scattering parameters in logarithm scale of hole (left) and disk arrays (right) under vertically polarized illumination. (solid line) Transmission coefficient and (dashed line) reflection coefficient

To put this pedagogically in an example within the context of the topic of the chapter, let us show the response of hole and disk arrays infinitely extended transversally and illuminated by a plane-wave under normal incidence. These results have been obtained numerically by the commercial software CST Microwave Studio™. This example will make it easier the description and understanding of the perfect linear polarizer of next section. As Fig. 6 displays, the complementariness of the two structures is completely satisfied in terms of transmission and reflection. Notice how the transmission coefficient of one is the reflection coefficient of the other and the other way around.

If we now take a step further and apply Babinet's principle to a self-complementary screen like the one shown in Fig. 5, we can immediately realize that in the same screen we have both EM responses shown at the bottom of Fig. 4, and thus, a linear polarizer is successfully achieved since for a given polarization there is high transmittance within a certain frequency range, whereas for its orthogonal, there is high reflectance within the same bandwidth.

4. Self-complementary extraordinary transmission layer: perfect linear polarizer

So far, we have given a description of arbitrary screens, but the Babinet's principle may hold evidently for SHAs (Beruete et al., 2007c) such as the doubly periodic arrangement shown in Fig. 7(a). Indeed, by a slight modification of the doubly periodic SHAs, a self-complementary screen is achievable, see Fig. 7(b) where $d_x = a + s$ and $d_y = 2a$ (to fulfil the self-complementary criterion), which should exhibit the linear polarization response described in the previous paragraph. This assumption is confirmed in Fig. 8, where transmittance and reflectance of both SHAs and self-complementary SHAs alongside the SSAs case are plotted. These results have been generated by CST Microwave Studio™ where $s = 0.04d_y$ (needless to say, $d_x = 0.54d_y$, $a = d_y/2$, t infinitely thin and metal modelled as PEC) and we have considered a subwavelength array of slits and holes with or without slits with the periodicity in the x-direction and extending all the way in the y-direction. The remaining z-direction is perpendicular to the plane of array. Apart from the clear self-complementariness of the EM response of the modified SHAs (notice this fact in both transmission and reflection), prominent in Fig. 8(a) is that the transmittance for vertical polarization (E_y) remains unaltered when narrow vertical slits connecting holes are included in the SHAs (note that the solid red trace of Fig. 8(a) is superimposed to the solid black one), whereas the transmittance of the orthogonal polarization (E_x) suffers a strong modification in order to accommodate itself to the self-complementariness principle.

That the vertical polarization remains unaltered by the inclusion of the subwavelength slits may be the manifestation that the current distribution in the vicinity of the hole responsible for the ET suffers no perturbation. Let us then compare the current distribution of simple SHAs and self-complementary ones. As we have already pointed out in this text, the ET is a self-resonance of the hole array, which can be modelled as an LC tank (Beruete et al., 2006; Medina et al., 2008; Ulrich, 1967) and has inductive character for large λ, hence transmittance is negligible when $\lambda \rightarrow \infty$ [see solid red trace of Fig. 8(a)]. Besides, the inset of Fig. 8(a) reinforces the statement about the inductive character of hole arrays in the very subwavelength regime since the phase for vertical polarization tends to + 90 deg as the SSAs (well-known inductive grid) do. Notice in Fig. 9(a) how the conduction currents at ET, which are the main contribution to the inductance – the inductance and the capacitance are balanced, at least in a first order approach (Ulrich, 1967), at the resonance peak –, are concentrated in the narrow wire between holes. Moreover, the current lines rendered in Fig. 9(a) are closed through displacement currents connecting the upper and lower sides of the hole. This description keeps valid when a narrow vertical slit is introduced connecting consecutive holes [see Fig. 7(b) and Fig. 9(b)], and thus, the assumption pointed out at the beginning of the paragraph is confirmed.

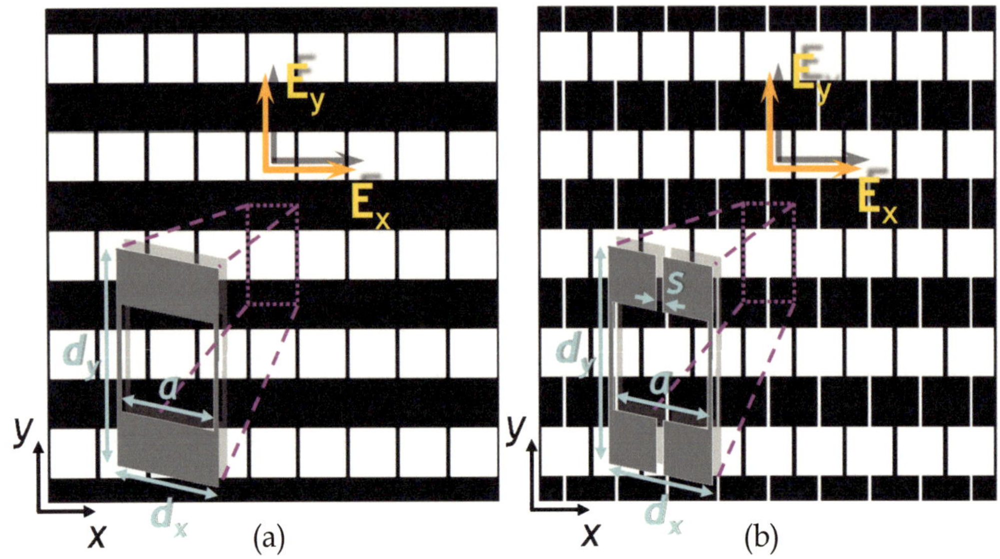

Fig. 7. Sketch of (a) doubly periodic subwavelength hole arrays, and (b) self-complementary subwavelength hole arrays. Insets: unit cell with parameters

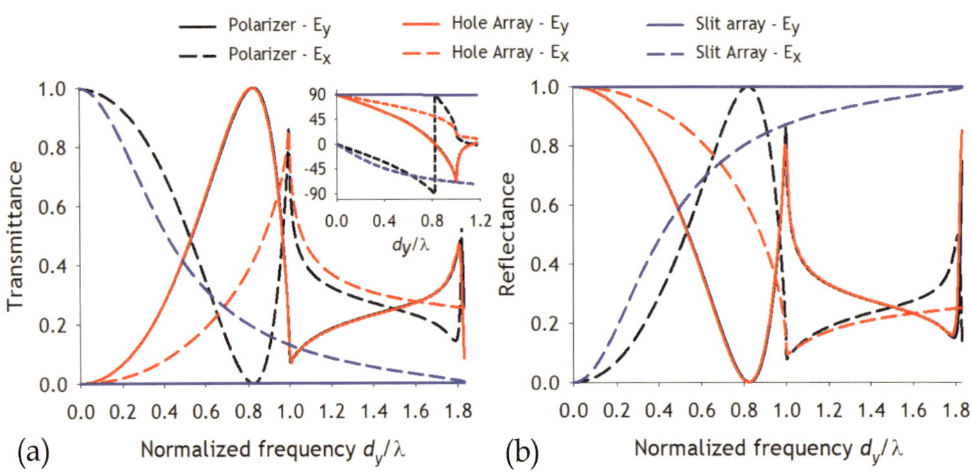

Fig. 8. Simulation results of the polarization dependent transmittance (a) and reflectance (b). Vertical, E_y, polarization and horizontal, E_x, polarization are represented by continuous and dashed traces respectively for vertical slit arrays (blue), hole array (red) and for the proposed polarizer (black). (inset) Transmittance phase

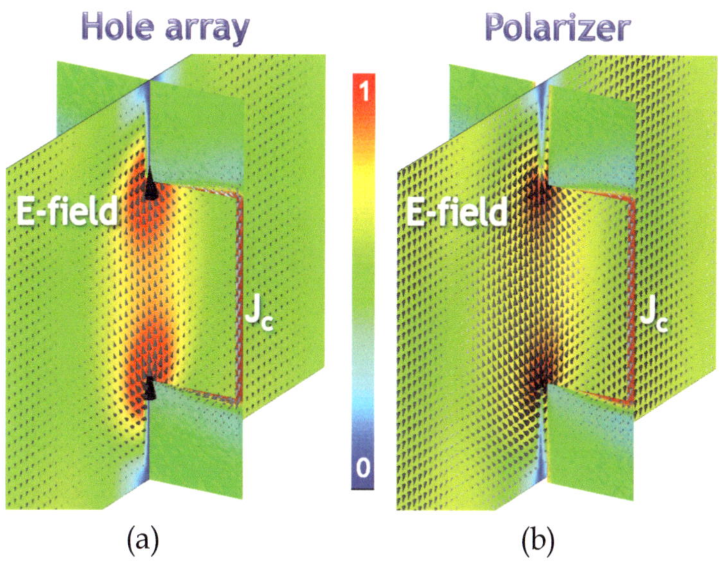

Fig. 9. Plot of the conduction and displacement currents at the ET resonance with vertically polarized (E_y) wave. Hole array case (a) and polarizer (b). The E-field is depicted in a perpendicular yz-cutting plane through the middle of a hole

On the other hand, under the horizontal polarization the EM response of the self-complementary SHAs can be seen as governed by small patch arrays - which are precisely the complementary screen of the SHAs and whose spectrum presents a resonant rejection band at the ET frequency (extraordinary reflection, ER) - connected with narrow wires. Following the same preceding analysis of the current distribution we may arrive at the conclusion that, like slits connecting holes, the connecting narrow wires cause in this case a negligible perturbation on the current distribution and thus, in the response of the small patch arrays.

4.1 Equivalent circuit analysis

An alternative and more quantitative approach to explain the electromagnetic response of the self-complementary SHAs is based on lumped element equivalent circuit and high order mode excitation. This approach is founded on the reduction of the problem of a plane wave normally incident on a periodic structure to a single unit cell in an artificial waveguide (Beruete et al., 2007). When the periodic screen has symmetric elements, some transversal periodic boundary conditions can be substituted by electric or magnetic walls/symmetry planes (assuming linear polarized plane wave normal incidence). With this approach it is possible to compare the performance of ET hole arrays and ET-based polarizers, see Fig. 7.

As it is well-known, transversal electric (TE) modes below cutoff store magnetic energy, whereas transversal magnetic (TM) modes store electric energy. It is precisely this reactive character of evanescent higher order modes added to the divergence obtained at the onset of propagation that gives rise to the ET, as explained in (Beruete et a., 2007; Medina et al., 2008) and in the text below. From this platform and invoking once again Babinet's principle, the ER can also be explained.

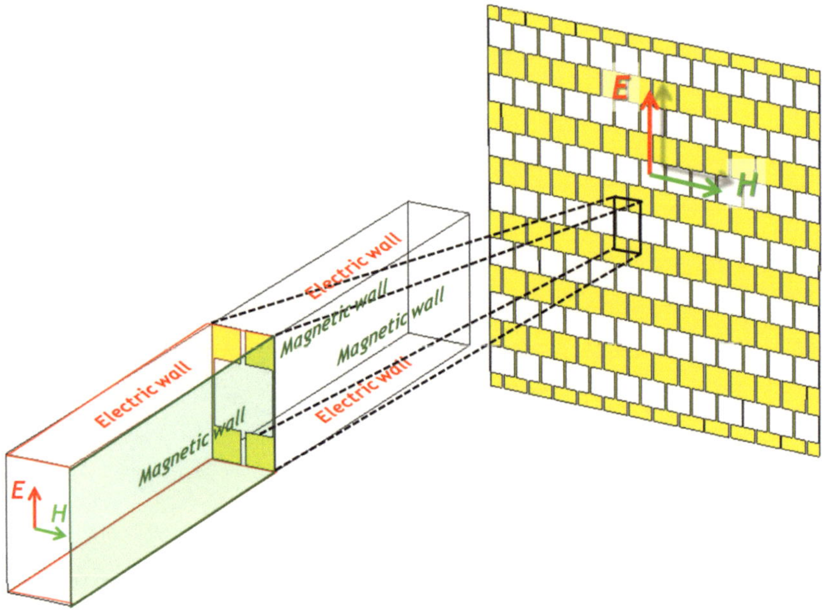

Fig. 10. Planes defining the boundary conditions of the artificial waveguide (vertical polarization)

Starting with the case of vertical polarization, where the polarizer presents an ET-like behaviour, and accepting that the slit does not fundamentally modifies the global performance, the equivalent circuit extracted for ET hole arrays (Beruete et al., 2007; Medina et al., 2008) can be applied without modification. It can be deduced from the dominant modes of the artificial waveguide. Before going to a comprehensive study of the modes, ET can be qualitatively explained using the concept of excess energy stored by evanescent modes (modes below cutoff). More specifically, applying the well-known procedure of component elimination from rotational Maxwell's equations (Jackson, 1999; Ramo et al., 1994) and imposing vertical polarization and the boundary conditions of Fig. 10, the functional expressions of the artificial waveguide modes are easily found:

- TE modes:

$$E_y(x,y) = \frac{j\omega\mu}{k_c^2}\frac{m\pi}{d_x}A^{mn}\cos(m\pi x/d_x)\cos(n\pi y/d_y) \tag{2a}$$

$$H_x(x,y) = \frac{-j\beta}{k_c^2}\frac{m\pi}{d_x}A^{mn}\cos(m\pi x/d_x)\cos(n\pi y/d_y) \tag{2b}$$

$$H_z(x,y) = A^{mn}\sin(m\pi x/d_x)\cos(n\pi y/d_y) \tag{2c}$$

- TM modes:

$$E_y(x,y) = \frac{j\beta}{k_c^2}\frac{n\pi}{d_y}B^{mn}\cos(m\pi x/d_x)\cos(n\pi y/d_y) \tag{3a}$$

$$H_x(x,y)=\frac{j\omega\varepsilon}{k_c^2}\frac{n\pi}{d_y}B^{mn}\cos(m\pi x/d_x)\cos(n\pi y/d_y) \tag{3b}$$

$$E_z(x,y)=B^{mn}\cos(m\pi x/d_x)\sin(n\pi y/d_y) \tag{3c}$$

where β is the propagation constant in the longitudinal direction, A^{mn}, B^{mn} are normalization factors that are found once the power carried by each mode is known, and k_c is the cutoff wavenumber calculated as:

$$k_c=\sqrt{\left(\frac{m\pi}{d_x}\right)^2+\left(\frac{n\pi}{d_y}\right)^2} \tag{4}$$

The symmetry of the problem only allows for even indices: m, n = ±2, ±4, ±6, and so forth. The TEM wave is directly deduced from Eqs. 2 or 3 imposing $m = n = 0$. It is evident here that TE (TM) modes do not admit a solution with $m = 0$ ($n = 0$), because otherwise the wave vanishes. With that restriction in view, it is deduced that the lowest high-order mode of a rectangular unit cell where $d_x < d_y$ is the TM_{02} mode with a cutoff frequency $f_c = c/d_y$, where c is the speed of light in vacuum.
The TM_{02} mode has the next equivalent impedance below cutoff (Medina et al., 2008):

$$Z_{TM_{02}}=-j\eta_0\sqrt{\left(\frac{c}{d_y f}\right)^2-1} \tag{5}$$

where it is explicitly shown the capacitive character of the reactance, which is the fingerprint of TM modes below cutoff. From here, the equivalent capacitance can be deduced (Medina, 2008):

$$C_{TM_{02}}=\frac{A_{TM}}{2\pi f\eta_0\sqrt{\left(\frac{c}{d_y f}\right)^2-1}} \tag{6}$$

The small central aperture below cutoff has an inductive character, as it is well-known from the times of radar development (Collin, 1991; Ramo et al., 1994; Schwinger & Saxon, 1968). This inductance is due to the TE modes below cutoff. A crucial feature of Eq. 6 is that the capacitance of the TM_{02} mode gradually diverges as the frequency approaches the cutoff frequency ($f_c = c/d_y$) and becomes infinite exactly at cutoff. This means that at the cutoff frequency a shunt short-circuit is introduced by the central element and transmission drops to zero. Additionally, as the capacitance diverges to infinite, there must be a frequency where it exactly balances the inductance of the hole and total transmission takes place. This agrees with the fact that at c/d_y there is always a zero of transmission, known as Rayleigh-Wood anomaly in grating theory and that ET resonance happens always at a frequency slightly below that condition (Medina et al., 2008).
A parallel analysis can be done for the polarization where ER occurs. However, Babinet's principle can be invoked to obtain the propagation characteristics of an infinitely thin patch array. As it has been indicated earlier, the reflection coefficient of an infinitely thin and

lossless small patch array obtained by interchanging air and metal in the previous hole array can be obtained from those of the hole array by considering the complementary excitation after the following changes (Marqués et al., 2005):

$$cB_{hole}^{inc} = E_{patch}^{inc} \tag{7a}$$

$$E_{hole}^{inc} = -cB_{patch}^{inc} \tag{7b}$$

So, the transmission coefficient of the patch array are obtained from those of the hole array for the complementary (orthogonal) polarization and the same incidence angle through (Marqués et al., 2005):

$$t_{hole} + t_{patch} = 1 \tag{8}$$

For the sake of completeness, the modal analysis is done in the following. Here, instead of rotating the polarization, we rotate the unit cell so that the previous boundary conditions can be applied to this problem and the same mode formulation is applicable. With this modification in mind, the lowest high-order mode now is transversal electric, the TE_{20} whose cutoff frequency is $f_c = c/d_x$ (note that, as d_x and d_y have interchanged their roles with respect to the previous case, the frequency is the same as before).
The TE_{20} mode has the next equivalent admittance below cutoff (Beruete et al., 2011; Medina et al., 2008):

$$Y_{TE_{20}} = -\frac{j}{\eta_0}\sqrt{\left(\frac{c}{d_x f}\right)^2 - 1} \tag{9}$$

where it is shown the inductive character of the reactance, which is the mark of TE modes below cutoff. From here, an equivalent inductance can be deduced (Beruete et al., 2011; Medina et al., 2008):

$$L_{TE_{20}} = \frac{A_{TM}\eta_0}{2\pi f\sqrt{\left(\frac{c}{d_y f}\right)^2 - 1}} \tag{10}$$

An array of disconnected apertures such as a patch array (recall that we have found that the small channel connecting patches has no effect on the resonance) has capacitive character. Observing that Eq. 10 predicts a divergence of the inductance as the frequency approaches cutoff, then there is a frequency where it fully compensates the capacitance of the patch array. So, apparently the previous argument applies, but there is a subtle modification: now, due to Babinet's principle, the equivalent circuit is a series LC so that when $f = f_c$ the patch array is an open circuit, and the transmission coefficient is unity and at the frequency where inductance and capacitance are matched it is equivalent to a short circuit and there is a null in transmission.

4.2 Toward the experiment

The primary obstacle for testing self-complementary SHAs is the difficulty of producing a free-standing structure while the robustness and stiffness are kept, because there are as many apertures as metal. To overcome this problem, the patterns were milled in the copper layer of the commercial microwave substrate ARLON CUCLAD 250 LX (metallization thickness t = 35 μm, and conductivity $\sigma_{Cu} = 5.8 \cdot 10^7$ S/m) and the structure was sandwiched between two dielectric slabs (dielectric constant ε_r = 2.43 and dielectric thickness h = 0.49 mm) to deal with symmetric geometries. Because of the dielectric embedding, the Babinet's principle is no longer entirely valid and thus, a fine tuning is required. The final in-plane periodicities of the prototype are d_x = 1.8 mm and d_y = 3.8 mm.

In Fig. 11, the simulated frequency response of a pure self-complementary structure (parameters d_x = 2.7 mm, d_y = 5.0 mm, a = 2.5 mm, s = 0.2 mm as shown on the right panel) embedded in the aforementioned dielectric slabs is shown (black traces). Apart from the fact that the ET peak/ER dip moves to 34 GHz, out of our experimental bandwidth that extends from 45 to 260 GHz, which does not diminish the generality of the following discussion, peaks and dips in the frequency response take place at different points. In general, the dielectric loading has different effects on each polarization and the structure is not necessarily self-complementary. This is due to the fact that the dielectric has a finite thickness resulting in different impedance for each polarization. Then, any structure loaded with finite dielectric slabs is not generally self-complementary (García-de-Abajo et al., 2005). Nevertheless, despite the movement from the strict scenario of Babinet's principle, we want to remark that the structure shows localized complementariness in the band of interest (ET/ER resonance) and that the principles governing the ET peak/ER dip are the same as for the ideal free-standing screen.

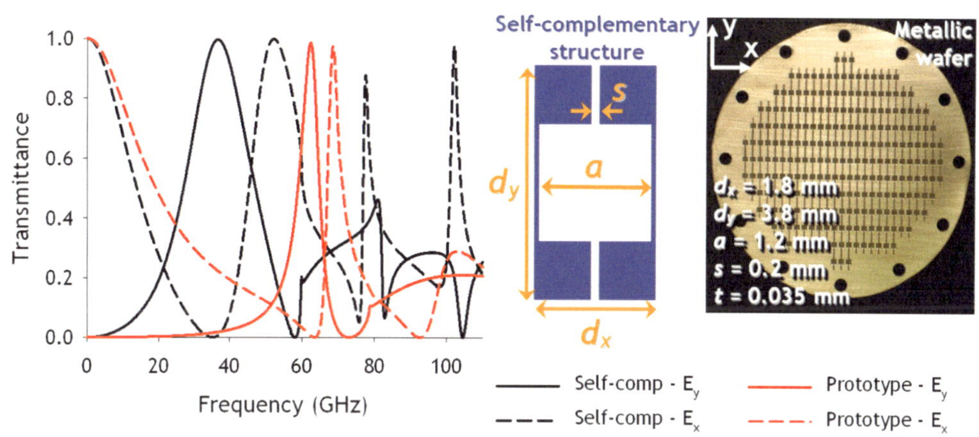

Fig. 11. (Left) Simulated polarization-dependent transmittance of self-complementary (black) and optimized (red) structures embedded in two dielectric slabs (ε_r = 2.43 and thickness of each slab h = 0.49 mm). (Right) Unit cell of a rectangular self-complementary ET polarizer with parameters: d_x = 2.7 mm, d_y = 5.0 mm, a = 2.5 mm, and s = 0.2 mm, and photograph of the fabricated prototype whose parameters are depicted in the figure

Taking the ideas of the previous paragraph into account, along with our fabrication tolerances and our interest in having the ET peak/ER dip and Wood's anomaly within our experimental bandwidth, the fine tuning of the hole array parameters leads to the final prototype shown on the right-hand side of Fig. 11, whose simulated EM response is plotted also in the graph of Fig. 11 (red lines). In addition to the desired frequency shift, note that the peaks have become sharper mainly because the filling factor (aperture-to-unit cell area ratio) has decreased from 0.5 to 0.29.

CST Microwave Studio™ allows launching a circularly polarized wave and then, visualizing the evolution of the electric field when it passes through the polarizer screen (assuming screen of infinite extension). Thus, we take advantage of this option to illustrate in Fig. 12 the polarization filtering of the designed prototype. The incident wave is circularly polarized, but because of the strong reflection of the horizontal component, the pattern seen at the input (top right corner in each panel of Fig. 12) becomes elliptical. This is nicely manifested by comparing both t_0 and $t_0+3\Delta t$ cases, where the horizontal component, E_x, seen in $t_0+3\Delta t$ exceeds in magnitude the electric field component y, E_y, of t_0. On the other hand, the outgoing wave has an ordinary linear (vertical) polarized pattern.

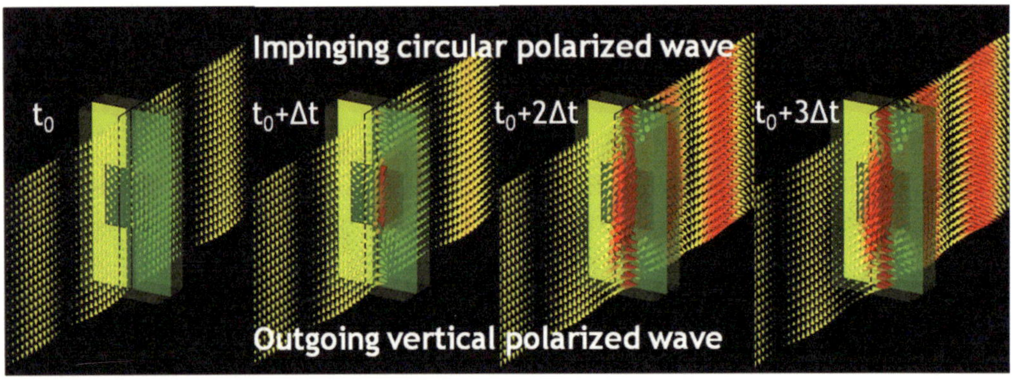

Fig. 12. Time evolution of the electric field for the self-complementary SHAs. From left to right t_0, $t_0+\Delta t$, $t_0+2\Delta t$, and $t_0+3\Delta t$

To finish with the numerical results, the effective parameters: electric permittivity ε, magnetic permeability μ and index of refraction n are retrieved from the complex scattering parameters, which is an established method for the experimental characterization of unknown materials (Ghodgaonkar et al., 1990; Nicolson & Ross, 1970). Notice that, for gratings, this is just an assignation of equivalent material parameters to a thin periodic pattern and lacks of truly physical meaning because they may be changed when several identical layers are stacked together (because of loading effect), for instance. Also, it should be noted that we adopt in this chapter a time dependence of e^{jwt}, which is followed in most engineering disciplines.

It can be shown that the scattering parameters are related to the transmission T and reflection coefficient Γ by the following equations:

$$S_{11} = \frac{\Gamma(1-T^2)}{1-\Gamma^2 T^2} = \frac{\Gamma(1-e^{-j2nk_0 d})}{1-\Gamma^2 e^{-j2nk_0 d}} \tag{11a}$$

$$S_{21} = \frac{T(1-\Gamma^2)}{1-\Gamma^2 T^2} = \frac{(1-\Gamma^2)e^{-jnk_0 d}}{1-\Gamma^2 e^{-j2nk_0 d}} \tag{11b}$$

After some mathematics it can be easily derived from these equations that:

$$z = \pm\sqrt{\frac{(1+S_{11})^2 - S_{21}^2}{(1-S_{11})^2 - S_{21}^2}} \tag{12}$$

$$n = \frac{1}{k_0 d}\left\{\left(\mathrm{Im}\left(\ln\left(e^{-jnk_0 d}\right)\right) + 2m\pi\right) + j\,\mathrm{Re}\left(\ln\left(e^{-jnk_0 d}\right)\right)\right\} \tag{13}$$

where m is an integer related to the branch index of Re(n). Finally, the electric permittivity $\varepsilon_{eff} = n/z$, and the magnetic permeability $\mu_{eff} = n \cdot z$.

The retrieved parameters are displayed in Fig. 13 at those frequencies nearby the ET peak/ER dip where the transmittance is above 0.5. The main features observed on the retrieval are:

- For E_y (vertical) polarization, the effective electric permittivity undergoes a Lorentzian-like dispersion around the ET peak arisen from the resonant character. Moreover, it comes from negative values - not appreciable in Fig. 13 - as it is expected because the structure behaves as a diluted metal (Drude dispersion) for low frequencies. The effective magnetic permeability in turn displays an anti-resonance whose origin is hypothesised to come from the finite size of the unit cell (Koschny et al., 2003). However, recently it has been associated to magneto-electric coupling (Alú, 2010, 2010b). Finally, the effective index of refraction presents a strong dispersion because of the resonant nature, reaching values from 0 up to 2.2.
- For the E_x (horizontal) polarization, the effective electric permittivity, magnetic permeability and index of refraction exhibits similar responses as for E_y, but at higher frequencies (at around Rayleigh-Wood's anomaly (just after the ER dip)). Nevertheless, it should be mentioned that the effective electric permittivity for this polarization cannot be described by a Drude dispersion for low frequencies since the structure is capacitive and supports a propagating mode at DC.

It should be noted that from the effective permittivity and permeability, we could find the parameters of the lumped elements modelling the response of the self-complementary-inspired hole array polarizer through the relationships $Z'=Z/l = j\omega\mu_0\mu_r$ and $Y'=Y/l = j\omega\varepsilon_0\varepsilon_r$, in which l is assumed to be the effective length of the slab. Nevertheless, this is left as an exercise for the reader.

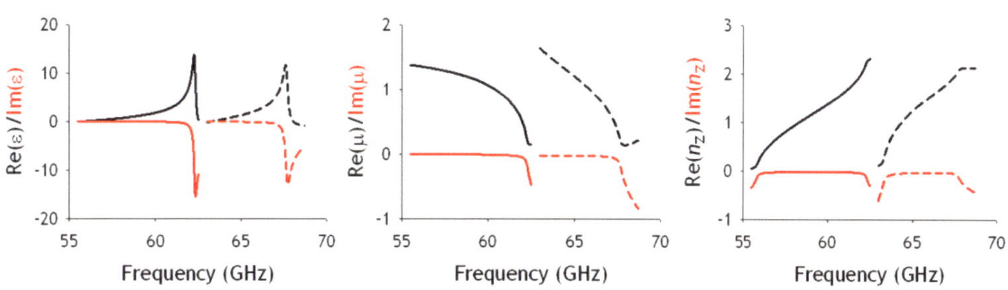

Fig. 13. From left to right: effective electric permittivity, magnetic permeability, and index of refraction. Solid and dashed lines account for E_y (vertical) and E_x (horizontal) polarization, respectively. Black and red lines represent real and imaginary part of the aforementioned constitutive parameters, respectively. The values are only computed within the -3 dB pass-band

Measurements were performed under Fresnel illumination with the AB-Millimetre™ quasi-optical (QO) vector network analyzer by using a QO bench, see Fig. 14. The AB-Millimetre™ is an all-solid-state electronics which generates millimetre-submillimetre waves by frequency multiplication, and detects them by harmonic mixing. The QO setup consists of a corrugated horn antenna that generates a very well linearly polarized Gaussian beam which, after two ellipsoidal mirrors, is focused over the sample under test with a beamwaist of around 28 mm at the frequencies of this work. There, the transmitted and reflected beams are obtained. The transmitted beam passes through another pair of identical mirrors and reaches the receiver antenna. This antenna is another corrugated horn well matched to the Gaussian beam.

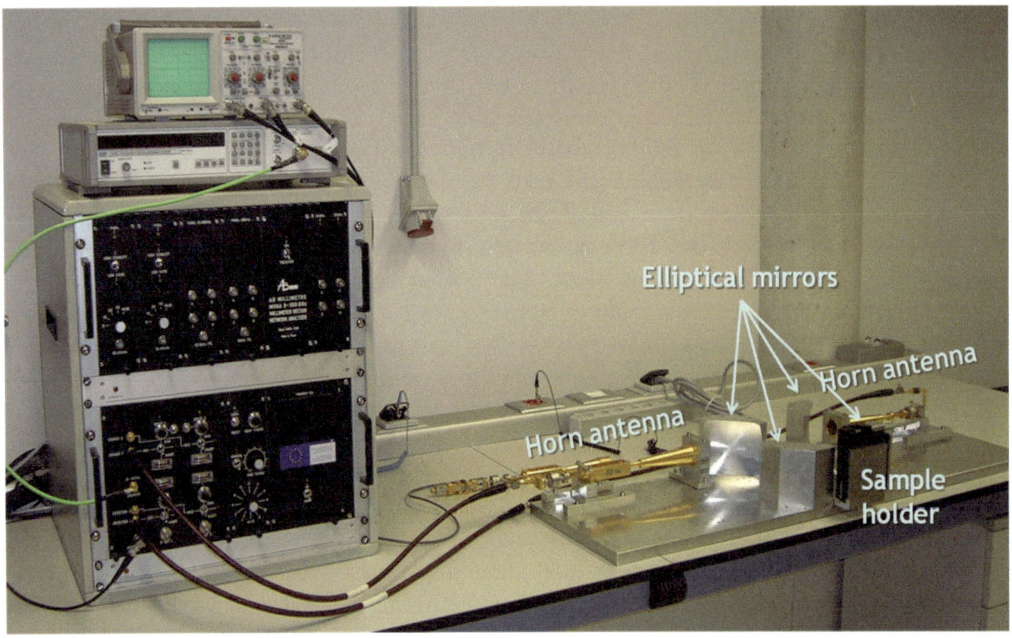

Fig. 14. Photograph of the experimental setup

The experimental results are shown in the black curves of Fig. 15 for the transmission [Fig. 15(a)] and reflection [Fig. 15(b)]. Both vertical polarization, E_y (solid lines), and the orthogonal, E_x (dashed lines) are measured. For incident E_y, a clear resonance around 61 GHz is detected with practically total transmission (0.95), followed by a second relative maximum at 74 GHz (0.12) related to the symmetric distribution of the surface waves at each interface (Kuznetsov et al., 2009) and the null corresponding to the Rayleigh-Wood's anomaly at around 79 GHz. The orthogonal level at 61 GHz is 0.029, whereas it reaches 0.002 at 63 GHz where the power intensity peak of E_y is still above 0.9. Notice also that at 69 GHz, the horizontal polarization achieves a transmittance of 0.81 while the vertical is 0.06, indicating that at this frequency we have as a by-product a polarization filtering behaviour too (for the orthogonal polarization).

As it is anticipated from the nearby Babinet's scenario, there is a relatively good complementary correspondence between the transmission and reflection measured results. The best we can expect with the given dielectric loaded structure is having the peak of transmission and the dip of reflection of the orthogonal components at almost the same frequency.

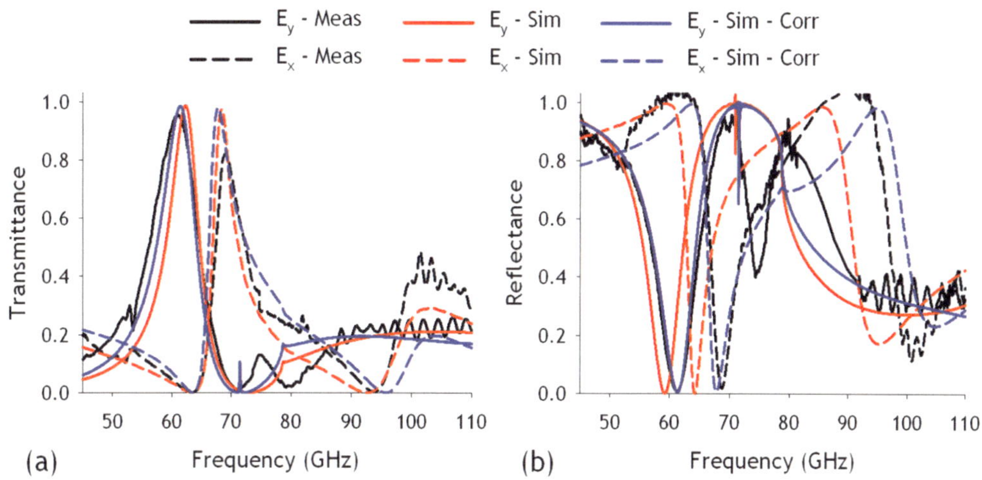

Fig. 15. Measurement (black) and simulation results of the prototype (red) and the prototype with an assembly defect modelled by an air slab of 0.1 mm between the perforated plate and one of the dielectric slabs (blue). Solid trace for E_y and dashed one for E_x: (a) Transmission spectrum for E_y and E_x and (b) Idem for reflection

Figure 15 renders also the numerical results given by CST Microwave Studio™ when the problem is treated as an infinite screen and under an ideal plane wave illumination, see red traces. Despite all the differences between the experimental setup (finite structure and Gaussian beam illumination to name the more prominent) and the simulation, the peak/dip positions and magnitudes are in reasonable agreement. To further approximate to the real problem, we have also explored numerically the possibility that the dielectric slab position is no longer ideal. The motivation of this study emerged after carefully analysing the manufacture and assembly procedures: firstly, the pattern is etched in one of the copper layer of ARLON CUCLAD 250 LX, whereas the other is removed; secondly, both metal layers of another wafer of ARLON are removed; finally, the single dielectric slab is screwed to the previous dielectric-backed metal pattern. Therefore, these steps permit only to ensure perfect contact between the metal layer with one of the dielectric slabs and good contact in the frame of the wafer with the other dielectric slab (see photograph of the prototype in Fig. 11, where there are 12 circular holes on the framework of the wafer for the screwing), but not at the centre. Then, a thin air layer is susceptible of existing between the metal and one slab. To account for this potential imperfection, we have included an air slab of 0.1 mm between the perforated plate and one of the dielectric slabs. The numerical results are shown by blue traces in Fig. 15 and they corroborate that the manufacture process can be the origin of the non-idealities measured.

The results presented up to now confirm the hypothesis of self-complementary structures as a guide method for designing dielectric embedded structures. Had it been a single plate metallic free-standing polarizer, the response would have been similar to the theoretical curves of Fig. 8.

5. Stacked self-complementariness-based extraordinary transmission layers

The next question to be answered is whether the subwavelength-stacking of these modified SHAs would still lead to backward wave propagation (effective negative refractive index) for the vertical polarization as it happens for the fishnet-like metamaterial (Beruete et al., 2006; Dolling et al., 2006; Zhang et al., 2005b). And also, whether the propagation of the horizontal polarization would be inhibited as it happens for the single layer configuration. To answer these question one should recall that Babinet's principle applies for infinitely thin metallic layers, as it was explained in section 3. Thus, the multi-layering of self-complementary extraordinary transmission layers leads certainly to backward wave propagation, but does not have to fulfil Babinet. Therefore, the direct stack of the previous prototypes may or may not arrive at a linear polarizer exhibiting effective negative refractive index. Nevertheless, the authors showed in a previous work that by a smart design one can positively have such linear polarizer (Beruete et al., 2007b).

We can further play around with the unit cell dimensions, and by a new clever design, overlap the first mode of each polarization, leading to a birefringent medium, where the ET axis behaves as an effective negative refractive index and the orthogonal axis as an effective positive refractive index (Beruete et al., 2008).

Here, we investigate in more detail this last design by constructing a wedge, which is a common geometrical approach to check the value of the effective index of refraction of a metamaterial (Navarro-Cía et al., 2008). If the stack behaves indeed as a single negative birefringent medium, one polarization undergoes negative refraction, whereas the orthogonal has positive refraction.

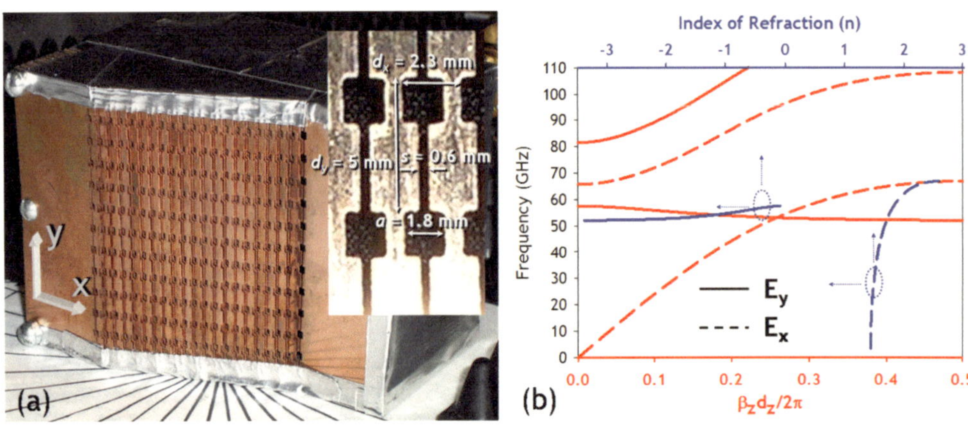

Fig. 16. (a) Picture of the prototype whose angle is 20.28 deg. (b) Dispersion diagram (red curve) and index of refraction (blue curve) - for the first propagating modes - calculated numerically for the infinite stack; solid lines correspond to electric field impinging vertically (E_y), whereas dashed lines represent the response under horizontal polarization (E_x); dotted blue arrows indicate the corresponding axes to the curve

To check this supposition, a prism made of modified SHAs was manufactured (Beruete et al., 2010; Navarro-Cía et al., 2010). A picture of the prototype and zoom-in of the unit cell whose parameters are: transversal periodicities d_x = 2.3 mm, d_y = 5 mm, longitudinal periodicity d_z = 0.85 mm, hole size a = 1.8 mm, slit width s = 0.6 mm and metal (copper)

thickness w = 0.35 mm; are shown in Fig. 16(a). The effective zone of the wedge comprises 21 slits (and the corresponding number of lines of holes) and 11 horizontal rows, which means a 55 mm × 48.3 mm size. On the other hand, the prism has total dimensions of 91 mm × 65 mm × 65 mm/45 mm (width × height × length).
As we have done in previous works (Beruete et al., 2006, 2006b, 2007, 2007b, 2007c, 2008, 2010; Navarro-Cía et al., 2008, 2010), our assumptions in the stack's properties rely on the dispersion diagram. In this case, the dispersion diagram of our present prism is plotted in Fig. 16(b), along with the index of refraction associated to each first couple of propagation modes (one corresponds to vertical polarization, E_y, whereas the other to the orthogonal one, E_x). Notice that in the frequency range between 52 and 57.5 GHz, an overlap of effective negative and positive refractive index propagation modes is expected.
The experimental setup is similar to the one used in preceding works (Beruete et al., 2010; Navarro-Cía et al. 2008, 2010). In short, the prism is illuminated from the back by a corrugated horn antenna and the outgoing beam deflection is scanned at a distance z = 400 mm by another identical corrugated horn antenna. Both antennas are connected to the AB-Millimetre™ QO vector network analyzer. The frequency band and effective aperture used impose that we are dealing with near field or Fresnel zone due to the fact that $2 \cdot D^2/\lambda \approx 1100$ mm, where D is the largest dimension of the prism (Balanis, 2005). The calibration or reference is done by using a face to face antennas configuration.

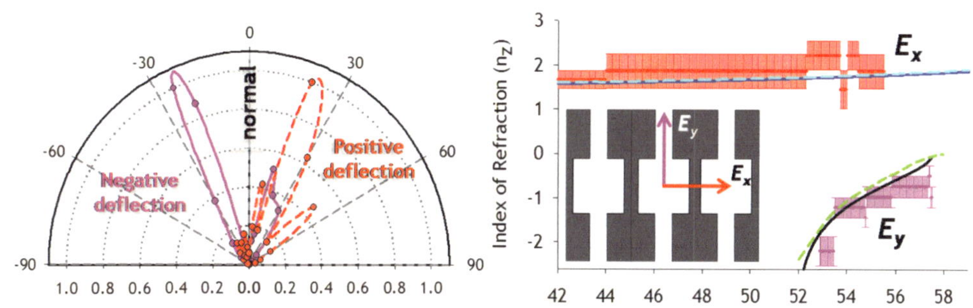

(a) (b)

Fig. 17. (a) Angular power distribution normalized to its maximum for both polarizations at 53.8 GHz. The spots correspond to the measurement results, whereas the curves are a smooth interpolation. (b) Evolution of the refractive index as a function of the frequency for both polarizations. Red and pink spots are the experimental points along with the corresponding error bars, the blue and black, and cyan and green curves are the equivalent simulation results derived from the dispersion diagram and from the standard S-parameter method respectively for each polarization

Figure 17(a) shows undoubtedly the negative and positive deflection experienced by the vertical and horizontal polarized beam respectively at 53.8 GHz, whereas Fig. 17(b) displays a quantitative comparison between numerical and experimental results in terms of effective index of refraction. There, the effective index of refraction derived from the experimental results via Snell's law for both polarizations alongside the experimental error bars are plotted, and the index computed with CST Microwave Studio™ by two different methods are also shown superimposed. Solid lines account for the effective index of refraction derived from the dispersion diagram, whereas dashed lines define the effective index of refraction retrieved from the S-parameters (Ghodgaonkar et al., 1990; Nicolson & Ross, 1970). In the latter procedure, the metal has been modelled by copper and 10 stacked layers have been considered as a representative number of stacked layers of the real prototype. Remarkable is the good agreement between numerical and experimental results despite the fact already pointed out regarding near-field measurement and finite structure.

6. Conclusion

The EM properties of SHAs arranged in rectangular lattice can be enhanced by applying Babinet's concepts. Indeed, by inserting vertical subwavelength slit arrays connecting holes, a self-complementary structure is obtained whose EM response in terms of transmittance and reflectance fulfil Babinet's principle. This feature leads to a perfect linear polarizer: the vertical polarization undergoes ET (total transmission in the lossless case), whereas the orthogonal polarization exhibits extraordinary reflection (total reflection in the lossless case) at the same frequency. The qualitatively explanation founded on Babinet's principle has been complemented with a description based on lumped element equivalent circuit and high order mode excitation. Furthermore, an inversion technique has been used to retrieve the effective constitutive EM parameters (effective electric permittivity, effective magnetic permeability and effective index of refraction) that model the response of the grating. In addition, the induced surface currents as well as the electric field evolution through the grating have been represented to pedagogically support the conclusions derived from Babinet's principle and self-complementariness. As a natural extension, it has been analysed briefly the propagation behaviour of stacked self-complementariness-based ET layers. It has been shown that for a certain design parameters it is possible to have an effective birefringent medium whose vertical and horizontal axis behaves as an effective negative and positive refractive index, respectively. Finally, the theoretical and numerical analyses for single- and multi-layer have been confirmed experimentally at millimetre-waves.

7. Acknowledgment

We wish to thank a whole host of our friends and colleagues for many useful discussions. They are Dr. Igor Campillo, Universidad de Deusto, Spain; Dr. Francisco Falcone, Universidad Pública de Navarra, Spain; Professor Vitaliy Lomakin, University of California San Diego, United States; and Dr. Philippe Goy, AB Millimetre, France.
Work supported by Spanish Government under contract Consolider "ENGINEERING METAMATERIALS" CSD2008-00066.

8. References

Alú, A. (December 2010). First-Principle Homogenization Theory for Periodic Metamaterial Arrays, In: *arXiv,* Available from: http://arxiv.org/abs/1012.1351

Alú, A. (December 2010). Restoring the Physical Meaning of Metamaterial Constitutive Parameters, In: *arXiv,* Available from: http://arxiv.org/abs/1012.1353

Balanis, C.A. (2005). *Antenna Theory Analysis and Design,* Wiley-Interscience, ISBN 978-047-1667-82-7, Hoboken (NJ), USA

Barnes, W.L., Murray, W.A., Dintinger, J., Devaux, E., & Ebbesen, T.W. (2004). Surface plasmon polaritons and their role in the enhanced transmission of light through periodic arrays of subwavelength holes in a metal film, *Physical Review Letters,* Vol. 92, No. 10, (March 2004), pp. 107401-1-4

Beruete, M., Sorolla, M., Campillo, I., Dolado, J.S., Martín-Moreno, L., Bravo-Abad, J., & García-Vidal, F.J. (2004). Enhanced millimetre-wave transmission through subwavelength hole arrays, *Optics Letters,* Vol. 29, No. 21, (November 2004), pp. 2500-2502

Beruete, M., Campillo, I., Dolado, J.S., Rodríguez-Seco, J.E., Pere, E., & Sorolla, M. (2004b). Enhanced microwave transmission and beaming using a subwavelength slot in corrugated plate, *IEEE Antennas and Wireless Propagation Letters,* Vol. 3, No. 16, (December 2004), pp. 328-331

Beruete, M., Sorolla, M., Campillo, I., & Dolado, J.S. (2005). Subwavelength slotted Corrugated plate with enhanced quasioptical millimeter wave transmission, *IEEE Microwave and Wireless Components Letters,* Vol. 15, No. 4, (April 2005), pp. 286-288

Beruete, M., Sorolla, M., & Campillo, I. (2006). Left-handed extraordinary optical transmission through a photonic crystal of subwavelength hole arrays, *Optics Express,* Vol. 14, No. 12, (June 2006), pp. 5445-5455

Beruete, M. (2006b). *Millimeter-Wave Extraordinary Transmission: Connection to Metamaterials and Technological Applications,* Doctoral thesis

Beruete, M., Campillo, I., Navarro-Cía, M., Falcone, F., & Sorolla, M. (2007). Molding Left- or Right-Handed Metamaterials by Stacked Cut-Off Metallic Hole Arrays, *IEEE Transactions on Antennas and Propagation, Special issue in honour or Prof. L.B. Felsen,* Vol. 55, No. 6, (June 2007), pp. 1514-1521

Beruete, M., Navarro-Cía, M., Sorolla, M., & Campillo, I. (2007b). Polarized left-handed extraordinary optical transmission of subterahertz waves, *Optics Express,* Vol. 15, No. 13, (June 2007), pp. 8125-8134

Beruete, M., Navarro-Cía, M., Campillo, I., Goy, P., & Sorolla, M. (2007c). Quasioptical Polarizer Based on Self-Complementary Sub-Wavelength Hole Arrays, *IEEE Microwave and Wireless Components Letters,* Vol. 17, No. 12, (December 2007), pp. 834-836

Beruete, M., Navarro-Cía, M., Sorolla, M., & Campillo, I. (2008). Polarization selection with stacked hole array metamaterial, *Journal of Applied Physics,* Vol. 103, No. 5, (March 2008), pp. 053102-1-4

Beruete, M., Navarro-Cía, M., Falcone, F., Campillo, I., & Sorolla, M. (2010). Single negative birefringence in stacked spoof plasmon metasurfaces by prism experiment, *Optics Letters,* Vol. 35, No. 5, (March 2010), pp. 643-645

Beruete, M., Navarro-Cía, M., Kuznetsov, S.A., & Sorolla, M. (2011). Circuit approach to the minimal configuration of terahertz anomalous extraordinary transmission, *Applied Physics Letters,* Vol. 98, No. 1, (January 2011), pp. 014106-1-3

Bethe, H.A. (1944). Theory of diffraction by small holes, *Physical Review,* Vol. 66, No. 7-8, (October 1944), pp. 163-182

Betzig, R.E., Lewis, A., Harootunian, A., Isaacson, M., & Kratschmer, E. (1986). Near-field scanning optical microscopy (NSOM): Development and Biophysical Applications, *Biophysical Journal,* Vol. 49, No. 1, (January 1986), pp. 269-279

Betzig, R.E., (1988). *Nondestructive optical imaging of surfaces with 500 angstrom resolution,* Doctoral thesis

Born, M. & Wolf, E. (1999). *Principles of Optics: Electromagnetic Theory of Propagation, Interference and Diffraction of Light,* Cambridge University Press, ISBN 978-052-1642-22-4, Cambridge, UK

Bouwkamp, C.J. (1950). On Bethe's theory of diffraction by small holes, *Philips Research Reports,* Vol. 5, No. 5, (1950), pp. 321-332

Bouwkamp, C.J. (1950b). On the diffraction of electromagnetic waves by small circular disks and holes, *Physical Review,* Vol. 5, No. 6, (1950), pp. 401-422

Bouwkamp, C.J. (1954). Diffraction theory, *Reports on Progress in Physics,* Vol. 17, No. 1, (January 1954), pp. 35-100

Brown, J. (1953). Artificial dielectric having refractive indices less than unity, *Proceedings of the IEE,* Vol. 100, No. 5, (October 1953), pp. 51-62

Chen, C.C. (1971). Diffraction of electromagnetic waves by a conducting screen perforated periodically with circular holes, *IEEE Transactions on Microwave Theory and Techniques,* Vol. 19, No. 5, (May 1971), pp. 475-481

Chen, C.C. (1973). Transmission of microwave through perforated flat plates of finite thickness, *IEEE Transactions on Microwave Theory and Techniques,* Vol. 21, No. 1, (January 1973), pp. 1-7

Collin, R.E. (1991). *Field Theory of Guided Waves,* Wiley-IEEE Press, ISBN 978-087-9422-37-0, Piscataway (NJ), USA

Degiron, A., Lezec, H.J., Barnes, W.L., & Ebbesen, T.W. (2002). Effects of hole depth on enhanced light transmission through subwavelength hole arrays, *Applied Physics Letters,* Vol. 81, No. 23, (October 2002), pp. 4327- 4329

Dolling, G., Enkrich, C., Wegener, M., Soukoulis, C.M., & S. Linden. (2006). Simultaneous negative phase and group velocity of light in a metamaterial, *Science,* Vol. 312, No. 5775, (May 2006), pp. 892- 894

Ebbesen, T.W., Lezec, H.J., Ghaemi, H., Thio, T., & Wolf, P.A. (1998). Extraordinary optical transmission through sub-wavelength hole arrays, *Science,* Vol. 391, No. 6668, (February 1998), pp. 667-669

Falcone, F., Lopetegi, T., Laso, M.A.G., Baena, J.D., Bonache, J., Beruete, M., Marqués, R., Martín, F., & Sorolla, M. (2004). Babinet principle applied to metasurface and metamaterial design, *Physical Review Letters,* Vol. 93, No. 19, (November 2004), pp. 197401-1-4

Fano, U. (1941). The theory of anomalous diffraction grating and of quasistationary waves on metallic surfaces (Sommerfield's waves), *Journal of Optical Society of America,* Vol. 31, No. 3, (March 1941), pp. 213-222

García-de-Abajo, F.J., Gómez-Medina, R., & Sáenz, J.J. (2005). Full transmission through perfect-conductor subwavelength hole arrays, *Physical Review E,* Vol. 72, No. 1, (July 2005), pp. 016608-1-4

García-Vidal, F.J., Lezec, H.J., Ebbesen, T.W., & Martín-Moreno, L. (2003). Multiple paths to enhance optical transmission through a single subwavelength slit, *Physical Review Letters*, Vol. 90, No. 21, (May 2003), pp. 213901-1-4

García-Vidal, F.J., Martín-Moreno, L., & Pendry, J.B. (2005). Surfaces with holes in them: new plasmonic metamaterials, *Journal of Optics A: Pure and Applied Optics*, Vol. 7, No. 2, (January 2005), pp. S97-S101

Genet, C., van Exter, M.P., & Woerdman, J.P. (2003). Fano-type interpretation of red shifts and red tails in hole array transmission spectra, *Optics Communications*, Vol. 225, No. 4-6, (October 2003), pp. 331-336

Ghaemi, H.F., Thio, T., Grupp, D.E., Ebbesen, T.W., & Lezec, H.J. (1998). Surface plasmons enhance optical transmission through sub-wavelength holes, *Physical Review B*, Vol. 58, No. 11, (September 1998), pp. 6779-6782

Ghodgaonkar, D.K., Varadan, V.V., & Varadan, V.K. (1990). Free-Space Measurement of Complex Permittivity and Complex Permeability of Magnetic Materials at Microwave Frequencies, *IEEE Transactions on Instrumentation and Measurements*, Vol. 39, No. 2, (April 1990), pp. 387-394

Goldsmith, P.F. (1998). *Quasioptical Systems: Gaussian Beam, Quasioptical Propagation, and Applications*, Wiley-IEEE Press, ISBN 978-078-0334-39-7, Piscataway (NJ), USA

Grupp, D.E., Lezec, H.J., Thio, T., & Ebbesen, T.W. (1999). Beyond the Bethe limit: Tunable enhanced light transmission through a single subwavelength aperture. *Advanced Materials*, Vol. 11, No. 10, (July 1999), pp. 860-862

Hessel, A. & Oliner A.A. (1965). A new theory of Wood's anomalies on optical gratings. *Applied Optics*, Vol. 4, No. 10, (October 1965), pp. 1275-1297

Ishimaru, A. (1990). *Electromagnetic Wave Propagation, Radiation, and Scattering*, Prentice hall, ISBN 978-013-2490-53-5, Boston, USA

Jackson, J.D. (1999). *Classical Electrodynamics*, John Wiley & Sons, ISBN 978-047-1309-32-1, New York, USA

Jackson, D.R., Oliner, A.A., Zhao, T., & Williams, J.T. (2005). The beaming of light at broadside through a subwavelength hole: Leaky-wave model and open stopband effect. *Radio Science*, Vol. 40, No. 6, (September 2005), pp. 1-7

Koschny, Th, Markos, P., Smith, D.R., & Soukoulis, C.M. (2001). Resonant and antiresonant frequency dependence of the effective parameters of metamaterials. *Physical Review E*, Vol. 68, No. 6, (December 2003), pp. 065602-1-4

Krishnan, A., Thio, T., Kim, T.J., Lezec, H.J., Ebbesen, T.W., Wolff, P.A., Pendry, J.B., Martín-Moreno, L., & García-Vidal, F.J. (2001). Evanescently coupled resonance in surface plasmon enhanced transmission. *Optics Communications*, Vol. 200, No. 1-6, (October 2001), pp. 1-7

Kuznetsov, S.A., Navarro-Cía, M., Kubarev, V.V., Gelfand, A.V., Beruete, M., Campillo, I., & Sorolla, M. (2009). Regular and anomalous Extraordinary Optical Transmission at the THz-gap. *Optics Express*, Vol. 17, No. 14, (July 2009), pp. 11730-11738

Lezec, H.J., Degiron, A., Deveaux, E., Linke, R.A., Martín-Moreno, L., García-Vidal, F.J., & Ebbesen, T.W. (2002). Beaming light from a sub-wavelength aperture. *Science*, Vol. 297, No. 5582, (August 2002), pp. 820-822

Lezec, H.J., & Thio, T. (2004). Diffracted evanescent wave model for enhanced and suppressed optical transmission through subwavelength hole arrays. *Optics Express,* Vol. 12, No. 16, (August 2004), pp. 3629-3651

Lomakin, V., Chen, N.W., Li, S.Q., & Michielssen, E. (2004). Enhanced transmission through two-period arrays of sub-wavelength holes. *IEEE Microwave and Wireless Components Letters,* Vol. 14, No. 7, (July 2004), pp. 355-357

Lomakin, V., & Thio, T. (2005). Enhanced transmission through metallic plates perforated by arrays of subwavelength holes and sandwiched between dielectric slabs. *Physical Review B,* Vol. 71, No. 23, (June 2005), pp. 235117-1-10

Marqués, R., Baena, J.D., Beruete, M., Falcone, F., Lopetegi, T., Sorolla, M., Martín, F., & García, J. (2005). Ab initio Analysis of Frequency Selective Surfaces Based on Conventional and Complementary Split Ring resonators. *Journal of Optics A: Pure and Applied Optics,* Vol. 7, No. 2, (January 2005), pp. S38-S43

Martín-Moreno, L., García-Vidal, F.J., Lezec, H.J., Pellerin, K.M., Thio, T., Pendry, J.B., & Ebbesen, T.W. (2001). Theory of extraordinary optical transmission through subwavelength hole arrays. *Physical Review Letters,* Vol. 86, No. 6, (February 2001), pp. 1114-1117

Medina, F., Mesa, F., & Marqués, R. (2008). Extraordinary Transmission Through Arrays of Electrically Small Holes From a Circuit Theory Perspective, *IEEE Transactions on Microwave Theory and Techniques,* Vol. 56, No. 12, (December 2008), pp. 3108-3120

Medina, F., Ruiz-Cruz, J.A., Mesa, F., Rebollar, J.M., Montejo-Garai, J.R., & Marqués, R. (2009). Experimental verification of extraordinary transmission without surface plasmons, *Applied Physics Letters,* Vol. 95, No. 7, (August 2009), pp. 071102-1-3

Mumcu, G., Sertel, K., & Volakis, J.L. (2006). Miniature antennas and arrays embedded within magnetic photonic crystals, *IEEE Antennas and Wireless Propagation Letters,* Vol. 5, No. 1, (December 2006), pp. 168-171

Navarro-Cía, M., Beruete, M., Sorolla, M., & Campillo, I. (2008). Negative refraction in a prism made of stacked subwavelength hole arrays. *Optics Express,* Vol. 16, No. 2, (January 2008), pp. 560-566

Navarro-Cía, M., Beruete, M., Falcone, F., Sorolla, M., & Campillo, I. (2010). Polarization-tunable negative or positive refraction in self-complementariness-based extraordinary transmission prism. *Progress In Electromagnetics Research,* Vol. 103, (April 2010), pp. 101-114

Nicolson, A.M., & Ross, G.F. (1970). Measurement of the Intrinsic Properties of Materials by Time-Domain Techniques, *IEEE Transactions on Instrumentation and Measurements,* Vol. 19, No. 4, (November 1970), pp. 377-382

Notomi, M. (2002). Negative refraction in photonic crystals, *Optical and Quantum Electronics,* Vol. 34, No. 1-3, (January 2002), pp. 133-143

Pendry, J.B., Martín-Moreno, L., & García-Vidal, F.J. (2004). Mimicking surface plasmons with structured surfaces. *Science,* Vol. 305, No. 5685, (August 2004), pp. 847-848

Porto, J.A., García-Vidal, F.J., & Pendry, J.B. (1999). Transmission resonances on metallic gratings with very narrow slits. *Physical Review Letters,* Vol. 83, No. 14, (October 1999), pp. 2845-2848

Ramo, S., Whinnery, J.R., & Van Duzer, T. (1994). *Fields and Waves in Communication Electronics,* John Wiley & Sons, ISBN 978-047-1585-51-3, New York, USA

Rayleigh, Lord (1907). On the dynamical theory of gratings. *Proceedings of the Royal Society of London A*, Vol. 79, No. 532, (August 2007), pp. 399-416

Robinson, L.A. (1960). Electrical properties of metal-loaded radomes. *Wright air Develop. Div. Rep. WADD-TR-60-84*, Vol. 60-84, (February 1960), pp. 1-114

Salomon, L., Grillot, F.D., Zayats, A.V., & de Fornel, F. (2001). Near-field distribution of optical transmission of periodic subwavelength holes in a metal film, *Physical Review Letters,* Vol. 86, No. 6, (February 2001), pp. 1110- 1113

Sarrazin, M., Vigneron, J.P., & Vigoureux, J.M. (2003). Role of Wood anomalies in optical properties of thin metallic films with a bidimensional array of subwavelength holes, *Physical Review B,* Vol. 67, No. 8, (February 2003), pp. 085415-1-8

Sarrazin, & M., Vigneron, J.P. (2005). Light transmission assisted by Brewster-Zenneck modes in chromium films carrying a subwavelength hole array, *Physical Review B,* Vol. 71, No. 7, (February 2005), pp. 075404-1-5

Schröter, U., & Heitmann, D. (1998). Surface-plasmon enhanced transmission through metallic gratings, *Physical Review B,* Vol. 58, No. 23, (December 1998), pp. 15419-15421

Schwinger, J., & Saxon, D. (1968). *Discontinuities In Waveguides – Notes on Lectures by Julian Schwinger,* Gordon and Breach, ISBN 978-067-7018-40-9, New York, USA

Synge, E.H. (1928). A suggested model for extending microscopic resolution into the ultra-microscopic region, *Philosophical Magazine,* Vol. 6, No. 35, (August 1928), pp. 356-362

Thio, T., Pellerin, K.M., Linke, R.A., Ebbesen, T.W., & Lezec, H.J. (2001). Enhanced light transmission through a single sub-wavelength aperture, *Optics Letters,* Vol. 26, No. 24, (December 2001), pp. 1972-1974

Thio, T., Lezec, H.J., Ebbesen, T.W., Pellerin, K.M., Lewen, G.D., Nahata, A., Linke, R.A. (2002). Giant optical transmission of sub-wavelength apertures: physics and applications, *Nanotechnology,* Vol. 13, No. 3, (June 2002), pp. 429-432

Treacy, M.M.J. (1999). Dynamical diffraction in metallic optical gratings, *Applied Physics Letters,* Vol. 75, No. 5, (June 1999), pp. 606-608

Treacy, M.M.J. (2002). Dynamical diffraction explanation of the anomalous transmission of light through metallic gratings, *Physical Review B,* Vol. 66, No. 19, (November 2002), pp. 195105-1-11

Ulrich, R. (1967). Far-infrared properties of metallic mesh and its complementary structure, *Infrared Physics,* Vol. 7, No. 1, (March 1967), pp. 37-55

Ulrich, R., & Tacke, M. (1972). Submillimeter waveguide on periodic metal structure, *Applied Physics Letters,* Vol. 22, No. 5, (March 1973), pp. 251-253

Ulrich, R. (1974). Modes of propagation on an open periodic waveguide for the far infrared, *Proceedings of Symposium Optical and Acoustical Microelectronics,* pp. 359-376, New York, USA, April 16-18, 1974

Wood, R.W. (1902). On a remarkable case of uneven distribution of light in a diffraction grating spectrum, *Proceedings of the Physical Society of London,* Vol. 18, No. 1, (June 1902), pp. 269-275

Ye, Y.H., & Zhang, J.Y. (2005). Enhanced light transmission through cascaded metal films perforated with periodic hole arrays, *Optics Letters,* Vol. 30, No. 12, (June 2005), pp. 1521-1523

Zhang, S., Fan, W., Malloy, K.J., & Brueck, S.R.J. (2005). Near-infrared double negative metamaterials, *Optics Express,* Vol. 13, No. 13, (June 2005), pp. 4922-4930

Zhang, S., Fan, W., Panoiu, N.C., Malloy, K.J., Osgood, R.M., & Brueck, S.R.J. (2005b). Experimental demonstration of near-infrared negative-index metamaterials, *Physical Review Letters,* Vol. 95, No. 13, (September 2005), pp. 137404-1-4

17

The Influence of Vacuum Electromagnetic Fluctuations on the motion of Charged Particles

Guozhong Wang
Wenzhou University
China

1. Introduction

Lorentz-Dirac equation (LDE) is the widely accepted classical equation to describe the motion of a scalar point charge acted by external electromagnetic fields and its own radiating fields. Because LDE is highly nonlinear, the ubiquitous electromagnetic fluctuating fields of the vacuum would produce a nonzero contribution to the motion of charged particles, which provides a promising way to understand the century-old problems and puzzles associated with uniformly accelerating motion of a point charge. The vacuum fluctuations also have an intimate relationship with the Unruh effect. In this chapter we will restrict our main stamina to investigate the influence of the vacuum electromagnetic fluctuating fields on the motion of a point charge. Because the discussion upon problems, such as the Unruh effect etc., would greatly digress from the motif of this book, we just disperse brief remarks on these problems at suitable places.

2. New reduction of order form of LDE

In 1938, Dirac for the first time systematically deduced the relativistic equation of motion for a radiating point charge in his classical paper [Dirac, 1938]. Being a singular third order differential equation, the controversy about the validity of LDE has never ceased due to its intrinsic pathological characteristics, such as violation of causality, nonphysical runaway solutions and anti-damping effect etc. [Wang et al., 2010]. All these difficulties of LDE can be traced to the fact that its order reduces from three to two as the Schott term is neglected. However, LDE derived by using the conservation laws of momentum and energy is quite elegant in mathematics and is of Lorentz invariance. Furthermore, many different methods used to derive the equation of motion for a radiating point charge lead to the same equation, and all pathological characteristics of LDE would disappear in its reduction of order form. Plass invented the backward integration method for scattering problems, and H. Kawaguchi et al. constructed a precise numerical integrator of LDE using Lorentz group Lie algebra property [Plass, 1961; Kawaguchi et al., 1997]. These methods are enough to numerically study the practical problems. On the other hand, Landau and Lifshitz obtained the reduction of order form of LDE [Landau & Lifshitz, 1962], which fully meets the requirements for dynamical equation of motion and is even recommended to substitute for the LDE. But one should keep in mind that Landau and Lifshitz equation (LLE) gaining the advantages over LDE is at the price of losing the orthogonality of four-velocity and four-acceleration of point charges. This complexion means that LDE is still the most qualified equation of motion for a radiating point charge. To be clear, we make the assumption that LDE is the exact equation of motion for a radiating point charge.

2.1 Description of reduction of order form of LDE

For a point charge of mass m and charge e, LDE reads

$$\ddot{x}^{\mu} = \frac{e}{m} F^{\mu\nu}\dot{x}_{\nu} + \tau_0(\dddot{x}^{\mu} + \ddot{x}^2\dot{x}^{\mu}), \tag{1}$$

where $x^{\mu}(\tau)$ is the spacetime coordinates of the charge at proper time τ, $\tau_0 = 2e^2/3m$ the characteristic time of radiation reaction which approximately equals to the time for a light to transverse across the classical radius of a massive charge. The upper dots denote the derivative with respect to the proper time, Greek indices μ, ν etc. run over from 0 to 3. Repeated indices are summed tacitly, unless otherwise indicated. The diagonal metric of Minkowski spacetime is $(1,\ -1,\ -1,\ -1)$. For simplicity, we work in relativistic units, so that the speed of light is equal to unity. The second term of the right side of Eq. (1) is referred to as the radiation reaction force, and $\dddot{x}^{\mu}$ is the so called Schott term.

We have assumed in Eq. (1) that charged particles interact only with electromagnetic fields $F^{\mu\nu}$ which has the matrix expression:

$$\begin{bmatrix} 0 & -E_1 & -E_2 & -E_3 \\ E_1 & 0 & -B_3 & B_2 \\ E_2 & B_3 & 0 & -B_1 \\ E_3 & -B_2 & B_1 & 0 \end{bmatrix}.$$

The Lorentz force is

$$eF^{\mu\nu}\dot{x}_{\nu} = e(\vec{E}\cdot\dot{\vec{x}},\ \dot{x}_0\vec{E} + \dot{\vec{x}}\times\vec{B}) = e\gamma(\vec{E}\cdot\frac{d\vec{x}}{dt},\ \vec{E} + \frac{d\vec{x}}{dt}\times\vec{B}), \tag{2}$$

where $\gamma = (1-\dot{x}^2)^{-1/2}$ is the relativistic factor.

By replacing the acceleration in the radiation reaction force with that produced only by external force, Landau and Lifshitz obtained the reduction of order form of LDE

$$\ddot{x}^{\mu} = f^{\mu}(\tau, x, \dot{x}) + \tau_0[\frac{df^{\mu}}{d\tau} + f^2\dot{x}^{\mu}], \tag{3}$$

where $f^{\mu}(\tau, x, \dot{x}) = eF^{\mu\nu}\dot{x}_{\nu}/m$. Eq. (3) is nonsingular and gets rid of most pathological characteristics of LDE, but the applicable scope is also slightly reduced. LLE is quite convenient to numerically study macroscopic motions of a point charge.

If one does not care about the complexity, there exists another more accurate reduction of order form of LDE than LLE, which also implies a corresponding reduction of order series form of LDE. In this section, we will present this reduction of order form of LDE. To do so, the most important step is using the acceleration produced only by external forces to approximate the Schott term, namely

$$\ddot{x}^\mu = f^\mu(\tau,\vec{x},\dot{\vec{x}}) + \tau_0[\dddot{x}^\mu + \ddot{x}^2\dot{x}^\mu] \approx F^\mu + \tau_0[\ddot{x}^\nu \frac{\partial f^\mu}{\partial \dot{x}^\nu} + \ddot{x}^2\dot{x}^\mu], \tag{4}$$

where

$$F^\mu = f^\mu + \tau_0(\frac{\partial f^\mu}{\partial \tau} + \dot{x}^\nu \frac{\partial f^\mu}{\partial x^\nu})$$

is the analytical function of proper time, spacetime coordinates and four-velocity. Letting $k = \ddot{x}^2$, Eq. (4) becomes

$$\ddot{x}^\mu - \tau_0\ddot{x}^\nu \partial f^\mu / \partial \dot{x}^\nu = F^\mu + \tau_0 k \dot{x}^\mu ,$$

which can be put into the matrix form $A\ddot{x} = F + \tau_0 k\dot{x}$, with $F = (F^0,\ F^1,\ F^2,\ F^3)^T$ and $\dot{x} = (\dot{x}^0,\ \dot{x}^1,\ \dot{x}^2,\ \dot{x}^3)^T$. The symbol " T " denotes transpose operation of a matrix. The explicit expression of matrix A is

$$\begin{bmatrix} 1-\tau_0\partial f^0/\partial\dot{x}^0 & -\tau_0\partial f^0/\partial\dot{x}^1 & -\tau_0\partial f^0/\partial\dot{x}^2 & -\tau_0\partial f^0/\partial\dot{x}^3 \\ -\tau_0\partial f^1/\partial\dot{x}^0 & 1-\tau_0\partial f^1/\partial\dot{x}^1 & -\tau_0\partial f^1/\partial\dot{x}^2 & -\tau_0\partial f^1/\partial\dot{x}^3 \\ -\tau_0\partial f^2/\partial\dot{x}^0 & -\tau_0\partial f^2/\partial\dot{x}^1 & 1-\tau_0\partial f^2/\partial\dot{x}^2 & -\tau_0\partial f^2/\partial\dot{x}^3 \\ -\tau_0\partial f^3/\partial\dot{x}^0 & -\tau_0\partial f^3/\partial\dot{x}^1 & -\tau_0\partial f^3/\partial\dot{x}^2 & 1-\tau_0\partial f^3/\partial\dot{x}^3 \end{bmatrix},$$

each element of matrix A is the analytical function of τ , x and $\dot{x}$. We define generalized four-velocity and four-acceleration vectors as

$$\dot{X}^\mu = (\frac{\Delta_0^{\dot{x}}}{\Delta}, \frac{\Delta_1^{\dot{x}}}{\Delta}, \frac{\Delta_2^{\dot{x}}}{\Delta}, \frac{\Delta_3^{\dot{x}}}{\Delta}),\ D^\mu = (\frac{\Delta_0^F}{\Delta}, \frac{\Delta_1^F}{\Delta}, \frac{\Delta_2^F}{\Delta}, \frac{\Delta_3^F}{\Delta}),$$

where Δ is the determinant of matrix A , and Δ_μ^F and $\Delta_\mu^{\dot{x}}$ are determinants of matrices obtained by replacing the μ-*th* column of A with column matrices F and $\dot{x}$ respectively. So four-acceleration can be expressed as

$$\ddot{x}^\mu = D^\mu + \tau_0 k \dot{X}^\mu . \tag{5}$$

Because the square of the four-acceleration k is involved, Eq. (5) is still not the explicit expression of acceleration. However, k can be expressed as

$$k = (D^\mu + \tau_0 k\dot{X}^\mu)(D_\mu + \tau_0 k \dot{X}_\mu) = \tau_0^2 \dot{X}^2 k^2 + 2\tau_0 k D^\mu \dot{X}_\mu + D^2 , \tag{6}$$

which is a quadratic algebra equation of k . The physical solution of k is

$$k = \frac{2D^2}{(1-2\tau_0 D^\mu \dot{X}_\mu) + \sqrt{(1-2\tau_0 D^\mu\dot{X}_\mu)^2 - 4\tau_0^2\dot{X}^2 D^2}} . \tag{7}$$

Thus we obtained the expression of acceleration, and the result is

$$\ddot{x}^\mu = D^\mu + \frac{2\tau_0 D^2 \dot{X}^\mu}{(1-2\tau_0 D^\mu\dot{X}_\mu) + \sqrt{(1-2\tau_0 D^\mu \dot{X}_\mu)^2 - 4\tau_0^2 \dot{X}^2 D^2}} , \tag{8}$$

which is now the explicit function of proper time τ, spacetime coordinates x and velocity $\dot{x}$. As a corollary, we can discuss the applicable scope of LDE from the existing condition of the solution of k, namely, the quantity under the square root appeared in Eq. (7) must be nonnegative. It is often taken for granted that LDE would be invalid at the scale of the Compton wavelength of the charge.

Following the above procedure, we can construct an iterative reduction of order form of LDE, which is a more accurate approximation to the original LDE. As the first step, we approximately expressed LDE as

$$\ddot{x}^{\mu} = f^{\mu}(\tau, x, \dot{x}) + \tau_0[\dddot{x}^{\mu} + \ddot{x}^2\dot{x}^{\mu}] \approx G^{\mu} + \tau_0 k\dot{x}^{\mu},$$

where

$$G^{\mu} = f^{\mu} + \tau_0(\frac{\partial\ddot{x}^{\mu}}{\partial\tau} + \dot{x}^{\nu}\frac{\partial\ddot{x}^{\mu}}{\partial x^{\nu}} + \ddot{x}^{\nu}_{known}\frac{\partial\ddot{x}^{\mu}}{\partial\dot{x}^{\nu}})$$

is the function of τ, x and $\dot{x}$ owing to the four-acceleration $\ddot{x}^{\nu}_{known}$ being taken as that given by Eq. (8). From its definition, the square of four-acceleration k can be worked out

$$k = \frac{2G^2}{(1-2\tau_0 G\cdot\dot{x}) + \sqrt{(1-2\tau_0 G\cdot\dot{x})^2 - 4\tau_0^2 G^2}},$$

and the four-acceleration is

$$\ddot{x}^{\mu} = G^{\mu} + \frac{2\tau_0 G^2\dot{x}^{\mu}}{(1-2\tau_0 G\cdot\dot{x}) + \sqrt{(1-2\tau_0 G\cdot\dot{x})^2 - 4\tau_0^2 G^2}}.$$

By repeating this procedure, we can obtain the *n-th* iterative expression of four-acceleration:

$$\ddot{x}^{\mu}_n = G^{\mu}_n + \frac{2\tau_0 G_n^2\dot{x}^{\mu}}{(1-2\tau_0 G_n\cdot\dot{x}) + \sqrt{(1-2\tau_0 G_n\cdot\dot{x})^2 - 4\tau_0^2 G_n^2}}, \tag{9}$$

where

$$G^{\mu}_n = f^{\mu} + \tau_0\frac{d\ddot{x}^{\mu}_{n-1}}{d\tau} = f^{\mu} + \tau_0(\frac{\partial\ddot{x}^{\mu}_{n-1}}{\partial\tau} + \dot{x}^{\nu}\frac{\partial\ddot{x}^{\mu}_{n-1}}{\partial x^{\nu}} + \ddot{x}^{\nu}_{n-1}\frac{\partial\ddot{x}^{\mu}_{n-1}}{\partial\dot{x}^{\nu}}).$$

We emphasize again that $\ddot{x}^{\mu}_0$ is taken as that given by Eq. (8). Hereto we have obtained the iterative self-contained reduction of order form of LDE.

As an example, we apply the new reduction of order iterative form of LDE to a special case, a point charge undergoing one-dimensional uniformly accelerating motion along x^1 direction acted by a constant electric field E. Assuming that the ratio of charge e to mass m is one, the equations of motion are

$$\begin{aligned} \ddot{x}^0 &= E\dot{x}^1 + \tau_0[\dddot{x}^0 + \ddot{x}^2\dot{x}^0] \approx E\dot{x}^1 + \tau_0[E\ddot{x}^1 + k\dot{x}^0] \\ \ddot{x}^1 &= E\dot{x}^0 + \tau_0[\dddot{x}^1 + \ddot{x}^2\dot{x}^1] \approx E\dot{x}^0 + \tau_0[E\ddot{x}^0 + k\dot{x}^1] \end{aligned}.$$

It is easy to obtain the expressions of $\dot{x}_0$ and $\dot{x}_1$, they are

$$\ddot{x}^0 = \frac{E\dot{x}^1 + \tau_0 E^2 \dot{x}^0 + \tau_0 k(\dot{x}^0 + \tau_0 E\dot{x}^1)}{1-(\tau_0 E)^2},$$
$$\ddot{x}^1 = \frac{E\dot{x}^0 + \tau_0 E^2 \dot{x}^1 + \tau_0 k(\dot{x}^1 + \tau_0 E\dot{x}^0)}{1-(\tau_0 E)^2}.$$

From its definition $k = \ddot{x}^2 = (\ddot{x}^0)^2 - (\ddot{x}^1)^2$, we obtain $k = -E^2$. The expression of four-acceleration is extremely simple, namely $\ddot{x}^0 = E\dot{x}^1,\ \ddot{x}^1 = E\dot{x}^0$. It is obvious that further iterative procedures will not bring any changes, which completely coincides with LDE for this motion. It is astonishing that a point charge undergoing one-dimensional uniformly accelerating motion emitting energy and momentum does not suffer radiation reaction force at all, which induces a series of puzzling problems lasted for over one hundred years since the radiation fields of this motion had been calculated [Born, 1909, as cited in Fulton & Rohrlich, 1960; Lyle, 2008].

We would like to make a brief remark on Eq. (8). Our approximate procedure contains all effects of the second term of radiation reaction force, so our result is more accurate than LLD. This conclusion is also embodied by its Taylor series form on τ_0 which includes infinite terms, while LLE only includes the linear term of τ_0.

It is easy for one to utilize the method of Landau and Lifshitz to construct a reduction of order iterative form of LDE [Aguirregabiria, 1997]. To compare two different reduction of order forms of LDE is the main content of the next subsection.

2.2 Reduction of order form of LDE up to τ_0^2 term

We know that the quantity τ_0 characterizing the radiation reaction effect is an extremely small time scale $(10^{-24}s)$, so every piece involved in Eq. (8) could be expanded as power series of the parameter τ_0. We are just interested in the first three terms of this series form of LDE, which is accurate enough to study practical problems and making the comparison between two series forms of LDE obtained respectively by Landau and Lifshitz's method and ours meaningful. To get this series form up to τ_0^2 term, we first expand matrix A to τ_0^2 term, and the result is

$$\Delta \approx 1 - \tau_0 \frac{\partial f^\mu}{\partial \dot{x}^\mu} + \tau_0^2 \left(\frac{\partial f^\mu}{\partial \dot{x}^\mu}\frac{\partial f^\nu}{\partial \dot{x}^\nu} - \frac{\partial f^\mu}{\partial \dot{x}^\nu}\frac{\partial f^\nu}{\partial \dot{x}^\mu} \right) + \cdots. \tag{10}$$

For the calculation of the four-vector D^μ, it is adequate to calculate its zero-th component and retain the result to τ_0^2 term

$$D^0 \approx F^0 + \tau_0 F^\mu \frac{\partial f^0}{\partial \dot{x}^\mu} + \tau_0^2 F^\mu \frac{\partial f^\nu}{\partial \dot{x}^\mu}\frac{\partial f^0}{\partial \dot{x}^\nu} + \cdots.$$

The space component expressions of four-vector D^μ can be obtained from the Lorentz covariance, and the result is

$$D^\mu \approx F^\mu + \tau_0 F^\alpha \frac{\partial f^\mu}{\partial \dot{x}^\alpha} + \tau_0^2 F^\alpha \frac{\partial f^\beta}{\partial \dot{x}^\alpha}\frac{\partial f^\mu}{\partial \dot{x}^\beta} + \cdots. \tag{11}$$

Due to the same footing as D^{μ}, the generalized four-velocity vector $\dot{X}^{\mu}$ in Eq. (8) can be immediately written out by changing F^{μ} in Eq. (11) to $\dot{x}^{\mu}$, namely

$$\dot{X}^{\mu}=\dot{x}^{\mu}+\tau_0\dot{x}^{\alpha}\frac{\partial f^{\mu}}{\partial\dot{x}^{\alpha}}+\tau_0^2\dot{x}^{\alpha}\frac{\partial f^{\beta}}{\partial\dot{x}^{\alpha}}\frac{\partial f^{\mu}}{\partial\dot{x}^{\beta}}+\cdots. \tag{12}$$

Then we need to calculate the square of four-acceleration k. Eqs. (8) and (9) show that it is enough for the expansion of k to retain the τ_0 term,

$$\begin{aligned}k&=\frac{2D^2}{(1-2\tau_0D^{\mu}\dot{X}_{\mu})+\sqrt{(1-2\tau_0D^{\mu}\dot{X}_{\mu})^2-4\tau_0{}^2\dot{X}^2D^2}}\\&\approx\frac{D^2}{(1-2\tau_0D^{\mu}\dot{X}_{\mu})}\approx D^2(1+2\tau_0D^{\mu}\dot{X}_{\mu})\\&=f^2+2\tau_0[f^2f^{\mu}\dot{x}_{\mu}+f_{\mu}(\frac{\partial f^{\mu}}{\partial\tau}+\dot{x}^{\nu}\frac{\partial f^{\mu}}{\partial x^{\nu}}+f^{\nu}\frac{\partial f^{\mu}}{\partial\dot{x}^{\nu}})]\end{aligned}\quad. \tag{13}$$

So we obtian the four-acceleration accurate to τ_0 term is

$$\begin{aligned}\ddot{x}^{\mu}&=D^{\mu}+\tau_0k\dot{X}^{\mu}=f^{\mu}+\tau_0(\frac{\partial f^{\mu}}{\partial\tau}+\dot{x}^{\nu}\frac{\partial f^{\mu}}{\partial x^{\nu}}+f^{\nu}\frac{\partial f^{\mu}}{\partial\dot{x}^{\nu}})+\tau_0f^2\dot{x}^{\mu}\\&=f^{\mu}+\tau_0(\frac{\partial f^{\mu}}{\partial\tau}+\dot{x}^{\nu}\frac{\partial f^{\mu}}{\partial x^{\nu}}+f^{\nu}\frac{\partial f^{\mu}}{\partial\dot{x}^{\nu}}+f^2\dot{x}^{\mu})+\cdots\end{aligned}, \tag{14}$$

which is the same as LLE.

We need to calculate the result of once iteration to obtain the τ_0^2 term of the series form of the acceleration. All involved calculation is straightforward but cumbersome. Retaining to the τ_0^2 term, the four-vector G^{μ} is

$$\begin{aligned}G^{\mu}&=f^{\mu}+\tau_0(\frac{\partial\ddot{x}^{\mu}}{\partial\tau}+\dot{x}^{\nu}\frac{\partial\ddot{x}^{\mu}}{\partial x^{\nu}}+\ddot{x}^{\nu}\frac{\partial\ddot{x}^{\mu}}{\partial\dot{x}^{\nu}})\\&=f^{\mu}+\tau_0[\frac{\partial(f^{\mu}+\tau_0\ddot{x}_1^{\mu})}{\partial\tau}+\dot{x}^{\nu}\frac{\partial(f^{\mu}+\tau_0\ddot{x}_1^{\mu})}{\partial x^{\nu}}+(f^{\nu}+\tau_0\ddot{x}_1^{\nu})\frac{\partial(f^{\mu}+\tau_0\ddot{x}_1^{\mu})}{\partial\dot{x}^{\nu}}]\\&=f^{\mu}+\tau_0[\frac{\partial f^{\mu}}{\partial\tau}+\dot{x}^{\nu}\frac{\partial f^{\mu}}{\partial x^{\nu}}+f^{\nu}\frac{\partial f^{\mu}}{\partial\dot{x}^{\nu}}]+\tau_0^2[\frac{\partial\ddot{x}_1^{\nu}}{\partial\tau}+\dot{x}^{\nu}\frac{\partial\ddot{x}_1^{\mu}}{\partial x^{\nu}}+f^{\nu}\frac{\partial\ddot{x}_1^{\mu}}{\partial\dot{x}^{\nu}}+\ddot{x}_1^{\nu}\frac{\partial f^{\mu}}{\partial\dot{x}^{\nu}}]\end{aligned}\quad. \tag{15}$$

We then calculate the square of four-acceleration k and the result is

$$\begin{aligned}k&=G^2(1+2\tau_0G\cdot\dot{x})=(f^{\mu}+\tau_0[\frac{\partial f^{\mu}}{\partial\tau}+\dot{x}^{\nu}\frac{\partial f^{\mu}}{\partial x^{\nu}}+\frac{\partial f^{\mu}}{\partial\dot{x}^{\nu}}f^{\nu}])^2(1+2\tau_0G\cdot\dot{x})\\&=f^2+2\tau_0[f_{\mu}\frac{\partial f^{\mu}}{\partial\tau}+\dot{x}^{\nu}\frac{\partial f^{\mu}}{\partial x^{\nu}}f_{\mu}+f_{\mu}\frac{\partial f^{\mu}}{\partial\dot{x}^{\nu}}f^{\nu}+f^2f\cdot\dot{x}]\end{aligned}\quad. \tag{16}$$

At last, we arrive at the targeted expression of $\ddot{x}^{\mu}$, which can be expressed as

$$\ddot{x}^{\mu}=G^{\mu}+\tau_0k\dot{x}^{\mu}=\ddot{x}_0^{\mu}+\tau_0\ddot{x}_1^{\mu}+\tau_0^2\ddot{x}_2^{\mu}, \tag{17}$$

and the quantities $\ddot{x}_0^\mu$, $\ddot{x}_1^\mu$ and $\ddot{x}_2^\mu$ are

$$\ddot{x}_0^\mu = f^\mu$$

$$\begin{aligned} \ddot{x}_1^\mu &= \frac{\partial f^\mu}{\partial \tau} + \dot{x}^\nu \frac{\partial f^\mu}{\partial x^\nu} + f^\nu \frac{\partial f^\mu}{\partial \dot{x}^\nu} + f^2 \dot{x}^\mu \\ \ddot{x}_2^\mu &= \frac{\partial \ddot{x}_1^\nu}{\partial \tau} + \dot{x}^\nu \frac{\partial \ddot{x}_1^\mu}{\partial x^\nu} + f^\nu \frac{\partial \ddot{x}_1^\mu}{\partial \dot{x}^\nu} + \ddot{x}_1^\nu \frac{\partial f^\mu}{\partial \dot{x}^\nu} \\ &+2(f_\alpha \frac{\partial f^\alpha}{\partial \tau} + \dot{x}^\nu \frac{\partial f^\alpha}{\partial x^\nu} f_\alpha + f^\nu \frac{\partial f^\alpha}{\partial \dot{x}^\nu} f_\alpha + f^2 f \cdot \dot{x})\dot{x}^\mu \end{aligned} , \tag{18}$$

which is quite forbidding on the first face compared with the original LDE!
As stated before, by repeating the method of Landau and Lifshitz to reduce the order of LDE, one can also construct the series form of order reduced approximation of LDE, which is quite simple and the result is

$$\begin{aligned} \ddot{x}_0^\mu &= f^\mu \\ \ddot{x}_i^\mu &= f^\mu + \tau_0 (\frac{d\ddot{x}_{i-1}^\mu}{d\tau} + \ddot{x}_{i-1}^2 \dot{x}^\mu) \end{aligned} . \tag{19}$$

One needs to iterate two times to get the series expression up to τ_0^2 term using Eq. (19), and the result is

$$\ddot{x}_L^\mu = f^\mu + \tau_0 [\frac{df^\mu}{d\tau} + f^2 \dot{x}^\mu] + \tau_0^2 [\frac{d}{d\tau}(\frac{df^\mu}{d\tau} + f^2 \dot{x}^\mu) + \frac{df^2}{d\tau}\dot{x}^\mu] . \tag{20}$$

We just point out that Eqs. (17) and (20) obtained through two different ways are completely consistent with each other, which shows that the reduction of order form of LDE has unique expression so long as the external force is orthogonal to four-velocity and depends only on proper time, spacetime coordinate and four-velocity. This fact also supports that LDE is the correct equation describing the motion of charged particles.
The general reduction of order series form of LDE can be expressed as

$$\ddot{x}^\mu = \sum_{n=0}^{N} \tau_0^n \ddot{x}_n^\mu + R^\mu = \ddot{x}_N^\mu + R^\mu , \tag{21}$$

where R^μ is of the order τ_0^{N+1} . Apart from the first term $\ddot{x}_0^\mu$, all other single term is not orthogonal to $\dot{x}^\mu$ as the original LDE does, which can be seen by the calculation

$$\begin{aligned} (\frac{d\ddot{x}_N^\mu}{d\tau} + \ddot{x}_N^2 \dot{x}^\mu)\dot{x}_\mu &= \dot{x}_\mu \frac{d\ddot{x}_N^\mu}{d\tau} + \ddot{x}_N^2 = \dot{x}_\mu \frac{d(\ddot{x}^\mu - R^\mu)}{d\tau} + \ddot{x}_N^2 = \dot{x}_\mu \frac{d\ddot{x}^\mu}{d\tau} + \ddot{x}_N^2 - \dot{x}_\mu \frac{dR^\mu}{d\tau} \\ &= -\ddot{x}^\mu \frac{d\dot{x}_\mu}{d\tau} + \ddot{x}_N^2 - \dot{x}_\mu \frac{dR^\mu}{d\tau} = -\ddot{x}^2 + (\ddot{x} - R)^2 - \dot{x}_\mu \frac{dR^\mu}{d\tau} = R^2 - 2\ddot{x} \cdot R - \dot{x}_\mu \frac{dR^\mu}{d\tau} \end{aligned} , \tag{22}$$

which is of order τ_0^{N+1} . To guarantee the orthogonal property between four-velocity and four-acceleration, the reduction of order series form of LDE must contain infinite terms. The present method reducing the order of LDE contains infinite terms of τ_0 at each iterative process indicating that it is more accurate than that of Landau and Lifshitz.

In subsection 1, we have known that the uniformly accelerating motion of a point charge is quite special. According to the Maxwell's electromagnetic theory, the accelerated charge would for certain radiate electromagnetic radiation which would dissipate the charge's energy and momentum. W. Pauli observed that at $t=0$ when the charge is instantaneously at rest [Pauli, 1920, as cited in Fulton & Rohrlich, 1960], the magnetic field of its radiating field is zero everywhere in the corresponding inertial frame shown by Born's original solution which means that the Poynting vector is zero everywhere in the rest frame of the charge at that instant of time and came to the conclusion that the uniformly accelerated charge does not radiate at all. Whether or not a charge undergoing uniformly accelerating motion emits electromagnetic radiation is still an open question. Nowadays, most authors of this area think that it does emit radiation. But one is faced another difficult question, namely what physical processes taking place near the neighborhood of the charge are able to give precise zero radiation reaction force. These problems have been extensively studied for a long time, but the situation still remains in a controversial status [Ginzburg, 1970; Boulware, 1980].

If one measures the macroscopic external field at the vicinity of the charge performing one-dimensional motion, what conclusions would he/she obtain? Just as one can not distinguish the gravitation field from a uniform acceleration by local experiments done in a tiny box, which is called Einstein equivalence principle, any macroscopic external fields felt by the charge are almost constant and the radiation reaction force would vanish according to LDE for charge's one-dimensional macroscopic motions! What is the mechanism of radiation reaction? There are various points of view for charges' radiation reaction. Lorentz regarded charges as rigid spheres of finite size, and the radiation reaction force comes from the interactions of the retarded radiation fields of all parts of the charge; Dirac regarded electrons as point charges and regarded the radiation reaction field as half the difference of its retarded radiation field minus its advanced field; Teitelboim obtained the radiation reaction force just using retarded radiating field of accelerated charges [Teitelboim , 1970; Teitelboim & Villarroel, 1980]; while Feynman and Wheeler thought that the radiation reaction of accelerated charges comes from the advanced radiating fields of all other charges in the Universe coherently superimposed at the location of the radiating charge [Wheeler & Feynman, 1945; 1949]. The common point of these different viewpoints is that the radiation reaction is represented by the variation of external force field with time, which gives zero radiation reaction for one-dimensional uniformly accelerating motion of charges showing that the general mechanism of radiation reaction is not complete. It has been proved that the electron of a hydrogen atom would never collapse using the nonrelativistic version of LDE which is quite contrary to the conventional idea of classical electrodynamics that the atoms would fall into the origin within rather a short time interval, which is called the theorem [Eliezer, 1947; Carati,2001]. Cole and Zou studied the stability of hydrogen atoms using LLE and obtained similar results with that of quantum mechanics [Cole & Zou, 2003].

It seems that the problems associated with the radiation reaction have little possibility to be resolved without introducing new factors. According to quantum field theory, the vacuum is not empty but full of all kinds of fluctuating fields. To investigate how the electromagnetic fluctuating fields of the vacuum influence the motion of a charge is the main content of this chapter.

3. Electromagnetic fluctuating fields of the vacuum

The development of quantum field theory (QFT) shows that the vacuum is not an empty space but an extremely complicated system. All kinds of field quanta are created and then annihilated or vice versa. The effects of electromagnetic fluctuating fields of the vacuum have been verified by experiments， such as Lamb shift of hydrogen atoms and Casimir force etc.. Unruh effect, which has been a very active area of physics and has not been verified by experiments as yet, shows that the vacuum state is dependent on the motion of observers. The quantized free fields of QFT are operator expressions, which can not be used directly in classical calculation. T. W. Marshall proposed a Lorentz invariant random classical radiation to model the corresponding fluctuating fields of the vacuum which was carried forward by T. H. Boyer [Boyer, 1980]. A number of phenomena associated with the vacuum of quantum electrodynamics can be understood in purely classical electrodynamics, provided we change the homogeneous boundary conditions on Maxwell's equations to include random classical radiation with a Lorentz invariant spectrum. This section briefly introduces this classical model of the vacuum fluctuating fields, which is borrowed heavily from T. H. Boyer's paper.

3.1 Random classical radiation fields for massless scalar cases

The introduced random radiation is not connected with temperature radiation but exist in the vacuum at the absolute zero of temperature; hence it is termed classical zero-point radiation, which is treated just as fluctuations of classical thermal radiation. The only special aspect of zero-point radiation is Lorentz invariant indicating there is no preferred frame. Thermal radiation involves radiation above the zero-point spectrum and involves a finite amount of energy and singles out a preferred frame of reference.

A spectrum of random classical radiation can be written as a sum over plane waves of various frequencies and wave vectors with random phases. For the massless scalar field, the spatially homogeneous and isotropic distribution in empty space can be written as an expansion in plane waves with random phases

$$\begin{aligned}\varphi(\vec{r},t) &= \int d\vec{k}\; f(\omega)\cos[\vec{k}\cdot\vec{r}-\omega t-\theta(\vec{k})] \\ &= \int d\vec{k}\; \frac{1}{2} f(\omega)[a(\vec{k})\exp(-ik\cdot x)+a^{*}(\vec{k})\exp(ik\cdot x)]\end{aligned}, \tag{23}$$

where the $\theta(\vec{k})$ is the random phase distributed uniformly on the interval $(0,\ 2\pi)$ and independently for each wave vector $\vec{k}$. The Lorentz invariance of Eq. (23) requires that $f(\omega)$ must be proportional to $1/\sqrt{\omega}$. The average over the random phases are

$$\begin{aligned}&< \cos[\theta(\vec{k})]\cos[\theta(\vec{k}')] > = < \sin[\theta(\vec{k})]\sin[\theta(\vec{k}')] > = \frac{1}{2}\delta(\vec{k}-\vec{k}') \\ &< \cos[\theta(\vec{k})]\sin[\theta(\vec{k}')] > = 0\end{aligned}. \tag{24}$$

Eqs. (23) and (24) have immediate connection with the counterparts of QFT, which are free field

$$\phi(x) = \int \frac{d\vec{k}}{2\pi}\sqrt{\frac{\hbar}{\omega}}[\hat{a}(\vec{k})\exp(-ik\cdot x)+\hat{a}^{\dagger}(\vec{k})\exp(-ik\cdot x)],$$

and the commutators for creation and annihilation operators $\hat{a}^{\dagger}(\vec{k})$ and $\hat{a}(\vec{k})$ satisfying

$$\begin{aligned}&[\hat{a}(\vec{k}),\hat{a}^{\dagger}(\vec{k}')]=\delta(\vec{k}-\vec{k}')\\&[\hat{a}(\vec{k}),\hat{a}(\vec{k}')]=[\hat{a}^{\dagger}(\vec{k}),\hat{a}^{\dagger}(\vec{k}')]=0\end{aligned}.$$

Compared with the free field of QFT one can immediately get the Lorentz invariant spectral function, that is

$$f(\omega)=\frac{1}{\pi}\sqrt{\frac{\hbar}{2\omega}}.$$

These similarities make the quantitative calculations using random classical radiation fields comparable to the results given by full quantum theory.

As an example, we will deduce the Unruh effect using random classical radiation field. For a charge undergoing one-dimensional uniformly accelerating motion, the world line is

$$x=\frac{1}{a}\cosh(a\tau),\quad t=\frac{1}{a}\sinh(a\tau),\tag{25}$$

where a is the constant acceleration seen at charge's rest frame. The uniform accelerating motion of a charge will be investigated in section 4.

We would like to evaluate the average value

$$<\varphi(\vec{r},s-t/2)\varphi(\vec{r},s+t/2)>,$$

which characterizes the random classical field. By a Lorentz transformation, the point $\vec{r}$ can be changed to the origin of another inertial frame, which will be taken as the laboratory frame. Thus we just need to calculate

$$<\varphi(0,\sigma-\tau/2)\varphi(0,\sigma+\tau/2)>,\tag{26}$$

where $\sigma\pm\tau/2$ have two different interpretations; the results of Lorentz transformation or just two different proper times. The charge undergoes the uniformly accelerating motion along the x direction of the laboratory frame, and we calculate Eq. (26) at two positions of the charge at proper times $\sigma\pm\tau/2$, namely

$$\varphi(0,\sigma\pm\tau/2)=\varphi(\frac{\cosh(a(\sigma\pm\tau/2))}{a},\frac{\sinh(a(\sigma\pm\tau/2))}{a}).\tag{27}$$

By the expression of field φ, the correlation function Eq. (26) becomes

$$\begin{aligned}<\varphi(0,\sigma-\tau/2)\varphi(0,\sigma+\tau/2)>=\int d^3k\frac{\hbar}{4\pi^2\omega}\cos\{\frac{k_x}{a}[\cosh(a(\sigma-\tau/2))\\-\cosh(a(\sigma+\tau/2))]-\frac{\omega}{a}[\sinh(a(\sigma-\tau/2))-\sinh(a(\sigma+\tau/2))]\}\end{aligned}.\tag{28}$$

The stationary character of the correlation function is not exhibited since the free parameter σ is included. However, the physical argument that there is no preferred time for uniformly accelerating motions indicates that Eq. (28) must be independent of σ. In fact, by performing the Lorentz transformation

$$\omega' = \omega\cosh(a\sigma) - k_x \sinh(a\sigma),$$
$$k_x' = k_x \cosh(a\sigma) - \omega\sinh(a\sigma) \quad ,$$

from unprimed laboratory frame to the primed frame in which the charge is instantaneously at rest at proper time σ, the correlation function becomes

$$< \varphi(0,\sigma - \tau/2)\varphi(0,\sigma + \tau/2) >$$
$$= \int \frac{d^3k'}{4\pi^2}\frac{\hbar}{\omega'}\cos[\frac{2\omega'}{a}[\sinh(\frac{a\tau}{2})]] = -\frac{\hbar}{\pi}(\frac{a}{2})^2 \operatorname{csc}h^2(\frac{a\tau}{2}) , \tag{29}$$

which is the function of the proper time τ.
The result Eq. (29) will be compared with the correlation function

$$< \varphi_T(0,s - t/2)\varphi_T(0,s + t/2) > \tag{30}$$

for a charge at the origin of an inertial frame where the random thermal radiation is

$$\pi^2 f_T^2(\omega) = \frac{\hbar}{\omega}(\frac{1}{2} + \frac{1}{\exp(\hbar\omega/kT) - 1}) = \frac{\hbar c^2}{2\omega}\coth(\frac{\hbar\omega}{2kT}),$$

where k is the Boltzmann constant, T the temperature. The calculation of correlation function Eq. (30) is straightforward and the result is

$$< \phi_T(0,s - t/2)\phi_T(0,s + t/2) >$$
$$= \frac{\hbar}{\pi c}\int_0^\infty d\omega\, \omega\coth(\frac{\hbar\omega}{2kT})\cos\omega t = -\frac{\hbar}{\pi}(\frac{\pi k_B T}{\hbar})^2 \operatorname{csc}h^2(\frac{\pi k_B T t}{\hbar}) . \tag{31}$$

The time parameter s is automatically counteracted without employing the Lorentz transformation.
If we compare the correlation function of the charge accelerated through the vacuum Eq. (29) with the correlation function Eq. (31) for a stationary charge in random classical scalar zero point radiation plus a Plank thermal spectrum, we find that they are identical in functional form provided that the acceleration and the temperature are related by

$$T = \frac{\hbar a}{2\pi k} . \tag{32}$$

It is in this sense that one speaks of an observer accelerated through the inertial vacuum as finding himself in a thermal bath which is called the Unruh effect [Unruh, 1976].

3.2 Random classical radiation fields for electromagnetic fields case

We just list the expressions used in the following sections of the classical model of the vacuum electromagnetic fluctuating fields, which can be written as

$$\vec{e}(\vec{r},t)=\sum_{\lambda=1,2}\int d^3k\sqrt{\frac{\hbar\omega}{2\pi^2}}\hat{\varepsilon}(\vec{k},\lambda)\cos[\vec{k}\cdot\vec{r}-\omega t-\theta(\vec{k},\lambda)] \\ \vec{b}(\vec{r},t)=\sum_{\lambda=1,2}\int d^3k\sqrt{\frac{\hbar\omega}{2\pi^2}}\hat{k}\times\hat{\varepsilon}(\vec{k},\lambda)\cos[\vec{k}\cdot\vec{r}-\omega t-\theta(\vec{k},\lambda)] \tag{33}$$

where λ denotes the polarization degree of freedom and $\hat{\varepsilon}(\vec{k},\lambda)$ the polarization vector which satisfies the sum rule

$$\sum_{\lambda=1}^{2}\varepsilon_i(\vec{k},\lambda)\varepsilon_j(\vec{k},\lambda)=\delta_{ij}-k_ik_j/k^2 . \tag{34}$$

The average over random phases satisfies the rules:

$$<\sin[\theta(\vec{k},\lambda)]\sin[\theta(\vec{k}',\lambda')]>=<\cos[\theta(\vec{k},\lambda)]\cos[\theta(\vec{k}',\lambda')]>=\frac{1}{2}\delta_{\lambda\lambda'}\delta(\vec{k}-\vec{k}') \\ <\cos[\theta(\vec{k},\lambda)]\sin[\theta(\vec{k}',\lambda')]>=0 \tag{35}$$

The average values of two components of random classical electromagnetic radiation fields can be obtained by using Eqs. (33)-(35). For example, the average value of two electric components is

$$\begin{aligned} e_ie_j &= \sum_{\lambda=1,2}\int d^3k\sqrt{\frac{\hbar\omega}{2\pi^2}}\varepsilon_i(\vec{k},\lambda)\sum_{\lambda'=1,2}\int d^3k'\sqrt{\frac{\hbar\omega'}{2\pi^2}}\hat{\varepsilon}_j(\vec{k}',\lambda') \\ &\times<\cos[k\cdot x-\theta(\vec{k},\lambda)]\cos[k'\cdot x-\theta(\vec{k}',\lambda')]> \\ &= \sum_{\lambda,\lambda'=1,2}\int d^3k\sqrt{\frac{\hbar\omega}{2\pi^2}}\varepsilon_i(\vec{k},\lambda)\sum_{\lambda'=1,2}\int d^3k'\sqrt{\frac{\hbar\omega'}{2\pi^2}}\hat{\varepsilon}_j(\vec{k}',\lambda')\frac{1}{2}\delta_{\lambda\lambda'}\delta(\vec{k}-\vec{k}'), \\ &= \sum_{\lambda=1,2}\int d^3k\frac{\hbar\omega}{4\pi^2}\hat{\varepsilon}_i(\vec{k},\lambda)\hat{\varepsilon}_j(\vec{k},\lambda)=\frac{\hbar}{4\pi^2}\delta_{ij}\int d^3k\omega-\frac{\hbar}{4\pi^2}\int d^3k\frac{k_ik_j}{k} \\ &= \delta_{ij}\frac{\hbar}{6\pi^2}\int d^3k\cdot k=\delta_{ij}\frac{\hbar}{6\pi}k_{cutoff}^4 \end{aligned} \tag{36}$$

where k_{cutoff} is the cutoff of wave vector and is usually taken as that corresponding to the Compton wavelength of massive charges. The other two average values are

$$<b_ib_j>=<e_ie_j>=\delta_{ij}\frac{\hbar}{6\pi}k_{cutoff}^4 , \quad <e_ib_j>=0 . \tag{37}$$

The preparation of the vacuum electromagnetic fluctuating fields is finished. However, we strongly recommend serious-minded readers to read the original papers of T. H. Boyer. In the next section, we will investigate the possible effects of electromagnetic fluctuating fields of the vacuum on the radiation reaction of a radiating charge using the reduction of order series form of LDE obtained in section 2.

4. Nonzero contribution of vacuum fluctuations to radiation reaction

In section 2, we have obtained the reduction of order series form of LDE up to τ_0^2 term, and here we present it again

$$\begin{aligned}\ddot{x}^\mu &= f^\mu + \tau_0[\frac{\partial f^\mu}{\partial \tau} + \dot{x}^\nu \frac{\partial f^\mu}{\partial x^\nu} + f^\nu \frac{\partial f^\mu}{\partial \dot{x}^\nu} + f^2 \dot{x}^\mu] + \tau_0^2[\frac{\partial \ddot{x}_1^\mu}{\partial \tau} + \dot{x}^\nu \frac{\partial \ddot{x}_1^\mu}{\partial x^\nu} \\ &+ f^\nu \frac{\partial \ddot{x}_1^\mu}{\partial \dot{x}^\nu} + \ddot{x}_1^\nu \frac{\partial f^\mu}{\partial \dot{x}^\nu} + 2(f_\alpha \frac{\partial f^\alpha}{\partial \tau} + \dot{x}^\nu \frac{\partial f^\alpha}{\partial x^\nu} f_\alpha + f^\nu \frac{\partial f^\alpha}{\partial \dot{x}^\nu} f_\alpha + f^2 f \cdot \dot{x}) \dot{x}^\mu]\end{aligned}, \tag{38}$$

where

$$\ddot{x}_1^\mu = f^\mu + \tau_0(\frac{\partial f^\mu}{\partial \tau} + \dot{x}^\nu \frac{\partial f^\mu}{\partial x^\nu} + f^\nu \frac{\partial f^\mu}{\partial \dot{x}^\nu} + f^2 \dot{x}^\mu).$$

In getting this equation, we have assumed that the external fields are electromagnetic fields which can be expressed by the functions of the proper time and spacetime coordinates. Therefore the external force has the Lorentz-force form:

$$f^\mu = \frac{e}{m} F^{\mu\nu} \dot{x}_\nu = [(\vec{E} + \vec{e}) \cdot \dot{\vec{x}},\ \dot{x}_0(\vec{E} + \vec{e}) + \dot{\vec{x}} \times (\vec{B} + \vec{b})], \tag{39}$$

where $\vec{E}(\tau, x)$ and $\vec{B}(\tau, x)$ are external electromagnetic fields, $\vec{e}(x)$ and $\vec{b}(x)$ the vacuum electromagnetic fluctuating fields expressed in Eq. (33). For the sake of simplicity, we have these electromagnetic fields absorb the factor e/m. The component expressions of the force field are

$$\begin{aligned} f^0 &= \dot{\vec{x}} \cdot (\vec{E} + \vec{e}) = f_E^0 + f_e^0 \\ f^1 &= \dot{x}_0(E_1 + e_1) + [\dot{x}_2(B_3 + b_3) - \dot{x}_3(B_2 + b_2)] = f_E^1 + f_e^1 \\ f^2 &= \dot{x}_0(E_2 + e_2) + [\dot{x}_3(B_1 + b_1) - \dot{x}_1(B_3 + b_3)] = f_E^2 + f_e^2 \\ f^3 &= \dot{x}_0(E_3 + e_3) + [\dot{x}_1(B_2 + b_2) - \dot{x}_2(B_1 + b_1)] = f_E^3 + f_e^3 \end{aligned}. \tag{40}$$

We would like to emphasize again that Eq. (38) is accurate enough for any macroscopic motions of a charge due to τ_0 ($\sim 10^{-24}s$) being an extremely small time scale. Because the atomic nucleus is of finite size, Eq. (38) could even be used to study the electron's motion of a hydrogen atom.

It is impossible or meaningless to trace the effects of vacuum electromagnetic fluctuations by directly solve the equation of motion of charged particles due to its stochastic property. Therefore, we should perform the average calculation for each term of the acceleration over the random phases of vacuum fluctuating fields, which is a coarse grained process. The nonlinear terms of Eq. (38) will produce nonzero contribution of vacuum electromagnetic fluctuating fields to the radiation reaction. According to the rules of average manipulation over random phases Eq. (35), The result of the average for the first term of charge's acceleration is just the external force f_E^μ. We will focus on the calculation of the average results for the next two terms of the acceleration, which is quite cumbersome.

4.1 Effects of the vacuum fluctuations on the radiation reaction in the order of τ_0

The part of acceleration linear with τ_0 is

$$\ddot{x}_1^\mu = \frac{\partial f^\mu}{\partial \tau} + \dot{x}^\nu \frac{\partial f^\mu}{\partial x^\nu} + f^\nu \frac{\partial f^\mu}{\partial \dot{x}^\nu} + f^2 \dot{x}^\mu. \tag{41}$$

Because the average value of the product of odd number vacuum electromagnetic fluctuating components is zero, the average results of the first two terms of Eq. (41) are just

$$\frac{\partial f_E^\mu}{\partial \tau}+\dot{x}^\nu \frac{\partial f_E^\mu}{\partial x^\nu}.$$

The third term

$$f^\nu \frac{\partial f^\mu}{\partial \dot{x}^\nu}$$

is possible to produce nonzero contribution to radiation reaction. Firstly, we calculate the average value of its zero component

$$\begin{aligned}
&< f^\nu \frac{\partial f^0}{\partial \dot{x}^\nu} >=< (f_E^\nu + f_e^\nu)\frac{\partial (f_E^0 + f_e^0)}{\partial \dot{x}^\nu} >= f_E^\nu \frac{\partial f_E^0}{\partial \dot{x}^\nu} + < f_e^\nu \frac{\partial f_e^0}{\partial \dot{x}^\nu} > \\
&= f_E^\nu \frac{\partial f_E^0}{\partial \dot{x}^\nu} + < f_e^0 \frac{\partial f_e^0}{\partial \dot{x}^0} + f_e^1 \frac{\partial f_e^0}{\partial \dot{x}^1} + f_e^2 \frac{\partial f_e^0}{\partial \dot{x}^2} + f_e^3 \frac{\partial f_e^0}{\partial \dot{x}^3} > \\
&= f_E^\nu \frac{\partial f_E^0}{\partial \dot{x}^\nu} + < (\dot{x}_0 e_1 + \dot{x}_2 b_3 - \dot{x}_3 b_2) e_1 + (\dot{x}_0 e_2 + \dot{x}_3 b_1 - \dot{x}_1 b_3) e_2 + (\dot{x}_0 e_3 + \dot{x}_1 b_2 - \dot{x}_2 b_1) e_3 > \\
&= f_E^\nu \frac{\partial f_E^0}{\partial \dot{x}^\nu} + 3\dot{x}_0 < e_1 e_1 >
\end{aligned}.$$

Then we calculate the average value of the first space component and the result is

$$\begin{aligned}
&< f^\nu \frac{\partial f^1}{\partial \dot{x}^\nu} >=< (f_E^\nu + f_e^\nu)\frac{\partial (f_E^1 + f_e^1)}{\partial \dot{x}^\nu} >= f_E^\nu \frac{\partial f_E^1}{\partial \dot{x}^\nu} + < f_e^\nu \frac{\partial f_e^1}{\partial \dot{x}^\nu} > \\
&= f_E^\nu \frac{\partial f_E^1}{\partial \dot{x}^\nu} + < f_e^0 \frac{\partial f^1}{\partial \dot{x}^0} + f_e^1 \frac{\partial f^1}{\partial \dot{x}^1} + f_e^2 \frac{\partial f^1}{\partial \dot{x}^2} + f_e^3 \frac{\partial f^1}{\partial \dot{x}^3} > \\
&= f_E^\nu \frac{\partial f_E^1}{\partial \dot{x}^\nu} + < \dot{x}_i e_i e_1 + (\dot{x}_0 e_2 + \dot{x}_3 b_1 - \dot{x}_1 b_3) b_3 - (\dot{x}_0 e_3 + \dot{x}_1 b_2 - \dot{x}_2 b_1) b_2 > \\
&= f_E^\nu \frac{\partial f_E^1}{\partial \dot{x}^\nu} + < \dot{x}_1 e_1^2 - \dot{x}_1 b_3^2 - \dot{x}_1 b_2^2 >= f_E^\nu \frac{\partial f_E^1}{\partial \dot{x}^\nu} - \dot{x}_1 < e_1 e_1 >
\end{aligned}.$$

Because all three space components are equivalent, the final result is

$$< f^\nu \frac{\partial f^\mu}{\partial \dot{x}^\nu} >= f_E^\nu \frac{\partial f_E^\mu}{\partial \dot{x}^\nu} + [3\dot{x}_0,\ -\dot{\vec{x}}] < e_1 e_1 >. \tag{42}$$

The average value of the fourth term of Eq. (41) can be directly calculated and the result is

$$\begin{aligned}
&< f^2 > \dot{x}^\mu =< (f_E^\mu + f_e^\mu)(f_{E,\mu} + f_{e,\mu}) > \dot{x}^\mu = f_E^2 \dot{x}^\mu + < f_e^2 > \dot{x}^\mu \\
&= f_E^2 \dot{x}^\mu + < \dot{x}_i^2 e_i^2 - (\dot{x}_0 e_1 + \dot{x}_2 b_3 - \dot{x}_3 b_2)^2 - (\dot{x}_0 e_2 + \dot{x}_3 b_1 - \dot{x}_1 b_3)^2 \\
&-(\dot{x}_0 e_3 + \dot{x}_1 b_2 - \dot{x}_2 b_1)^2 > \dot{x}^\mu \\
&= f_E^2 \dot{x}^\mu + < \dot{x}_i^2 e_i^2 - (\dot{x}_0^2 e_1^2 + \dot{x}_2^2 b_3^2 + \dot{x}_3^2 b_2^2) - (\dot{x}_0^2 e_2^2 + \dot{x}_3^2 b_1^2 + \dot{x}_1^2 b_3^2) \\
&-(\dot{x}_0^2 e_3^2 + \dot{x}_1^2 b_2^2 + \dot{x}_2^2 b_1^2) > \dot{x}^\mu \\
&= f_E^2 \dot{x}^\mu - (3 + 4\dot{\vec{x}}^2)\dot{x}^\mu < e_1 e_1 >
\end{aligned}.$$

Therefore the complete expression linear with τ_0 of the contribution of vacuum fluctuations to the radiation reaction is

$$-4\tau_0(\dot{\ddot{x}}^2\dot{x}_0,\ \dot{x}_0^2\dot{\ddot{x}})<e_1e_1>. \tag{43}$$

4.2 Effects of the vacuum fluctuations on the radiation reaction in the order of τ_0^2

The part of acceleration that is proportional to τ_0^2 is

$$\begin{aligned}\ddot{x}_2^\mu &= \frac{\partial \ddot{x}_1^\mu}{\partial \tau} + \dot{x}^\nu \frac{\partial \ddot{x}_1^\mu}{\partial x^\nu} + f^\nu \frac{\partial \ddot{x}_1^\mu}{\partial \dot{x}^\nu} + \ddot{x}_1^\nu \frac{\partial f^\mu}{\partial \dot{x}^\nu} \\ &+2(f_\alpha \frac{\partial f^\alpha}{\partial \tau} + \dot{x}^\nu \frac{\partial f^\alpha}{\partial x^\nu} f_\alpha + f^\nu \frac{\partial f^\alpha}{\partial \dot{x}^\nu} f_\alpha + f^2 f\cdot \dot{x})\dot{x}^\mu\end{aligned} \tag{44}$$

We had better expand it into explicit expression of external electromagnetic fields and vacuum fluctuating fields and then collect the same terms together. The expansion of the first term of Eq. (44) is

$$\begin{aligned}\frac{\partial \ddot{x}_1^\mu}{\partial \tau} &= \frac{\partial}{\partial \tau}(\frac{\partial f^\mu}{\partial \tau} + \dot{x}^\nu \frac{\partial f^\mu}{\partial x^\nu} + f^\nu \frac{\partial f^\mu}{\partial \dot{x}^\nu} + f^2 \dot{x}^\mu) \\ &= \frac{\partial^2 f^\mu}{\partial^2 \tau} + \dot{x}^\nu \frac{\partial^2 f^\mu}{\partial \tau \partial x^\nu} + f^\nu \frac{\partial f^\mu}{\partial x^\nu} + \frac{\partial f^\nu}{\partial \tau}\frac{\partial f^\mu}{\partial \dot{x}^\nu} + f^\nu \frac{\partial^2 f^\mu}{\partial \tau \partial \dot{x}^\nu} + 2 f^\nu \frac{\partial f_\nu}{\partial \tau}\dot{x}^\mu + f^2 f^\mu\end{aligned}$$

The second term of Eq. (44) is

$$\begin{aligned}\dot{x}^\nu \frac{\partial \ddot{x}_1^\mu}{\partial x^\nu} &= \dot{x}^\nu \frac{\partial}{\partial x^\nu}(\frac{\partial f^\mu}{\partial \tau} + \dot{x}^\alpha \frac{\partial f^\mu}{\partial x^\alpha} + f^\alpha \frac{\partial f^\mu}{\partial \dot{x}^\alpha} + f^2 \dot{x}^\mu) \\ &= \dot{x}^\nu(\frac{\partial^2 f^\mu}{\partial \tau \partial x^\nu} + \dot{x}^\alpha \frac{\partial^2 f^\mu}{\partial x^\nu \partial x^\alpha} + \frac{\partial f^\alpha}{\partial x^\nu}\frac{\partial f^\mu}{\partial \dot{x}^\alpha} + f^\alpha \frac{\partial^2 f^\mu}{\partial x^\nu \partial \dot{x}^\alpha} + 2 f_\alpha \frac{\partial f^\alpha}{\partial x^\nu}\dot{x}^\mu)\end{aligned}$$

The third term of Eq. (44) is

$$\begin{aligned}f^\nu \frac{\partial \ddot{x}_1^\mu}{\partial \dot{x}^\nu} &= f^\nu \frac{\partial}{\partial \dot{x}^\nu}(\frac{\partial f^\mu}{\partial \tau} + \dot{x}^\alpha \frac{\partial f^\mu}{\partial x^\alpha} + f^\alpha \frac{\partial f^\mu}{\partial \dot{x}^\alpha} + f^2 \dot{x}^\mu) \\ &= f^\nu(\frac{\partial^2 f^\mu}{\partial \tau \partial \dot{x}^\nu} + \delta_\nu^\alpha \frac{\partial f^\mu}{\partial x^\alpha} + \dot{x}^\alpha \frac{\partial^2 f^\mu}{\partial x^\alpha \partial \dot{x}^\nu} + \frac{\partial f^\alpha}{\partial \dot{x}^\nu}\frac{\partial f^\mu}{\partial \dot{x}^\alpha} + f^\alpha \frac{\partial^2 f^\mu}{\partial \dot{x}^\alpha \partial \dot{x}^\nu} + 2 f_\alpha \frac{\partial f^\alpha}{\partial \dot{x}^\nu}\dot{x}^\mu + f^2 \delta_\nu^\mu) \\ &= f^\nu \frac{\partial^2 f^\mu}{\partial \tau \partial \dot{x}^\nu} + f^\alpha \frac{\partial f^\mu}{\partial x^\alpha} + \dot{x}^\alpha \frac{\partial^2 f^\mu}{\partial x^\alpha \partial \dot{x}^\nu} f^\nu + f^\nu \frac{\partial f^\alpha}{\partial \dot{x}^\nu}\frac{\partial f^\mu}{\partial \dot{x}^\alpha} + f^\alpha \frac{\partial^2 f^\mu}{\partial \dot{x}^\alpha \partial \dot{x}^\nu} f^\nu + 2 f_\alpha \frac{\partial f^\alpha}{\partial \dot{x}^\nu} f^\nu \dot{x}^\mu \\ &+ f^2 f^\mu\end{aligned},$$

and the fourth term is

$$\ddot{x}_1^\nu \frac{\partial f^\mu}{\partial \dot{x}^\nu} = \frac{\partial f^\nu}{\partial \tau}\frac{\partial f^\mu}{\partial \dot{x}^\nu} + \dot{x}^\alpha \frac{\partial f^\nu}{\partial x^\alpha}\frac{\partial f^\mu}{\partial \dot{x}^\nu} + f^\alpha \frac{\partial f^\nu}{\partial \dot{x}^\alpha}\frac{\partial f^\mu}{\partial \dot{x}^\nu} + \frac{\partial f^\mu}{\partial \dot{x}^\nu} f^2 \dot{x}^\nu .$$

Now we collected all these terms together and found that there are three terms contained in $\ddot{x}_2^\mu$ which are linear with force field:

$$\frac{\partial^2 f^\mu}{\partial^2 \tau} + 2\dot{x}^\nu \frac{\partial^2 f^\mu}{\partial \tau \partial x^\nu} + \dot{x}^\nu \dot{x}^\alpha \frac{\partial^2 f^\mu}{\partial x^\nu \partial x^\alpha}.$$

The average of these three terms over the random phases of vacuum fluctuating fields just leaves the contribution of the external force field, and the result is

$$< \frac{\partial^2 f^\mu}{\partial^2 \tau} + 2\dot{x}^\nu \frac{\partial^2 f^\mu}{\partial \tau \partial x^\nu} + \dot{x}^\nu \dot{x}^\alpha \frac{\partial^2 f^\mu}{\partial x^\nu \partial x^\alpha} >= \frac{\partial^2 f_E^\mu}{\partial^2 \tau} + 2\dot{x}^\nu \frac{\partial^2 f_E^\mu}{\partial \tau \partial x^\nu} + \dot{x}^\nu \dot{x}^\alpha \frac{\partial^2 f_E^\mu}{\partial x^\nu \partial x^\alpha}.$$

There are seven terms contained in $\ddot{x}_2^\mu$ which are quadratic in force fields. Because the vacuum fluctuating field just the function of spacetime coordinates, the derivatives with respect to proper time τ or four-velocity $\dot{x}^\mu$ equal to zero. Furthermore, the derivative with respect to spacetime coordinates would turn the cosine function into sine function, which combining with another cosine function of a vacuum fluctuating field would make the average value zero. Therefore, the average value of all quadratic terms of force fields contained in $\ddot{x}_2^\mu$ over the random phases of vacuum fluctuating fields is

$$\begin{aligned} &< 2f^\nu \frac{\partial f^\mu}{\partial x^\nu} + 2\frac{\partial f^\nu}{\partial \tau}\frac{\partial f^\mu}{\partial \dot{x}^\nu} + 2f^\nu \frac{\partial^2 f^\mu}{\partial \tau \partial \dot{x}^\nu} + 4f^\nu \frac{\partial f_\nu}{\partial \tau}\dot{x}^\mu \\ &+2\dot{x}^\nu \frac{\partial f^\alpha}{\partial x^\nu}\frac{\partial f^\mu}{\partial \dot{x}^\alpha} + 2\dot{x}^\nu f^\alpha \frac{\partial^2 f^\mu}{\partial x^\nu \partial \dot{x}^\alpha} + 4\dot{x}^\nu f_\alpha \frac{\partial f^\alpha}{\partial x^\nu}\dot{x}^\mu > \\ &= 2f_E^\nu \frac{\partial f_E^\mu}{\partial x^\nu} + 2\frac{\partial f_E^\nu}{\partial \tau}\frac{\partial f_E^\mu}{\partial \dot{x}^\nu} + 2f_E^\nu \frac{\partial^2 f_E^\mu}{\partial \tau \partial \dot{x}^\nu} + 4f_E^\nu \frac{\partial f_E^\nu}{\partial \tau}\dot{x}^\mu \\ &+2\dot{x}^\nu \frac{\partial f_E^\alpha}{\partial x^\nu}\frac{\partial f_E^\mu}{\partial \dot{x}^\alpha} + 2\dot{x}^\nu f_E^\alpha \frac{\partial^2 f_E^\mu}{\partial x^\nu \partial \dot{x}^\alpha} + 4\dot{x}^\nu f_{E,\alpha} \frac{\partial f_E^\alpha}{\partial x^\nu}\dot{x}^\mu \end{aligned}.$$

There are six terms contained in $\ddot{x}_2^\mu$ which are trinomial of force fields

$$2f^2 f^\mu + 2f^\nu \frac{\partial f^\alpha}{\partial \dot{x}^\nu}\frac{\partial f^\mu}{\partial \dot{x}^\alpha} + f^\alpha \frac{\partial^2 f^\mu}{\partial \dot{x}^\alpha \partial \dot{x}^\nu} f^\nu + 4f_\alpha \frac{\partial f^\alpha}{\partial \dot{x}^\nu} f^\nu \dot{x}^\mu + \frac{\partial f^\mu}{\partial \dot{x}^\nu} f^2 \dot{x}^\nu + 2\dot{x}^\mu f^2 f \cdot \dot{x}.$$

These terms are sequently denoted as (a), (b), (c), (d), (e) and (f). We just present the calculation process for the first term, only final results of other five terms are given.
We first expand term (a) into

$$\begin{aligned} 2 < f^2 f^\mu >&= 2 < (f_E^\nu + f_e^\nu)(f_{E,\nu} + f_{e,\nu})(f_E^\mu + f_e^\mu) > \\ &= 2 < (f_E^\nu + f_e^\nu)(f_{E,\nu} + f_{e,\nu})f_E^\mu + (f_E^\nu + f_e^\nu)(f_{E,\nu} + f_{e,\nu})f_e^\mu > \\ &= 2 < (f_E^2 + f_e^2)f_E^\mu + 2f_E^\nu f_{e,\nu} f_e^\mu > \\ &= 2f_E^2 f_E^\mu + 2 < f_e^2 > f_E^\mu + 4f_E^\nu < f_{e,\nu} f_e^\mu > \end{aligned},$$

and last two terms are denoted as $(a.2)$ an $(a.3)$. We will calculate the average values of these two terms separately.
For the average value of term $(a.2)$, the result is

$$2f_E^\mu < f_e^2 >= 2f_E^\mu < f_{e,0}^2 - f_{e,1}^2 - f_{e,2}^2 - f_{e,3}^2 >$$

$$
\begin{aligned}
&= 2f_E^{\mu} < \dot{x}_i e_i \dot{x}_j e_j - (\dot{x}_0 e_1 + \dot{x}_2 b_3 - \dot{x}_3 b_2)^2 - (\dot{x}_0 e_2 + \dot{x}_3 b_1 - \dot{x}_1 b_3)^2 - (\dot{x}_0 e_3 + \dot{x}_1 b_2 - \dot{x}_2 b_1)^2 > \\
&= 2f_E^{\mu} < \dot{x}_i e_i \dot{x}_i e_i - (\dot{x}_0^2 + \dot{x}_2^2 + \dot{x}_3^2) e_1 e_1 - (\dot{x}_0^2 + \dot{x}_1^2 + \dot{x}_3^2) e_1 e_1 - (\dot{x}_0^2 + \dot{x}_2^2 + \dot{x}_1^2) e_1 e_1 > \\
&= 2f_E^{\mu} [\dot{x}_i \dot{x}_i - (\dot{x}_0^2 + \dot{x}_2^2 + \dot{x}_3^2) - (\dot{x}_0^2 + \dot{x}_1^2 + \dot{x}_3^2) - (\dot{x}_0^2 + \dot{x}_2^2 + \dot{x}_1^2)] < e_1 e_1 > \\
&= -2(4\dot{x}_0^2 - 1) f_E^{\mu} < e_1 e_1 >
\end{aligned}
$$

As for the term $(a.3)$, we first calculate its zero-th component and the result is

$$
\begin{aligned}
& 4f_E^{\nu} < f_{e,\nu} f_e^0 >= 4f_E^0 < f_{e,0} f_e^0 > -4f_E^1 < f_{e,1} f_e^0 > -4f_E^2 < f_{e,2} f_e^0 > -4f_E^3 < f_{e,3} f_e^0 > \\
&= 4f_E^0 < \dot{x}_i \dot{x}_i e_1 e_1 > -4f_E^1 < \dot{x}_0 e_1 \dot{x}_i e_i > -4f_E^2 < \dot{x}_0 e_2 \dot{x}_i e_i > -4f_E^3 < \dot{x}_0 e_3 \dot{x}_i e_i > \\
&= 4f_E^0 \dot{x}_i \dot{x}_i < e_1 e_1 > -4f_E^1 \dot{x}_0 \dot{x}_1 < e_1 e_1 > -4f_E^2 \dot{x}_0 \dot{x}_2 < e_1 e_1 > -4f_E^3 \dot{x}_0 \dot{x}_3 < e_1 e_1 > \\
&= 4(f_E^0 \dot{\vec{x}}^2 - \dot{x}_0 \vec{f}_E \cdot \dot{\vec{x}}) < e_1 e_1 >
\end{aligned},
$$

then we calculate its first space component, and the result is

$$
\begin{aligned}
& 4f_E^{\nu} < f_{e,\nu} f_e^1 >= 4f_E^0 < f_{e,0} f_e^1 > -4f_E^1 < f_{e,1} f_e^1 > -4f_E^2 < f_{e,2} f_e^1 > -4f_E^3 < f_{e,3} f_e^1 > \\
&= 4f_E^0 < \dot{x}_i \dot{x}_0 e_i e_1 > -4f_E^1 < (\dot{x}_0^2 + \dot{x}_2^2 + \dot{x}_3^2) e_1 e_1 > +4f_E^2 < \dot{x}_1 \dot{x}_2 e_1 e_1 > +4f_E^3 < \dot{x}_1 \dot{x}_3 e_1 e_1 > \\
&= [4f_E^0 \dot{x}_0 \dot{x}_1 - 4f_E^1 (2\dot{x}_0^2 - 1) + 4f_E^1 \dot{x}_1 \dot{x}_1 + 4f_E^2 \dot{x}_1 \dot{x}_2 + 4f_E^3 \dot{x}_1 \dot{x}_3] < e_1 e_1 > \\
&= [4(f_E^0 \dot{x}_0 + \vec{f}_E \cdot \dot{\vec{x}}) \dot{x}_1 - 4f_E^1 (2\dot{x}_0^2 - 1)] < e_1 e_1 >
\end{aligned}.
$$

The other two space components of term $(a.3)$ can be obtained by the isotropy of the vacuum, and the final average result of term $(a.3)$ is

$$
\begin{aligned}
& 2 < f^2 f^{\mu} >= 2f_E^2 f_E^{\mu} + 2 < f_e^2 > f_E^{\mu} + 4f_E^{\nu} < f_{e,\nu} f_e^{\mu} > \\
&= 2f_E^2 f_E^{\mu} - 2(4\dot{x}_0^2 - 1) f_E^{\mu} < e_1 e_1 > +[4(f_E^0 \dot{\vec{x}}^2 - \dot{x}_0 \vec{f}_E \cdot \dot{\vec{x}}),\ 4(f_E^0 \dot{x}_0 + \vec{f}_E \cdot \dot{\vec{x}}) \dot{\vec{x}} - 4\vec{f}_E (2\dot{x}_0^2 - 1)] < e_1 e_1 > \\
&= 2f_E^2 f_E^{\mu} - [(4\dot{x}_0^2 + 2) f_E^0 + 4\dot{x}_0 \vec{f}_E \cdot \dot{\vec{x}}),\ -4(f_E^0 \dot{x}_0 + \vec{f}_E \cdot \dot{\vec{x}}) \dot{\vec{x}} + (16\dot{x}_0^2 - 6) \vec{f}_E] < e_1 e_1 > \\
&= 2f_E^2 f_E^{\mu} - [(8\dot{x}_0^2 + 2) f_E^0,\ -8f_E^0 \dot{x}_0 \dot{\vec{x}} + (16\dot{x}_0^2 - 6) \vec{f}_E] < e_1 e_1 >
\end{aligned}.
$$

In the following, we just list the final results for terms $(b)-(f)$, which would facilitate the check processes for serious-minded readers.

The average result for term (b) is

$$
2 < f^{\alpha} \frac{\partial f^{\nu}}{\partial \dot{x}^{\alpha}} \frac{\partial f^{\mu}}{\partial \dot{x}^{\nu}} >= 2f_E^{\alpha} \frac{\partial f_E^{\nu}}{\partial \dot{x}^{\alpha}} \frac{\partial f_E^{\mu}}{\partial \dot{x}^{\nu}} + 6[f_E^0,\ 2\dot{x}_0 \vec{E} - \vec{f}_E] < e_1 e_1 >,
$$

and the result for term (c) is

$$
< f^{\alpha} \frac{\partial^2 f^{\mu}}{\partial \dot{x}^{\alpha} \partial \dot{x}^{\nu}} f^{\nu} >= 0 ,
$$

which is the result of the facts that external electromagnetic fields are linear function of four-velocity and vacuum fluctuating fields are the functions of spacetime coordinates.

The average result of term (d) is

$$
4 < f_{\alpha} \frac{\partial f^{\alpha}}{\partial \dot{x}^{\nu}} f^{\nu} > \dot{x}^{\mu} = 0 ,
$$

which is closely dependent on the structure of the Lorentz force , and the average result for term (e) is

$$<\dot{x}^{\nu}\frac{\partial f^{\mu}}{\partial \dot{x}^{\nu}}f^{2}>=\dot{x}^{\nu}\frac{\partial f_{E}^{\mu}}{\partial \dot{x}^{\nu}}f_{E}^{2}+[-f_{E}^{0}(4\dot{x}_{0}^{2}+1),\ 4\dot{x}_{0}f_{E,0}\dot{\vec{x}}-\vec{f}_{E}(5\dot{x}_{0}^{2}+3\dot{\vec{x}}^{2})]<e_{1}e_{1}>.$$

The last term (f) vanishes due to the fact that the Lorentz force is orthogonal to four-velocity vector. Collecting these pieces together, we arrive at the final expression proportional to τ_0^2 of the contribution of vacuum fluctuating fields to the radiation reaction of a radiating charge:

$$\tau_{0}^{2}[(-12\dot{x}_{0}^{2}+3)f_{E}^{0},\ 12f_{E,0}\dot{x}_{0}\dot{\vec{x}}+12\dot{x}_{0}\vec{E}-(24\dot{x}_{0}^{2}-3)\vec{f}_{E})]<e_{1}e_{1}>. \quad (45)$$

It is easy to check that the contribution of vacuum fluctuating fields to the radiation reaction is orthogonal to the four-velocity of the radiating charge. The only undetermined quantity contained in Eqs. (43) and (45) is the average value of two electric components of vacuum fluctuating fields, which can not be determined precisely. However, the cutoff of wave vector is often taken as that corresponding to the Compton wavelength of the massive charge. For electrons, this quantity is $<e_1e_1>=\hbar k_C^4/6\pi=\Omega=9.3378\times10^{-5}T^2$, here " T " denotes the unit of magnetic field-Tesla.

It is necessary to illustrate the justification of our procedure, first performing the average calculation before solving the equation of motion, used to treat Eq. (38). It seems that the opposite procedure is more in line with general ideas. However, if one obtained the solution by directly solving Eq. (38) without performing the average calculation over random phases beforehand, we are justified to do the Taylor expansion about the vacuum fluctuations and then do average calculation for each term of this series. We can expect that two Taylor series obtained from two opposite procedures would coincide with each other at least for the first several terms.

In the following two sections, we will study the one-dimensional uniformly accelerating motion and the planar motion using the equation obtained in this section, which includes the radiation reaction up to τ_0^2 term with the contribution of the external field and the vacuum fluctuating fields.

5. One-dimensional motion of a charge acted by a constant electric field

In section 1, we have obtained the equation of motion of the one-dimensional uniformly accelerating motion of a point charge in a constant electric field. Although the radiation is emitted out during the charge undergoing this motion, the radiation reaction is bewilderingly zero! Much effort has been devoted to this problem, but there is no reasonable interpretation of this problem as yet. We propose a viewpoint that the radiation reaction of one-dimensional macroscopic motions of charged particles mainly comes from vacuum electromagnetic fluctuations expressed by Eqs. (43) and (45). Otherwise, it is very hard to understand the zero radiation reaction for the uniformly accelerating motion of a charge.

Assuming the constant electric field E is along x-axis, the radiation reaction of external field is zero, and the equation of motion of this case can be easily written out

$$\begin{aligned}\ddot{x}^\mu &= f_E^\mu - \tau_0\Omega[4\dot{\vec{x}}^2\dot{x}_0, 4\dot{x}_0^2\dot{\vec{x}}] \\ &-\tau_0^2\Omega[(12\dot{x}_0^2-3)f_E^0, -12\dot{x}_0\vec{E} - 12 f_E^0\dot{x}_0\dot{\vec{x}} + (24\dot{x}_0^2-3)\vec{f}_E]'\end{aligned} \tag{46}$$

where all the derivatives are with respect to proper time τ, and Ω denotes $< e_1 e_1 >$. In this case, the Lorentz force is $f_E^\mu = (\dot{x}E,\ \dot{x}_0 E,\ 0,\ 0)$. Eq. (46) shows that the magnitude of transverse motion exponentially decreases with proper time, namely $\dot{x}_{2,3}^2 \sim \exp(-8\tau_0\Omega\dot{x}_0^2\tau)$, which is reasonable because the mass of the charge increased with its velocity making it harder and harder for vacuum fluctuations to jolt it. However, the decay time $\sim (8\tau_0\Omega\dot{x}_0^2)^{-1}$ is very large showing that transverse motion induced by vacuum fluctuations could remain considerable magnitude for a long time and triggers extra radiation which is interpreted as Unruh radiation by some authors [Chen & Tajima, 1999; Thirolf et al., 2009; Iso et al., 2010]. The transverse motion of the charge is neglected in the following calculation.
The four-velocity and acceleration are $\dot{x}^\mu = (\dot{x}_0,\ \dot{x}),\quad \ddot{x}^\mu = (\ddot{x}_0,\ \ddot{x})$ respectively, and the component form of equation of motion can be expressed as

$$\begin{aligned}\ddot{x}_0 &= \dot{x}E - 4\tau_0\Omega\dot{x}^2\dot{x}_0 - \tau_0^2\Omega E(12\dot{x}_0^2-3)\dot{x} \\ \ddot{x} &= \dot{x}_0 E - 4\tau_0\Omega\dot{x}_0^2\dot{x} - \tau_0^2\Omega E(12\dot{x}_0^2-3)\dot{x}_0\end{aligned}. \tag{47}$$

Letting $\dot{x}_0 = \cosh\beta(\tau),\ \dot{x} = \sinh\beta(\tau)$, Eq. (47) is transformed into

$$\dot{\beta} = E(1-3\tau_0^2\Omega) - 2\tau_0\Omega\sinh(2\beta) - 6\tau_0^2\Omega E\cosh(2\beta), \tag{48}$$

which can be further expressed as

$$\begin{aligned}\tau - \tau_o &= \int_{\beta_0}^{\beta}\frac{d\beta}{E(1-3\tau_0^2\Omega) - 2\tau_0\Omega\sinh(2\beta) - 6\tau_0^2\Omega E\cosh(2\beta)} \\ &= \int_{\beta_0}^{\beta}\frac{exp(2\beta)d\beta}{E(1-3\tau_0^2\Omega)exp(2\beta) - \tau_0\Omega[exp(4\beta)-1] - 3\tau_0^2\Omega E[exp(4\beta)+1]} \\ &= \int_{\beta_0}^{\beta}\frac{ydy}{-(\tau_0\Omega + 3\tau_0^2\Omega E)y^4 + E(1-3\tau_0^2\Omega)y^2 + \tau_0\Omega - 3\tau_0^2\Omega E} \\ &= \int_{\beta_0}^{\beta}\frac{ydy}{-My^4 + Ny^2 + L}\end{aligned}, \tag{49}$$

where $y = exp(\beta) > 1$, and $M = \tau_0\Omega + 3\tau_0^2\Omega E,\ N = E(1-3\tau_0^2\Omega),\ L = \tau_0\Omega - 3\tau_0^2\Omega E$. The initial value of proper time τ_o should not be confused with the characteristic time of radiation reaction τ_0.
The time coordinate $x_0(\tau)$ can be obtained from $\dot{x}_0 = \cosh\beta$, namely

$$\begin{aligned}x_0(\tau) - x_0(0) &= \int_0^\tau \cosh\beta d\tau = \int_0^\tau \frac{\cosh\beta}{\dot{\beta}}d\beta \\ &= \int_0^\tau \frac{\cosh\beta d\beta}{E(1-3\tau_0^2\Omega) - 2\tau_0\Omega\sinh(2\beta) - 6\tau_0^2\Omega E\cosh(2\beta)}\end{aligned}$$

$$
\begin{aligned}
&=\frac{1}{2}\int_0^{\tau}\frac{[\exp(3\beta)+\exp(\beta)]d\beta}{E(1-3\tau_0^2\Omega)\exp[2\beta]-\tau_0\Omega[\exp(4\beta)-1]-3\tau_0^2\Omega E[\exp(4\beta)+1]} \\
&=\frac{1}{2}\int_0^{\tau}\frac{y^2+1}{-My^4+Ny^2+L}dy
\end{aligned}
\tag{50}
$$

Similarly, the space coordinate $x(\tau)$ has the expression:

$$x(\tau)-x(0)=\int_0^{\tau}\sinh\beta d\tau=\int_0^{\tau}\frac{\sinh\beta}{\dot{\beta}}d\beta=\frac{1}{2}\int_0^{\tau}\frac{y^2-1}{-My^4+Ny^2+L}dy\,. \tag{51}$$

Now all interesting quantities are ascribed to the calculation of three integrals:

$$\int_{y0}^{y}\frac{ydy}{-My^4+Ny^2+L},\quad \int_{y0}^{y}\frac{y^2+1}{-My^4+Ny^2+L}dy,\quad \int_{y0}^{y}\frac{y^2-1}{-My^4+Ny^2+L}dy\,.$$

Before we cope with these complicated integrals, let us first see two simple cases: (a) the vacuum fluctuating fields are neglected, namely the conventional uniformly accelerating motion, thus $M=L=0,\ N=E$, and Eq. (49) becomes

$$\tau-\tau_o=\frac{1}{E}\int_{y0}^{y}\frac{dy}{y}=\frac{1}{E}\ln(\frac{y}{y_0}),$$

which leads to the result $\beta-\beta_0=E(\tau-\tau_o)$. The expressions of spacetime coordinates are

$$
\begin{aligned}
x_0(\tau)-x_0(0)&=\frac{1}{2E}\int_{y0}^{y}\frac{y^2+1}{y^2}dy=\frac{1}{E}[\sinh\beta-\sinh\beta_0] \\
x_1(\tau)-x_1(0)&=\frac{1}{2E}\int_{y0}^{y}\frac{y^2-1}{Ey^2}dy=\frac{1}{E}[\cosh\beta-\cosh\beta_0]
\end{aligned},
$$

which are extensively used in the study of Unruh effect.
Case (b): the external field is absent, so $M=L=\tau_0\Omega$ and $N=E=0$. The solutions are

$$
\begin{aligned}
\tau-\tau_o&=\frac{1}{2M}\int_{y0}^{y}\frac{dy^2}{1-y^4}=\frac{1}{2M}\int_{y_0^2}^{y^2}\frac{dy}{1-y^2}=\frac{1}{2M}[\tanh^{-1}(y^{-2})-\tanh^{-1}(y_0^{-2})] \\
x_0(\tau)-x_0(0)&=\frac{1}{2M}\int_{y0}^{y}\frac{y^2+1}{1-y^4}dy=\frac{1}{2M}\int_{y0}^{y}\frac{1}{1-y^2}dy=\frac{1}{2M}[\tanh^{-1}(y^{-1})-\tanh^{-1}(y_0^{-1})], \\
x(\tau)-x(0)&=\frac{1}{2M}\int_{y0}^{y}\frac{y^2-1}{1-y^4}dy=\frac{-1}{2M}\int_{y0}^{y}\frac{1}{1+y^2}dy=\frac{1}{2M}[\tan^{-1}(y_0)-\tan^{-1}(y)]
\end{aligned}
$$

in getting these results, we used the integral formula

$$\int\frac{1}{1-x^2}dx=\begin{cases}\tanh^{-1}(x) & (0\le|x|<1)\\ \tanh^{-1}(x^{-1}) & (|x|>1)\end{cases},$$

where functions $\tanh^{-1}(y)$ and $\tan^{-1}(y)$ are inverse functions of $\tanh(x)$ and $\tan(x)$. From the expression of the proper time, we can see that $\beta(\infty)\to 0$ as $\tau\to\infty$ showing that a free charge moving in vacuum would exhaust its energy ending up with a quivering state.
At last we deal with the general case (c), namely Eqs. (49)-(51) which are expressed by three integrals. We just give the results of these integrals, the calculating process is not difficult. The results of these integrals are

$$\int\frac{ydy}{-My^4+Ny^2+L}=\frac{1}{\sqrt{4ML+N^2}}\tanh^{-1}[(\frac{2My^2-N}{\sqrt{4ML+N^2}})^{\pm}]$$

$$\int\frac{y^2}{-My^4+Ny^2+L}dy=\sqrt{\frac{\sqrt{4ML+N^2}+N}{2M\sqrt{4ML+N^2}}}\tanh^{-1}[(y\sqrt{\frac{2M}{\sqrt{4ML+N^2}+N}})^{\pm}]$$
$$-\sqrt{\frac{\sqrt{4ML+N^2}-N}{2M\sqrt{4ML+N^2}}}\tan^{-1}(y\sqrt{\frac{2M}{\sqrt{4ML+N^2}-N}}) \quad ,$$

$$\int\frac{1}{-My^4+Ny^2+L}dy=\sqrt{\frac{\sqrt{4ML+N^2}+N}{2L(4ML+N^2)}}\tan^{-1}[\sqrt{\frac{2M}{\sqrt{4ML+N^2}-N}}y]$$
$$+\sqrt{\frac{\sqrt{4ML+N^2}-N}{2L(4ML+N^2)}}\tanh^{-1}[(\sqrt{\frac{2M}{\sqrt{4ML+N^2}+N}}y)^{\pm}]$$

where the symbol " $\pm$ " is involved. When the argument of function $\tanh^{-1}(x)$ is larger than one, we take minus symbol, otherwise we take plus symbol.
It is not necessary to write out the expressions of proper time and spacetime coordinates as the functions of $y=\exp(\beta)$. We have already gained some insight into this kind of motion from cases (a) and (b). But the general case (c) could provide further insight into the effects of vacuum on the charge's motion. Let us investigate some information contained in Eq. (40). Its explicit expression is

$$\begin{aligned}\tau-\tau_o&=\frac{1}{\sqrt{4ML+N^2}}\tanh^{-1}[(\frac{2My^2-N}{\sqrt{4ML+N^2}})^{\pm}]\Big|_{y0}^{y}\\&\sim\frac{1}{E}\tanh^{-1}[(\frac{2\tau_0\Omega y^2-E}{E})^{\pm}]\Big|_{y0}^{y}\end{aligned}\quad , \tag{52}$$

in the second line we have used the approximation $M=L\sim\tau_0\Omega$, $N\sim E$, and we also assumed that external field $E<<\tau_0^{-1}$ (complete expression is $E<<mc/e\tau_0$). If we choose the " $+$ " as the superscript of the argument of function $\tanh^{-1}(x)$, we tacitly use the condition $y<\sqrt{E/(\tau_0\Omega)}$ and arrive at the expression

$$y^2\sim\frac{E}{2\tau_0\Omega}[1+\tanh(\tau E)],$$

with the initial value $y_0^2=y^2(0)=E/2\tau_0\Omega$, which corresponds to an extremely large energy of the charge. when $\tau\to\infty$, we have $y\to\sqrt{E/(\tau_0\Omega)}$ and the velocity $\tanh\beta$ of charge observed in laboratory frame approaches to the value $1-2\tau_0\Omega/E$. All these results and conditions are self-consistent.

When the initial value y_0^2 is small and does not meet the condition $y_0^2 = E / 2\tau_0\Omega$, we have approximate expression

$$y^2 = \frac{E(1+Y)}{2\tau_0\Omega}\frac{1+\tanh(E\tau)}{1+Y\tanh(E\tau)},$$

where $Y = (2\tau_0\Omega y_0^2 - E)/E$. As $\tau \to \infty$, the value of y increases to $\sqrt{E/(\tau_0\Omega)}$ which is close to what happens in linear accelerators. The vacuum fluctuations and the external electric field E together put an upper limit to its final energy.

Similarly we can discuss the opposite situation $y > \sqrt{E/(\tau_0\Omega)}$, Eq. (52) becomes

$$E(\tau - \tau_o) \sim \tanh^{-1}[\frac{E}{2\tau_0\Omega y^2 - E}]\Big|_{y0}^{y} .$$

For initial value $y_0 = y(0) = \infty$, the solution can be expressed as

$$y^2 = \frac{E}{2\tau_0\Omega}[\frac{1}{\tanh(E\tau)} + 1] .$$

when $\tau \to \infty$, y^2 decreases to $\sqrt{E/(\tau_0\Omega)}$.

If the initial value $y_0 > \sqrt{E/(\tau_0\Omega)}$ is finite, the solution becomes

$$y^2 = \frac{E(1+Y)}{2\tau_0\Omega}\frac{1+\tanh(E\tau)}{Y+\tanh(E\tau)},$$

where $Y = E/(2\tau_0\Omega y_0^2 - E) < 1$. As $\tau \to \infty$, y^2 also decreases to $\sqrt{E/(\tau_0\Omega)}$. This situation can be used to estimate the energy of charged cosmic rays and the electric field of the universe. In fact the limit value $\sqrt{E/(\tau_0\Omega)}$ of y^2 can be obtained directly from Eq. (47) by using the condition that the total force of the charge equals to zero.

We would like to briefly discuss about what we obtian in this section. If our starting point Eq. (38) is right, then the existence of the limit value of y^2 shows that the conventional uniformly accelerating motion can not be strictly realized in the real world, and the radiation reaction originated from the vacuum fluctuations could solve the puzzles associated with this motion, such as whether or not charges uniformly accelerated radiate with respect to inertial observers and the problems associated with a freely falling charge in a gravitation field etc. [Dewitt & Brehme, 1960; Singal, 1995, 1997; Lyle, 2008; Spallicci, 2010]. As seen in section 3, uniformly accelerating motion has an intimate relationship with the Unruh effect. Although B. L. Hu based on the model for a spinless charge interacting with a scalar field objects to relating the radiation reaction to vacuum fluctuation [Johnson, & Hu, 2005], we still insist on the viewpoint that the radiation reaction of the charge's one-dimensional macroscopic motions mainly comes from the vacuum electromagnetic fluctuating fields, which is another source besides external fields responsible for the radiation reaction of charges. As expressed by Eqs. (34) and (36), the radiation reaction produced by vacuum fluctuations exists in all motions of charges, but only for macroscopic linear motions it preponderates over that of external fields.

6. Planar motion of a charge in a constant magnetic field

Since the LDE is nonlinear, it is hard to solve it even in simple external field configurations. The planar motion of a charge in a constant magnetic field is another extremely important case except for the uniformly accelerating motion. D. J. Endres has proved that in this situation LDE admits a unique physical solution [Endres, 1993; Wang, 2010]. Many authors studied this problem and obtained various approximate solutions under certain conditions [Lubart, 1974]. To our knowledge no precise solution of LDE for this case has been found yet. J. S. Bell and J. M. Leinaas suggested that the residue polarization of electrons in storage rings could be served as a measure of Unruh temperature [Bell, & Leinaas, 1983, 1987; Unruh, 1998].

Classically electrons are regarded as scalar point charges, and all effects associated with spin are neglected [Jackson, 1976]. A charge moving in the plane with a constant magnetic field along its normal direction should execute an inward spiral motion radiating away energy and momentum according to Larmor law. Using the equation obtained in section 3, which contains radiation reaction effects produced by external field and vacuum fluctuations up to τ_0^2 term and is accurate enough for any macroscopic motions of charges. The solution of electrons' planar motion also means a way to experimentally detect the contribution of vacuum electromagnetic fluctuating fields to the radiation reaction.

The component form of the equation of motion is

$$\begin{aligned} \ddot{x}_0 &= -\tau_0(B^2+4\Omega)(\dot{x}_0^2-1)\dot{x}_0 \\ \ddot{x}_1 &= \dot{x}_2 B - \tau_0(B^2+4\Omega)\dot{x}_0^2\dot{x}_1 - \tau_0^2(B^3+B\Omega)(26\dot{x}_0^2-3)\dot{x}_2 \ , \\ \ddot{x}_2 &= -\dot{x}_1 B - \tau_0(B^2+4\Omega)\dot{x}_0^2\dot{x}_2 + \tau_0^2(B^3+B\Omega)(26\dot{x}_0^2-3)\dot{x}_1 \end{aligned} \tag{53}$$

which contains the radiation reaction up to τ_0^2 term. The external magnetic field $\vec{B}$ is along z-axis, and the electron moves in x-y plane.

The first expresion of Eq. (53) describes the time component of four-velocity changing with the proper time, which can be written as

$$\int \frac{d\dot{x}_0^2}{(\dot{x}_0^2-1)\dot{x}_0^2} = \int -2\tau_0(B^2+4\Omega)d\tau \ .$$

This expression can be integrated out immediately, and the result is

$$\dot{x}_0^2(\tau) = \frac{\dot{x}_0^2(0)}{\dot{x}_0^2(0)-[\dot{x}_0^2(0)-1]exp[-2\tau_0(B^2+4\Omega)\tau]} \ . \tag{54}$$

When $\tau \to \infty$, we can see that $\dot{x}_0^2(\tau) \to 1$ indicating that the electron's kinetic energy was exhausted by radiation reaction force after some time. The total energy of the electron changing with proper time τ is described by

$$m\dot{x}_0 = \frac{m\dot{x}_0(0)}{\sqrt{\dot{x}_0^2(0)-[\dot{x}_0^2(0)-1]exp[-2\tau_0(B^2+4\Omega)\tau]}} \ .$$

To solve the equations of $\dot{x}_1$ and $\dot{x}_2$, we introduce the complex variable $\dot{Z} = \dot{x}_1 + i\dot{x}_2$, which satisfies the equation:

$$\ddot{Z} = -[iB + \tau_0(B^2 + 4\Omega)\dot{x}_0^2 - i\tau_0^2(B^3 + B\Omega)(26\dot{x}_0^2 - 3)]\dot{Z}\,. \tag{55}$$

This equation can be easily solved and the result is

$$\dot{Z} = \dot{Z}_0 exp[-\int_0^\tau [iB + \tau_0(B^2 + 4\Omega)\dot{x}_0^2 - i\tau_0^2(B^3 + B\Omega)(26\dot{x}_0^2 - 3)]d\tau]$$

$$= \dot{Z}_0 exp[-iB\tau - 3i\tau_0^2(B^3 + B\Omega)\tau]exp\{[-\tau_0(B^2 + 4\Omega) + 26i\tau_0^2(B^3 + B\Omega)]\int_0^\tau \dot{x}_0^2(\tau)d\tau\}$$

Using Eq. (54), the integral of $\dot{x}_0^2(\tau)$ can be carried out

$$\int_0^\tau \dot{x}_0^2(\tau)d\tau = \int_0^\tau \frac{\dot{x}_0^2(0)}{\dot{x}_0^2(0) - [\dot{x}_0^2(0) - 1]exp[-2\tau_0(B^2 + 4\Omega)\tau]}d\tau$$

$$= \int_0^\tau \frac{exp[2\tau_0(B^2 + 4\Omega)\tau]}{exp[2\tau_0(B^2 + 4\Omega)\tau] - [1 - 1/\dot{x}_0^2(0)]}d\tau \qquad .$$

$$= \frac{Ln[\dot{x}_0^2(0)exp[2\tau_0(B^2 + B_v^2)\tau] - \dot{x}_0^2(0) + 1]}{2\tau_0(B^2 + 4\Omega)}$$

Now, the expression of $\dot{Z}$ becomes

$$\dot{Z} = \dot{Z}_0 exp[-iB\tau - 3i\tau_0^2(B^3 + B\Omega)\tau]exp\{[-\tau_0(B^2 + 4\Omega) + 26i\tau_0^2(B^3 + B\Omega)]$$

$$\times\frac{Ln[\dot{x}_0^2(0)exp[2\tau_0(B^2 + 4\Omega)\tau] - \dot{x}_0^2(0) + 1]}{2\tau_0(B^2 + 4\Omega)}\} \qquad , \tag{56}$$

$$= \dot{Z}_0\dot{x}_0^{2\alpha}(0)exp[(-iB - \tau_0(B^2 + 4\Omega) + 23i\tau_0^2(B^3 + B\Omega))\tau][1 - \left|\dot{Z}_0\right|^2 exp[-2\tau_0(B^2 + 4\Omega)\tau]/\dot{x}_0^2(0)]^\alpha$$

where

$$\alpha = -\frac{1}{2} + 13i\tau_0\frac{B^3 + B\Omega}{B^2 + 4\Omega},$$

and $\dot{x}_0^2(0) = 1 + \left|\dot{Z}_0\right|^2$. It is very hard to find the exact expressions for $x_{1,2}(\tau)$, or they do not exist at all. However, Eq. (56) is the product of two parts; the first part is

$$\dot{Z}_0 exp[-iB\tau - 3i\tau_0^2(B^3 + B\Omega)\tau + 2\alpha\tau_0(B^2 + 4\Omega)\tau],$$

which varies quickly with the proper time compared with the second part

$$[\dot{x}_0^2(0) - \left|\dot{Z}_0\right|^2 exp[-2\tau_0(B^2 + 4\Omega)\tau]]^\alpha .$$

Therefore, $Z(\tau)$ can be expressed as the series in the quantity

$$\frac{2\tau_0(B^2 + 4\Omega)}{-iB - \tau_0(B^2 + 4\Omega) + 23i\tau_0^2(B^3 + B\Omega)}.$$

By the method of integration by parts, $Z(\tau)$ can be expressed as

$$Z(\tau) = Z_0 + \frac{\dot{Z}_0 \dot{x}_0^{2\alpha}(0)}{-iB - \tau_0(B^2 + 4\Omega) + 23i\tau_0^2(B^3 + B\Omega)}$$
$$\times\{exp[(-iB - \tau_0(B^2 + 4\Omega) + 23i\tau_0^2(B^3 + B\Omega))\tau][1 - |\dot{Z}_0|^2 exp[-2\tau_0(B^2 + 4\Omega)\tau] / \dot{x}_0^2(0)]^\alpha - 1\} + \cdots \quad , (57)$$

where only the first term is presented. This expression is of high accuracy for any proper time and any initial value of velocity. It is easy to prove that the complete series of Eq. (57) is absolutely convergent.

For the inward spiral motion of electrons, the magnitude r of the radius vector $\vec{r}(\tau) = x_1(\tau)\vec{e}_x + x_2(\tau)\vec{e}_y$ changing with proper time is an interested quantity. Due to $dr = \sqrt{dx_1^2 + dx_2^2} = |dZ|$ and with Eq. (56), we have

$$-\frac{dr}{d\tau} = |\dot{Z}_0[\dot{x}_0^2(0) exp[2\tau_0(B^2 + 4\Omega)\tau] - \dot{x}_0^2(0) + 1]^\alpha| = \frac{|\dot{Z}_0|}{\sqrt{\dot{x}_0^2(0) exp[2\tau_0(B^2 + 4\Omega)\tau] - |\dot{Z}_0|^2}}, \quad (58)$$

which can be further expressed as

$$r(\tau) = r(0) + \frac{|\dot{Z}_0|}{\tau_0(B^2 + 4\Omega)} \int_0^\tau \frac{d exp[-\tau_0(B^2 + 4\Omega)\tau]}{\sqrt{\dot{x}_0^2(0) - |\dot{Z}_0|^2 exp[-2\tau_0(B^2 + 4\Omega)\tau]}}, \quad (59)$$

where $r(0)$ is the initial distance of the electron away from the origin.

Using the integral formulas

$$\int \frac{dx}{\sqrt{1 - x^2}} = \sin^{-1}(x), \quad \int \frac{-1}{\sqrt{1 - x^2}} dx = \cos^{-1}(x),$$

the calculation of Eq. (59) can be carried out and the result is

$$\int_0^\tau \frac{d exp[-\tau_0(B^2 + 4\Omega)\tau]}{\sqrt{\dot{x}_0^2(0) - |\dot{Z}_0|^2 exp[-2\tau_0(B^2 + 4\Omega)\tau]}}$$
$$= \frac{1}{|\dot{Z}_0|}\{\sin^{-1}[\frac{|\dot{Z}_0|}{\dot{x}_0(0)} exp(-\tau_0(B^2 + 4\Omega)\tau)] - \sin^{-1}[\frac{|\dot{Z}_0|}{\dot{x}_0(0)}]\}$$

Thus we obtain the expression of $r(\tau)$:

$$r(\tau) = r(0) + \frac{1}{\tau_0(B^2 + 4\Omega)}\{\sin^{-1}[\frac{|\dot{Z}_0|}{\dot{x}_0(0)} exp(-\tau_0(B^2 + 4\Omega)\tau)] - \sin^{-1}[\frac{|\dot{Z}_0|}{\dot{x}_0(0)}]\}, \quad (60)$$

which shows that the electron's distance away from the origin slowly decreases with the proper time. The electron really undergoes an inward spiral motion.

We prefer to express physical quantities by laboratory time x_0 rather than proper time τ. To do so, we first calculate

$$\frac{dt}{d\tau}=\frac{\dot{x}_0(0)}{\sqrt{\dot{x}_0^2(0)-\left|\dot{Z}_0\right|^2 exp[-2\tau_0(B^2+4\Omega)\tau]}}$$
$$=\frac{\dot{x}_0(0)exp[\tau_0(B^2+4\Omega)\tau]}{\sqrt{\dot{x}_0^2(0)exp[2\tau_0(B^2+4\Omega)\tau]-\left|\dot{Z}_0\right|^2}} \tag{61}$$

from Eq. (54). Using integral formula $\int(x^2-1)^{-1/2}dx=\cosh^{-1}(x)$,we obtain

$$t=\int_0^\tau\frac{\dot{x}_0(0)exp[\tau_0(B^2+4\Omega)\tau]d\tau}{\sqrt{\dot{x}_0^2(0)exp[2\tau_0(B^2+4\Omega)\tau]-\left|\dot{Z}_0\right|^2}},$$
$$=\frac{\dot{x}_0(0)}{\tau_0(B^2+4\Omega)\left|\dot{Z}_0\right|}\{\cosh^{-1}[\frac{\dot{x}_0(0)}{\left|\dot{Z}_0\right|}exp[\tau_0(B^2+4\Omega)\tau]]-\cosh^{-1}[\frac{\dot{x}_0(0)}{\left|\dot{Z}_0\right|}]\} \tag{62}$$

in getting this result, we have assumed that initial laboratory time $t=0$ corresponds to initial proper time $\tau=0$.

The distance of electron away from the origin varies with the laboratory time t can be written out according to Eqs. (54) and (58) and the result is

$$-\frac{dr}{dt}=-\frac{dr/d\tau}{dt/d\tau}=\frac{\left|\dot{Z}_0\right|}{\dot{x}_0(0)}exp[-\tau_0(B^2+4\Omega)\tau]. \tag{63}$$

If we can express the right hand side of this equation by the expression of t, then we get the equation describing r changing with laboratory time t. From Eq. (62), after some calculation we can obtain a quadratic equation of $exp[\tau_0(B^2+4\Omega)\tau]$, which is

$$exp[2\tau_0(B^2+4\Omega)\tau]-2exp[\tau_0(B^2+4\Omega)\tau]\cosh[\tau_0(B^2+4\Omega)t]$$
$$+\frac{1}{\dot{x}_0^2(0)}+\frac{\left|\dot{Z}_0\right|^2}{\dot{x}_0^2(0)}\cosh^2[\tau_0(B^2+4\Omega)t]=0 .$$

It is easy to solve this equation and the result is

$$exp[\tau_0(B^2+4\Omega)\tau]=\cosh[\tau_0(B^2+4\Omega)t]\pm\frac{1}{\dot{x}_0(0)}\sinh[\tau_0(B^2+4\Omega)t]. \tag{64}$$

We know that when the electron is stationary, $\dot{x}_0(0)=1$ and $t=\tau$. So we should take the " + " in Eq. (64), and Eq. (63) becomes

$$-\frac{dr}{dt}=\frac{\left|\dot{Z}_0\right|}{\dot{x}_0(0)\cosh[\tau_0(B^2+4\Omega)t]+\sinh[\tau_0(B^2+4\Omega)t]}.$$

This equation can be easily solved and the result is

$$
\begin{aligned}
r(0)-r(t) &= \int_0^t \frac{|\dot{Z}_0|\,dt}{\dot{x}_0(0)\cosh[\tau_0(B^2+4\Omega)t]+\sinh[\tau_0(B^2+4\Omega)t]} \\
&= \frac{2|\dot{Z}_0|}{\tau_0(B^2+4\Omega)}\int_0^t \frac{dexp(y)}{[\dot{x}_0(0)+1]exp(2y)+[\dot{x}_0(0)-1]} \\
&= \frac{2}{\tau_0(B^2+4\Omega)}\{\tan^{-1}[\sqrt{\frac{\dot{x}_0(0)+1}{\dot{x}_0(0)-1}}exp[\tau_0(B^2+4\Omega)t]]-\tan^{-1}[\sqrt{\frac{\dot{x}_0(0)+1}{\dot{x}_0(0)-1}}]\}
\end{aligned}
.
$$

So the final expression of the electron's distance away from the origin expressed by the laboratory time is

$$
r(t)=r(0)-\frac{2}{\tau_0(B^2+4\Omega)}\{\tan^{-1}[\sqrt{\frac{\dot{x}_0(0)+1}{\dot{x}_0(0)-1}}exp[\tau_0(B^2+4\Omega)t]]-\tan^{-1}(\sqrt{\frac{\dot{x}_0(0)+1}{\dot{x}_0(0)-1}})\}, \quad (65)
$$

which explicitly shows the inward spiral characteristic of the electron's planar motion with a constant magnetic field along its normal direction. Classically the final destiny of electrons performing such a motion is falling into the origin.

For ultrarelativistic electrons, $\dot{x}_0(0)$ is very large and $\tan^{-1}[\sqrt{(\dot{x}_0(0)+1)/(\dot{x}_0(0)-1)}]\doteq\pi/4$. Using Eq. (65), We can estimate the laboratory time needed for an ultrarelativistic electron to decrease its the distance from $r(0)$ to $r(0)/2$, and the result is

$$
t_{half}=\frac{1}{\tau_0(B^2+4\Omega)}\ln\frac{1+\tan[\tau_0(B^2+4\Omega)r(0)/4]}{1-\tan[\tau_0(B^2+4\Omega)r(0)/4]}.
$$

The initial value $r(0)$ can be approximated by that of the situation without considering the radiation reaction effects, namely $r(0)\approx\dot{x}_0/B$, so t_{haff} can be written as

$$
t_{half}=\frac{1}{\tau_0(B^2+4\Omega)}\ln\frac{1+\tan[\tau_0(B^2+4\Omega)\dot{x}_0(0)/4B]}{1-\tan[\tau_0(B^2+4\Omega)\dot{x}_0(0)/4B]}. \quad (66)
$$

However, due to τ_0 being a very small quantity, we are justified to further simplify this expression and the result is

$$
t_{half}\sim\frac{\dot{x}_0(0)}{2}\frac{mc}{eB}\sim 10^{-3}\frac{\dot{x}_0(0)}{B}
$$

seconds for electrons which shows that t_{half} has little relationship with vacuum fluctuations.
For ultrarelativistic electrons, according to Eq. (64), we have approximate expression

$$
exp[\tau_0(B^2+4\Omega)\tau]\doteq\cosh[\tau_0(B^2+4\Omega)t],
$$

so the energy of electrons can be simplified as

$$
\frac{\dot{x}_0(\tau)}{\dot{x}_0(0)}\sim\frac{\cosh[\tau_0(B^2+4\Omega)t]}{\sqrt{\dot{x}_0^2(0)\cosh^2[\tau_0(B^2+4\Omega)t]-|\dot{Z}_0|^2}}\sim\frac{\cosh[\tau_0(B^2+4\Omega)t]}{\dot{x}_0(0)\sinh[\tau_0(B^2+4\Omega)t]}.
$$

The ratio of two $\dot{x}_0(\tau)/\dot{x}_0(0)$'s respectively with and without the considerations of vacuum fluctuations is

$$\frac{[\dot{x}_0(\tau)/\dot{x}_0(0)]_{with}}{[\dot{x}_0(\tau)/\dot{x}_0(0)]_{without}} \sim 1 - 4\Omega/B^2 , \tag{67}$$

which shows that the effect of the vacuum fluctuations. We look forward to seeing that this result would be tested in future experiments. The nonzero effects of vacuum fluctuations had been recognized in microscopic world long time ago, such as the Lamb shift and the Casmir effect etc. However, whether or not the vacuum fluctuations has a relationship with the radiation reaction for the motion of charges is still an open question. The planar motion of high energy electrons with a constant magnetic field perpendicular to its moving plane provides a possible experimental scheme to test this viewpoint.

7. Conclusion

In this chapter, we presented a new reduction of order form of LDE, which coincides with that obtained by the method of Landau and Lifshitz in its Taylor series form. Using the classical version of zero-point electromagnetic fluctuating fields of the vacuum, we obtained the contributions of vacuum fluctuations to radiation reaction of a radiating charge up to the τ_0^2 term. Then we use the obtained reduction of order equation of LDE including the radiation reaction induced by external force and vacuum fluctuations up to the τ_0^2 term, which is accurate enough for any macroscopic motions of charges and even applicable to the electron's motion of a hydrogen atom due to τ_0 being extremely small, to study the one-dimensional uniformly accelerating motion produced by a constant electric field and the planar motion produced by a constant magnetic field. Our calculations show that for any one-dimensional uniformly accelerating motion the velocity of charges has a limit value and almost all puzzles associated with this special motion disappear; while the planar motion of electrons provides an experimental scheme to test the conjecture that the interaction between charged particles and the vacuum electromagnetic fluctuations is anther mechnisim for the charge's radiation reaction, which plays a dominant role only for one-dimensional macroscopic motions of charged particles.

8. Acknowledgment

This work was supported by the postdoc foundation of Shanghai city, China. The author would like to thank Dr Hui Li for reading the manuscript and proposing fruitful comment on the content of this chapter.

9. References

Aguirregabiria, J. M., (1997), Solving Forward Lorentz-Dirac-like Equations, *J. Phys. A: Math. Gen*, Vol. 30, pp. 2391-2402.

Boulware, D., (1980), Radiation from a Uniformly Accelerated Charge, *Ann. Phys*, Vol. 124, pp. 169-188.

Boyer, T. H., (1980), Thermal Effects of Acceleration through Random Classical Radiation. *Phys. Rev. D*, Vol. 21, pp. 2137-2148.

Bell, J. S., & Leinaas, J. M., (1983), Electrons as Accelerated Thermometers, *Nucl. Phys B*, Vol. 212, pp.131-149; (1987), The Unruh Effect and Quantum Fluctuaions of Electrons in Storage Ring, *Nucl. Phys. B*, Vol. 284, pp. 488-508.

Chen, P., & Tajima, T., (1999). Testing Unruh Radiation with Ultraintense Lasers, *Phys. Rev. Lett*, Vol. 83, pp. 256-259.

Carati, A., (2001). An extension of Eliezer's theorem on the Abraham-Lorentz-Dirac Equation, *J. Phys. A. Gen*, Vol. 34, pp. 5937-5944.

Cole, D. C., & Zou, Y., (2003), Quantum mechanical Ground State of Hydrogen Obtained from Classical Electrodynamics, *Phys. Lett. A*, Vol. 317, pp. 14-20.

Dirac, P. A. M., (1938). Classical Theory of Radiating Electrons, *Proc. R. Soc. London A*, Vol. 167, pp. 148-168.

Dewitt, B. S., & Brehme, R., (1960), Radiation Damping in a Gravitational Field, *Ann. Phys*, Vol. 9, pp. 220-259.

Eliezer, C. J., (1947). The Interaction of Electrons and an Electromagnetic Field, *Rev. Mod. Phys*, Vol. 19, pp. 147-184.

Endres, D. J., (1993), The Physical Solution to the Lorentz-Dirac Equation for Planar Motion in a Constant Magnetic Field, *Nonlinearity*, Vol. 6, pp. 953-971.

Fulton, T., & Rohrlich, F., (1960), Classical Radiation from a Uniformly Accelerated charge, *Ann. Phys*, Vol. 9, pp. 499-517.

Ginzberg, V. L., (1970), Radiation and Radiation Friction Force in Uniformly Accelerated Motion of a Charge, *Sov. Phys. Usp*, Vol. 12, pp. 569-585.

Iso, S., Yamamoto, Y, & Zhang, S., (2010), Does an Accelerated Electron Radiate Unruh Radiation, available from, arXiv:1011.419lvl.

Johnson, P. R., & Hu, B. L., (2005), Uniformly Accelerated Charge in a Quantum Field: from Radiation Reaction to Unruh Effect, *Foundations of Physics*, Vol. 35, pp. 1117-1147.

Jackson, J. D., (1976), On Understanding Spin-Flip Synchrotron Radiation and the Transverse Polarization of Electrons in Storage Rings, *Rev. Mod. Phys*, Vol. 48, pp. 417-433.

Kawaguchi, H., Tsubota, K., & Honma, T., Lorentz Group Lie Algebra Map of Ultra-Relativistic Radiating Electron, availabel from http://epaper.kek.jp/pac97/papers/pdf/8P086.PDF.

Landau, L. D., & Lifshitz, E. M., (1962), *The Classical Theory of Fields*, Pergamon, Oxford.

Lyle, S. N., (2008), *Uniformly Accelerating Charged Particles*, Springer, ISBN. 978-3-540-68469-5, Berlin, Heidelberg.

Lubart, N. D., (1974), Solution of Radiation Reaction Problem for the Uniform Magnetic field, *Phys. Rev. D*, Vol. 9, pp. 2717-2722.

Plass, G. N., (1961). Classical Electrodynamic Equations of Motion with Radiative Reaction, *Rev. Mod. Phys*, Vol. 33, pp. 37-62.

Spallicci, A., (2010), Free Fall and Self-Force: an Historical Perspective, available from, arXiv:1005.0611v1.

Singal, A. K., (1995), The Equivalence Principle and an Electric Charge in a Gravitational Field, *Gen. Rel. Grav*, Vol. 27, pp. 953-967; (1997), The Equivalence Principle and an Electric Charge in a Gravitational Field II. A Uniformly Accelerated Charge does not Radiate, *Gen. Rel. Grav*, Vol. 29, pp. 1371-1390.

Thirolf, P. G., Habs, D., Henig, A., Jung, D., Kiefer, D., Lang, C., Schrerber, J., Mais, C., Schaller, G., Schutzhold, R., & Tajima, T., (2009), Signatures of the Unruh Effect via High-Power, Short-Pulse Lasers, *Eur. Phys. J. D*, Vol. 55, pp. 379-389.

Teitelboim, C., (1970), Splitting of the Maxwell Tensor: Radiation Reaction without advanced Field, *Phys. Rev. D*, Vol. 1, pp. 1572-1582.

Teitelboim, C., Villarroel, D., & Weert, C. G., (1980), Classical Electrodynamics of Retarded Fields and Point Particles, *Riv. Nuovo Cimento*, Vol. 3, pp. 1-64.

Unruh, W. G., (1998), Accleration Radiation for Orbiting Electrons, available from, arXiv:hep-th/9804158v1;

Unruh, W. G., (1976), Notes on Black Hole Evaporation, *Phys. Rev. D*, Vol. 14, pp. 870-892.

Wang, Guozhong., Li, Hui., Shen, Yifeng., Yuan, Xianzhang., & Zi, Jian., (2010). Anti-Dampimg Effect of Radiation Reaction, *Phys. Scr*, Vol. 81, pp. 1-10.

Wheeler, J. A., & Feynman, R. P., (1945), Interaction with the Absorber as the Mechanism of Radiation, *Rev. Mod. Phys*, Vol. 17, pp. 157-181; (1949), Classical Electrodynamics in Terms of Diract Interparticle Action, *Rev. Mon. Phys*, Vol. 21, pp. 425-433.

18

Observation of Cavity Interface and Mechanical Stress in Opaque Material by THz Wave

Tsuguhiro Takahashi
Central Research Institute of Electric Power Industry
Japan

1. Introduction

One of the recent topics of optical measurement techniques is the generation and utilization of the "Terahertz wave (THz wave)". In the long history of the research on electromagnetic waves, it remains as the last unrevealed area because its generation and detection techniques have not been sufficiently developed. In recent years, some useful and convenient devices have become commercially available, and much related research is being carried out. One important feature of THz wave is the high transmission probability through some opaque materials, many of which are dielectric materials through which a visible light beam cannot pass. Solid insulating materials are usually opaque dielectric materials, and it is difficult to measure their interior by conventional optical measurement techniques. If THz wave can pass through an insulating material sufficiently, the same measurement technique as with a visible light beam, already developed and practically applied, such as the technique of utilizing polarization, will become applicable. In this chapter, the research work on the introduction of THz wave techniques to the internal measurement of solid insulating materials is reviewed [1,2].

2. Applicability of THz wave to internal measurement of solid insulating materials

The dielectric strength of insulating materials generally increases in the order of gas < liquid < solid, but from the viewpoints of the selfrecovery of insulation after discharge, the cooling method and so on, SF6 gas and insulating oil are generally adopted in high voltage electric power equipment. On the other hand, polyethylene is successfully utilized in high voltage XLPE cables with well controlled manufacturing techniques. In the research work on the "All Solid Insulated Substation" [3], high voltage moulded transformer, compact connector and bus system are being developed. Such equipment is expected to show high insulation performance, if there is no fault in the manufacturing process and it is operated with the monitoring of aging. Cost reduction will be realized by compactification and long time use, if solid insulating materials can be sufficiently utilized. Practical problems of solid insulating materials are, internal cavities (voids) occurrence, electrical and mechanical distortions and so on. In order to utilize solid insulating materials effectively, these faults should be monitored during production and operation, but there is no established non-destructive measurement method for opaque insulating materials.

In order to obtain internal information from the outside, the detecting "probe" must be inserted. For a solid material, it is not possible to insert some objects, but sonic or electromagnetic waves can be utilized. These waves should have high transmission probability and sufficient interaction with the target to detect physical quantities. An example of the classifications of an electromagnetic waves is shown in Fig. 1.

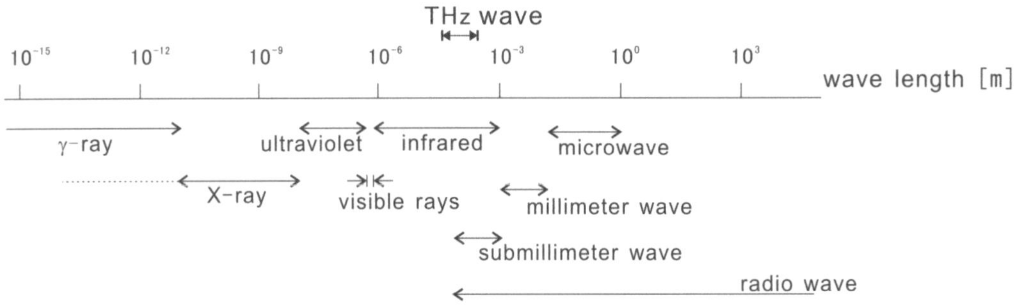

Fig. 1. Names and wave lengths of electromagnetic waves

"THz wave" means an electromagnetic wave with a "terahertz" frequency band. Generally, research work on electromagnetic waves of several100 GHz - several10 THz in frequency is being carried out recently. This frequency band has remained an "unrevealed area", but because of the recent development of several related techniques, such as the femtosecond laser, the generation and detection techniques of THz wave have been developed (Table 1 and Table 2), and several related applications are being examined. THz wave techniques are expected to be applied in a wide range of areas in the future.

2.1 Transmission probability of THz wave for solid insulating materials

Most solid insulating materials are opaque dielectric materials. "Opaque" means, visible light of about 400 nm - 750 nm in wavelength and electro magnetic waves of around this band cannot be sufficiently transmitted. It is said that electromagnetic waves of longer wavelengths (far red) can be transmitted, and it is known that microwaves and millimeter waves, which have much longer wave lengths, can also be transmitted. In order to utilize THz waves for transmission measurement of solid insulating materials, it should be clarified, which frequency is suitable for target materials. But there is not yet a systematic database of such data.

Examples of measurement results of transmission probability for PE (polyethylene), Epoxy, silicone rubber, and EPDM (ethylene propylene diene Monomer) rubber, which are typical insulating materials, are shown in Fig. 2, 3, 4 and 5. Except for PE, there are large and almost flat absorptions for the indicated frequency range.

Other transmission probability measurements for several kinds of solid materials have been made, as shown in Fig. 6 as an example[1]. It was obtained with the BWO[2] spectroscope (Fig. 7), which is commercially available. In Fig. 6, adjacent points have been averaged, but there

---

[1] These characteristics are those of author's samples. There is a possibility the same material with other filler and manufacturing process will show other characteristics.

[2] Back-ward Wave Oscillator Tube; it generates single frequency electromagnetic waves, which can be adjusted in the range of millimeter wave to sub-millimeter waves by changing the applied voltage. The changeable range depends on the model number of the tube.

still remains oscillation, which is called the "etalon effect", which is decided by the thickness of samples (5mm) and frequency. According to this data, the transmission probability of THz waves is not high for many solid materials. From the next section, actual measurement made for polyethylene, through which THz waves can transmit well, are described, but the same measurement can be applied for other materials with lower frequency waves (, in such a case, the spatial resolution of measurement should be worse).

Method		frequency	power	source material
optical mixing [4,5]		--3THz 1.5--7THz 1−10THz	several 10nW(CW) 11mW(PW) 0.1μW(PW)	GaAs organic DAST crystal InGaP/InGaAs/GaAs
Parametric oscillation [6,7]		0.65--2.6THz	200mW(PW)	$LiNbO_3$ $LiTaO_3$
femto second plus +	photo conduction switch [8]	--3THz	1μW(PW)	GaAs
	super conductive optical switch [9,10]	--2THz	0.5μW(PW)	YBCO
	semiconductor (by surface electric field) [11,12]	--37THz	(PW)	InP, GaAs
	semiconductor (by coherent phonon) [13]	--4THz	(PW)	Te, PbTe, CdTe
	semiconductor in high magnetic field [14]	--2THz	100mW(PW)	InAs, GaAs
	single quantum well [13,15]	1.4−2.6THz	(PW)	GaAs/ $Al_{0.3}Ga_{0.7}As$ (quantum well structure)
	double quantum well [13,16]	1.5THz	(PW)	$Al_{0.2}Ga_{0.8}As$
gas laser		0.3--7THz	50mW(CW)	CH_3OH, CH_2F_2, CH_3Cl
quantum cascade laser [17]		1.5--THz	数10mW(PW)	GaAs/AlGaAs, InGaAs/InAlAs
millimeter wave + frequency multiplier [18]		2.7THz	(CW)	GaAs schottky diode
backward wave oscillation from electron beam in high magnetic field [19]		--1.4THz	--several mW(CW)	backward wave oscillator tube

Table 1. Generation methods of THz wave (CW: continuous wave, PW: pulsed wave)

method	notes
photo conduction switch [13,16]	The same switch as in the generation method is utilized. Frequency spectrum is obtained from a waveform in the time domain by Fourier transform.
thermal absorbtion	Temperature change of materials (gas cell) absorbing THz wave is detected (, applicable for wide frequency range).
EO sampling [13,16]	THz wave intensity is converted to visible light beam intensity by the Pockels effect. Imaging with a CCD camera is realized.
photon counting [20]	Single electron transistor of quantum dot structure is utilized. Intensity is measured directly as numbers of photons.
Superconductive tunnel junction device [21]	Photon-assisted tunneling effect is detected. There is the possibility of being arrayed.

Table 2. Detection methods of THz wave

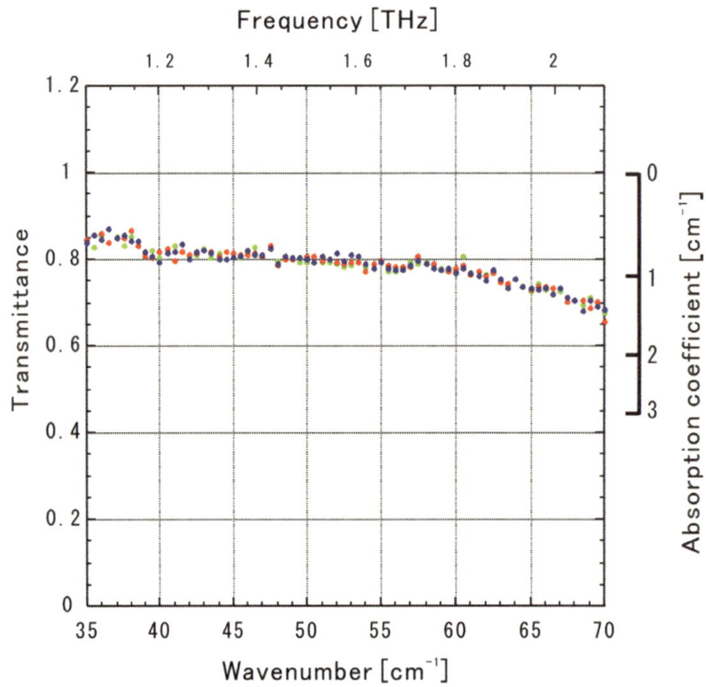

Fig. 2. Transmission probability measurement for PE (thickness: 2.7mm; dot colours represent measurement ID numbers)

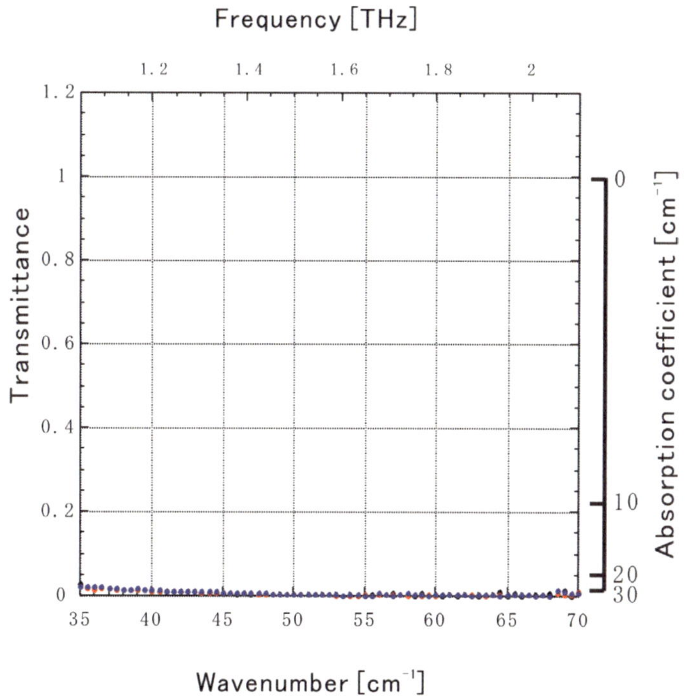

Fig. 3. Transmission probability measurement for Epoxy (thickness: 1.5mm)

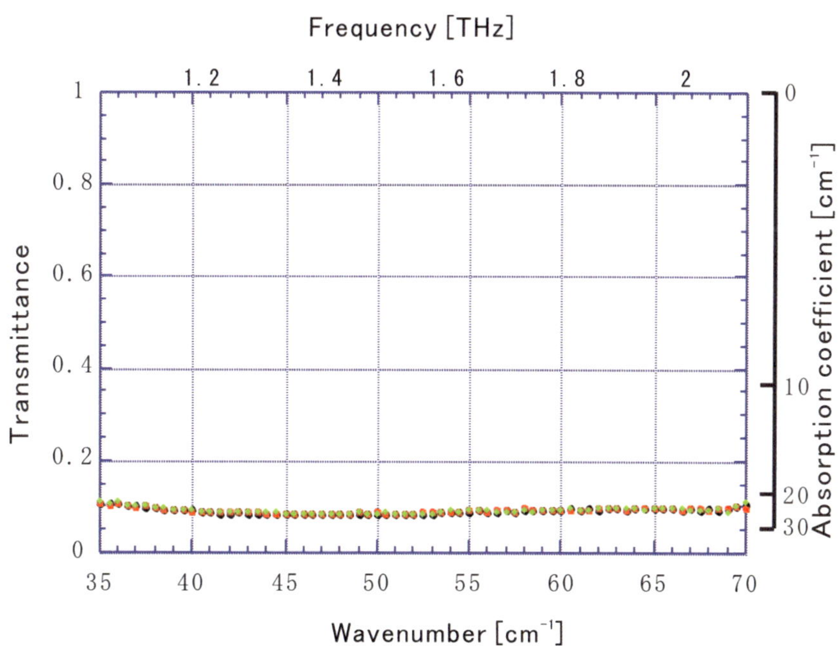

Fig. 4. Transmission probability measurement for silicone rubber (thickness: 1mm)

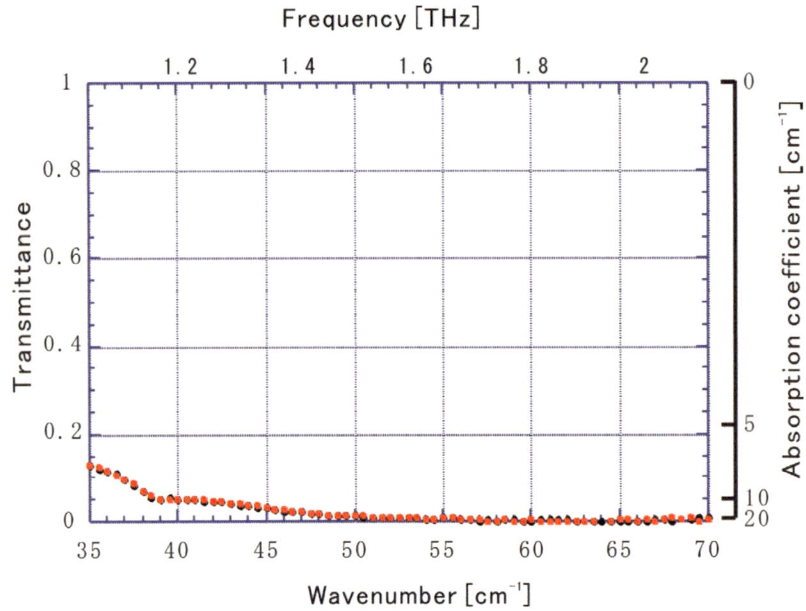

Fig. 5. Transmission probability measurement for EPDM rubber (thickness: 3mm)

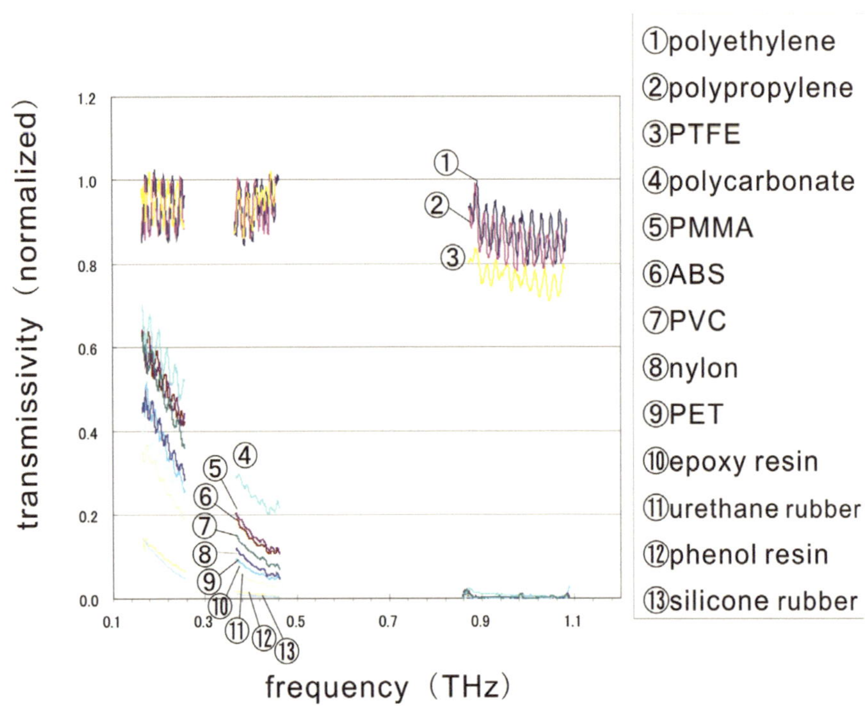

Fig. 6. Transmission probability of solid materials

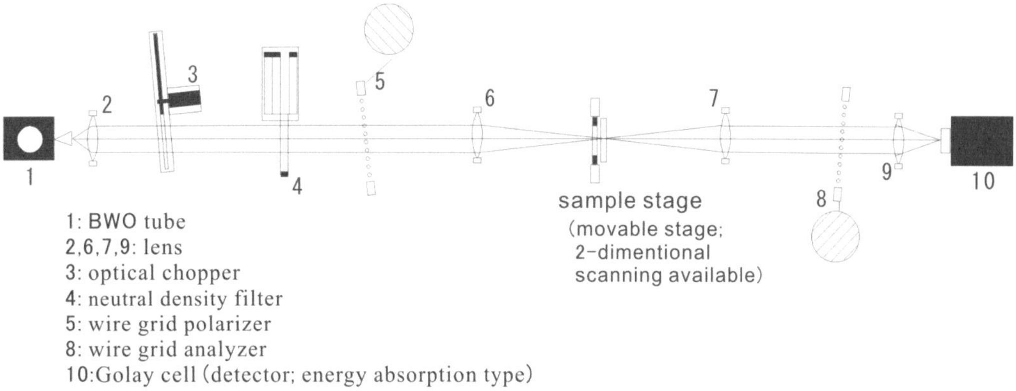

Fig. 7. Schematic of BWO experimental setup

2.2 Possibility of internal physical quantity measurement of solid insulating materials

<Cavity> For an internal cavity of solid materials, such as a void or a crack, all transmittable electromagnetic waves can be applied[3]. The key point is spatial resolution, which is decided by the wave length. By considering both its spatial resolution and transmission probability, a suitable wave length should be selected. Other than electromagnetic waves, ultra sonic waves are also applicable for elastic interfaces. It is said that the spatial resolution of ultra sonic methods reaches about 10 μm.

<Electric field> When electric field is applied to a material, an electron cloud of the constituent atoms is deformed, and as a consequence, its polarization, that is, dielectric constant (, or refractive index), changes. Among such phenomenon, the electro-optical Kerr effect (quadratic electro-optical effect) exists in all materials (, its quantity depends on the material)[4]. Therefore, there is a possibility that the applied electric field can be detected through a change in the refractive index. Moreover, Maxwell stress is expected to occur with the application of electric field, and the refractive index also changes through the photoelastic effect. If a light beam is transmitted to the material, such a change of the refractive index is converted to a detectable polarization change of the incident light beam. The electric field in a solid material is expected to be detected by utilizing the polarized light beam (electromagnetic wave) as the detecting "probe". Its wave length (frequency) should be selected considering its transmission probability.

<Mechanical stress> When mechanical stress is applied to a material, its dielectric constant (refractive index) changes owing to the photoelastic effect. Therefore, similarly to the electric field, it is expected to be detected by utilizing a polarized light beam (electro magnetic wave). Such measurement is already practically utilized for transparent plastic material, such as poly methyl methacrylate, using a visible laser beam.

<Thermal strain> When there is thermal strain in a material, such as an abnormal localized temperature rise in a solid insulating material, the distribution of the dielectric constant

---

[3] Electromagnetic waves of very short wave length (high energy), such as X-rays, should be handled with care as they may influence the target materials.

[4] Pockels effect (linear electro-optical effect) exists in only crystalline materials without centrosymmetry. Generally, cubic and higher electro-optical effects are said to be very small.

changes because of the photoelastic effect, similarly to mechanical stress above mentioned. Therefore, a polarized light beam (electro magnetic wave) is expected to function as a detecting "probe", as for electric field and mechanical stress.

From the viewpoints of transmission probability and the possibility of an internal physical quantity in solid materials, as described, THz wave is expected to be a suitable detecting "probe" for insulating materials, when a suitable frequency is selected. The internal measurement technique using a visible light beam has already been utilized for transparent materials. This technique tends to be regarded for only transparent materials, but here, "transparent" means the "probe" light beam can be transmitted with sufficient probability. Therefore, by selecting a suitable frequency, the applicable areas of the internal measurement technique can be widely expanded.

3. Study of internal measurement for a solid insulating material

By utilizing the measurement system in Fig. 7, internal measurement for polyethylene has been examined.

3.1 Detection of cavity interface

One of the most frequent abnormal features is a cavity, such as a void or a crack. It is expected that the transmission intensity of the "probe" light beam differs between the normal part and the cavity, but there is a possibility the difference is insufficient from the viewpoint of the S/N ratio. On the other hand, if there is a cavity, there must be an interface at which refraction occurs and light beam intensity changes. As a simple example, a polyethylene plate (5 mm thick) with holes (10 mm and 8 mm in diameter) has been measured (2-dimensional scanning) with THz wave (1.3 THz). The results are shown in Fig. 8. The light beam intensity at the hole is the highest, and a decrease in the light beam intensity is also prominent. This decrease is caused by refraction at the interface[5]. This result suggests that even if the transmission probability difference is not sufficiently large between the solid material and the cavity (normally, air), its interface should be clearly distinguishable. Another measurement example with a smaller hole (1.5 mm) is shown in Fig. 9. There is an area of light beam intensity decrease, which should depend on the thickness of the sample; therefore there is no area of light intensity increase in this case.

In these results, detected shapes of the cavity interface are deformed. This means that the measurement does not have high spatial resolution. From the viewpoint of cavity (void and crack) detection in a solid insulating material (detection of abnormal condition), there should be sufficient worth in establishing its existence or nonexistence, without an accurate shape measurement.

3.2 Detection of mechanical stress

In order to examine the measurement technique for mechanical stress in a solid material, a polyethylene sample (5 mm thick) pinched by a metal clamp has been measured using a 1.0

---

[5] The light beam is condensed by lens. The refraction occurs according to relation between cavity interface angle and light beam direction. It is supposed, decreases of the light beam intensity at right and left side of holes in Fig. 4 differs according to the this relation.

THz light beam. One measurement example is shown in Fig. 10. Fig. 10(a) is a simple scanned image of the transmission light beam intensity, and Fig. 10(b) was obtained with a polarizer and an analyzer placed form a cross-Nicol system. Fig. 10(a) shows the difference in the transmission probabilities of each part. In Fig. 10(b), the incident THz wave is linearly polarized by a polarizer with a tilted angle 45 degrees to the vertical direction, and the output THz wave transmitted from the sample is passed through an analyzer with a transmission direction tilted at an angle of 90 degrees to the polarizer. Therefore, Fig. 10(b) shows the polarization change of the transmitting THz wave in the sample. At the region of air and the polyethylene part with no mechanical stress, there is no polarization change and the detected signal intensity is very low. On the other hand, at the mechanically stressed polyethylene part, there occurs anisotropy of the dielectric constant because of the photoelastic effect, and the polarization of the transmitting THz wave changes from linear to elliptic. The elliptically polarized THz wave can be partially transmitted through the analyzer. Therefore, only the mechanically stressed polyethylene part is detected[6].

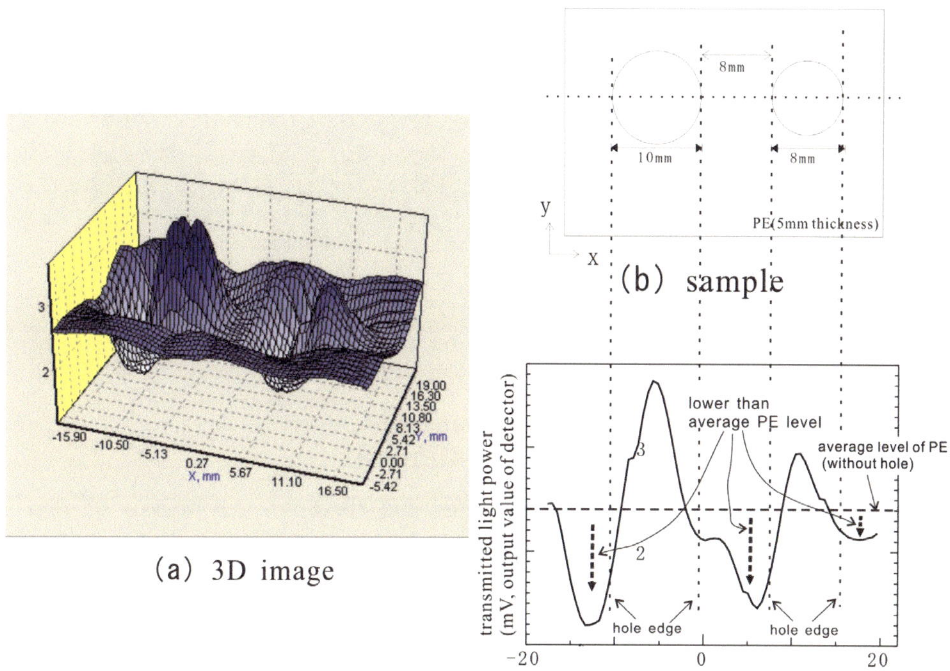

(a) 3D image

(b) sample

(c) scan result along X-axis on dotted line in Fig (b)

Fig. 8. Result of scanning perforated PE panel
(scanning step: 0.54mm, scanning area: 37.3×28.2mm)

---

[6] With this cross-Nicol configuration, mechanical stress along horizontal direction in Fig.6(b) is detected.

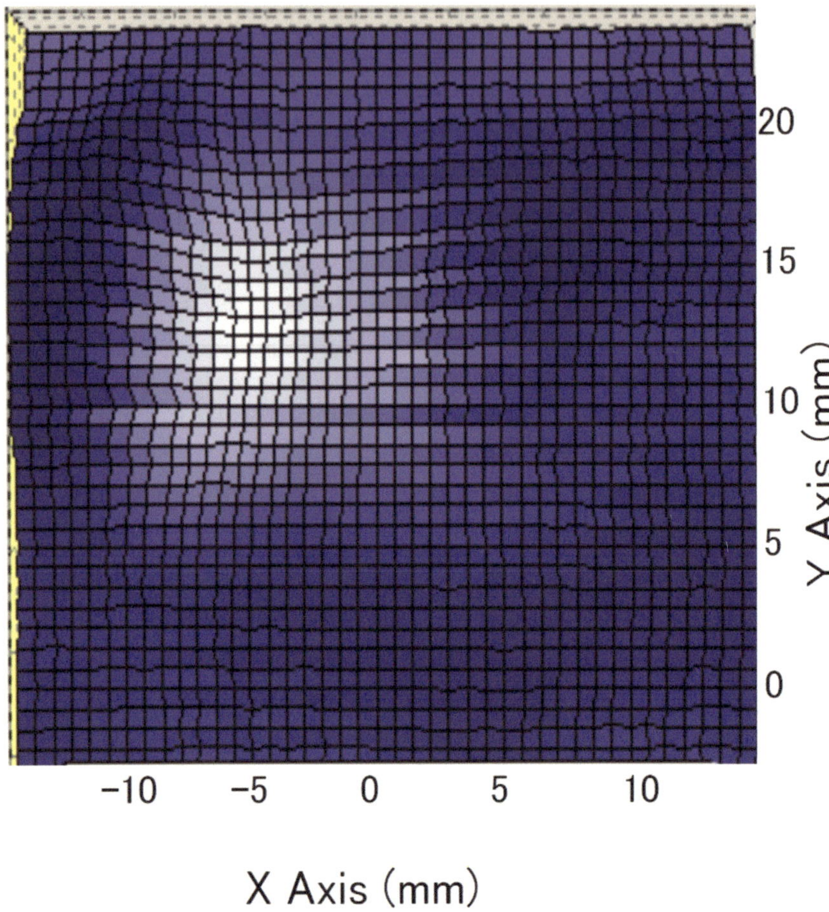

Fig. 9. Scanning image of PE panel with 1.5mm hole
(scanning step: 0.54mm, scanning area: 22.1×21.1mm)

When two orthogonally crossed coordinates and their normal stresses σ1 and σ2 are considered, the photoelestic effect is expressed by phase difference θ between two electric field components of the incident light beam along those coordinates as

$$\theta = \frac{2\pi \cdot h}{\lambda} \cdot C(\sigma_1 - \sigma_2) \tag{1}$$

where λ is the wave length of the incident light beam, *h* is the thickness of the sample, *C* is the photoelastic constant. A phase difference such as that in Eq. (1) appears in materials with dielectric anisotropy. It can be detected by the "probe" light beam, as shown in Fig. 11. The dielectric anisotropy (refractive index anisotropy) affects the linearly polarized incident light beam, and the polarization of the output light beam becomes elliptical, the ellipticity of which is governed by the phase difference θ. Generally a polarizer is utilized to make the polarization of the incident light beam linear, and an analyzer (, which is the same optical parts as a polarizer) is utilized to detect the ellipticity of the output light beam. As in the experiment corresponding to Fig. 10, when a mechanically stressed sample is set in a cross-Nicol system and the direction of normal stresses are set to the angle of 45 degrees to the transmitting direction of the polarizer and analyzer, the output light beam intensity from the analyzer I is expressed as

$$I = \frac{I_0}{2}\{1 + \cos(\Delta\theta + \theta_0)\} \tag{2}$$

where I_0 is the incident light intensity, and θ_0 is the phase difference cased by the natural birefringence (, birefringence when σ1=σ2). The relationship between I and θ is shown in Fig.12.

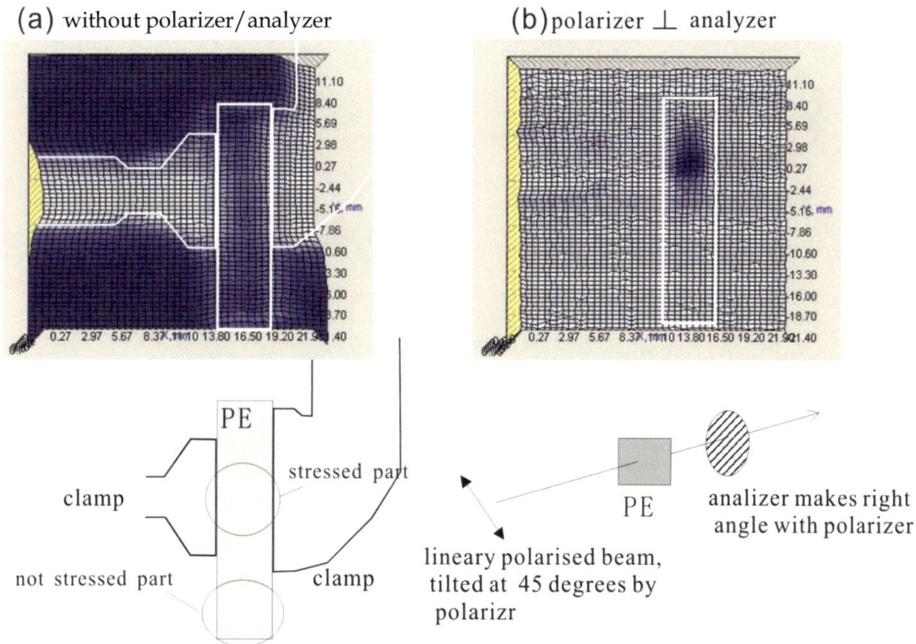

Fig. 10. Scanning result of stressed PE panel
(detected stress direction: horizontal, scanning step: 0.54mm, scanning area: 23.2×33.1mm)

In order to confirm the positive relationship between applied mechanical stress and the difference in the refractive index, and to estimate the photoelastic constant of PE in the THz frequency range, the output THz wave strength from the analyzer has been measured by changing the applied mechanical stress. The stress was applied with a hydraulic jack, which was replaced with the sample stage in Fig. 7. The transmitted THz wave was 1.0 THz in frequency, and the dimensions of the measured PE sample block were 70.4 mm in width, 30.5 mm in thickness (direction of mechanical stress application) and 30.9 mm in length (direction of THz wave transmission). The area of applying mechanical stress with the hydraulic jack was 70.4mm x 30.9mm. Under these conditions, the relationship between the applied mechanical stress and the phase difference θ in Eq. (1) was obtained, as shown in Fig. 13. In this measurement, very large stress was applied to the PE sample; therefore, some distortion was observed[7]. By measuring the sample size during this experiment, the relationship between applied mechanical stress and distortion was obtained, as shown in Fig. 14. The range in which mechanical stress is almost proportional to distortion can be regarded as the elastic region, and when stress is larger than the compressive yield strength of PE, it can be said to be the plastic region and the distortion increases rapidly[8]. In Fig. 13, a linear approximation was made in the elastic region, and the photoelastic constant under these experimental conditions was estimated.

$$C = 4\times10^{-11} \tag{3}$$

The estimation of the photoelastic constant in this frequency range seems to be unique.

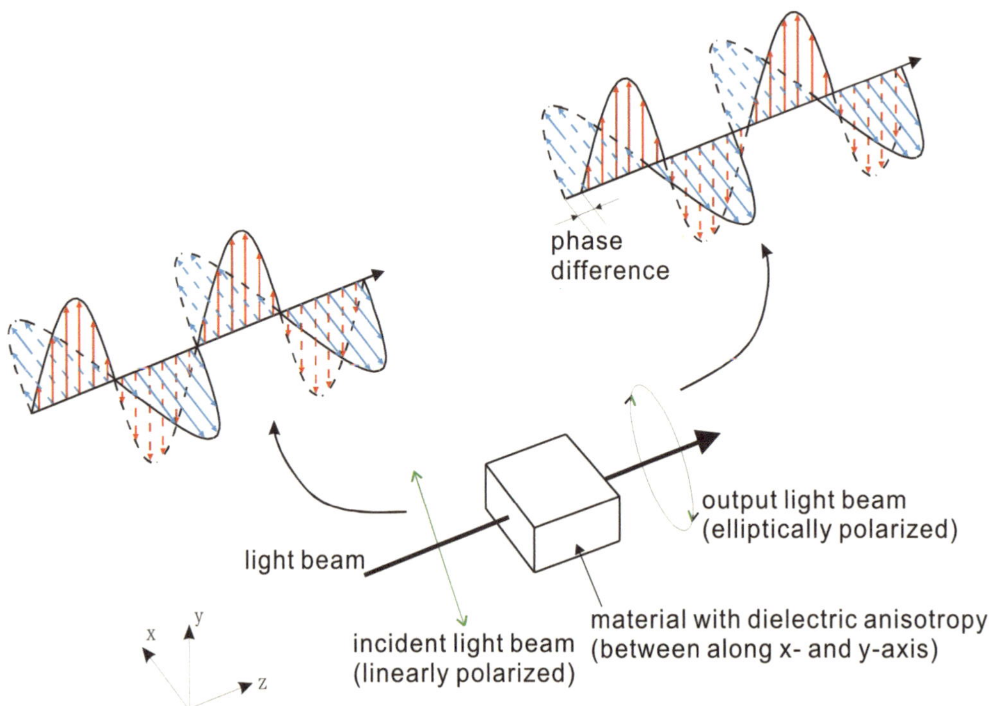

Fig. 11. Polarization change of "probe" light beam in material with dielectric anisotropy

[7] The distortion of PE changed gradually after mechanical stress was applied. In this experiment, values of applied mechanical stress were recorded after the change was almost saturated (, about 30 - 60 minutes after), from the stress value indicator of the hydraulic jack.
[8] In this experiment, the compressive yield strength was obtained to be about 18 - 23 MPa, which is in agreement with some references (e.g., The Society of Polymer Science Japan, "Polymer Data Handbook" (1986)).

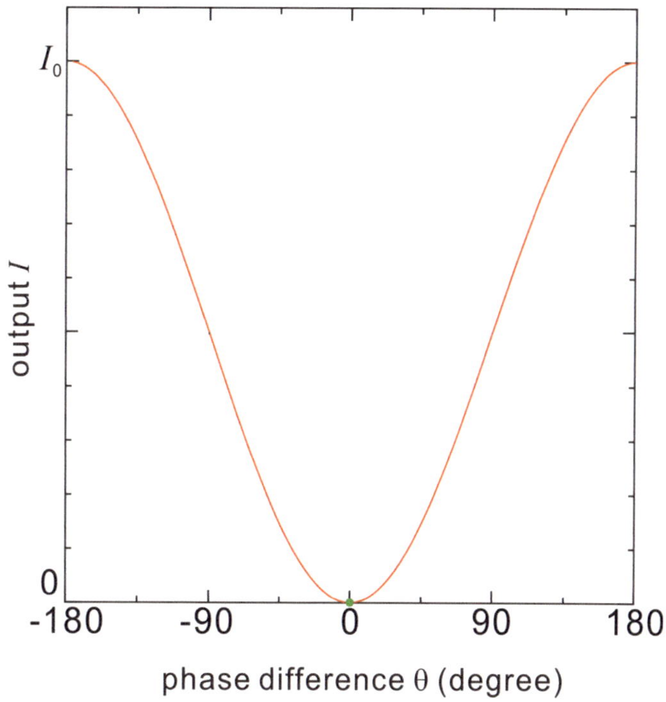

Fig. 12. Relationship between output light intensity from analyzer and phase difference θ

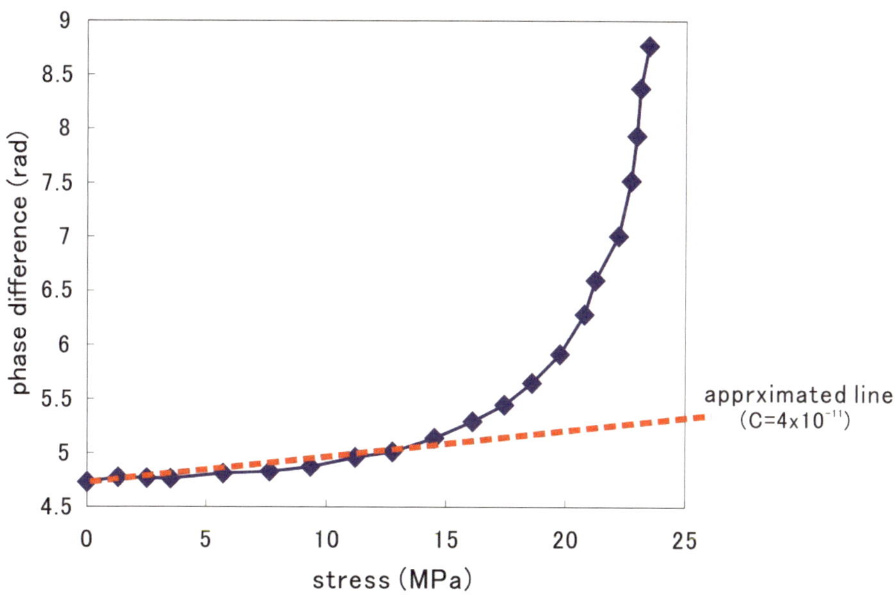

Fig. 13. Relationship between phase difference θ and applied mechanical stress

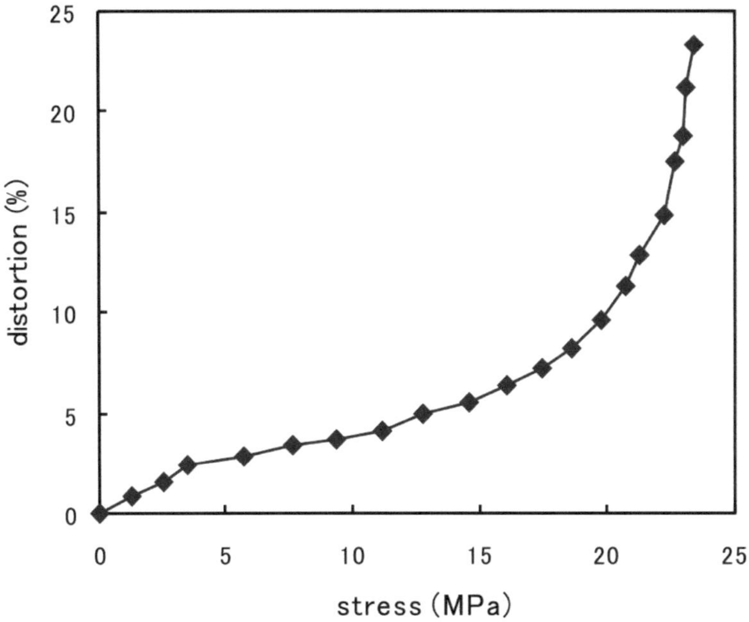

Fig. 14. Relationship between distortion of PE sample block and applied mechanical stress

4. Conclusion

THz wave technology is outstanding in recent optical engineering. Much development and application research is being carried out. In this chapter, one such research, the study of its application in the internal measurement of insulating materials, has been reviewed. Some measurements have been carried out with polyethylene, in which THz waves show high transmission probability, and the internal cavity interface and mechanical stress were detected. The applicability of such internal measurement in polyethylene was presented. A calibration (and estimation) method should be investigated for it.
Other internal problems, such as locally high electric field and temperature rise, are also expected to be detected by the same transmitting measurement of THz wave in the future. Moreover, surface (and near surface) conditions of materials are expected to be detected by the measurement of reflected THz wave.

5. Acknowledgment

The author would like to thank Prof. Hidaka and his laboratory members of the University of Tokyo for their instruction and collaboration.

6. References

[1] T. Takahashi et al., Application of Measurement Technique for Polyethylene with Mechanical Stress/a Cavity Interface using THz Wave, *Trans. Of IEE Japan*-A, Vol. 127-A, No. 10, pp. 593-598 (2007) [in Japanese]

[2] T. Takahashi, T. Takahashi and T. Okamoto, `Study on measurement technique for polyethylene by using THz beam --detection of vacant gap interface of mm order and mechanichal stress--', CRIEPI Report W03014 (2004) [in Japanese]
[3] M. Shibuya et al., Proposition of all solid insulated substation', CRIEPI Report W00047 (2000) (in Japanese)
[4] S. Matsuura et al., A compact terahertz radiation source for high-resolution spectroscopy, *IEICE Technical Report*, No. LQE97-5 (1997) [in Japanese]
[5] T. Taniuchi et al., Widely Tunable Terahertz Electromagnetic Radiation by Nonlinear Optical Effect, *The Review of Laser Engineering (The Laser Society of Japan)*, Vol. 30, No. 7, pp. 365-369 (2002) [in Japanese]
[6] K. Kawase et al., Widely Tunable Coherent THz-Wave Generation Using Nonlinear Optical Effect, *Trans. Of IEICE*, Vol. J81-C-I, No. 2, pp. 66-73 (1998) [in Japanese]
[7] K. Kawase et al., "Injection-seeded terahertz-wave parametric generator with wide tenability", *Appl. Phys. Lett.*, Vol. 80, No. 2, pp. 195-197 (2002)
[8] J. T. Darrow et al., Saturation Properties of Large-Aperture Photoconducting Antennas, *IEEE J. Quantum Electron.*, 28, pp. 1607-1616 (1992)
[9] M. Hangyo et al., Terahertz Radiation from High-Temperature Superconductors, *The Review of Laser Engineering (The Laser Society of Japan)*, Vol. 26, No. 7, pp. 536-540 (1998) [in Japanese]
[10] M. Tonouchi ,THz radiation from high-Tc superconductor, *OYOBUTURI(JSAP)*, Vol. 66, No. 9, pp. 988-989 (1997) [in Japanese]
[11] X. -C. Zhang et al., Generation of femtosecond electromagnetic pulses from semiconductor surfaces, *Appl. Phys. Lett.*, Vol. 56(11), pp. 1011-1013 (1990)
[12] X. -C. Zhang et al., Optoelectronic measurement of semiconductor surfaces and interfaces with femtosecond optics, *J. Appl. Phys.*, Vol. 71(1), pp. 326-338 (1992)
[13] K. Sakai et al., Terahertz Electromagnetic Waves: Generation and Applications, *The Review of Laser Engineering (The Laser Society of Japan)*, Vol. 26, No. 7, pp. 515-521 (1998)
[14] X. -C. Zhang et al., Influence of electric and magnetic fields on THz radiation, *Appl. Phys. Lett.*, Vol. 62(20), pp. 2477-2479 (1993)
[15] H. Ohtake et at., Intense THz Radiation from Semiconductors in a Magnetic Field and Its ApplicationA33, *The Review of Laser Engineering (The Laser Society of Japan)*, Vol. 30, No. 7, pp. 360-364 (2002) [in Japanese]
[16] K. Sakai et al., Terahertz optoelectronics, *OYOBUTURI(JSAP)*, Vol. 70, No. 2, pp. 149-155 (2001) [in Japanese]
[17] Optoelectronic Industry and Technology Development Association, Optoelectronics Technology Report (2007) [in Japanese]
[18] F. Maiwald et al., Design and performance of a 2.7 THz waveguide tripler, *Proc. of 12th Int. Symp. on Space terahertz Technology*, pp. 320-329 (2001)
[19] A. Volkov, Submillimeter BWO spectroscopy of solids, *Int. J. Infrared Millim. Waves* , Vol. 8, No. 1, pp. 55-61 (1987)
[20] S. Komiyama et al., Ultrahigh Sensitivity Detection of Terahertz Waves Using Quantum Dots, *The Review of Laser Engineering (The Laser Society of Japan)*, Vol. 30, No. 7, pp. 385-390 (2002) [in Japanese]

[21] S. Ariyoshi et al., Terahertz Imaging with a Superconducting Detector Array, *IEICE Technical Report*, Vol. 106, No. 403(ED2006 184-199), pp. 59-63 (2006) [in Japanese]

19

Reciprocity in Nonlocal Optics and Spectroscopy

Huai-Yi Xie
Research Center for Applied Sciences, Academia Sinica, Taipei
Taiwan

1. Introduction

In any wave propagation, the wave can go through various scattering processes through interaction with target in the environment from a source to a detector. In such a process, reciprocity refers to the equality in the signal received when the source and the detector are reversed, that is; their respective positions switched (Potton, 2004). We can find many interesting applications which based on either its validity or its breakdown, in the large number of areas involving transmission of signals ranging of classical optical problems. In classical optics, reciprocity is a powerful result which finds applications in many problems in optics (Potton, 2004) and spectroscopy (Hill et al., 1997). For example, we can establish relations between far fields and near fields from different sources as well as spectroscopic analysis of surface enhanced Raman scattering (SERS) at metallic structure (Kahl & Voges 2000; Ru & Etchegoin, 2006). However, in the previous literature, the optical reciprocity always has been discussed under local optics (Potton, 2004). Now we try to describe the optical reciprocity from electrostatics to electrodynamics under nonlocal optics in order to consider some quantum effects of the particles. Our goal is that the general conditions to determine that the optical reciprocity remains or breaks down will be constructed under nonlocal optics. Some examples and applications will also be discussed.

2. Reciprocity in electrostatics (Green reciprocity)

If we consider an object whose size is much smaller than the wavelength of the incident light, then the effect of retardation can be neglected. Hence we can simply use electrostatics to discuss the interaction between the light and the material. In mathematics, we usually use two popular forms to describe optical reciprocity. One is the Lorentz lemma in electrostatics and the other is the symmetry of the scalar Green function.

2.1 Lorentz lemma in electrostatics

Lorentz lemma in electrostatics form is well-known with local optical response of the medium in the literature. We will extend to consider nonlocal optical response of the medium, since it is known that such response is rather significant with metallic nano structures due to the large surface-to-volume ratio of these systems. First we write the mathematical form of the Lorentz lemma in electrostatics as follows (Griffiths 1999; Jackson, 1999):

$$\int \rho_1(\vec{r})\Phi_2(\vec{r})d^3\vec{r} = \int \rho_2(\vec{r})\Phi_1(\vec{r})d^3\vec{r}\,, \tag{1}$$

where Φ_1 (Φ_2) is the electric potential resulting from the total charge density ρ_1 (ρ_2). Here we will derive this lemma in two different kinds of circumstances.

2.1.1 Anisotropic local response

In the beginning, we start from the Poisson equations with two different distributions of charge density ρ_1 and ρ_2:

$$\left.\begin{array}{l}\vec{\nabla}\cdot\left[\ddot{\varepsilon}(\vec{r})\cdot\vec{\nabla}\Phi_1\right] = -4\pi\rho_1 \\ \vec{\nabla}\cdot\left[\ddot{\varepsilon}(\vec{r})\cdot\vec{\nabla}\Phi_2\right] = -4\pi\rho_2\end{array}\right\}, \tag{2}$$

where $\ddot{\varepsilon}(\vec{r})$ is a dielectric tenser. Next we use Eq. (A1) and put the tensor $\ddot{\lambda} = \ddot{\varepsilon}$, the value $\Phi = \Phi_1$ and $\Psi = \Phi_2$. Thus we have the following equality:

$$\int\left[\Phi_1\vec{\nabla}\cdot\left(\ddot{\varepsilon}\cdot\vec{\nabla}\Phi_2\right) - \Phi_2\vec{\nabla}\cdot\left(\ddot{\varepsilon}\cdot\vec{\nabla}\Phi_1\right)\right]d^3\vec{r} = \oint_S \hat{n}\cdot\left[\Phi_1\ddot{\varepsilon}\cdot\vec{\nabla}\Phi_2 - \Phi_2\ddot{\varepsilon}\cdot\vec{\nabla}\Phi_1\right]da\,, \tag{3}$$

under the symmetric condition of a dielectric tensor $\varepsilon_{ij} = \varepsilon_{ji}$. Combining with Eq. (2) and extending the finite volume to all space ($\mathbb{R}^3$), we can remove the surface integral in Eq. (3) and obtain Eq. (1). Hence we prove the Lorentz lemma in electrostatics under the symmetry condition of the dielectric tensor, that is; the optical reciprocity does not break down under the symmetry condition of the dielectric tensor ($\varepsilon_{ij} = \varepsilon_{ji}$) in the case of anisotropic local response of the medium.

2.1.2 Anisotropic nonlocal response

In this case, we will extend to consider the nonlocal response. Here we write the Poisson equations with two different charge densities ρ_1 and ρ_2:

$$\left.\begin{array}{l}\vec{\nabla}\cdot\left[\int\ddot{\varepsilon}(\vec{r},\vec{r}')\cdot\vec{\nabla}'\Phi_1(\vec{r}')d^3\vec{r}'\right] = -4\pi\rho_1(\vec{r}) \\ \vec{\nabla}\cdot\left[\int\ddot{\varepsilon}(\vec{r},\vec{r}')\cdot\vec{\nabla}'\Phi_2(\vec{r}')d^3\vec{r}'\right] = -4\pi\rho_2(\vec{r})\end{array}\right\}, \tag{4}$$

and we use Eq. (A5) with $\ddot{\lambda}(\vec{r},\vec{r}_1) = \ddot{\varepsilon}(\vec{r},\vec{r}_1)$, the value $\Phi = \Phi_1$ and $\Psi = \Phi_2$. Thus we get the following identity:

$$\begin{array}{l}\int d^3\vec{r}\int d^3\vec{r}_1\left\{\Phi_1(\vec{r})\vec{\nabla}\cdot\left[\ddot{\varepsilon}(\vec{r},\vec{r}_1)\cdot\vec{\nabla}_1\Phi_2(\vec{r}_1)\right] - \Phi_2(\vec{r})\vec{\nabla}\cdot\left[\ddot{\varepsilon}(\vec{r},\vec{r}_1)\cdot\vec{\nabla}_1\Phi_1(\vec{r}_1)\right]\right\} \\ = \oint_S da\int d^3\vec{r}_1\left\{\hat{n}\cdot\left[\Phi_1(\vec{r})\ddot{\varepsilon}(\vec{r},\vec{r}_1)\cdot\vec{\nabla}_1\Phi_2(\vec{r}_1) - \Phi_2(\vec{r})\ddot{\varepsilon}(\vec{r},\vec{r}_1)\cdot\vec{\nabla}_1\Phi_1(\vec{r}_1)\right]\right\}\end{array}, \tag{5}$$

under the condition $\varepsilon_{ij}(\vec{r},\vec{r}') = \varepsilon_{ji}(\vec{r}',\vec{r})$. Next we combine Eq. (4) and extend the finite volume to $\mathbb{R}^3$; thus we can remove the surface integral in Eq. (5) and get Eq. (1) again. Hence we prove the Lorentz lemma in electrostatics under the symmetry condition of the dielectric tensor $\varepsilon_{ij}(\vec{r},\vec{r}') = \varepsilon_{ji}(\vec{r}',\vec{r})$. Thus the optical reciprocity does not break down under the symmetry condition of the dielectric tensor in the case of anisotropic nonlocal response of the medium.

2.2 Scalar Green function

Another method to describe the optical reciprocity is the symmetry of the scalar Green function. The mathematical form is (Jackson, 1999):

$$G(\vec{r}'',\vec{r}')=G(\vec{r}',\vec{r}'')\,. \tag{6}$$

We divide into two cases, and consider two kinds of boundary conditions to discuss the symmetry of the scalar Green function. One is the Dirichlet boundary condition and the other is the Neumann boundary condition.

2.2.1 Anisotropic local response

Referring to Eq. (2), the corresponding two equations of scalar Green function are in the following forms:

$$\left.\begin{aligned}\vec{\nabla}\cdot\left[\ddot{\varepsilon}(\vec{r})\cdot\vec{\nabla}G(\vec{r},\vec{r}')\right]=-4\pi\delta(\vec{r}-\vec{r}')\\ \vec{\nabla}\cdot\left[\ddot{\varepsilon}(\vec{r})\cdot\vec{\nabla}G(\vec{r},\vec{r}'')\right]=-4\pi\delta(\vec{r}-\vec{r}'')\end{aligned}\right\}, \tag{7}$$

where $\vec{r}$ and $\vec{r}'$ ($\vec{r}''$) are the positions of the field and source, respectively. δ denotes the Dirac delta function. Let us apply Eq. (A1) and put $\ddot{\lambda}=\ddot{\varepsilon}(\vec{r})$, $\Phi=G(\vec{r},\vec{r}')$ and $\Psi=G(\vec{r},\vec{r}'')$. Hence we have the following equality:

$$\begin{aligned}&\int\left\{G(\vec{r},\vec{r}')\vec{\nabla}\cdot\left[\ddot{\varepsilon}(\vec{r})\cdot\vec{\nabla}G(\vec{r},\vec{r}'')\right]-G(\vec{r},\vec{r}'')\vec{\nabla}\cdot\left[\ddot{\varepsilon}(\vec{r})\cdot\vec{\nabla}G(\vec{r},\vec{r}')\right]\right\}d^3\vec{r}\\ &=\oint_S\hat{n}\cdot\left[G(\vec{r},\vec{r}')\ddot{\varepsilon}(\vec{r})\cdot\vec{\nabla}G(\vec{r},\vec{r}'')-G(\vec{r},\vec{r}'')\ddot{\varepsilon}(\vec{r})\cdot\vec{\nabla}G(\vec{r},\vec{r}')\right]da\end{aligned}, \tag{8}$$

under the condition $\varepsilon_{ij}(\vec{r})=\varepsilon_{ji}(\vec{r})$ and we also combine Eq. (6) to get the following result:

$$\begin{aligned}&-4\pi G(\vec{r}'',\vec{r}')+4\pi G(\vec{r}',\vec{r}'')\\ &=\oint_S\hat{n}\cdot\left[G(\vec{r},\vec{r}')\ddot{\varepsilon}(\vec{r})\cdot\vec{\nabla}G(\vec{r},\vec{r}'')-G(\vec{r},\vec{r}'')\ddot{\varepsilon}(\vec{r})\cdot\vec{\nabla}G(\vec{r},\vec{r}')\right]da\end{aligned}. \tag{9}$$

Next we will divide into two different boundary conditions to discuss. In the case of the Dirichlet boundary condition, we have:

$$G(\vec{r},\vec{r}')=G(\vec{r},\vec{r}'')=0\,, \tag{10}$$

with $\vec{r}\in S$. Substitute this into Eq. (9) and we obtain Eq. (6), establishing the symmetry of the scalar Green function with the Dirichlet boundary condition under the symmetry condition of the dielectric tensor.

In the case of the Neumann boundary condition, let us generalize the results in Kim et al (Kim, 1993) to introduce the following Neumann boundary conditions (Xie, 2010):

$$\begin{aligned}\hat{n}\cdot\left[\ddot{\varepsilon}(\vec{r})\cdot\vec{\nabla}G_N(\vec{r},\vec{r}')\right]\Big|_{\vec{r}\in S}&=-\frac{4\pi}{A}\\ \hat{n}\cdot\left[\ddot{\varepsilon}(\vec{r})\cdot\vec{\nabla}G_N(\vec{r},\vec{r}'')\right]\Big|_{\vec{r}\in S}&=-\frac{4\pi}{A}\end{aligned}, \tag{11}$$

where A is the area of the closed boundary S. Eq. (9) then becomes:

$$-4\pi G_N(\vec{r}'',\vec{r}')+4\pi G_N(\vec{r}',\vec{r}'')=-\frac{4\pi}{A}\oint_S G_N(\vec{r},\vec{r}')da+\frac{4\pi}{A}\oint_S G_N(\vec{r},\vec{r}'')da\,. \tag{12}$$

We can then follow our pervious work (Xie, 2010) to define the following symmetrized Green function:

$$G_N^S(\vec{r}'',\vec{r}')=G_N(\vec{r}'',\vec{r}')-\frac{1}{A}\oint_S G_N(\vec{r},\vec{r}')da\,, \tag{13}$$

which can be shown explicitly to lead to the same solution for the potential with no contributions from the additional surface term.

2.2.2 Anisotropic nonlocal response

In this case, we will consider the anisotropic nonlocal response in the material. The Poisson equations with two different distributions of charge density as given in Eq. (4) will have the corresponding scalar Green functions satisfying:

$$\left.\begin{aligned}\int d^3\vec{r}_1\vec{\nabla}\cdot\left[\ddot{\varepsilon}(\vec{r},\vec{r}_1)\cdot\vec{\nabla}_1G(\vec{r}_1,\vec{r}')\right]=-4\pi\delta(\vec{r}-\vec{r}')\\ \int d^3\vec{r}_1\vec{\nabla}\cdot\left[\ddot{\varepsilon}(\vec{r},\vec{r}_1)\cdot\vec{\nabla}_1G(\vec{r}_1,\vec{r}'')\right]=-4\pi\delta(\vec{r}-\vec{r}'')\end{aligned}\right\}. \tag{14}$$

Next we use Eq. (A5) and let $\ddot{\lambda}=\ddot{\varepsilon}$, $\Phi(\vec{r})=G(\vec{r},\vec{r}')$ and $\Psi(\vec{r})=G(\vec{r},\vec{r}'')$, leading to:

$$\begin{aligned}&\int d^3\vec{r}\int d^3\vec{r}_1\left\{G(\vec{r},\vec{r}')\vec{\nabla}\cdot\left[\ddot{\varepsilon}(\vec{r},\vec{r}_1)\cdot\vec{\nabla}_1G(\vec{r}_1,\vec{r}'')\right]-G(\vec{r},\vec{r}'')\vec{\nabla}\cdot\left[\ddot{\varepsilon}(\vec{r},\vec{r}_1)\cdot\vec{\nabla}_1G(\vec{r}_1,\vec{r}')\right]\right\}\\&=\oint_S da\int d^3\vec{r}_1\left\{\hat{n}\cdot\left[G(\vec{r},\vec{r}')\ddot{\varepsilon}(\vec{r},\vec{r}_1)\cdot\vec{\nabla}_1G(\vec{r}_1,\vec{r}'')-G(\vec{r},\vec{r}'')\ddot{\varepsilon}(\vec{r},\vec{r}_1)\cdot\vec{\nabla}_1G(\vec{r}_1,\vec{r}')\right]\right\}\end{aligned}, \tag{15}$$

under the condition $\varepsilon_{ij}(\vec{r},\vec{r}')=\varepsilon_{ji}(\vec{r}',\vec{r})$, combining Eqs. (14) and (15) leads to:

$$\begin{aligned}&-4\pi G(\vec{r}'',\vec{r}')+4\pi G(\vec{r}',\vec{r}'')\\&=\oint_S da\int d^3\vec{r}_1\left\{\hat{n}\cdot\left[G(\vec{r},\vec{r}')\ddot{\varepsilon}(\vec{r},\vec{r}_1)\cdot\vec{\nabla}_1G(\vec{r}_1,\vec{r}'')-G(\vec{r},\vec{r}'')\ddot{\varepsilon}(\vec{r},\vec{r}_1)\cdot\vec{\nabla}_1G(\vec{r}_1,\vec{r}')\right]\right\}\end{aligned}. \tag{16}$$

Now we will divide into two different kinds of the boundary conditions. First we consider the Dirichlet boundary condition (Eq. (10)). The RHS of Eq. (16) becomes zero and thus we obtain the optical reciprocity under the symmetry of the dielectric tensor. Next we consider the Neumann boundary condition G_N. Again, introducing the following generalized Neumann conditions (Xie, 2010) :

$$\begin{aligned}\hat{n}\cdot\int d^3\vec{r}_1\left[\ddot{\varepsilon}(\vec{r},\vec{r}_1)\cdot\vec{\nabla}_1G_N(\vec{r}_1,\vec{r}')\right]\Big|_{\vec{r}\in S}&=-\frac{4\pi}{A}\\ \hat{n}\cdot\int d^3\vec{r}_1\left[\ddot{\varepsilon}(\vec{r},\vec{r}_1)\cdot\vec{\nabla}_1G_N(\vec{r}_1,\vec{r}'')\right]\Big|_{\vec{r}\in S}&=-\frac{4\pi}{A}\end{aligned}, \tag{17}$$

the Green function can be symmetrized in the form as in Eq. (13) under the following symmetric condition for the dielectric tensor: $\varepsilon_{ij}(\vec{r},\vec{r}')=\varepsilon_{ji}(\vec{r}',\vec{r})$. Note further that just like the local case, this symmetrized Green function can be shown to lead to the same solution for the potential. Now we demonstrate explicitly that the newly-constructed symmeterized

Green functions in Eq. (13) yield the same solution for the potential with no contribution from the additional surface term. To achieve this, we start with Eq. (A5) and set $\Phi=\Phi(\vec{r})$, $\Psi(\vec{r})=G_N^S(\vec{r},\vec{r}')$, and $\vec{\lambda}(\vec{r},\vec{r}_1)=\vec{\varepsilon}(\vec{r},\vec{r}_1)$ to obtain

$$\begin{aligned}&\oint_S da\int d^3\vec{r}_1\left\{\hat{n}\cdot\left[\Phi(\vec{r})\vec{\varepsilon}(\vec{r},\vec{r}_1)\cdot\vec{\nabla}_1 G_N^S(\vec{r}_1,\vec{r}')-G_N^S(\vec{r},\vec{r}')\vec{\varepsilon}(\vec{r},\vec{r}_1)\cdot\vec{\nabla}_1\Phi(\vec{r}_1)\right]\right\}\\&=\int d^3\vec{r}\int d^3\vec{r}_1\left\{\Phi(\vec{r})\vec{\nabla}\cdot\left[\vec{\varepsilon}(\vec{r},\vec{r}_1)\cdot\vec{\nabla}_1 G_N^S(\vec{r}_1,\vec{r}')\right]-G_N^S(\vec{r},\vec{r}')\vec{\nabla}\cdot\left[\vec{\varepsilon}(\vec{r},\vec{r}_1)\cdot\vec{\nabla}_1\Phi(\vec{r}_1)\right]\right\}\end{aligned}. \quad (18)$$

With Eq. (13) into the G_N^S of Eq. (18) and using Eq. (14) together with the Poisson equation for Φ, we obtain the following result:

$$\begin{aligned}\Phi(\vec{r}')&=\frac{1}{A}\oint_S\Phi(\vec{r})da+\frac{1}{4\pi}\oint_S da\int d^3\vec{r}_1 G_N(\vec{r},\vec{r}')\hat{n}\cdot\left[\vec{\varepsilon}(\vec{r},\vec{r}_1)\cdot\vec{\nabla}_1\Phi(\vec{r}_1)\right]\\&+\int\rho(\vec{r})G_N(\vec{r},\vec{r}')d^3\vec{r}+F(\vec{r}')\left\{\int\rho(\vec{r})d^3\vec{r}+\frac{1}{4\pi}\oint_S da\int d^3\vec{r}_1\hat{n}\cdot\left[\vec{\varepsilon}(\vec{r},\vec{r}_1)\cdot\vec{\nabla}_1\Phi(\vec{r}_1)\right]\right\}\\&=\frac{1}{A}\oint_S\Phi(\vec{r})da+\frac{1}{4\pi}\oint_S da\int d^3\vec{r}_1 G_N(\vec{r},\vec{r}')\hat{n}\cdot\left[\vec{\varepsilon}(\vec{r},\vec{r}_1)\cdot\vec{\nabla}_1\Phi(\vec{r}_1)\right]+\int\rho(\vec{r})G_N(\vec{r},\vec{r}')d^3\vec{r}\end{aligned}, \quad (19)$$

where the surface term $F(\vec{r}')$ has no contribution since the term {....} in Eq. (18) vanishes based on the nonlocal version of the Gauss law for Φ [See Eq. (4)]. Thus the symmetrized G_N^S leads to the same potential as the one obtained from using G_N.

2.3 Some examples for scalar Green functions

We will give some examples to display the explicit forms of the Green function in both Dirichlet and Neumann boundary conditions (Chang 2008; Xie, 2010). Consider a metal sphere (radius a) with an isotropic (for simplicity) but nonlocal dielectric response $\varepsilon(k,\omega)$. For the case of Dirichlet condition, we have previously applied the model of the nonlocal polarizability by Fuchs and Claro (Fuchs & Claro, 1987) to obtain the following symmetric Green function:

$$G(\vec{r},\vec{r}')=4\pi\sum_{\ell=0}^{\infty}\sum_{m=-\ell}^{\ell}\frac{1}{2\ell+1}\left[\frac{r_<^\ell}{r_>^{\ell+1}}-\frac{\alpha_\ell^{NL}}{(rr')^{\ell+1}}\right]Y_{\ell m}^*(\theta',\varphi')Y_{\ell m}(\theta,\varphi), \quad (20)$$

where $(r_<,r_>)$ denote the smaller or greater of (r,r'), and

$$\alpha_\ell^{NL}=\frac{\xi_\ell-1}{\xi_\ell+\left(\frac{\ell+1}{\ell}\right)}a^{2\ell+1}, \quad (21)$$

with the "effective dielectric function" given by:

$$\xi_\ell(\omega)=\left[\frac{2(2\ell+1)a}{\pi}\int_0^\infty\frac{j_\ell^2(ka)}{\varepsilon(k,\omega)}dk\right]^{-1}. \quad (22)$$

For the same problem under the Neumann condition, first we get the following Green function by solving the corresponding boundary value problem:

$$G_N(\vec{r},\vec{r}')=\frac{1}{r_>}-\frac{1}{r}+4\pi\sum_{\ell=1}^{\infty}\sum_{m=-\ell}^{\ell}\frac{1}{2\ell+1}\left[\frac{r_<^{\ell}}{r_>^{\ell+1}}+\frac{\ell}{\ell+1}\frac{\alpha_\ell^{NL}}{(rr')^{\ell+1}}\right]Y_{\ell m}^*(\theta',\varphi')Y_{\ell m}(\theta,\varphi), \tag{23}$$

which does not satisfy Eq. (6). However, we can use the pervious methods to symmetrize the above asymmetric Green function. Applying Eq. (13) to calculate the surface term using only the first two terms in Eq. (23) and obtain the following:

$$\frac{1}{A}\oint_S G_N(\vec{r},\vec{r}')da=\frac{1}{r'}-\frac{1}{a}. \tag{24}$$

We thus obtain the final symmetrized Neumann Green function for the region outside a nonlocal metal sphere in the following form:

$$G_N^S(\vec{r},\vec{r}')=\left[\frac{1}{r_>}-\left(\frac{1}{r}+\frac{1}{r'}\right)+\frac{1}{a}\right] \\ +4\pi\sum_{\ell=1}^{\infty}\sum_{m=-\ell}^{\ell}\frac{1}{2\ell+1}\left[\frac{r_<^{\ell}}{r_>^{\ell+1}}+\frac{\ell}{\ell+1}\frac{\alpha_\ell^{NL}}{(rr')^{\ell+1}}\right]Y_{\ell m}^*(\theta',\varphi')Y_{\ell m}(\theta,\varphi). \tag{25}$$

In our examples, we can find that the optical reciprocity does not break down since the dielectric function satisfies $\varepsilon(\vec{r},\vec{r}')=\varepsilon(|\vec{r}-\vec{r}'|)$.

3. Reciprocity in electrodynamics

In the pervious discussion, we have restricted our problem to the "long wavelength approximation" in which electrostatics has been applied (Chang, 2008). In problems with very high frequency source (e.g. scattering between X-rays and a nanoparticle (Ruppin, 1975)) in which electrostatics breaks down and nonlocal effects can become even more significant due to the large value of the wavevector, the previous formulation (Chang, 2008) becomes inadequate. Here we use the method of exact electrodynamics to study the optical reciprocity. Again, we shall refer to both the Lorentz lemma in electrodynamics and the symmetry of dyadic Green function.

3.1 Lorentz lemma in electrodynamics

The mathematical form of the Lorentz lemma in electrodynamics is as follows (Landau et al., 1984):

$$\int\vec{J}_1(\vec{r})\cdot\vec{E}_2(\vec{r})d^3\vec{r}=\int\vec{J}_2(\vec{r})\cdot\vec{E}_1(\vec{r})d^3\vec{r}, \tag{26}$$

where $\vec{E}_1$ ($\vec{E}_2$) is the electric field resulting from the current density $\vec{J}_1$ ($\vec{J}_2$). Next we will derive this formula in two different cases.

3.1.1 Anisotropic local response

In the beginning, let us consider time harmonic Maxwell's equations ($\sim e^{-i\omega t}$):

$$\left.\begin{aligned}\vec{\nabla}\cdot\vec{D}=4\pi\rho \quad & \vec{\nabla}\times\vec{E}=\frac{i\omega}{c}\vec{B}\\ \vec{\nabla}\cdot\vec{B}=0 \quad & \vec{\nabla}\times\vec{H}=\frac{4\pi}{c}\vec{J}-\frac{i\omega}{c}\vec{D}\end{aligned}\right\}. \tag{27}$$

Now we consider two different sources $\vec{J}_1$ and $\vec{J}_2$ which correspond to two electric fields $\vec{E}_1$ and $\vec{E}_2$ and two magnetic fields $\vec{H}_1$ and $\vec{H}_2$, respectively. We obtain from Eq. (27) the following result:

$$\begin{aligned}&\vec{H}_2\cdot\vec{\nabla}\times\vec{E}_1-\vec{E}_1\cdot\vec{\nabla}\times\vec{H}_2+\vec{E}_2\cdot\vec{\nabla}\times\vec{H}_1-\vec{H}_1\cdot\vec{\nabla}\times\vec{E}_2\\&=\frac{i\omega}{c}\left[\left(\vec{B}_1\cdot\vec{H}_2-\vec{H}_1\cdot\vec{B}_2\right)+\left(\vec{E}_1\cdot\vec{D}_2-\vec{D}_1\cdot\vec{E}_2\right)\right]+\frac{4\pi}{c}\left(\vec{J}_1\cdot\vec{E}_2-\vec{J}_2\cdot\vec{E}_1\right)'\end{aligned} \tag{28}$$

which can be simplified to the following form:

$$\vec{\nabla}\cdot\left(\vec{H}_1\times\vec{E}_2-\vec{E}_1\times\vec{H}_2\right)=\frac{4\pi}{c}\left(\vec{J}_1\cdot\vec{E}_2-\vec{J}_2\cdot\vec{E}_1\right), \tag{29}$$

if the dielectric tensor satisfies the symmetry condition $\varepsilon_{ij}=\varepsilon_{ji}$ and the permeability satisfies the symmetry condition $\mu_{ij}=\mu_{ji}$. Furthermore, let us integrate over $\mathbb{R}^3$ in Eq. (29) and use the divergence theorem to convert the left side of Eq. (29) into a surface integral which can be removed. Hence we obtain the Lorentz lemma in electrodynamics (i.e. Eq. (26)).

3.1.2 Anisotropic nonlocal response

Next we will extend the local response to the nonlocal response where the relation between the auxiliary field and the magnetic field is described by

$$\vec{H}(\vec{r})=\int d^3\vec{r}'\ddot{\mu}^{-1}(\vec{r},\vec{r}')\cdot\vec{B}(\vec{r}') \ . \tag{30}$$

Thus Eq. (28) will become:

$$\begin{aligned}&\vec{\nabla}\cdot\left(\vec{H}_1\times\vec{E}_2-\vec{E}_1\times\vec{H}_2\right)\\&=\frac{i\omega}{c}\int d^3\vec{r}'\left\{\vec{B}_1(\vec{r})\cdot\ddot{\mu}^{-1}(\vec{r},\vec{r}')\cdot\vec{B}_2(\vec{r}')-\ddot{\mu}^{-1}(\vec{r},\vec{r}')\cdot\vec{B}_1(\vec{r}')\cdot\vec{B}_2(\vec{r})\right\}\\&+\frac{i\omega}{c}\int d^3\vec{r}'\left\{\vec{E}_1(\vec{r})\cdot\ddot{\varepsilon}(\vec{r},\vec{r}')\cdot\vec{E}_2(\vec{r}')-\ddot{\varepsilon}(\vec{r},\vec{r}')\cdot\vec{E}_1(\vec{r}')\cdot\vec{E}_2(\vec{r})\right\}+\frac{4\pi}{c}\left(\vec{J}_1\cdot\vec{E}_2-\vec{J}_2\cdot\vec{E}_1\right)\end{aligned} \ . \tag{31}$$

We integrate over $\mathbb{R}^3$ and use the divergence theorem to remove the surface term again. Thus Eq. (31) becomes:

$$\begin{aligned}&\frac{4\pi}{c}\int\left(\vec{J}_1\cdot\vec{E}_2-\vec{J}_2\cdot\vec{E}_1\right)d^3\vec{r}\\&=\frac{i\omega}{c}\int d^3\vec{r}\int d^3\vec{r}'\left\{\vec{B}_1(\vec{r})\cdot\ddot{\mu}^{-1}(\vec{r},\vec{r}')\cdot\vec{B}_2(\vec{r}')-\ddot{\mu}^{-1}(\vec{r},\vec{r}')\cdot\vec{B}_1(\vec{r}')\cdot\vec{B}_2(\vec{r})\right\},\\&+\frac{i\omega}{c}\int d^3\vec{r}\int d^3\vec{r}'\left\{\vec{E}_1(\vec{r})\cdot\ddot{\varepsilon}(\vec{r},\vec{r}')\cdot\vec{E}_2(\vec{r}')-\ddot{\varepsilon}(\vec{r},\vec{r}')\cdot\vec{E}_1(\vec{r}')\cdot\vec{E}_2(\vec{r})\right\}\end{aligned} \tag{32}$$

which can be simplified to get the Lorentz lemma in electrodynamics under the symmetry condition of the dielectric tensor $\varepsilon_{ij}(\vec{r},\vec{r}')=\varepsilon_{ji}(\vec{r}',\vec{r})$ and the permeability tensor $\mu_{ij}(\vec{r},\vec{r}')=\mu_{ji}(\vec{r}',\vec{r})$.

3.2 Dyadic Green function

The other method to describe the optical reciprocity is the symmetry of dyadic Green function and the mathematical form is the following form (Tai, 1993; Xie, 2009a, 2009b):

$$\left[\ddot{G}_e(\vec{r}',\vec{r}'')\right]^T=\ddot{G}_e(\vec{r}'',\vec{r}'), \tag{33}$$

where T is the transport. We shall establish our results in two steps and restrict ourselves to the case of boundary conditions for the dyadic Green function.

3.2.1 Anisotropic local response

First we consider only local response which is simpler and sets the framework for the treatment of the more complicated nonlocal case. Thus we assume the following constitutive relations: $\vec{D}(\vec{r})=\ddot{\varepsilon}(\vec{r})\cdot\vec{E}(\vec{r})$ and $\vec{B}(\vec{r})=\ddot{\mu}(\vec{r})\cdot\vec{H}(\vec{r})$. For fields with time harmonic dependence, we have the following vector identity:

$$\vec{\nabla}\times\ddot{\mu}^{-1}(\vec{r})\cdot\vec{\nabla}\times\vec{E}(\vec{r})-\frac{\omega^2}{c^2}\ddot{\varepsilon}(\vec{r})\cdot\vec{E}(\vec{r})=i\omega\frac{4\pi}{c^2}\vec{J}(\vec{r}), \tag{34}$$

which implies the following differential equation for the electric dyadic Green function of the problem:

$$\vec{\nabla}\times\ddot{\mu}^{-1}(\vec{r})\cdot\vec{\nabla}\times\ddot{G}_e(\vec{r},\vec{r}')-\frac{\omega^2}{c^2}\ddot{\varepsilon}(\vec{r})\cdot\ddot{G}_e(\vec{r},\vec{r}')=\frac{4\pi}{c}\ddot{I}\delta(\vec{r}-\vec{r}'). \tag{35}$$

Using Eq. (A9), we set $\ddot{A}=\ddot{G}_e(\vec{r},\vec{r}')$, $\ddot{B}=\ddot{G}_e(\vec{r},\vec{r}'')$ and $\ddot{\lambda}=\ddot{\mu}^{-1}(\vec{r})$ to obtain the following identity:

$$\begin{aligned}&\int\left\{\left[\vec{\nabla}\times\ddot{\mu}^{-1}\cdot\vec{\nabla}\times\ddot{G}_e(\vec{r},\vec{r}'')\right]^T\cdot\ddot{G}_e(\vec{r},\vec{r}')-\left[\ddot{G}_e(\vec{r},\vec{r}'')\right]^T\cdot\vec{\nabla}\times\ddot{\mu}^{-1}\cdot\vec{\nabla}\times\ddot{G}_e(\vec{r},\vec{r}')\right\}d^3\vec{r}\\&=\oint_S\left\{\left[\hat{n}\times\ddot{G}_e(\vec{r},\vec{r}'')\right]^T\cdot\left(\ddot{\mu}^{-1}\cdot\vec{\nabla}\times\ddot{G}_e(\vec{r},\vec{r}')\right)-\left[\ddot{\mu}^{-1}\cdot\vec{\nabla}\times\ddot{G}_e(\vec{r},\vec{r}'')\right]^T\cdot\hat{n}\times\ddot{G}_e(\vec{r},\vec{r}')\right\}da\end{aligned}. \tag{36}$$

Hence from either the dyadic Dirichlet condition:

$$\hat{n}\times\ddot{G}_e(\vec{r},\vec{r}')=0, \tag{37}$$

or the dyadic Neumann condition[1]:

$$\hat{n}\times\left[\ddot{\mu}^{-1}\cdot\vec{\nabla}\times\ddot{G}_e(\vec{r},\vec{r}')\right]=0, \tag{38}$$

[1] Note that the condition introduced in Eq. (38) is a generalization of $\hat{n}\times\left[\vec{\nabla}\times\ddot{G}_e(\vec{r},\vec{r}')\right]=0$ with the explicit inclusion of the magnetic permeability tensor.

Eq. (36) leads to:

$$\int\left\{\left[\vec{\nabla}\times\ddot{\mu}^{-1}\cdot\vec{\nabla}\times\ddot{G}_e(\vec{r},\vec{r}'')\right]^T\cdot\ddot{G}_e(\vec{r},\vec{r}')-\left[\ddot{G}_e(\vec{r},\vec{r}'')\right]^T\cdot\vec{\nabla}\times\ddot{\mu}^{-1}\cdot\vec{\nabla}\times\ddot{G}_e(\vec{r},\vec{r}')\right\}d^3\vec{r}=0\,. \quad (39)$$

Substituting Eq. (35) into Eq. (39), we have:

$$\begin{aligned}&\int\left\{\frac{\omega^2}{c^2}\left[\ddot{\varepsilon}(\vec{r})\cdot\ddot{G}_e(\vec{r},\vec{r}'')\right]^T+\frac{4\pi}{c}\ddot{I}\delta(\vec{r}-\vec{r}'')\right\}\cdot\ddot{G}_e(\vec{r},\vec{r}')d^3\vec{r}\\&-\int\left[\ddot{G}_e(\vec{r},\vec{r}'')\right]^T\cdot\left\{\frac{\omega^2}{c^2}\ddot{\varepsilon}(\vec{r})\cdot\ddot{G}_e(\vec{r},\vec{r}')+\frac{4\pi}{c}\ddot{I}\delta(\vec{r}-\vec{r}')\right\}d^3\vec{r}=0\end{aligned}, \quad (40)$$

which implies the symmetry of the dyadic Green function. Eq. (39) provides that the dielectric tensor is symmetric: $\varepsilon_{ij}(\vec{r})=\varepsilon_{ji}(\vec{r})$. We remark that the validity of Eq. (36) has already required $\mu_{ij}(\vec{r})=\mu_{ji}(\vec{r})$ as explained before.

3.2.2 Anisotropic nonlocal response

Now we consider both electric and magnetic nonlocal responses and thus generalize the dyadic differential equation in Eq. (35) to:

$$\nabla\times\int\ddot{\mu}^{-1}(\vec{r},\vec{r}_1)\cdot\vec{\nabla}_1\times\ddot{G}_e(\vec{r}_1,\vec{r}')d^3\vec{r}_1-\frac{\omega^2}{c^2}\int\ddot{\varepsilon}(\vec{r},\vec{r}_1)\cdot\ddot{G}_e(\vec{r}_1,\vec{r}')d^3\vec{r}_1=\frac{4\pi}{c}\ddot{I}\delta(\vec{r}-\vec{r}')\,. \quad (41)$$

Using Eq. (A15), we set $\begin{cases}\ddot{A}(\vec{r})=\ddot{G}_e(\vec{r},\vec{r}')\\ \ddot{A}(\vec{r}_1)=\ddot{G}_e(\vec{r}_1,\vec{r}')\end{cases}$, $\begin{cases}\ddot{B}(\vec{r})=\ddot{G}_e(\vec{r},\vec{r}'')\\ \ddot{B}(\vec{r}_1)=\ddot{G}_e(\vec{r}_1,\vec{r}'')\end{cases}$ and $\ddot{\lambda}=\ddot{\mu}^{-1}$ and obtain the following form:

$$\begin{aligned}&\int d^3\vec{r}\int d^3\vec{r}_1\left\{\left[\vec{\nabla}\times\ddot{\mu}^{-1}(\vec{r},\vec{r}_1)\cdot\vec{\nabla}_1\times\ddot{G}_e(\vec{r}_1,\vec{r}'')\right]^T\cdot\ddot{G}_e(\vec{r},\vec{r}')\right\}\\&-\int d^3\vec{r}\int d^3\vec{r}_1\left\{\left[\ddot{G}_e(\vec{r},\vec{r}'')\right]^T\cdot\vec{\nabla}\times\ddot{\mu}^{-1}(\vec{r},\vec{r}_1)\cdot\vec{\nabla}_1\times\ddot{G}_e(\vec{r}_1,\vec{r}')\right\}\\&=\oint_S da\int d^3\vec{r}_1\left\{\left[\hat{n}\times\ddot{G}_e(\vec{r},\vec{r}'')\right]^T\cdot\left[\ddot{\mu}^{-1}(\vec{r},\vec{r}_1)\cdot\vec{\nabla}_1\times\ddot{G}_e(\vec{r}_1,\vec{r}')\right]\right\}\\&-\oint_S da\int d^3\vec{r}_1\left\{\left[\ddot{\mu}^{-1}(\vec{r},\vec{r}_1)\cdot\vec{\nabla}_1\times\ddot{G}_e(\vec{r}_1,\vec{r}'')\right]^T\cdot\hat{n}\times\ddot{G}_e(\vec{r},\vec{r}')\right\}\end{aligned}. \quad (42)$$

Again, with either Dirichlet or Neumann conditions[2], we obtain from Eq. (42):

$$\begin{aligned}&\int d^3\vec{r}\int d^3\vec{r}_1\left\{\left[\vec{\nabla}\times\ddot{\mu}^{-1}(\vec{r},\vec{r}_1)\cdot\vec{\nabla}_1\times\ddot{G}_e(\vec{r}_1,\vec{r}'')\right]^T\cdot\ddot{G}_e(\vec{r},\vec{r}')\right\}\\&-\int d^3\vec{r}\int d^3\vec{r}_1\left\{\left[\ddot{G}_e(\vec{r},\vec{r}'')\right]^T\cdot\vec{\nabla}\times\ddot{\mu}^{-1}(\vec{r},\vec{r}_1)\cdot\vec{\nabla}_1\times\ddot{G}_e(\vec{r}_1,\vec{r}')\right\}=0\end{aligned}. \quad (43)$$

---

[2] The Neumann condition in Eq. (38) in the nonlocal case has also to be generalized to the following form: $\int\hat{n}\times\left[\ddot{\mu}^{-1}(\vec{r},\vec{r}_1)\cdot\vec{\nabla}_1\times\ddot{G}_e(\vec{r}_1,\vec{r}'')\right]d^3\vec{r}=0$

Substituting Eq. (41) into Eq. (43), we have the following identity:

$$\begin{aligned}&\int\left[\frac{\omega^2}{c^2}\int\ddot{\varepsilon}(\vec{r},\vec{r}_1)\cdot\ddot{G}_e(\vec{r}_1,\vec{r}'')d^3\vec{r}_1+\frac{4\pi}{c}\ddot{I}\delta(\vec{r}-\vec{r}'')\right]^T\cdot\ddot{G}_e(\vec{r},\vec{r}')d^3\vec{r}\\&=\int\left[\ddot{G}_e(\vec{r},\vec{r}'')\right]^T\cdot\left\{\frac{\omega^2}{c^2}\int\ddot{\varepsilon}(\vec{r},\vec{r}_1)\cdot\ddot{G}_e(\vec{r}_1,\vec{r}')d^3\vec{r}_1+\frac{4\pi}{c}\ddot{I}\delta(\vec{r}-\vec{r}')\right\}d^3\vec{r}\end{aligned}, \quad (44)$$

which again implies Eq. (33) in a way that is similar to the above case for local response.

4. The equivalence of Lorentz lemma and Green function formulation

So far, we have shown two different mathematical formulations for discussing the optical reciprocity. Now the question is: are these two statements equivalent? Now we give a proof.

4.1 Electrostatics

First we demonstrate the equivalence between Lorentz lemma and the symmetry of the scalar Green function in electrostatics, by starting with a slightly more general form of Eq. (1) *with the surface terms retained*:

$$\int(\rho_1\Phi_2-\rho_1\Phi_2)d^3\vec{r}=\frac{1}{4\pi}\oint_S\hat{n}\cdot\left(\Phi_1\ddot{\varepsilon}\cdot\vec{\nabla}\Phi_2-\Phi_2\ddot{\varepsilon}\cdot\vec{\nabla}\Phi_1\right)da. \quad (45)$$

Note that the above can be applied to the finite boundary region. To demonstrate the equivalence between Eq. (1) and Eq. (6), let us consider two unit point charge distribution as follows:

$$\rho_1=\delta(\vec{r}-\vec{r}''),\ \rho_2-\delta(\vec{r}-\vec{r}'), \quad (46)$$

and the potentials at each of their locations are then given by the scalar Green function:

$$\Phi_1=G(\vec{r},\vec{r}''),\Phi_2=G(\vec{r},\vec{r}')\ . \quad (47)$$

Substituting Eqs. (46) and (47) into Eq. (45) leads to the following result[3]:

$$\begin{aligned}&G(\vec{r}'',\vec{r}')-G(\vec{r}',\vec{r}'')\\&=\frac{1}{4\pi}\oint_S\hat{n}\cdot\left[G(\vec{r},\vec{r}'')\ddot{\varepsilon}\cdot\vec{\nabla}G(\vec{r},\vec{r}')-G(\vec{r},\vec{r}')\ddot{\varepsilon}\cdot\vec{\nabla}G(\vec{r},\vec{r}'')\right]da\end{aligned}. \quad (48)$$

[3] Note that the proof of the equivalence between the two versions of the reciprocity principle in the previous section remains valid for the case with nonlocal response, with Eq. (48) generalized to the following form:

$$\begin{aligned}&G(\vec{r}'',\vec{r}')-G(\vec{r}',\vec{r}'')\\&=\frac{1}{4\pi}\oint_S da\int d^3\vec{r}_1\left\{\hat{n}\cdot\left[G(\vec{r},\vec{r}'')\ddot{\varepsilon}(\vec{r},\vec{r}_1)\cdot\vec{\nabla}_1G(\vec{r}_1,\vec{r}')-G(\vec{r},\vec{r}')\ddot{\varepsilon}(\vec{r},\vec{r}_1)\cdot\vec{\nabla}_1G(\vec{r}_1,\vec{r}'')\right]\right\}\end{aligned}.$$

Here we separate into two different kinds of the boundary conditions to discuss:
First, with the Dirichlet boundary condition given in Eq. (10) substituted into Eq. (48), we obtain Eq. (6). Thus we have demonstrated the equivalence between the Lorentz lemma in electrostatics and the scalar Green function under the Dirichlet boundary condition.
Second, the Neumann boundary condition is given by Eq. (11) and thus Eq. (48) becomes the following form:

$$G_N(\vec{r}'',\vec{r}') - G_N(\vec{r}',\vec{r}'') = -\frac{1}{A}\oint_S G_N(\vec{r},\vec{r}'')da + \frac{1}{A}\oint_S G_N(\vec{r},\vec{r}')da . \tag{49}$$

By the pervious method, we can establish the symmetry of the scalar Green function as shown in Eq. (6).

4.2 Electrodynamics

Next we will show that the equivalence between these two statements which are the optical reciprocity in the form of Lorentz lemma in electrodynamics and of the symmetry of the dyadic Green function. To demonstrate this equivalence, first we start from Lorentz lemma in electrodynamics *by retaining the surface terms (Xie, 2009b)*:

$$\frac{4\pi}{c}\int\left(\vec{J}_1\cdot\vec{E}_2 - \vec{J}_2\cdot\vec{E}_1\right)d^3\vec{r} = \oint_S \hat{n}\cdot\left(\vec{E}_1\cdot\vec{H}_2 - \vec{E}_2\cdot\vec{H}_1\right)da . \tag{50}$$

Note that Eq. (50) is a direct consequence from Maxwell's equations and the surface terms are kept to allow for the presence of finite boundaries and nontrivial material with both permittivity and permeability. Although these surface terms are usually discarded (Kahl & Voges, 2000; Ru & Etchegoin, 2006; Landau et al., 1984; Iwanaga et al., 2007), they have also been considered in some studies in the literatures (Xie et al., 2009; Porto et al., 2000; Joe et al., 2008). Hence we must keep them to demonstrate the exact equivalence between the two versions of optical reciprocity.
In the beginning, let us consider two unit point current sources due to electric dipole (with moment p) as follows:

$$\begin{aligned}\vec{J}_1 &= -i\omega p\delta(\vec{r}-\vec{r}'')\hat{e}_i \\ \vec{J}_2 &= -i\omega p\delta(\vec{r}-\vec{r}')\hat{e}_j\end{aligned}, \tag{51}$$

and the electric fields at each of their locations are given in terms of the column component of the dyad as follows:

$$\vec{E}_1 = \frac{\omega^2 p}{c}\vec{G}_{ei}(\vec{r},\vec{r}''),\ \vec{E}_2 = \frac{\omega^2 p}{c}\vec{G}_{ej}(\vec{r},\vec{r}') . \tag{52}$$

Substituting Eqs. (51) and (52) into Eq. (50) leads to the following result:

$$\begin{aligned}&-\frac{4\pi i\omega p}{c}\left[\hat{e}_i\cdot\vec{G}_{ej}(\vec{r}'',\vec{r}') - \hat{e}_j\cdot\vec{G}_{ei}(\vec{r}',\vec{r}'')\right] \\ &= \oint_S \hat{n}\cdot\left[\vec{G}_{ei}(\vec{r},\vec{r}'')\times\vec{H}_2(\vec{r}) - \vec{G}_{ej}(\vec{r},\vec{r}')\times\vec{H}_1(\vec{r})\right]da\end{aligned}. \tag{53}$$

Hence using Maxwell's equations and the vector triple product, we obtain[4]:

$$
\begin{aligned}
&-\frac{4\pi i\omega p}{c}\left\{\left[\ddot{G}_e(\vec{r}'',\vec{r}')\right]_{ij}-\left[\ddot{G}_e(\vec{r}',\vec{r}'')\right]_{ji}\right\}\\
&=\oint_S\left\{\vec{H}_2(\vec{r})\cdot\left[\hat{n}\times\vec{G}_{ei}(\vec{r},\vec{r}'')\right]-\vec{H}_1(\vec{r})\cdot\left[\hat{n}\times\vec{G}_{ej}(\vec{r},\vec{r}')\right]\right\}da\\
&=\frac{\omega}{i}\oint_S\left\{\left(\ddot{\mu}^{-1}\cdot\vec{\nabla}\times\vec{E}_2(\vec{r})\right)\cdot\left[\hat{n}\times\vec{G}_{ei}(\vec{r},\vec{r}'')\right]-\left(\ddot{\mu}^{-1}\cdot\vec{\nabla}\times\vec{E}_1(\vec{r})\right)\cdot\left[\hat{n}\times\vec{G}_{ej}(\vec{r},\vec{r}')\right]\right\}da\\
&=\frac{\omega p}{i}\oint_S\left\{\left(\ddot{\mu}^{-1}\cdot\vec{\nabla}\times\vec{G}_{ej}(\vec{r},\vec{r}')\right)\cdot\left[\hat{n}\times\vec{G}_{ei}(\vec{r},\vec{r}'')\right]-\left(\ddot{\mu}^{-1}\cdot\vec{\nabla}\times\vec{G}_{ei}(\vec{r},\vec{r}'')\right)\cdot\left[\hat{n}\times\vec{G}_{ej}(\vec{r},\vec{r}')\right]\right\}da
\end{aligned}
\quad . \quad (54)
$$

Hence we have:

$$
\begin{aligned}
&\frac{4\pi}{c}\left\{\left[\ddot{G}_e(\vec{r}'',\vec{r}')\right]_{ij}\hat{e}_i-\left[\ddot{G}_e(\vec{r}',\vec{r}'')\right]_{ji}\hat{e}_i\right\}\\
&=\oint_S\left\{\left[\hat{n}\times\ddot{G}_e(\vec{r},\vec{r}'')\right]^T\cdot\left(\ddot{\mu}^{-1}\cdot\vec{\nabla}\times\vec{G}_{ej}(\vec{r},\vec{r}')\right)-\left[\ddot{\mu}^{-1}\cdot\vec{\nabla}\times\ddot{G}_e(\vec{r},\vec{r}'')\right]^T\cdot\left[\hat{n}\times\vec{G}_{ej}(\vec{r},\vec{r}')\right]\right\}da
\end{aligned}
\quad . \quad (55)
$$

We can rewrite Eq. (55) in dyadic form as follows:

$$
\begin{aligned}
&\frac{4\pi}{c}\left\{\ddot{G}_e(\vec{r}'',\vec{r}')-\left[\ddot{G}_e(\vec{r}',\vec{r}'')\right]^T\right\}\\
&=\oint_S\left\{\left[\hat{n}\times\ddot{G}_e(\vec{r},\vec{r}'')\right]^T\cdot\left(\ddot{\mu}^{-1}\cdot\vec{\nabla}\times\ddot{G}_e(\vec{r},\vec{r}')\right)-\left[\ddot{\mu}^{-1}\cdot\vec{\nabla}\times\ddot{G}_e(\vec{r},\vec{r}'')\right]^T\cdot\left[\hat{n}\times\ddot{G}_e(\vec{r},\vec{r}')\right]\right\}da
\end{aligned}
\quad . \quad (56)
$$

By imposing on S either the dyadic Dirichlet condition (Eq. (37)) or the dyadic Neumann condition (Eq. (38)), the surface integral in Eq. (56) can be made vanished by applying the dyadic triple product in the Neumann case. Hence under either one of these boundary conditions, Eq. (56) will lead to the symmetric property of the dyadic Green function in Eq. (33).

5. Some examples

We have established the general conditions for optical reciprocity to hold in nonlocal optics from the method of electrostatics to electrodynamics. The general conditions are:

[4] Note that the proof of the equivalence between the two versions of the reciprocity principle in the previous section remains valid for the case with nonlocal response, with Eq. (54) generalized to the following form:

$$
\begin{aligned}
&-\frac{4\pi i\omega p}{c}\left\{\left[\ddot{G}_e(\vec{r}'',\vec{r}')\right]_{ij}-\left[\ddot{G}_e(\vec{r}',\vec{r}'')\right]_{ji}\right\}\\
&=\frac{\omega p}{i}\oint_S da\int d^3\vec{r}_1\left(\ddot{\mu}^{-1}(\vec{r},\vec{r}_1)\cdot\vec{\nabla}_1\times\vec{G}_{ej}(\vec{r}_1,\vec{r}')\right)\cdot\left[\hat{n}\times\vec{G}_{ei}(\vec{r},\vec{r}'')\right]\\
&-\frac{\omega p}{i}\oint_S da\int d^3\vec{r}_1\left(\ddot{\mu}^{-1}(\vec{r},\vec{r}_1)\cdot\vec{\nabla}_1\times\vec{G}_{ei}(\vec{r}_1,\vec{r}'')\right)\cdot\left[\hat{n}\times\vec{G}_{ej}(\vec{r},\vec{r}')\right]
\end{aligned}
.
$$

$$\begin{aligned}\varepsilon_{ij}(\vec{r},\vec{r}') &= \varepsilon_{ji}(\vec{r}',\vec{r}) \\ \mu_{ij}(\vec{r},\vec{r}') &= \mu_{ji}(\vec{r}',\vec{r})\end{aligned}, \tag{57}$$

which are *the extension conditions of local optics*. This reduces to the well-known local limit which requires only a symmetric local dielectric tensor for the validity of reciprocity (Chang, 2008; Iwanaga, 2007). It also reduces to the isotropic nonlocal case which is known to be valid for most of the well-known nonlocal quantum mechanical models for a homogeneous electron gas, such as the Linhard-Mermin function in which $\varepsilon(\vec{r},\vec{r}') = \varepsilon(|\vec{r}-\vec{r}'|)$ (Chang, 2008). Moreover, we also give two interesting examples that *may* lead to the *breakdown* of the reciprocity in linear optics. One example is that the following dielectric tensor:

$$\ddot{\varepsilon} = \begin{pmatrix} \varepsilon_x & -ig & 0 \\ ig & \varepsilon_x & 0 \\ 0 & 0 & \varepsilon_z \end{pmatrix}, \tag{58}$$

which is hermitian but not symmetric (Vlokh & Adamenko, 2008). Another example is to refer to the case studied in the literature (Malinowski et al., 1996) which involved the propagation of light along a cubic axis in a crystal of 23 point group. In this case, the nonlocality tensor γ_{ijk} may be asymmetric in the sense that $\gamma_{ijk} \neq \gamma_{jik}$, which can be shown to imply an asymmetric dielectric tensor $\varepsilon_{ij} \neq \varepsilon_{ji}$. Here we give a proof. With the dielectric function becoming a tensor, we have:

$$D_i(\vec{r}) = \int \varepsilon_{ij}(\vec{r},\vec{r}') \cdot E_j(\vec{r}') d^3\vec{r}'. \tag{59}$$

Next we change the variable $\vec{r}' = \vec{r} + \vec{a}$ and use a Taylor series for the electric field to obtain the following form:

$$\begin{aligned} D_i(\vec{r}) &= \int \varepsilon_{ij}(\vec{r},\vec{r}+\vec{a}) E_j(\vec{r}+\vec{a}) d^3\vec{a} \\ &= \int \varepsilon_{ij}(\vec{r},\vec{r}+\vec{a}) \left[E_j(\vec{r}) + \vec{a}\cdot\vec{\nabla}E_j(\vec{r}) + \frac{(\vec{a}\cdot\vec{\nabla})^2}{2} E_j(\vec{r}) + O(a^3) \right] d^3\vec{a} \end{aligned}. \tag{60}$$

For case of weak nonlocality, where $\varepsilon_{ij}(\vec{r},\vec{r}+\vec{a}) \neq 0$ only for $\vec{a}$ within a small neighborhood of $\vec{r}$, higher order terms in Eq. (60) can be neglected, and we recover the identity which has occurred in Eq. (1) of the literature (Malinowski et al., 1996):

$$\begin{aligned} D_i(\vec{r}) &= E_j(\vec{r}) \int \varepsilon_{ij}(\vec{r},\vec{r}+\vec{a}) d^3\vec{a} + \partial_k E_j(\vec{r}) \int \varepsilon_{ij}(\vec{r},\vec{r}+\vec{a}) a_k d^3\vec{a} \\ &\equiv \beta_{ij} E_j + \gamma_{ijk} \partial_k E_j(\vec{r}) \end{aligned}, \tag{61}$$

where the first term and second term of the above equation denote the contribution of locality and nonlocality, respectively. Since the nonlocality tensor γ_{ijk} satisfies the relation $\gamma_{ijk} \neq \gamma_{jik}$, we conclude that the electric tensor ε_{ij} satisfies the relation $\varepsilon_{ij} \neq \varepsilon_{ji}$. However, this is the same with what was studied in the literature (Malinowski et al., 1996), where nonlocality through the field gradient dependent response is required to break reciprocity symmetry for the rotation of the polarization plane of the transmitted wave. In their statement, if the nonlicality γ_{ijk} satisfies the relation $\gamma_{ijk} = \gamma_{jik}$, optical reciprocity breaks down. In our viewpoint, From Eq. (61), the relation $\gamma_{ijk} = \gamma_{jik}$ implies the relation $\varepsilon_{ij}(\vec{r},\vec{r}+\vec{a}) = \varepsilon_{ji}(\vec{r},\vec{r}+\vec{a})$ where violates Eq. (57). Thus the reciprocity *may* break down. Hence our mathematical formulations provide a general examination to determine if the optical reciprocity remain or break down initially.

6. Application to spectroscopic analysis

In this secton, we demonstrate the application of the reciprocity symmetry in the form of the Lorentz lemma for two dipolar sources (in obvious notations):

$$\vec{p}_1 \cdot \vec{E}_2 = \vec{p}_2 \cdot \vec{E}_1, \tag{62}$$

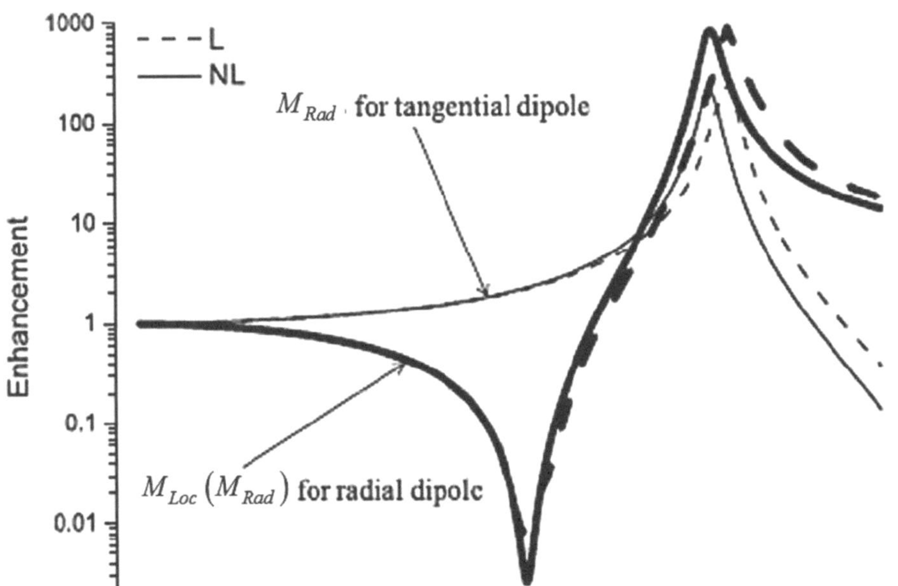

Fig. 1. Spectrum of the local field and radiation enhancement factors, with the latter plotted for both radial and tangential molecular dipoles, according to both the local (dashed lines) and nonlocal (solid lines) SERS models. The molecular dipole is located at a distance of 1 nm from a silver nanosphere of 5 nm radius

to the calculation of the various surface-enhanced Raman scattering (SERS) enhancement factors from a molecule adsorbed on a metallic nanoparticle following the recent work of Le Ru and Etchegoin. As pointed out by Le Ru and Etchegoin (Ru & Etchegoin, 2006), in any SERS analysis, one *must* distinguish carefully between the local field and the radiation enhancement since '. . . the induced molecular Raman dipole is not necessarily aligned

parallel to the electric field of the pump beam . . .'. Based on this distinction, it was proposed in the literature (Ru & Etchegoin, 2006) that the more correct SERS enhancement ratio should be a product of these two enhancement factors: $\vec{M}_{SERS} = \vec{M}_{Loc} \cdot \vec{M}_{Rad}$ with the latter enhancement calculable from an application of Eq. (62). This formulation has then corrected a conventional misconception in the literature of SERS theory with models exclusively based on the fourth power dependence of the local field.
In Fig. 1, we have essentially reproduced the key features in the corresponding Fig. 1 of the literature (Ru & Etchegoin, 2006), but for a much smaller metal sphere (radius = 5 nm) so that nonlocal effects are more pronounced. Note that in this figure, Eq. (21) has been used to calculate the various quantities represented by solid lines and we note that, with the nonlocal response of the metal particle, the sharp differences between $\vec{M}_{Loc}$ and $\vec{M}_{Rad}$ remain for the tangentially oriented dipoles, as was first observed in the literature (Ru & Etchegoin, 2006). The radially oriented dipole, however, gives very similar results for both the enhancement factors in both our nonlocal calculation and the local one as reported in the literature (Ru & Etchegoin, 2006). Note that the nonlocal effects are most significant in the vicinity of the plasmon resonance frequency, with the peaks slightly blueshifted due mainly to the semiclassical infinite barrier (SCIB) approximation adopted in this model (Fuchs & Claro, 1987).

7. Conclusion

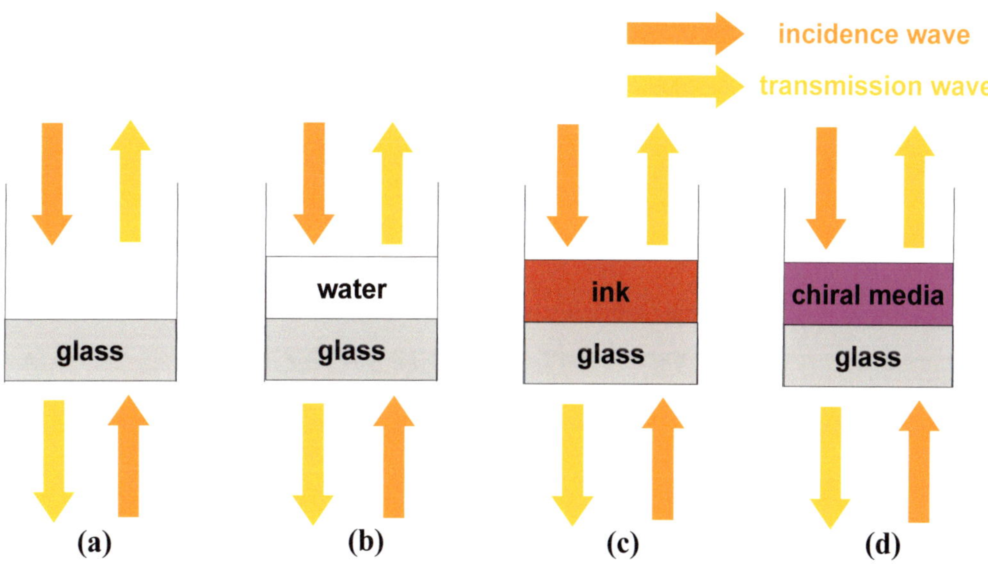

Fig. 2. The description of optical reciprocity in four different distributions of the material media

We have constructed the conditions for optical reciprocity in the case with a nonlocal anisotropic magnetic permeability and electric permittivity, motivated by the recent explosion in the research with metamaterials according to two different mathematical viewpoints (Lorentz lemma and Green function formulation) furthermore that are equivalent. These results reduce to the well-known conditions in the case of local response. Note that while the symmetry in $\vec{r}$ and $\vec{r}'$ will be valid for must materials on a macroscopic scale (Jenkins & Hunt, 2003), that in the tensorial indices will not be valid in general for complex materials such as bianisotropic or chiral materials (Kong, 2003). Importantly, our mathematical formulations provide a general examination to determine if the optical reciprocity remain or break down initially. However, it will be of interest to design some optical experiment to observe the breakdown of reciprocity symmetry with these systems in the study of metamaterials. One possible way is to observe transmission asymmetry in the light propagating through these materials as shown in Fig. 2 which shows this interesting process and lists four different distributions of the material media. According to our pervious mathematical prediction, we will have optical reciprocity still remains valid in (a), (b) and (c); but it may break down in (d).

8. Appendix Give a proof of some useful mathematical identities (Chang, 2008; Xie, 2009a, 2009b)

[1]
$$\int_V \left[\Phi\nabla\cdot\left(\ddot{\lambda}\cdot\nabla\Psi\right)-\Psi\nabla\cdot\left(\ddot{\lambda}\cdot\nabla\Phi\right)\right]d^3\vec{r}=\int_S \hat{n}\cdot\left(\Phi\ddot{\lambda}\cdot\nabla\Psi-\Psi\ddot{\lambda}\cdot\nabla\Phi\right)da\,, \tag{A1}$$

under the condition $\lambda_{ij}=\lambda_{ji}$.
To prove Eq. (A1), we will first prove the following identity:

$$\vec{\nabla}\cdot\left[\ddot{\lambda}\cdot\left(\Phi\vec{\nabla}\Psi-\Psi\vec{\nabla}\Phi\right)\right]=\Phi\vec{\nabla}\cdot\left(\ddot{\lambda}\cdot\vec{\nabla}\Psi\right)-\Psi\vec{\nabla}\cdot\left(\ddot{\lambda}\cdot\vec{\nabla}\Phi\right), \tag{A2}$$

under the condition $\lambda_{ij}=\lambda_{ji}$. Using the Einstein notation to express Eq. (A2), we have for the LHS:

$$\begin{aligned}\vec{\nabla}\cdot\left[\ddot{\lambda}\cdot\left(\Phi\vec{\nabla}\Psi-\Psi\vec{\nabla}\Phi\right)\right]&=\Phi\partial_i\lambda_{ij}\partial_j\Psi-\Psi\partial_i\lambda_{ij}\partial_j\Phi\\&\quad+\lambda_{ij}\left[(\partial_i\Phi)(\partial_j\Psi)-(\partial_i\Psi)(\partial_j\Phi)\right]\end{aligned}, \tag{A3}$$

and for the RHS of Eq. (A2):

$$\Phi\vec{\nabla}\cdot\left(\ddot{\lambda}\cdot\vec{\nabla}\Psi\right)-\Psi\vec{\nabla}\cdot\left(\ddot{\lambda}\cdot\vec{\nabla}\Phi\right)=\Phi\partial_i\lambda_{ij}\partial_j\Psi-\Psi\partial_i\lambda_{ij}\partial_j\Phi\,. \tag{A4}$$

Thus Eqs. (A3) and (A4) are equal under the condition $\lambda_{ij}=\lambda_{ji}$ and hence we prove Eq. (A1).

[2]
$$\begin{aligned}&\int d^3\vec{r}\int d^3\vec{r}_1\left\{\Phi(\vec{r})\nabla\cdot\left[\ddot{\lambda}(\vec{r},\vec{r}_1)\cdot\nabla_1\Psi(\vec{r}_1)\right]-\Psi(\vec{r})\nabla\cdot\left[\ddot{\lambda}(\vec{r},\vec{r}_1)\cdot\nabla_1\Phi(\vec{r}_1)\right]\right\}\\&=\int_S da\int d^3\vec{r}_1\left\{\hat{n}\cdot\left[\Phi(\vec{r})\ddot{\lambda}(\vec{r},\vec{r}_1)\cdot\nabla_1\Psi(\vec{r}_1)-\Psi(\vec{r})\ddot{\lambda}(\vec{r},\vec{r}_1)\cdot\nabla_1\Phi(\vec{r}_1)\right]\right\}\end{aligned}, \tag{A5}$$

under the condition $\lambda_{ij}(\vec{r},\vec{r}')=\lambda_{ji}(\vec{r}',\vec{r})$.
First we prove the following identity:

$$\begin{aligned}&\int d^3\vec{r}\int d^3\vec{r}_1\left\{\Phi(\vec{r})\vec{\nabla}\cdot\left[\ddot{\lambda}(\vec{r},\vec{r}_1)\cdot\vec{\nabla}_1\Psi(\vec{r}_1)\right]-\Psi(\vec{r})\vec{\nabla}\cdot\left[\ddot{\lambda}(\vec{r},\vec{r}_1)\cdot\vec{\nabla}_1\Phi(\vec{r}_1)\right]\right\}\\&=\int d^3\vec{r}\int d^3\vec{r}_1\left\{\vec{\nabla}\cdot\left[\Phi(\vec{r})\ddot{\lambda}(\vec{r},\vec{r}_1)\cdot\vec{\nabla}_1\Psi(\vec{r}_1)-\Psi(\vec{r})\ddot{\lambda}(\vec{r},\vec{r}_1)\cdot\vec{\nabla}_1\Phi(\vec{r}_1)\right]\right\}\end{aligned}, \tag{A6}$$

under the condition $\lambda_{ij}(\vec{r},\vec{r}')=\lambda_{ji}(\vec{r}',\vec{r})$. Again we express the left side as:

$$\begin{aligned}&\int d^3\vec{r}\int d^3\vec{r}_1\left\{\Phi(\vec{r})\vec{\nabla}\cdot\left[\ddot{\lambda}(\vec{r},\vec{r}_1)\cdot\vec{\nabla}_1\Psi(\vec{r}_1)\right]-\Psi(\vec{r})\vec{\nabla}\cdot\left[\ddot{\lambda}(\vec{r},\vec{r}_1)\cdot\vec{\nabla}_1\Phi(\vec{r}_1)\right]\right\}\\&=\int d^3\vec{r}\int d^3\vec{r}_1\left\{\Phi(\vec{r})\partial_i^{\vec{r}}\lambda_{ij}(\vec{r},\vec{r}_1)\partial_j^{\vec{r}_1}\Psi(\vec{r}_1)-\Psi(\vec{r})\partial_i^{\vec{r}}\lambda_{ij}(\vec{r},\vec{r}_1)\partial_j^{\vec{r}_1}\Phi(\vec{r}_1)\right\}\end{aligned}, \quad (A7)$$

and the right side as:

$$\begin{aligned}&\int d^3\vec{r}\int d^3\vec{r}_1\left\{\vec{\nabla}\cdot\left[\Phi(\vec{r})\ddot{\lambda}(\vec{r},\vec{r}_1)\cdot\vec{\nabla}_1\Psi(\vec{r}_1)-\Psi(\vec{r})\ddot{\lambda}(\vec{r},\vec{r}_1)\cdot\vec{\nabla}_1\Phi(\vec{r}_1)\right]\right\}\\&=\int d^3\vec{r}\int d^3\vec{r}_1\left\{\Phi(\vec{r})\partial_i^{\vec{r}}\lambda_{ij}(\vec{r},\vec{r}_1)\partial_j^{\vec{r}_1}\Psi(\vec{r}_1)-\Psi(\vec{r})\partial_i^{\vec{r}}\lambda_{ij}(\vec{r},\vec{r}_1)\partial_j^{\vec{r}_1}\Phi(\vec{r}_1)\right\}\\&+\int d^3\vec{r}\int d^3\vec{r}_1\left\{\lambda_{ij}(\vec{r},\vec{r}_1)\left[\left[\partial_i^{\vec{r}}\Phi(\vec{r})\right]\left[\partial_j^{\vec{r}_1}\Psi(\vec{r}_1)\right]-\left[\partial_i^{\vec{r}}\Psi(\vec{r})\right]\left[\partial_j^{\vec{r}_1}\Phi(\vec{r}_1)\right]\right]\right\}\end{aligned}. \quad (A8)$$

Thus Eqs. (A7) and (A8) are equal under the condition $\lambda_{ij}(\vec{r},\vec{r}')=\lambda_{ji}(\vec{r}',\vec{r})$ and hence Eq. (A6) is established. We can again use the divergence theorem to establish Eq. (A5).

[3]

$$\begin{aligned}&\int\left\{\left[\vec{\nabla}\times\ddot{\lambda}\cdot\vec{\nabla}\times\ddot{B}\right]^T\cdot\ddot{A}-\left[\ddot{B}\right]^T\cdot\vec{\nabla}\times\ddot{\lambda}\cdot\vec{\nabla}\times\ddot{A}\right\}d^3\vec{r}\\&=\oint_S\left\{\left[\hat{n}\times\ddot{B}\right]^T\cdot\left(\ddot{\lambda}\cdot\vec{\nabla}\times\ddot{A}\right)-\left[\ddot{\lambda}\cdot\vec{\nabla}\times\ddot{B}\right]^T\cdot\hat{n}\times\ddot{A}\right\}da\end{aligned}, \quad (A9)$$

under the condition $\lambda_{ij}=\lambda_{ji}$.
Let us first establish the following simpler vector identity:

$$\vec{\nabla}\cdot\left[\vec{B}\times\ddot{\lambda}\cdot\vec{\nabla}\times\vec{A}-\vec{A}\times\ddot{\lambda}\cdot\vec{\nabla}\times\vec{B}\right]=\vec{A}\cdot\vec{\nabla}\times\ddot{\lambda}\cdot\vec{\nabla}\times\vec{B}-\vec{B}\cdot\vec{\nabla}\times\ddot{\lambda}\cdot\vec{\nabla}\times\vec{A}, \quad (A10)$$

under the condition $\lambda_{ij}=\lambda_{ji}$. In explicit Einstein's summation convention, we have for the LHS of Eq. (A10):

$$\begin{aligned}\vec{\nabla}\cdot\left[\vec{B}\times\ddot{\lambda}\cdot\vec{\nabla}\times\vec{A}-\vec{A}\times\ddot{\lambda}\cdot\vec{\nabla}\times\vec{B}\right]&=\varepsilon_{ijk}\varepsilon_{lmn}\left(B_j\partial_i\lambda_{kl}\partial_m A_n-A_j\partial_i\lambda_{kl}\partial_m B_n\right)\\&+\varepsilon_{ijk}\varepsilon_{lmn}\lambda_{kl}\left[\left(\partial_i B_j\right)\left(\partial_m A_n\right)-\left(\partial_i A_j\right)\left(\partial_m B_n\right)\right]\end{aligned}, \quad (A11)$$

and the RHS of Eq. (A10):

$$\vec{A}\cdot\vec{\nabla}\times\ddot{\lambda}\cdot\vec{\nabla}\times\vec{B}-\vec{B}\cdot\vec{\nabla}\times\ddot{\lambda}\cdot\vec{\nabla}\times\vec{A}=\varepsilon_{ijk}\varepsilon_{lmn}\left(A_i\partial_j\lambda_{kl}\partial_m B_n-B_i\partial_j\lambda_{kl}\partial_m A_n\right). \quad (A12)$$

Thus Eqs. (A11) and (A12) are equal under the condition $\lambda_{ij}=\lambda_{ji}$ and hence Eq. (A10) is established.
From Eq. (A10) application of the divergence theorem leads to:

$$\begin{aligned}&\int\left[\vec{A}\cdot\vec{\nabla}\times\ddot{\lambda}\cdot\vec{\nabla}\times\vec{B}-\left(\vec{\nabla}\times\ddot{\lambda}\cdot\vec{\nabla}\times\vec{A}\right)\cdot\vec{B}\right]d^3\vec{r}\\&=-\oint_S\left\{\left[\hat{n}\times\left(\ddot{\lambda}\cdot\vec{\nabla}\times\vec{A}\right)\right]\cdot\vec{B}+\left(\hat{n}\times\vec{A}\right)\cdot\left(\ddot{\lambda}\cdot\vec{\nabla}\times\vec{B}\right)\right\}da\end{aligned}, \quad (A13)$$

and thus we get the following form by generalizing $\vec{B}$ to a second rank tensor $\ddot{B}$:

$$\begin{aligned}&\int\left\{\left[\vec{\nabla}\times\overleftrightarrow{\lambda}\cdot\vec{\nabla}\times\overleftrightarrow{B}\right]^{T}\cdot\vec{A}-\left[\overleftrightarrow{B}\right]^{T}\cdot\vec{\nabla}\times\overleftrightarrow{\lambda}\cdot\vec{\nabla}\times\vec{A}\right\}d^{3}\vec{r}\\&=\oint_{S}\left\{\left[\hat{n}\times\overleftrightarrow{B}\right]^{T}\cdot\left(\overleftrightarrow{\lambda}\cdot\vec{\nabla}\times\overleftrightarrow{A}\right)-\left[\overleftrightarrow{\lambda}\cdot\vec{\nabla}\times\overleftrightarrow{B}\right]^{T}\cdot\hat{n}\times\overleftrightarrow{A}\right\}da\end{aligned}. \quad (A14)$$

Hence we repeat this step for $\vec{A}$ leads to the result in Eq. (A9).

[4]

$$\begin{aligned}&\int d^{3}\vec{r}\int d^{3}\vec{r}_{1}\left\{\left[\vec{\nabla}\times\overleftrightarrow{\lambda}(\vec{r},\vec{r}_{1})\cdot\vec{\nabla}_{1}\times\overleftrightarrow{B}(\vec{r}_{1})\right]^{T}\cdot\vec{A}(\vec{r})-\left[\overleftrightarrow{B}(\vec{r})\right]^{T}\cdot\vec{\nabla}\times\overleftrightarrow{\lambda}(\vec{r},\vec{r}_{1})\cdot\vec{\nabla}_{1}\times\vec{A}(\vec{r}_{1})\right\}\\&=\oint_{S}da\int d^{3}\vec{r}_{1}\left\{\left[\hat{n}\times\overleftrightarrow{B}(\vec{r})\right]^{T}\cdot\left[\overleftrightarrow{\lambda}(\vec{r},\vec{r}_{1})\cdot\vec{\nabla}_{1}\times\overleftrightarrow{A}(\vec{r}_{1})\right]-\left[\overleftrightarrow{\lambda}(\vec{r},\vec{r}_{1})\cdot\vec{\nabla}_{1}\times\overleftrightarrow{B}(\vec{r}_{1})\right]^{T}\cdot\hat{n}\times\overleftrightarrow{A}(\vec{r})\right\}\end{aligned}, \quad (A15)$$

under the condition $\lambda_{ij}(\vec{r},\vec{r}')=\lambda_{ji}(\vec{r}',\vec{r})$.
Let us first establish the following identity:

$$\begin{aligned}&\int d^{3}\vec{r}\int d^{3}\vec{r}_{1}\vec{\nabla}\cdot\left[\vec{B}(\vec{r})\times\overleftrightarrow{\lambda}(\vec{r},\vec{r}_{1})\cdot\vec{\nabla}_{1}\times\vec{A}(\vec{r}_{1})-\vec{A}(\vec{r})\times\overleftrightarrow{\lambda}(\vec{r},\vec{r}_{1})\cdot\vec{\nabla}_{1}\times\vec{B}(\vec{r}_{1})\right]\\&=\int d^{3}\vec{r}\int d^{3}\vec{r}_{1}\left[\vec{A}(\vec{r})\cdot\vec{\nabla}\times\overleftrightarrow{\lambda}(\vec{r},\vec{r}_{1})\cdot\vec{\nabla}_{1}\times\vec{B}(\vec{r}_{1})-\vec{B}(\vec{r})\cdot\vec{\nabla}\times\overleftrightarrow{\lambda}(\vec{r},\vec{r}_{1})\cdot\vec{\nabla}_{1}\times\vec{A}(\vec{r}_{1})\right]\end{aligned}. \quad (A16)$$

Again we express the left side as:

$$\begin{aligned}&\int d^{3}\vec{r}\int d^{3}\vec{r}_{1}\vec{\nabla}\cdot\left[\vec{B}(\vec{r})\times\overleftrightarrow{\lambda}(\vec{r},\vec{r}_{1})\cdot\vec{\nabla}_{1}\times\vec{A}(\vec{r}_{1})-\vec{A}(\vec{r})\times\overleftrightarrow{\lambda}(\vec{r},\vec{r}_{1})\cdot\vec{\nabla}_{1}\times\vec{B}(\vec{r}_{1})\right]\\&=\varepsilon_{ijk}\varepsilon_{lmn}\int d^{3}\vec{r}\int d^{3}\vec{r}_{1}\left\{\left[B_{j}(\vec{r})\partial_{i}^{\vec{r}}\lambda_{kl}(\vec{r},\vec{r}_{1})\partial_{m}^{\vec{r}_{1}}A_{n}(\vec{r}_{1})-A_{j}(\vec{r})\partial_{i}^{\vec{r}}\lambda_{kl}(\vec{r},\vec{r}_{1})\partial_{m}^{\vec{r}_{1}}B_{n}(\vec{r}_{1})\right]\right\},\\&+\varepsilon_{ijk}\varepsilon_{lmn}\int d^{3}\vec{r}\int d^{3}\vec{r}_{1}\lambda_{kl}(\vec{r},\vec{r}_{1})\left\{\left[\partial_{i}^{\vec{r}}B_{j}(\vec{r})\right]\left[\partial_{m}^{\vec{r}_{1}}A_{n}(\vec{r}_{1})\right]-\left[\partial_{i}^{\vec{r}}A_{j}(\vec{r})\right]\left[\partial_{m}^{\vec{r}_{1}}B_{n}(\vec{r}_{1})\right]\right\}\end{aligned} \quad (A17)$$

and the right side as:

$$\begin{aligned}&\int d^{3}\vec{r}\int d^{3}\vec{r}_{1}\left[\vec{A}(\vec{r})\cdot\vec{\nabla}\times\overleftrightarrow{\lambda}(\vec{r},\vec{r}_{1})\cdot\vec{\nabla}_{1}\times\vec{B}(\vec{r}_{1})-\vec{B}(\vec{r})\cdot\vec{\nabla}\times\overleftrightarrow{\lambda}(\vec{r},\vec{r}_{1})\cdot\vec{\nabla}_{1}\times\vec{A}(\vec{r}_{1})\right]\\&=\varepsilon_{ijk}\varepsilon_{lmn}\int d^{3}\vec{r}\int d^{3}\vec{r}_{1}\left[A_{i}(\vec{r})\partial_{j}^{\vec{r}}\lambda_{kl}(\vec{r},\vec{r}_{1})\partial_{m}^{\vec{r}_{1}}B_{n}(\vec{r}_{1})-B_{i}(\vec{r})\partial_{j}^{\vec{r}}\lambda_{kl}(\vec{r},\vec{r}_{1})\partial_{m}^{\vec{r}_{1}}A_{n}(\vec{r}_{1})\right]\end{aligned}. \quad (A18)$$

Hence Eq. (A17) is equal to Eq. (A18) by imposing $\lambda_{ij}(\vec{r},\vec{r}')=\lambda_{ji}(\vec{r}',\vec{r})$ and the result in Eq. (A15) can again be obtained by the same method as that in proving Eq. (A9).

9. Acknowledgment

I thank Prof. Pui-Tak Leung and Prof. Din Ping Tsai for fruitful discussion.

10. References

[1] R. J. Potton (2004). Reciprocity in optics. *Reports on Progress in Physics*, Vol.67, No.5, pp. 717-754, ISSN 0034-4885

[2] S. C. Hill, G. Videen and J. D. Pendleton (1997). Reciprocity method for obtaining the far fields generated by a source inside or near a scattering object. *Journal of the Optical Society of America B*, Vol.14, No.10, pp. 2522-2529, ISSN 0740-3224

[3] M. Kahl and E. Voges (2000). Analysis of plasmon resonance and surface-enhanced Raman scattering on periodic silver structures. *Physical Review B*, Vol.61, No. 20, pp. 14078-14088, ISSN 0163-1829

[4] E. C. Le Ru and P. G. Etchegoin (2006). Rigorous justification of the $|E|^4$ enhancement factor in Surface Enhanced Raman Spectroscopy. *Chemical Physical Letter*, Vol. 423, pp. 63- 66, ISSN 0009-2614

[5] G. J. Griffiths (1999). *Introduction to Electrodynamics* 3rd, Prentice-Hall International, ISBN 0-13-919960-8, Upper Saddle River, New Jersey

[6] J. D. Jackson (1999). *Classical Electrodynamics* , 3rd , New York: Wiley, ISBN 0-471-30932-X

[7] R. Chang and P. T. Leung (2008). Reciprocity in nonlocal nano-optics. *Journal of Optics A: Pure and Applied Optics*, Vol. 10, No. 7, pp. 075201, ISSN 1464-4258

[8] K.-J. Kim and J. D. Jackson (1993). Proof that the Neumann Green's function in electrostatics can be symmetrized. *American Journal of Physics,* Vol. 61, No. 12, pp. 1144-1146, ISSN 0002-9505

[9] H. Y. Xie, P. T. Leung and D. P. Tsai (2010). Reciprocity theorem for nonlocal optics: completion of proof and application to spectroscopic analysis. *Journal of Optics,* Vol. 12, No. 3, pp. 035006, ISSN 2040-8978

[10] R. Fuchs and F. Claro (1987). Multipolar response of small metallic spheres: Nonlocal theory. *Physical Review B,* Vol. 35, No. 8, (March 1987), pp. 3722-3727, ISSN 1550 235x

[11] R. Ruppin (1975). Optical properties of small metal spheres. *Physical Review B,* Vol. 11, No. 8, (April 1975), pp. 2871-2876, ISSN 1550 235x

[12] L. D. Landau, E. M. Lifshitz and L. P. Pitaevskii (1984), *Electrodynamics of Continuous Media* 2nd. Butterworth-heinemann, ISBN 0750626348

[13] C. T. Tai (1993). *Dyadic Green Functions in Electromagnetic Theory* 2nd New York, IEEE Press, ISBN 0700223452

[14] H. Y. Xie, P. T. Leung and D. P. Tsai (2009). General proof of optical reciprocity for nonlocal electrodynamics. Vol. 42, No. 4, (December 2008), pp. 045402, ISSN 1751-8113

[15] H. Y. Xie, P. T. Leung and D. P. Tsai (2009). Clarification and extension of the optical reciprocity theorem. *Journal of Mathematical Physics,* Vol. 50, No. 7 (July 2009), pp. 072901, ISSN 1089-7658

[16] M. Iwanaga, A. S. Vengurlekar, T. Hatano and T. Ishihara (2007). Reciprocal transmittances and reflectances: An elementary proof. *American Journal of Physics,* Vol. 75, No. 10, (October 2007), pp. 899- ISSN 0002-9505

[17] J. A. Porto, R. Carminati and J. J. Greffet (2000). Theory of electromagnetic field imaging and spectroscopy in scanning near-field optical microscopy. *Journal of Applied Physics,* Vol. 88, No. 8, pp. 4845-4851, ISSN 1089-7550

[18] Y. S. Joe, J. F. D. Essiben and E. M. Cooney (2008). Radiation characteristics of waveguide antennas located on the same impedance plane. *Journal of Physics D: Applied Physics* , Vol. 41, No. 12, (April 2008), pp. 125503, ISSN 0022-3727

[19] R. Vlokh and D. Adamenko (2008). Non-reciprocity of Faraday rotation in gyrotropic crystals. *Ukrainian Journal of Physical Optics,* Vol. 9, No. 4, pp. 217-225, ISSN 1609-1833

[20] A. Malinowski, Yu P. Svirko and N. I. Zheludev (1996). Reciprocity in nonlocal optics. *Journal of the Optical Society of America B,* Vol. 13, No. 8 (October 1995) , pp. 1641-1644, ISSN 0740-3224

[21] O. S. Jenkins and K. L. C. Hunt (2003), Nonlocal dielectric functions on the nanoscale: Screened forces from unscreened potentials. *Journal of Chemical Physics,* Vol. 119, No. 16, (May 2003), pp. 8250-8255, ISSN 0021-9606

[22] J. A. Kong (1972), Theorems of bianisotropic media. *Proceedings of the IEEE,* Vol. 60, No. 9 (June 2005) pp. 1036-1046, ISSN 0018-9219

20

Focused Arrays Beamforming

Oleksandr Mazurenko and Yevhenii Yakornov
Institute of Telecommunication Systems,
National Technical University of Ukraine "Kyiv Polytechnic Institute"
Ukraine

1. Introduction

Over the past two decades many articles devoted to the antenna systems focused in it's near-field zone (NFZ) or intermediate-field zone (IFZ) and their application in medical engineering, geology, materials and environment sensing, RFID, energy transfer technologies were published. The development of this theory allows to improve the quality level of technique and to expand applicability of the focused antenna systems, for example, in the telecommunications engineering.

Development of the focused antennas theory began in the late 1950's. The first collection of papers that describe the properties of the focused antenna, edited by Hansen, was printed in 1964 (Hansen, 1964). Further development of this theory was not so active until the 1990's. Recent works in this area relates only to the practical application and realization of the focusing effect (Herben, 1999; Hristov, 2004; Karimkashi & Kishik, 2008; Rudolph & Grbic, 2008; etc.) and finding of the methods of improving the focused antennas performance (Hussain, 2004, 2008; Karimkashi & Kishik, 2009; etc.).

Reference materials for this paper are based on a current technical level, accordingly to URSI and IEEE papers, within the limits of knowledge of the near-field and the intermediate field diffraction theory, focal areas forming on the plane apertures radiation axis and signal processing methods of the focused arrays for various environments scanning.

The authors decided that the reference materials are insufficiently exploring the problem for wider and more flexible usage of the three-dimensionally directional signal transmission phenomenon due to an incompleteness of the focused antenna arrays (FAA) theory. The given incompleteness is revealing as a high level of calculations for obtaining the exact aperture phase distribution, inaccuracy and deficiency of theoretical models, that does not allow to use qualitatively the focused energy transmission to a certain area of space at a wide range of angles in azimuth and elevation planes.

The basis of this chapter is the results of research led for the purpose of improving FAA theory for its further usage in the telecommunication engineering that cannot be done without increasing of FAA performance. The research materials are devoted to a wide range of FAA structures with different types of radiator and to the methods of FAA directivity improving with a purpose to increase the 3-dimensional gain performance of antenna arrays at a wide range of angles in azimuth and elevation planes.

This chapter is organized as follows. Section 2 is devoted to a new approach that better reveals the principles of FAA radiation pattern forming, including FAA beamforming with various radiators types and allocation. FAA directivity improving methods are considered in Section 3. FAA possible applications for a short distance wireless communication are described in Section 4. Concluding remarks and future activities are collected in Section 5.

2. Focused Antenna Arrays radiation patterns

When writing this section the authors did not attempt to create a new huge mathematical model that would describe the distribution of field or power radiated by different types of antennas, but instead of it to find new approaches for better describing the characteristics of focused antennas. If the reader wants to see the detailed, but approximated by Fresnel description of a field radiated by focused aperture or its focusing properties, he can refer to an existing theory (Chu, 1971; Fenn, 2007; Graham, 1983; Hansen, 1964, 1985, 2009; Laybros et al., 2005; Malyuskin & Fusco, 2009; Narasimhan & Philips, 1987a, 1987b; Polk, 1956).
Generally consider the antenna arrays of linear structure as that is sufficient to study properties of FAA. Thus all tasks of study of FAA radiation pattern synthesis are sufficient to be done in its azimuth plane, while considering the linear antenna array.

2.1 Geometric models

In this subsection we present geometric models of different structures of arrays. In the next subsections we will describe radiation patterns of arrays based on this models.
For a start, consider the problem of finding an expression for a linear array with equivalent spaced radiators (LAESR) without any mathematical approximation (Fraunhofer or Fresnel), where phase shifts between array elements and the array phase center are determined by two exact components: the phase shift by angle and the phase shift by distance. The geometry model of LAESR is shown in Fig.1, where d - array element spacing between $2N+1$ radiators with number n, normal vector to the array is polar axis or the starting point of polar coordinates in which the problem is solved. An important factor is the location of the phase center, which contains the polar axis. Let phase center be located in the LAESR center element with $n = 0$. Then location of observation point is described by azimuth θ and distance R relatively to the array phase center. Thereby all equations related to the right side elements (RSE) with $n_{RSE} = 1 \ldots N = n$ and the left side elements (LSE) with $n_{LSE} = -1 \ldots -N = -n$ from the phase center differ by indexes and content. Then $\Delta(n_{LSE})$, $\Delta(n_{RSE})$ are spatial shifts between phase center and LSE, RSE respectively; $\upsilon(n_{LSE})$, $\upsilon(n_{RSE})$ are angles between phase center and LSE, RSE with number n respectively in observation point; $\theta(n_{LSE})$, $\theta(n_{RSE})$ are azimuths of observation point from LSE, RSE respectively.
Obtain the two equation systems using elementary trigonometry for LAESR (fig.1):

$$\begin{cases} \dfrac{dn_{LSE}}{\sin(\upsilon(n_{LSE}))} = \dfrac{R+\Delta(n_{LSE})}{\cos(\theta)} = \dfrac{R}{\cos(\theta+\upsilon(n_{LSE}))} \\ \dfrac{dn_{RSE}}{\sin(\upsilon(n_{RSE}))} = \dfrac{R+\Delta(n_{RSE})}{\cos(\theta)} = \dfrac{R}{\cos(\theta-\upsilon(n_{RSE}))} \end{cases}; \tag{1}$$

$$\begin{cases} \Delta(n_{LSE}) = \sqrt{R^2+(dn_{LSE})^2+2Rdn_{LSE}\sin(\theta)} - R \\ \Delta(n_{RSE}) = \sqrt{R^2+(dn_{RSE})^2-2Rdn_{RSE}\sin(\theta)} - R \end{cases}. \tag{2}$$

From equations (1), (2), difference between the expressions for the RSE and LSE is due to using the number sign of array element.

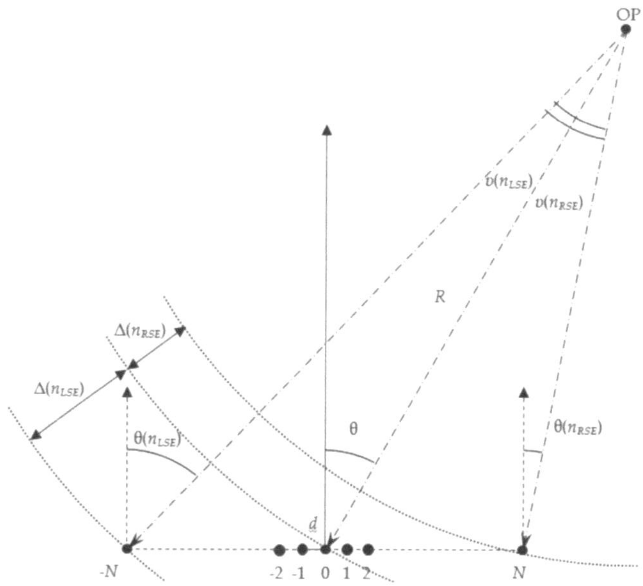

Fig. 1. The geometric model of LAESR

Expression (2) is regular for the spatial shift Δ (Fenn, 2007; Hansen, 1964), then using (1) to solve the problem mentioned before.
Obtain the equation systems for $\Delta(n_{LSE})$, $\Delta(n_{RSE})$ and $\upsilon(n_{LSE})$, $\upsilon(n_{RSE})$ from (1):

$$\begin{cases} \Delta(n_{LSE}) = \dfrac{dn_{LSE}\cos(\theta)}{\sin(\upsilon(n_{LSE}))} - R \\ \Delta(n_{RSE}) = \dfrac{dn_{RSE}\cos(\theta)}{\sin(\upsilon(n_{RSE}))} - R \end{cases}; \quad (3)$$

$$\begin{cases} \Delta(n_{LSE}) = dn_{LSE}\left(\sin(\theta) + \cos(\theta)\tan\left(\dfrac{\upsilon(n_{LSE})}{2}\right)\right) \\ \Delta(n_{RSE}) = dn_{RSE}\left(-\sin(\theta) + \cos(\theta)\tan\left(\dfrac{\upsilon(n_{RSE})}{2}\right)\right) \end{cases}; \quad (4)$$

$$\begin{cases} \upsilon(n_{LSE}) = \operatorname{arccot}\left(\dfrac{R}{dn_{LSE}\cos(\theta)} + \tan(\theta)\right), \upsilon(n_{LSE}) \in \left[0; \dfrac{\pi}{2}\right] \\ \upsilon(n_{RSE}) = \operatorname{arccot}\left(\dfrac{R}{dn_{RSE}\cos(\theta)} - \tan(\theta)\right), \upsilon(n_{RSE}) \in \left[0; \dfrac{\pi}{2}\right] \end{cases}. \quad (5)$$

Then equation (4) can be written as:

$$\begin{cases} \Delta(n_{LSE}) = dn_{LSE}\left(\sin(\theta)+\cos(\theta)\tan\left(0.5\operatorname{arccot}\left(\dfrac{R}{dn_{LSE}\cos(\theta)}+\tan(\theta)\right)\right)\right) \\ \Delta(n_{RSE}) = dn_{RSE}\left(-\sin(\theta)+\cos(\theta)\tan\left(0.5\operatorname{arccot}\left(\dfrac{R}{dn_{RSE}\cos(\theta)}-\tan(\theta)\right)\right)\right) \end{cases}. \quad (6)$$

Another important parameter for radiation pattern calculating is the observation point azimuth $\theta(n)$. The expressions for $\theta(n_{LSE})$, $\theta(n_{RSE})$ are:

$$\begin{cases} \theta(n_{LSE}) = \arcsin\left[\dfrac{R\sin(\theta)+dn_{LSE}}{\sqrt{R^2+(dn_{LSE})^2+2Rdn_{LSE}\sin(\theta)}}\right] \\ \theta(n_{RSE}) = \arcsin\left[\dfrac{R\sin(\theta)-dn_{RSE}}{\sqrt{R^2+(dn_{RSE})^2-2Rdn_{RSE}\sin(\theta)}}\right] \end{cases}. \quad (7)$$

From expressions (5), (6), (7) for left side and right side antenna elements, the phase center location affects the spatial shifts distribution in NFZ and IFZ can be concluded. General equation for $\Delta(n)$, $\upsilon(n)$ and $\theta(n)$, where $n\in[-N; N]$, $n\in \mathbf{Z}$, using (5), (6), (7) can be written as:

$$\Delta(n) = dn\left(\cos(\theta)\tan\left(0.5\operatorname{arccot}\left(\frac{R}{dn\cos(\theta)}-\tan(\theta)\right)\right)-\sin(\theta)\right); \quad (8)$$

$$\upsilon(n) = \operatorname{arccot}\left(\frac{R}{dn\cos(\theta)}-\tan(\theta)\right), \upsilon(n)\in\left[0;\frac{\pi}{2}\right]; \quad (9)$$

$$\theta(n) = \arcsin\left[\frac{R\sin(\theta)-dn}{\sqrt{R^2+(dn)^2-2Rdn\sin(\theta)}}\right]. \quad (10)$$

Expressions (8), (9), (10) are necessary in calculating LAESR radiation pattern.

Equation (8) of spatial shift obtained in the form of expression $\Delta = \Delta_R + \Delta_\theta$ is useful for FAA with separate phase steering by distance and by angular coordinates synthesis, it also helps to reduce the level of computational operations to calculate required phase distribution.

In elevation plane of spherical coordinates spatial shifts of LAESR mentioned before are equal at different elevation angles.

The linear structure is most often used in the construction of antenna arrays. However, there are other structures that as the best create the Rayleigh or the Fresnel diffraction field. There are polygonal structures where all elements are radiating inside the polygon. It should be noted that antenna arrays of polygonal structures have their own natural focus.

Next step is finding an expression for spatial shifts between phase center and elements of polygonal antenna array (PAA) using vector analysis theory. Place phase center of PAA in its natural focus to solve this task since the problem of obtaining equations between the PAA parameters is complex.

The geometric model of PAA is shown in Fig.2, where 1...n-th antenna location points, d_{ij} – distance between i-th and j-th antenna element of PAA, 0 – centre of the polygon and phase center of PAA, 0′ – observation point, **L** - polar axis of PAA in polar coordinate, where the problem is being solved; $\mathbf{R}_n(|\mathbf{R}_n|;\theta_{Rn})$ - vector of n–th antenna signal in phase center and its polar axis, where $|\mathbf{R}_n|$ - spatial shift between n–th antenna and phase center; $\boldsymbol{\Delta}(|\boldsymbol{\Delta}|;\theta_\Delta)$ - vector from phase center to observation point, where $|\boldsymbol{\Delta}|$ - spatial shift between phase center and observation point, θ_n – azimuth of observation point in n–th antenna. Then the spatial shift between n–th antenna and observation point:

$$\Delta_n = \sqrt{|\mathbf{R}_n|^2 + |\boldsymbol{\Delta}|^2 + |\mathbf{R}_n||\boldsymbol{\Delta}|\cos(\theta_{Rn} - \theta_\Delta)} - |\mathbf{R}_n| . \quad (11)$$

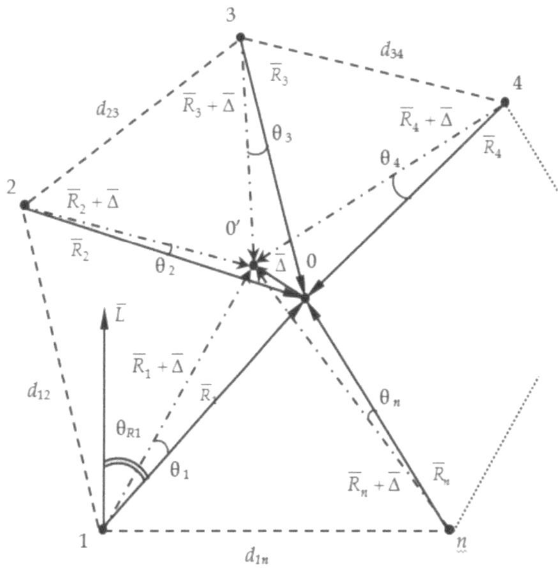

Fig. 2. Geometry model of PAA

Equations (11) can be written in form of $\Delta = \Delta_R + \Delta_\theta$, as follows:

$$\Delta_n = |\boldsymbol{\Delta}|\left(\sin(\theta_{Rn} - \theta_\Delta)\tan\left(0.5\operatorname{arccot}\left(\frac{|\mathbf{R}_n|}{|\boldsymbol{\Delta}|\sin(\theta_{Rn} - \theta_\Delta)} + \cot(\theta_{Rn} - \theta_\Delta)\right)\right) + \cos(\theta_{Rn} - \theta_\Delta)\right); \quad (12)$$

azimuth of observation point in n–th antenna:

$$\theta_n = \arctan\frac{|\mathbf{R}_n|\sin(\theta_{Rn}) + |\boldsymbol{\Delta}|\sin(\theta_\Delta)}{|\mathbf{R}_n|\cos(\theta_{Rn}) + |\boldsymbol{\Delta}|\cos(\theta_\Delta)} . \quad (13)$$

Search of θ_n, Δ_n expressions is possible with precise definition of the PAA polygon parameters $\mathbf{R}_n(|\mathbf{R}_n|;\ \theta_{Rn})$. In case of irregular polygonal structure, obtaining of its parameters which depend on d_{ij}, antenna and phase center positions is a hard analytical task. This task can be solved by empirical methods. Another way is predetermination of these parameters in the condition of preserving polygonal structure.

For regular polygonal structure with a side d, the task of obtaining of its parameters is much easier, because $\mathbf{R}_n$ is the bisecting line of each polygon corner and $|\mathbf{R}_n|$ is the radius of polygon escribed circle, where $n \in [1; N]$, $n \in \mathbf{Z}$, N – polygon sides quantity. Then:

$$|\mathbf{R}_n| = \frac{d}{2\sin(\pi / N)}; \tag{14}$$

$$\theta_{Rn} = \frac{\pi}{N}(2n-1). \tag{15}$$

Expression for spatial shifts between elements of antenna array with nonlinear structures, for example, parabola, hyperbola, can be obtained in the same way as for PAA.

2.2 Focused arrays radiation pattern

Radiation pattern of the near-field or the intermediate-field of FAA has the following features compared with common antenna arrays focused at the far-field zone (FFZ):

- signal attenuation by propagation $L(n,R) = 4\pi R^2(n) / \lambda^2$ is different from each FAA n-th antenna element due to different distance to observation point $R(n)$;
- each n-th element of FAA radiates signals to observation point at different angles;
- each element of FAA has the radiation pattern $\dot{f}_n(\theta(n), R)$ that depends on the distance;
- assumed that the near field of FAA doesn't include its elements reactive field;
- main lobe, side lobes, grating lobes that are expressed in angular values are being transformed to main, side and grating focal areas that are expressed in values of width, length and height, that causes the problem of transforming from polar (spherical) to Cartesian coordinates.

Then the radiation patterns of FAA which consist of phase shifters and delay lines according to a common theory (Hansen, 2009) are defined respectively as:

$$\dot{F}(\theta, R) = \sum_n \frac{\dot{f}_n(\theta(n), R)}{L(n,R)} e^{j(k(\omega)\Delta(n) - \Delta\Phi(n))}; \tag{16a}$$

$$\dot{F}(\theta, R) = \sum_n \frac{\dot{f}_n(\theta(n), R)}{L(n,R)} e^{j(\omega(\Delta(n)/c - \Delta T(n)))}; \tag{16b}$$

where $I(n)$ - n-th antenna element excitation amplitude;

$\Delta\varphi(n) = k\Delta(n)$ - the phase shift between phase center and the n-th antenna element;
$\Delta t(n) = \Delta(n)/c$ - time delay between the phase center and the n-th antenna element;
$\Delta\Phi(n)$, $\Delta T(n)$ - the phase and time distribution of n-th antenna element excitation;
ω - circular frequency;
$k = 2\pi/\lambda = \omega/c$ - coefficient of propagation;
λ – signal wavelength;
c – signal propagation speed in the environment.

If FAA is excited by wideband signal with multicomponent spectrum, for example, pulse signals, definition of its radiation pattern cannot be completed without taking into account the spectral structure of the signal $\dot{A}(j\omega)$, $\omega \in [\omega_{min}; \omega_{max}]$ – signal circular frequency band.

Therefore, this expression can be described in form of inverse Fourier transformation of radiation patterns at each spectrum component with their amplitude and phase as follows:

$$\dot{F}(\theta,R)=\sum_{\omega}\dot{A}(j\omega)\sum_{n}\frac{\dot{f}_n(\theta(n),R)}{L(n,R)}e^{j(k(\omega)\Delta(n)-\Delta\Phi(n,\omega))}\;; \tag{17a}$$

$$\dot{F}(\theta,R)=\sum_{\omega}\dot{A}(j\omega)\sum_{n}\frac{\dot{f}_n(\theta(n),R)}{L(n,R)}e^{j(\omega(\Delta(n)/v-\Delta T(n)))}\;; \tag{17b}$$

The concept of FAA excitation by wideband signal is introduced in papers (Wu, 1985; Kremer, 1984) and was used by several authors (Ishimaru et al.,2007 ; Hussain, 2004; Malyuskin & Fusco, 2009) to improve FAA performance, but they had not disclosed the nature of this method. Expression (17a) is used for a broadband signal in the time sense and (17b) in the space-time sense (Kremer, 1984).
For example, the expression for radiation pattern of focused LAESR can be written as:

$$\dot{F}(\theta,R)=\sum_{\omega}\dot{A}(j\omega)\sum_{n}\frac{\dot{f}_n(\theta(n),R)}{L(n,R)}e^{j\left(k(\omega)nd\left(\cos(\theta)\tan\left(0.5\operatorname{arccot}\left(\frac{R_F}{dn\cos(\theta)}-\tan(\theta)\right)\right)-\sin\theta\right)-\Delta\Phi(n,\omega)\right)}, \tag{18}$$

where $\Delta\Phi(n,\omega)=k(\omega)nd\left(\cos(\theta_F)\tan\left(0.5\operatorname{arccot}\left(\frac{R_F}{dn\cos(\theta_F)}-\tan(\theta_F)\right)\right)-\sin\theta_F\right);$

R_F, θ_F - coordinates of focus point.

Radiation patterns intersection in the azimuth plane of LAESR of nine omnidirectional in azimuth plane elements with $d = \lambda/2$ and octagonal structure PAA of patch antenna elements with radiation pattern in a form of cardioid which are radiating in the center of polygon with $d = 2\lambda$ without taking into account signal attenuation by the propagation are shown in Fig. 3(a) and Fig. 3(b) respectively, where the signal level is in dB.

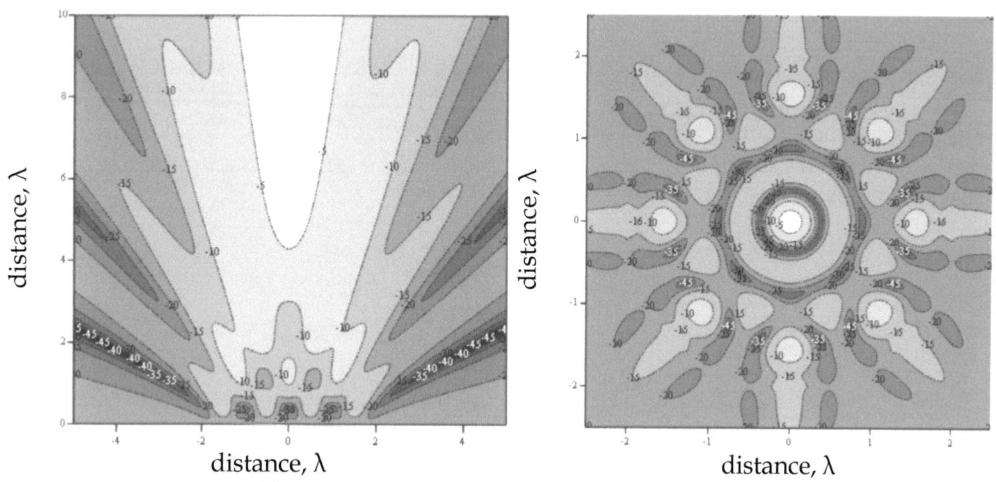

(a) (b)

Fig. 3. Radiation patterns intersection in the azimuth plane of NFZ and IFZ of LAESR, PAA

According to Fig.3 and sources (Fenn, 2007; Hansen, 1964; Graham, 1983), the features of radiation patterns by distance are the same as by angle coordinates. That is the level and the amount of mainlobe and sidelobes proportionally to the level and the amount of main focal area and side focal areas. Thus the methods of radiation pattern synthesis (Hansen,2009) in NFZ, IFZ and FFZ are similar.

2.3 Focusing properties of antenna arrays

According to a common theory (Hansen, 1964), focusing - the process of compensation of components of phase shifts between antenna elements and phase center of FAA, that depends on the distance in focus coordinates R_F, θ_F. Otherwise focusing is the process of inversion of the Fresnel and the Fraunhofer diffraction fields. Thereby FAA achieves a high directivity in its NFZ or IFZ and loses directivity in its FFZ.

The important factor is that the focus takes place exclusively in NFZ or IFZ (Hansen, 1964) limited for LAESR by the hyperfocal distance from its phase centre $R_{HF} < D^2\cos^2(\theta)/\lambda$, where D - the largest FAA dimensions in the plane where NFZ or IFZ is considered. Depth of focus and directivity are important parameters of FAA that describe its focusing or directional signal transmission properties.

According to (Hansen, 1964), depth of focus - the size of focal area in the dimension of distance $R \in [R_{near}, R_{far}]$. Expressions for R_{near} and R_{far} (Hansen, 1964), where the level of focused LAESR radiation patterns by distance becomes -0,1 dB are obtained by methods of theory of geometrical optics and are defined as:

$$R_{near} = \frac{D^2/\lambda}{1 + D^2/\lambda R_F(\theta_F)} \tag{19a}$$

$$R_{far} = \begin{cases} \dfrac{D^2/\lambda}{-1 + D^2/\lambda R_F(\theta_F)}, & R_F(\theta_F) < \dfrac{D^2\cos^2\theta_F}{\lambda} = R_{HF} \\ \infty, & R_F(\theta_F) \geq \dfrac{D^2\cos^2\theta_F}{\lambda} = R_{HF} \end{cases}. \tag{19b}$$

Graham (Graham, 1983) had studied the axial field radiation pattern synthesis for apertures focused in the Fresnel region with Fresnel approximation, and had obtained the equations for depth of focus ΔR, minima R_{min} and maxima R_{max} of field location. These equations can be used for focused LAESR and can be written for different angles of focus point in azimuth plane, respectively as:

$$\Delta R = \frac{7\lambda\left(R_F\cos(\theta_F)\right)^2}{D^2 - \left(3.5\lambda R_F\cos(\theta_F)/D\right)^2}; \tag{20}$$

$$R_{\min} = \frac{R_F\cos(\theta_F)}{1 - 8n\lambda R_F\cos(\theta_F)/D^2}, \text{ for } n < \frac{D^2}{8\lambda R_F\cos(\theta_F)}, \ n \in \mathbf{Z}; \tag{21}$$

$$R_{\max} = \frac{R_F\cos(\theta_F)}{1 - 8(n+0.5)\lambda R_F\cos(\theta_F)/D^2}, \text{ for } n < \frac{D^2}{8\lambda R_F\cos(\theta_F)} - 0.5, \ n \in \mathbf{Z}; \tag{22}$$

Equations (20), (21), (22) are exact only in IFZ.
The exact equations for LAESR parameters ΔR, R_{min} and R_{max} in NFZ can be obtained when the problem of finding an exact expression for its radiation pattern is solved. Equation for LAESR radiation pattern in azimuth plane can be written as follows:

$$F(\theta,R)=e^{-j\beta R}\int_{\upsilon_{LSE}}^{\upsilon_{RSE}} e^{j\beta R\left(\frac{\cos(\theta)}{\cos(\upsilon-\theta)}\right)}d\upsilon\,, \tag{23}$$

where $\upsilon_{LSE}=\operatorname{arccot}\left(\dfrac{R}{-dN\cos(\theta)}+\tan(\theta)\right)$, $\upsilon_{RSE}=\operatorname{arccot}\left(\dfrac{R}{dN\cos(\theta)}-\tan(\theta)\right)$,

For further description the parameters of FAA radiation patterns equations (20), (21), (22) are used since the expression (23) has no solution in the form of simple function.
Radiation patterns intersection in the azimuth plane of LAESR of twenty one patch antenna elements focused in polar coordinates at R_F = 50λ, θ_F = 0° with d = λ and phase centre in eleventh element with and without taking into account signal attenuation by the propagation are shown in Fig.5(a) and Fig.5(b) respectively, where the signal level is in dB. Axial radiation patterns on azimuth θ_F = 0° of this LAESR focused at R_F = 50λ with and without taking into account signal attenuation by the propagation are shown in Fig.6.

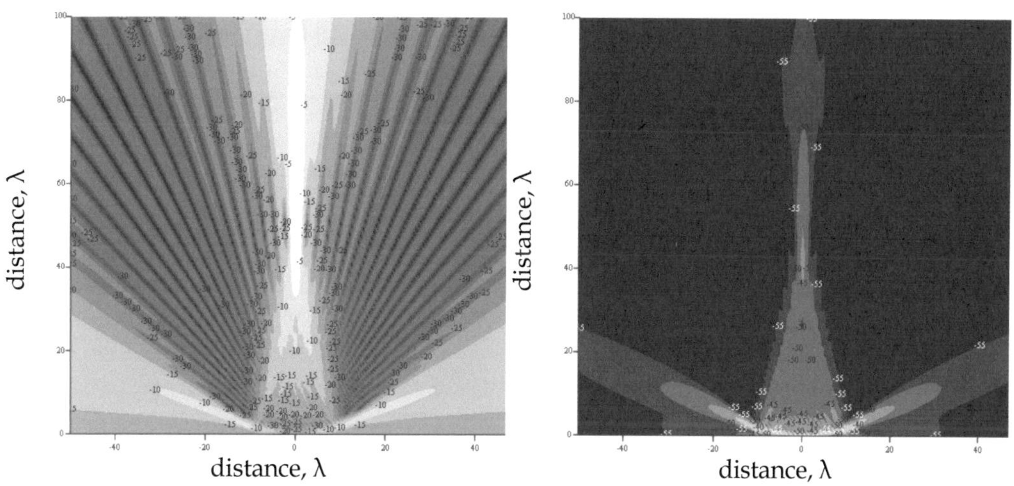

(a) (b)

Fig. 5. Radiation patterns intersection in the azimuth plane of LAESR

Fig.5, Fig.6 and source (Hansen, 1964) shows that the focal area or a certain segment of distance $R \in [0, R_F]$ where radiated power distribution is uniform can be the result of focusing process.
The level of radiation pattern in FFZ becomes the level of radiation pattern in point of NFZ or IFZ of LAESR focused on FFZ as the result of focusing on this point according to the fact that focusing is the process of inversion of the Fresnel and the Fraunhofer diffraction fields, as shown on Fig.7.
From Fig.7 and the materials (Hansen, 1964) it can be concluded that the minimum signal level in FFZ of FAA can be obtained since FAA is focused in points of minimum signal level

when it is focused in FFZ. These focus points are calculated from (21). The closer the focus point is placed then the weaker the level of signal in the FFZ is.

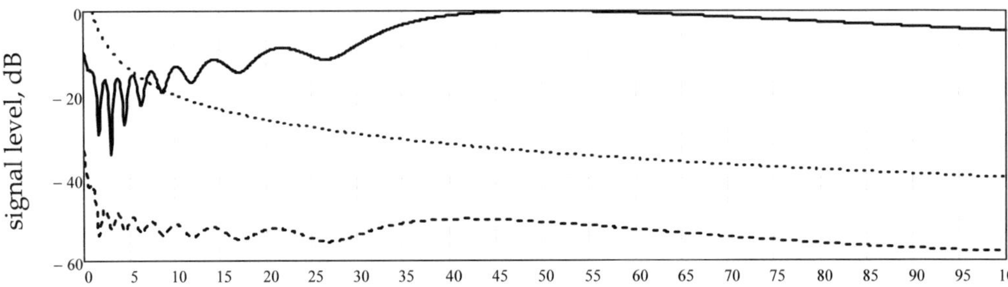

Fig. 6. Axial radiation patterns on azimuth $\theta_F = 0°$ of focused LAESR

Radiation patterns intersection in the azimuth plane of octagonal PAA of patch antenna elements which are radiating inside the structure and focused in the center of polygon with $d = 3\lambda$, $\lambda = 2$ m with and without taking into account signal attenuation by the propagation are shown in Fig.8(a) and Fig.8(b) respectively, where the signal level is in dB.

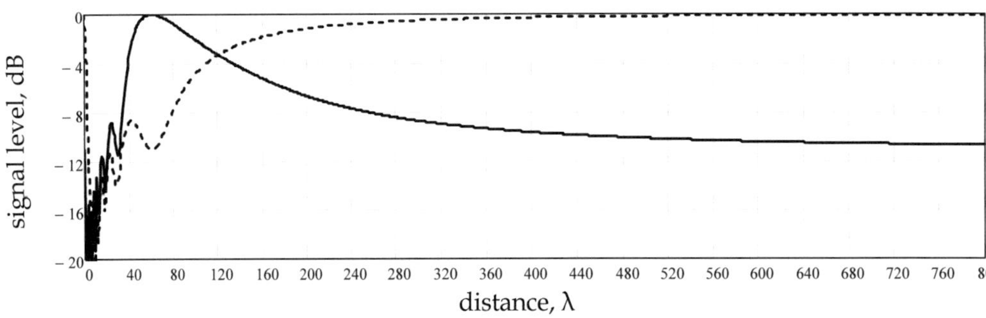

Fig. 7. Axial radiation patterns of LAESR focused in NFZ and FFZ

For antenna array of polygonal structures where all elements are radiating inside the polygon, radiation pattern is described as field in NFZ or IFZ of LAESR. Therefore, according to Fig.8 and the paper (Mazurenko & Yakornov, 2010) generally focal area has the form of ellipse, and for regular polygon array structures the form of circle with depth of focus $\Delta R \approx \lambda/3$.

The focus point of antenna array of regular polygonal structure with $N < 7$, $d < \lambda$ where $|\mathbf{R}_n| < \lambda$ is located in its reactive field zone that can be avoided by increasing the number of antenna elements N in the condition $|\mathbf{R}_n| > \lambda$ or increasing the element spacing $d >> \lambda$. Radiation of wideband signals by array or using the directional antennas is required for grating focal areas suppression since $d >> \lambda$. The method of radiation of wideband signals for grating focal areas suppression will be reviewed in the next section. The preliminary simulation result of this method in form of radiation pattern intersection in the azimuth plane of octagonal PAA of patch antenna elements which are radiating inside the structure periodic pulse signal with bandwidth $\Delta f = 300$ MHz, central frequency $f_C = 150$ MHz, period $T = 33.3$ ns, pulse duration $\tau_p = 6.6$ ns and focused in the center of polygon with $d = 4\lambda$, $\lambda = 2$ m with taking into account signal attenuation by the propagation is shown in Fig. 9.

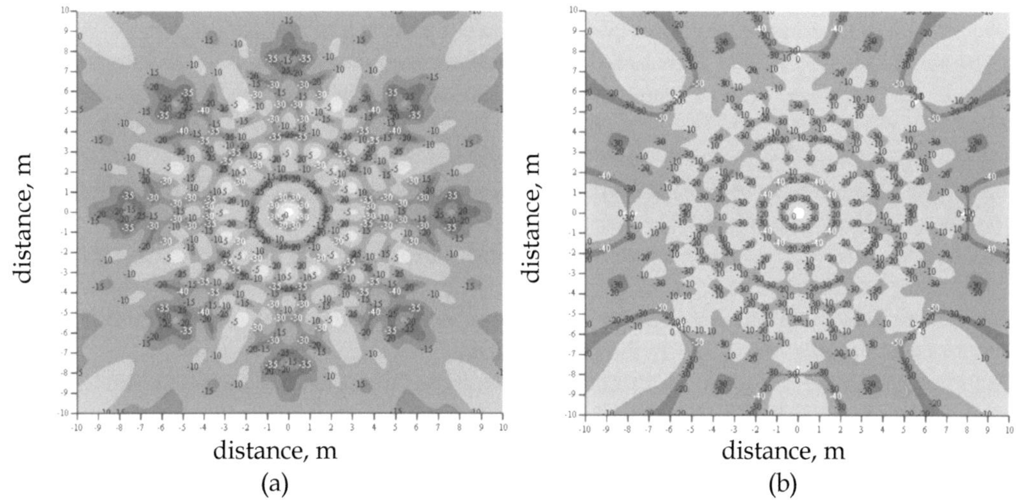

Fig. 8. Radiation patterns intersection in the azimuth plane of octagonal PAA

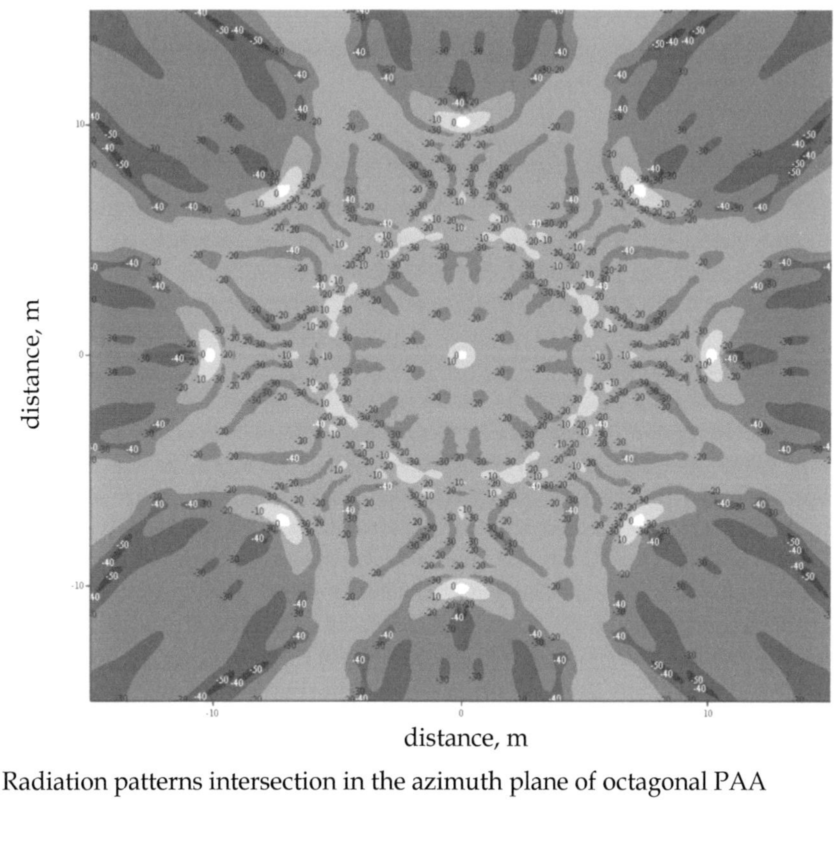

Fig. 9. Radiation patterns intersection in the azimuth plane of octagonal PAA

According to study results presented in the papers (Hansen, 1964; Polk, 1956), the directivity of focused linear antenna array in NFZ or IFZ cannot exceed directivity in its FFZ since assumed that the focusing is the process of inversion of Fresnel and Fraunhofer diffraction fields. Thus the methods of calculating the angular directivity of FAA are equal to methods of calculating the directivity of common antenna arrays in the FFZ (Hansen, 2009).
The polygonal structure array directivity is defined by degree of radiation inside it.
In common sense antenna directivity is relative to angular selectivity. For selectivity by distance the determination of antenna directivity is a complex task due to infinity of values of coordinate by distance. Thereby the directivity by distance can be determined only as angular directivity in focused point to angular directivity in FFZ ratio. The angular directivity as a function of distance with Fresnel approximation for focused apertures can be obtained from paper (Polk, 1956).
The overall FAA directivity as a multiplication of directivity by distance and angle coordinates can be obtained.
The overall antenna gain, obtained by focusing process cannot be calculated without taking into account signal attenuation by the propagation.
Thereby appliance of FAA is limited by distance R_{MD} due to signal attenuation by propagation that can be described as:

$$R_{MD} \to L\ (R_{MD}) \leq G_{AS}G_{AE}SINR, \tag{24}$$

where G_{AS} - gain of FAA;

G_{AE} – gain of FAA antenna element;

$SINR$ - necessary signal to interference and noise ratio at the appropriate detector.

The FAA ability to separate signals by distance as by angular coordinates can be concluded according to the materials of this section.

3. Focused Antenna Arrays directivity improving methods

Improving of FAA directivity is increasing its spatial resolution by distance and angular coordinates and increasing the hyperfocal distance. The first attempts of FAA directivity improvements are made in the papers (Karimkashi & Kishk, 2009; Hussain & Al-Zayed, 2008), but the authors used common methods of radiation pattern synthesis in the FFZ and in a case of array excitation by wideband signal (Hussain, 2004, 2008) the nature of improvement had not been disclosed.
According to the materials of previous section and the papers mentioned before in this section, FAA directivity can be improved by increasing element spacing jointly with using directed array elements (Hansen, 2009) or exciting wideband signal jointly with increasing array elements sparsity (Hussain & Al-Zayed, 2008). Also this task can be solved by using common methods of optimal radiation pattern synthesis for FFZ (Hansen, 2009).
Application of common methods of pattern synthesis by special amplitude-phase distributions (APD) of excited signal or spatial distributions of array elements despite the papers (Karimkashi & Kishk, 2009; Hussain & Al-Zayed, 2008) is not so effective for FAA as for conventional antenna arrays. Low efficiency of these methods may be linked to the fact that they are created for linear distribution of phase shifts between elements of array and the process of focusing in NFZ or IFZ based on creation of non-linear distribution. Thus application of specific APD (Hansen, 2009) created for far-field pattern synthesis, wideband signal radiation with linear distribution of spectrum components, for example, pulse signals

(Hussain & Al-Zayed, 2008) for FAA is leading to its partial defocusing in NFZ or IFZ. This defocusing occurs due to ignoring of phase shifts components that depend on distance when common APD and amplitude-frequency characteristic (AFC) of radiated signal had been created. Modification of specific antenna array APD or AFC of wideband signals, which is radiated by arrays is the solution for defocusing problem.

Use the concept of spectrum of spatial frequencies (SSF) for finding a solution of this defocusing problem. The SSF concept is presented from different sides in the books (Korostelev, 1987; Kremer, 1984). Also the SSF concept helps to reveal the nature of wideband signal excitation of FAA to improve its directivity.

The essence of SSF concept is the similarity of use of the inverse Fourier transformation for the synthesis signal in time domain with a limited frequency spectrum and radiation pattern in the spatial region using spatial frequency spectrum which is essentially the APD of array excitation.

Radiation patterns of antenna array of two elements excited by the wideband signal with m-components frequency spectrum and m-elements antenna array excited by the narrowband signal are equivalent that can be concluded from the use of the SSF concept, where space frequency $\omega_S = 2\pi d_0 n/\lambda_m = 2\pi d_0 n f_m/c = 2\pi d_0 n f_0 m/c$, $d_0 = l\lambda_0$, l – normalized array spacing in wavelength, $f_0 = c/d_0$, $f_m = f_0 m$, c – signal propagation speed in the environment. Thereby each frequency spectrum component f_m except f_0 of excited signal creates a SSF component or a virtual antenna array element located from its phase centre at distance $d_m = d_0 m$ that can be assumed. So SSF can be linked with AFC of radiated signal.

When array is excited by the signal with spectrum components $f_m < f_0$, $m = f_m/f_0 < 1$, the grating lobes and the side lobes are suppressing, the main lobe becomes wider. When array is excited by the signal with spectrum components $f_m > f_0$, $m = f_m/f_0 > 1$, the grating lobes and the side lobes are not suppressing, the main lobe becomes narrower. Similar result is obtained in the paper (Hussain & Al-Zayed, 2008).

Thus for increasing the directivity of antenna array by **the method of increasing array elements sparsity jointly with wideband signal excitation** is necessary to enhance the array element spacing $d_0 = l\lambda_0$ with $l > 1$ and to excite array by the signal with spectrum components $f_m < f_0$, $m = f_m/f_0 < 1$. The directivity increasing is much effective when created SSF components are located from its phase centre at distance $d_m = p\lambda_0/2$, $p \in \mathbf{Z}$ or $f_m = pf_0/2l$, $p/2l < 2$. So this method is effective when radio impulse signal is radiated by FAA. For example, excitation LAESR with element spacing $d_0 = 4\lambda_0$ by the periodic pulse signal with bandwidth Δf = 300 MHz, central frequency $f_C = f_0$ = 150 MHz, period T = 33.3 ns, pulse duration τ_p = 6.6 ns 4 times increases directivity and 16 times increases hyperfocal distance in comparison to the array with element spacing $d_0 = \lambda_0$ and narrowband signal excitation. Radiation patterns intersection in the azimuth plane of LAESR of 11 patch antennas focused in polar coordinates at R_F = 50 m, $\theta_F = 0°$ with $d_0 = 4\lambda_0$, λ_0 = 0.5 m excited by narrowband and by wideband signal are shown in Fig.10(a) and Fig.10(b) respectively.

For increasing the hyperfocal distance by **the method of wideband signal excitation** is necessary to excite array by the wideband or ultrawideband pulse signal with spectrum components $f_m > f_0$, $m = f_m/f_0 > 1$. The hyperfocal distance is linearly dependent on frequency band, so $R_{HFm} = mR_{HF0}$. Using of this method causes the grating focal area generating and the side focal area rising, so directivity by angular coordinate is decreased, but directivity by distance is increased, that are dependent on AFC of radiated signal as in previous method.

For increasing the directivity of antenna array by **the method of using special APD or AFC** is necessary to modify APD or AFC created for FFZ beamforming (original AFC can be taken from previous methods) by the way of transformation of SSF distributions from the linear $A(n\omega_s)$ to nonlinear form with Fresnel approximation $A([n - n^2/2R_F]\omega_S)$, where n - number of array elements; R_F - focusing distance. This transformation is performed by transferring the amplitude distribution of SSF with nonlinear distribution to SSF with linear distribution of its components, when the FAA elements are uniformly spaced, because SSF components distribution is equal to FAA elements spatial location distribution or AFD of signal radiated by the FAA.

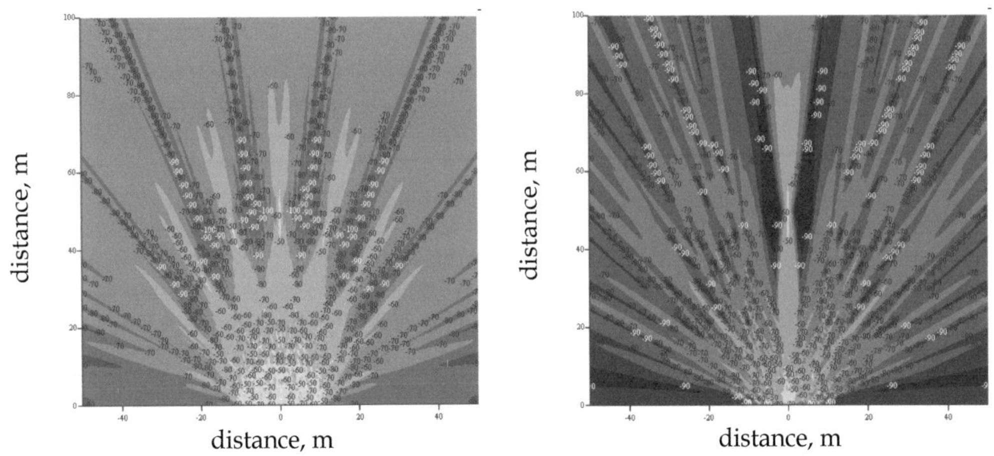

(a) (b)

Fig. 10. Radiation patterns intersection in the azimuth plane of focused LAESR

For example, the modification of Dolph-Chebyshev APD leads to APD of the same type, but for radiation pattern with higher side lobe level and narrower main lobe (Fig. 11), which essentially gives the best result. So, using Dolph-Chebyshev APD causes the defocusing versus side focal area suppressing in all cases as for FFZ beamforming.

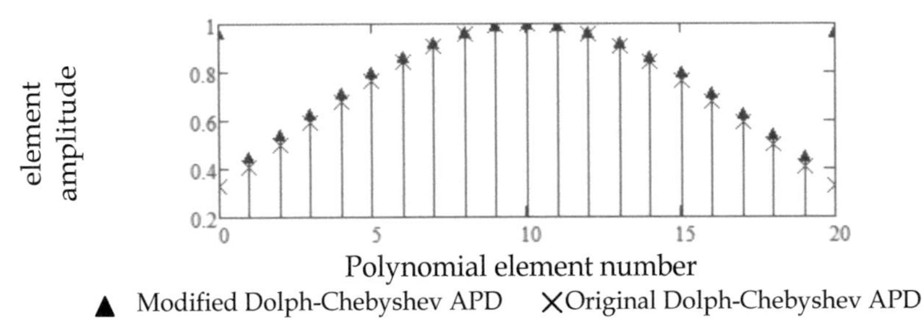

Fig. 11. Modification of Dolph-Chebyshev APD for LAESR of 21 antenna elements

Modification of AFC of wideband pulse signal emitted by FAA is done in same way and causes transformation of all parameters of radiated signal. This fact allows to consider the problem of optimal signals for high directivity of antenna arrays achieving. A comparison of radiation patterns by distance of LAESR which radiating wideband periodic radio impulse with modified and common AFC is shown in Fig.12(a) and comparison of radiation patterns by distance of LAESR with modified and common Dolph-Chebyshev APD which radiating wideband periodic radio impulse is shown in Fig.12(b). Fig.12 shows that this modification is effective, thus the defocusing is present indeed.

These modifications are effective only in NFZ. This method can be used only for linear FAA. The using of the methods of radiation pattern synthesis which are developed using SSF concept and studied by the authors of this chapter in this section is necessary for improve FAA performance according to materials mentioned before.

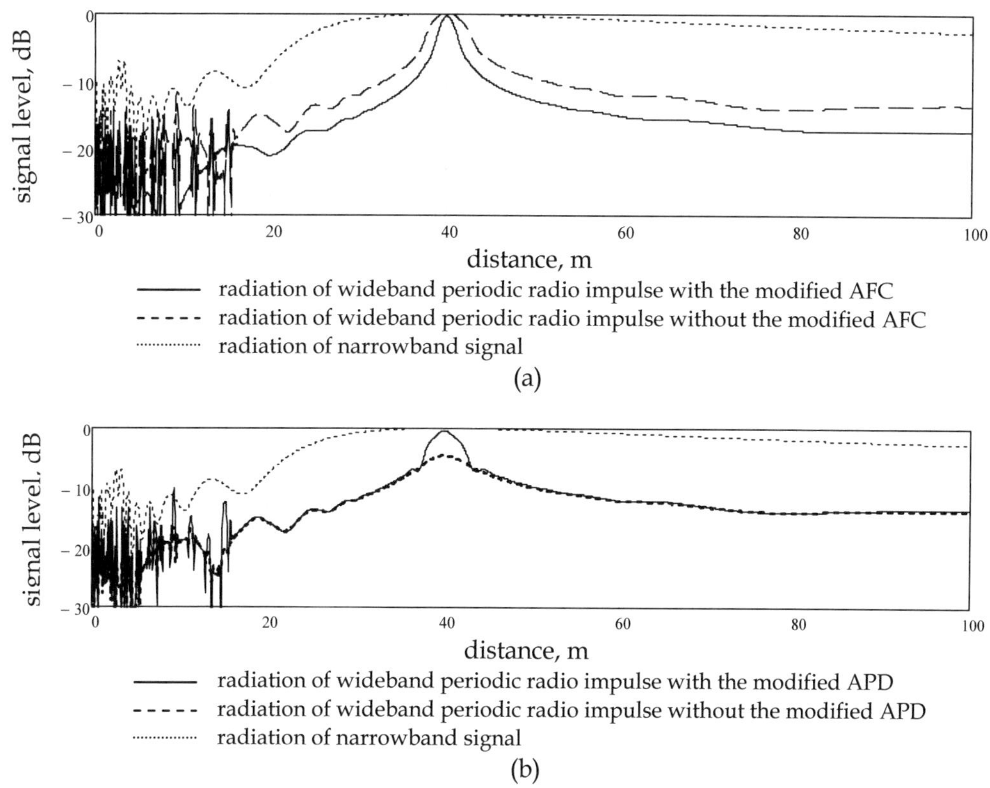

Fig.12. Axial radiation patterns of LAESR of 21 antenna elements excited by special signals

4. Focused Antenna Arrays in telecommunications engineering

Radio-frequency resource utilization efficiency in the spatial area is not enough because the spatial selectivity is limited by two dimensions: the azimuth and the elevation. This problem can be solved by use of antenna systems with controlled 3-dimensional, including distance, signal spatial division or FAA, which form operational area in their NFZ and IFZ.

Modern technics of focused antennas are presented by arrays (Buffi et al., 2009; Fenn, 2007; Malyuskin & Fusco, 2009), Fresnel zone plate and lens antennas (Garret & Wiltse, 1990; Herben, 1999; Hristov, 2004; Karimkashi & Kishik, 2008; Reid & Smith, 2009; Wang et al., 2007) and other (Rudolph & Grbic, 2008; Molotkov et al., 2009). The arrays and single radiators of these focused antennas can be used for 3-dimensional directional signal transmission.

Applicability of the antennas mentioned before for short range fixed or mobile communication is provided by its capability to perform the controlled 3-dimensional spatial signal division in necessary operational area of wireless telecommunication system.

According to the paper (Mazurenko & Yakornov, 2010), realization of the given FAA for achieving operation zone dimensions $R_{OA} = 0.1\ldots1$ km and focal area with depth of focus $\Delta R = 10\ldots100$m are limited by the hyperfocal distance. For FAA element spacing $d_0 = l\lambda_0$ the operation area dimension R_{OA} and array parameters are related by expression

$$R_{OA} < (N-1)^2 l^2 \lambda_0, \tag{25}$$

According to eq. (25) and materials of previous sections, FAA structures for some frequency ranges are:

- $f < 30$ MHz - element quantity $N < 10$, rarefied antenna array of polygonal structure with electrically small antenna elements.
- $f \in [30$ MHz; 30 GHz$]$ - element quantity $N \in [10; 100]$, rarefied antenna array of polygonal or linear structure with directional array elements and grating focal areas suppression methods application.
- $f > 30$ GHz - linear or planar antenna array with element quantity $N > 100$ or equivalent length continuous radiator, aperture, reflector.

Considering all achievements in FAA directivity improving and real condition of signal propagation the following antenna systems can be created:

- PAA with operation zone in the area inside structure of polygon. Radiation pattern of PAA is characterized by one focal area of circular form with power gain < 10 dB if maximum operational zone dimension not exceed 100 m, as shown on Fig.13.
- LAESR with operation zone in the area of its NFZ and IFZ. Radiation pattern of LAESR is characterized by one focal area of elliptical form with power gain 10-20 dB if maximum operational zone dimension not exceed 100 m, as shown on Fig.14, and by almost uniformly distributed power over the area with width equal to array maximum dimensions and length equal to focus point distance with power fluctuations < 5 dB if maximum operational zone dimension within the limits $R_{OA} \in [100\text{m}; 1000\text{m}]$, as shown on Fig.15.

Radiation patterns intersection in the azimuth plane of octagonal PAA of patch antennas with $d = 10\lambda$, $\lambda = 4$ m which are radiating inside the structure periodic radio impulse signal with $\Delta f = 150$ MHz, $f_C = 75$ MHz, $T = 173.3$ ns and $\tau_p = 13.3$ ns with taking into account signal attenuation by the propagation, where array focused in point with coordinates $R_F = 0$ m, $\theta_F = 0°$, $R_F = 15$ m, $\theta_F = 45°$ relatively to the center of polygon are shown in Fig.13(a) and Fig.13(b) respectively, where the signal level is in dB.

Radiation patterns intersection in the azimuth plane of LAESR of 401 patch antennas with $d = \lambda$ and $\lambda = 5$ mm or equivalent reflector with linear dimension $D = 2$ m with taking into account signal attenuation by the propagation, where array radiated narrowband signal in focal points with coordinates $R_F = 100$ m, $\theta_F = 0°$ and $R_F = 100$ m, $\theta_F = 30°$ are shown in Fig.14(a) and Fig.14(b) respectively, where the signal level is in dB.

Radiation patterns intersection in the azimuth plane of LAESR of 401 patch antennas with $d = \lambda$ and $\lambda = 5$ mm or equivalent reflector with linear dimension $D = 2$ m with taking into account signal attenuation by the propagation, where array radiated narrowband signal in focal points with coordinates $R_F = 800$ m, $\theta_F = 0°$ and $R_F = 800$ m, $\theta_F = 30°$ are shown in Fig.15(a) and Fig.15(b) respectively, where the signal level is in dB.

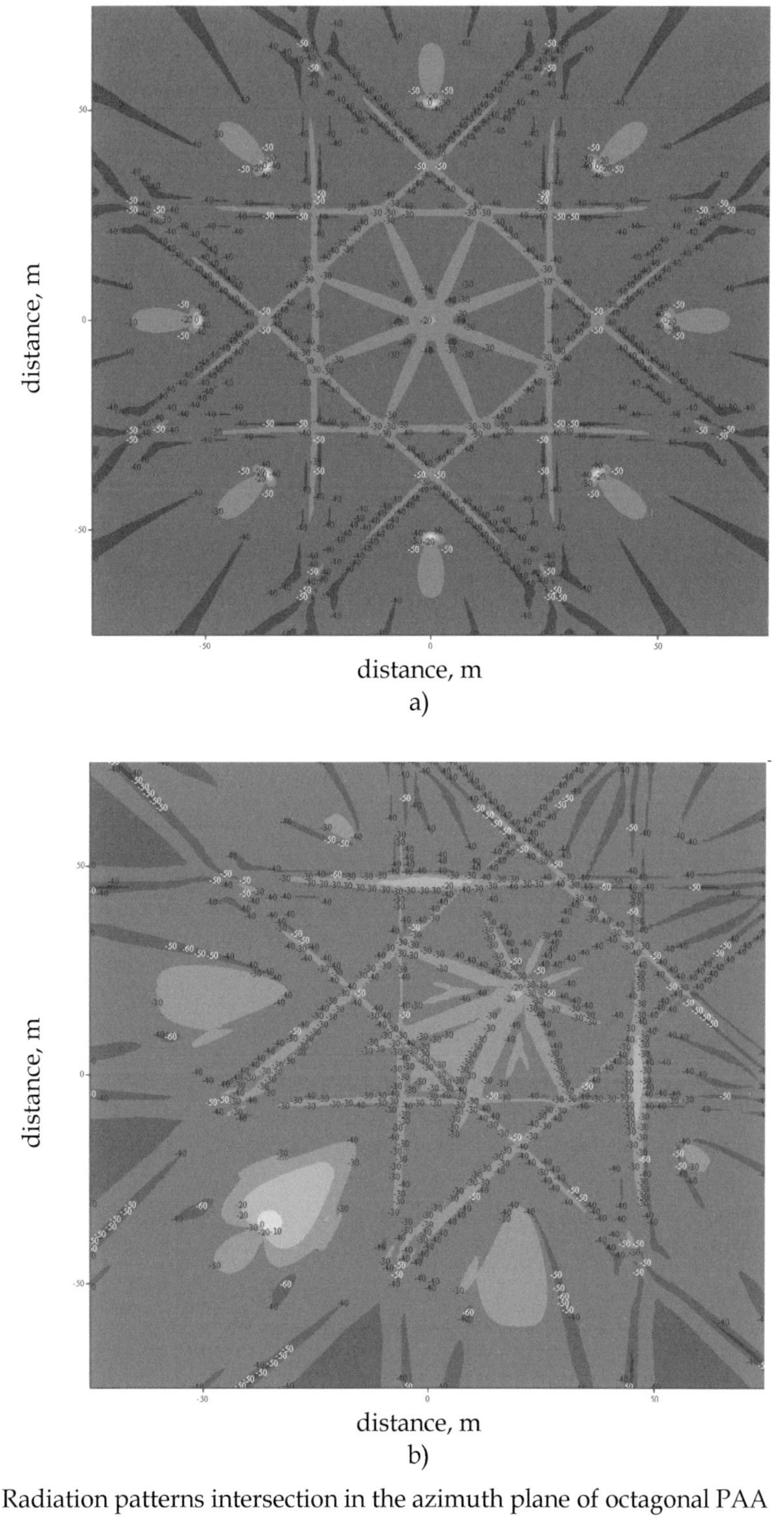

Fig. 13. Radiation patterns intersection in the azimuth plane of octagonal PAA

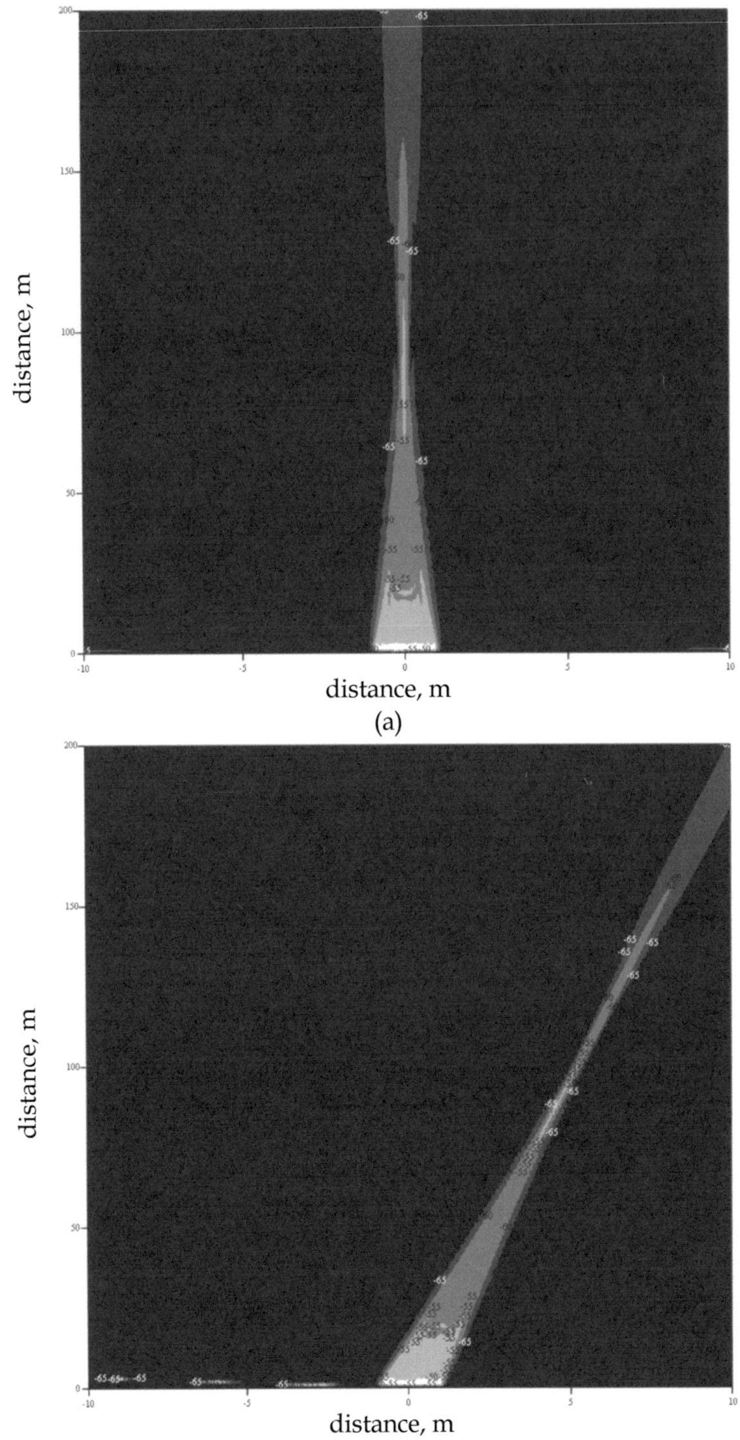

Fig. 14. Radiation patterns intersection in the azimuth plane of focused LAESR

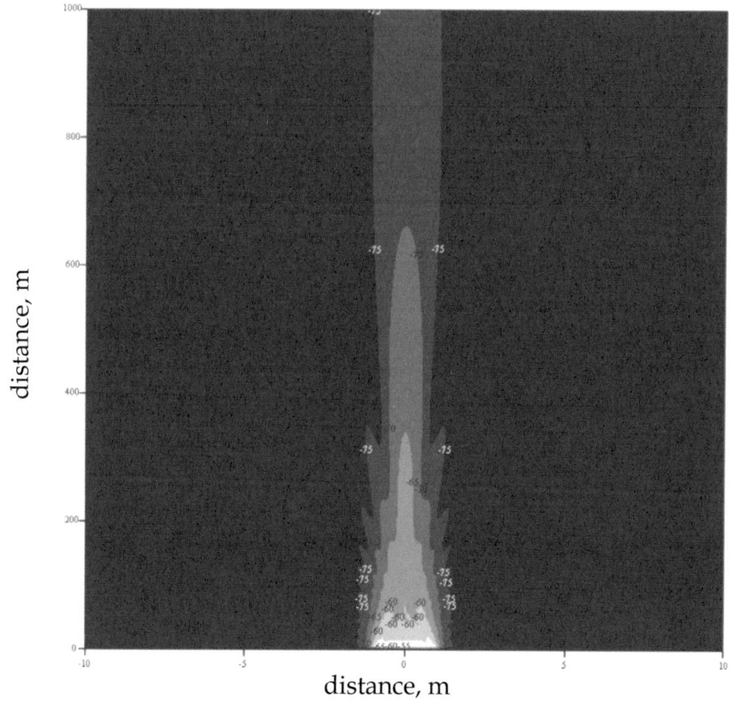

(a)

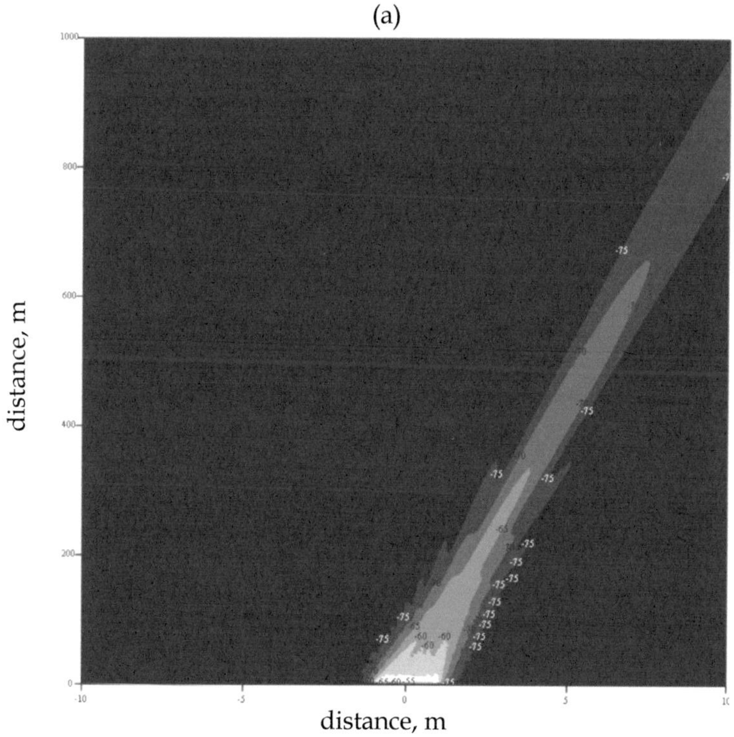

(b)

Fig. 15. Radiation patterns intersection in the azimuth plane of focused LAESR

5. Conclusion

In this chapter, beamforming of antenna arrays focused in NFZ or IFZ is studied and the new principles of this process are revealed with a purpose to increase the 3-dimensional gain performance of antenna arrays at a wide range of angles and to improve the quality level of this technique for expanding applicability of the focused antenna.

The antenna arrays of linear and polygonal structure with different location and types of array elements have been considered in this paper, so all tasks of study of FAA radiation pattern synthesis have done in its azimuth plane or in polar coordinates.

Expressions of spatial shifts between elements and phase center of arrays where spatial shifts by distance and spatial shifts by angular coordinate are obtained as two separate parts of one equation without any approximation (Fresnel or Fraunhofer) are introduced. These equations had been obtained first time and they can be useful for FAA with separate phase steering by distance and by angular coordinate synthesis, it also helps to reduce the level of computational operations to calculate required phase distribution in array elements.

Radiation pattern synthesis and focusing properties of focused arrays are considered in this work, which are based on obtained geometrical models of arrays of different structures. Location of array phase center, different signal attenuations by propagation from array elements to observation point, different angles of radiation from array elements to observation point and array elements radiation patterns dependence on the distance must be taken to account for precise beamforming of arrays focused in NFZ or IFZ. The principles of radiation pattern synthesis of arrays focused in NFZ or IFZ and arrays focused in FFZ are equal. The properties of radiation pattern by distance and angular coordinate are equal too.

The main focusing properties of antenna array are based on principle of inversion of the Fresnel and the Fraunhofer diffraction fields which is the result of focusing process. The applicability of arrays is limited by hyperfocal distance and signal attenuation by propagation. The focal area or a certain segment of distance $R \in [0, R_F]$ where radiated power distribution is uniform can be the result of focusing process.

FAA directivity by distance as angular directivity in focused point to angular directivity in FFZ ratio is introduced in this chapter.

FAA directivity can be improved by increasing array elements spacing jointly with grating focal area supressing or by excitation of array elements by wideband signals or by using of modified APD in array elements and modified AFC of excited wideband signal which are introduced in this chapter. The spectrum of spatial frequencies concept was used for FAA directivity improvement.

Investigation results of possible structures of FAA which capable to three-dimensional spatial division multiplexing in all frequency ranges used by a modern radio-technique are presented in this work. These results have shown the given FAA applicability in a radio and short range telecommunication engineering.

The task of obtaining a exact expression for radiation pattern of antenna array without any approximation still remain unsolved due to lack of solution in the form of a simple or a special function. Beamforming of planar array using study results achieved in this chapter have not considered. Methods of FAA directivity improving are considered more qualitatively than quantitatively. Mutual coupling effect among the array elements and its real dimensions is not assumed. These facts reveal the possibility of further research.

6. References

Buffi A.; Serra A.; Nepa P.; Manara G. & Luise M. (2009). Near field focused microstrip arrays for gate access control systems, *2009 IEEE International Antennas and Propagation Symp. Dig.*, vol. 47, (June 2009), pp. 1704 - 1707

Chu T. S. (1971). A note on simulating Fraunhofer radiation patterns in the Fresnel region, *Antennas and Propagation, IEEE Transactions on*, vol. 19, (Sept. 1971), pp. 691 - 692

Fenn A. J. (2007). *Adaptive Antennas and Phased Arrays for Radar and Communications*. Artech House Publishers, ISBN: 978-1-59693-273-9, Boston, USA

Garrett J. E. & Wiltse J. C. (1990). Antenna pattern characteristics of phase-correcting Fresnel zone plates, *1990 IEEE International Antennas and Propagation Symp. Dig.*, vol. 28, (May 1990), pp. 1906 - 1909

Graham W. J. (1983). Analysis and synthesis of axial field patterns of focused apertures, *Antennas and Propagation, IEEE Transactions on*, vol. 31, (July 1983), pp. 665 - 668

Hansen R. C. (1964). *Microwave Scanning Antennas. vol. 1, Apertures.*, Ed. Academic Press, New York, USA

Hansen R. C. (1985). Focal region characteristics of focused array antennas, *Antennas and Propagation, IEEE Transactions on*, vol. 33, (December 1985), pp. 1328 - 1337

Hansen R. C. (2009). *Phased Array Antennas, 2nd Ed.*, John Wiley & Sons, ISBN 978-0-470-40102-6, New Jersey, USA

Herben M. H. A. J. & Hristov H. D. (1999). Some developments in Fresnel zone plate lens antennas, *1999 IEEE International Antennas and Propagation Symp. Dig.*, vol. 37, (June 1999), pp. 726 - 729

Hristov H. D.; Feick R.; Grote W. & Fernández P. (2004). Indoor signal focusing by means of Fresnel zone plate lens attached to building wall, *Antennas and Propagation, IEEE Transactions on*, vol. 52, (April 2004), pp. 933 - 940

Hussain M. G. M. & Al-Zayed A.S. (2008). Aperture-sparsity analysis of ultrawideband two-dimensional focused array, *Antennas and Propagation, IEEE Transactions on*, vol. 56, (July 2008), pp. 1908 - 1918

Hussain M. G. M. (2004). Characteristics of ultrawideband electro-magnetic missiles generated by focused two-dimensional array. *Progress In Electromagnetics Research*, PIER 49, pp.143–159

Ishimaru A.; Jaruwatanadilok S. & Kuga Y. (2007). Imaging of a target through random media using a short-pulse focused beam, *Antennas and Propagation, IEEE Transactions on*, vol. 55, (June 2007), pp. 1622 - 1629

Karimkashi Sh. & Kishik A. A. (2008). A New Fresnel Zone Antenna with beam Focused in the Fresnel Region. *U.R.S.I. XXIX General Assembly 7-16 August 2008, Chicago, USA.* Available from:http://ursi-test.intec.ugent.be/files/URSIGA08/papers/BP7p7.pdf

Karimkashi, Sh. & Kishk, A.A. (2009). Focused Microstrip Array Antenna Using a Dolph-Chebyshev Near-Field Design, *Antennas and Propagation, IEEE Transactions on* , vol.57, no.12, (December 2009), pp.3813-3820

Korostelev A. A. (1987). *Space-time theory of radio system*. Radio i Svyaz, Moscow, USSR (in Russian)

Kremer I. Ya., Kremer A. I. & Petrov V. M. (1984). *Space-time signal processing*. Radio i Svyaz, Moscow, USSR (in Russian)